Foundation of Digital Electronics and Logic Design

Foundation of Digital Electronics and Logic Design

For B.Tech. IInd Year
(As per the New Syllabus prescribed by GBTU/MTU, formerly known as U.P.T.U.)

A. K. Singh
Dean (Academics) and HOD
Electronics and Communication Engineering
Goel Institute of Technology and Management
Lucknow (U.P.)

An Imprint of

NEW AGE INTERNATIONAL (P) LIMITED, PUBLISHERS

New Delhi • Bangalore • Chennai • Cochin • Guwahati
Hyderabad • Kolkata • Lucknow • Mumbai
Visit us at **www.newagepublishers.com**

Published by New Age International (P) Ltd., Publishers
First Edition: 2012

BRANCHES

- **Bangalore** 37/10, 8th Cross (Near Hanuman Temple), Azad Nagar, Chamarajpet, Bangalore-560 018
  Tel.: (080) 26756823, Telefax: 26756820, **E-mail: bangalore@newagepublishers.com**
- **Chennai** 26, Damodaran Street, T. Nagar, Chennai-600 017
  Tel.: (044) 24353401, Telefax: 24351463, **E-mail: chennai@newagepublishers.com**
- **Cochin** CC-39/1016, Carrier Station Road, Ernakulam South, Cochin-682 016
  Tel.: (0484) 2377004, Telefax: 4051303, **E-mail: cochin@newagepublishers.com**
- **Guwahati** Hemsen Complex, Mohd. Shah Road, Paltan Bazar, Near Starline Hotel, Guwahati-781 008
  Tel.: (0361) 2513881, Telefax: 2543669, **E-mail: guwahati@newagepublishers.com**
- **Hyderabad** 105, 1st Floor, Madhiray Kaveri Tower, 3-2-19, Azam Jahi Road, Nimboliadda, Hyderabad-500 027
  Tel.: (040) 24652456, Telefax: 24652457, **E-mail: hyderabad@newagepublishers.com**
- **Kolkata** RDB Chambers (Formerly Lotus Cinema) 106A, 1st Floor, S.N. Banerjee Road, Kolkata-700 014
  Tel.: (033) 22273773, Telefax: 22275247, **E-mail: kolkata@newagepublishers.com**
- **Lucknow** 16-A, Jopling Road, Lucknow-226 001
  Tel.: (0522) 2209578, 4045297, Telefax: 2204098, **E-mail: lucknow@newagepublishers.com**
- **Mumbai** 142C, Victor House, Ground Floor, N.M. Joshi Marg, Lower Parel, Mumbai-400 013
  Tel.: (022) 24927869, Telefax: 24915415, **E-mail: mumbai@newagepublishers.com**
- **New Delhi** 22, Golden House, Daryaganj, New Delhi-110 002
  Tel.: (011) 23262368, 23262370, Telefax: 43551305, **E-mail: sales@newagepublishers.com**

ISBN: 978-81-224-3375-3

₹ 299.00

C-12-06-6343

Printed in India at Print 'O' Pack, Delhi.

Typeset at Inhouse.

PUBLISHING FOR ONE WORLD

NEW AGE INTERNATIONAL (P) LIMITED, PUBLISHERS
7/30 A, Daryaganj, New Delhi-110002
Visit us at **www.newagepublishers.com**

This book is dedicated to my parents and teachers who have helped shape me into who I am today.

PREFACE

The objective of this book is to develop in the reader the ability to analyze and design the digital circuits. The increased use of digital technology in objects used for day-to-day life necessitates an in-depth knowledge of the subject for the professionals and engineers.

There are lots of references, available on Logic Design and Digital Electronics, which discuss various topics separately. But, through this text, our notion was to discover the topics rather than cover them. This comprehensive text fulfills the course requirement on the subject of Digital Electronics and Logic Design for B.Tech degree course in Electronics Engineering and Computer Science Engineering of different (GBTU and MTU) technical universities. This text is also bound to serve as a useful reference book for various competitive examinations.

There is no special pre-requisite before starting this book. Each chapter of the book starts with simple 'facts and concepts' and traversing through the examples and figures, it uncovers the advanced topics. The book has 10 well-organized chapters.

Chapter 1 deals with number systems and codes. It includes an exhaustive set of solved examples and exercise to clarify the concepts. *Chapter 2* introduces the basic building blocks of digital electronics. It starts with basic postulates. Boolean algebra and then introduces logic gates. It also deals with several types of implementation using logic gates. For beginners, we strongly recommend to work out this chapter twice before proceeding further.

Chapter 3 deals with the Boolean function minimization techniques using Postulates and Boolean Algebra, K-Map and Quine-McCluskey methods. *Chapter 4* presents various combinational logic design using the discrete logic gates and LSI and MSI circuits. This chapter also deals with hazards and fault detection. *Chapter 5* introduces the programmable logic devices. It also deals with basics of ROM, and then moves towards PLAs, PALs, CPLDs and FPGA.

Chapter 6 introduces the clocked (synchronous) sequential circuits. It starts with discussions on various flip-flops, their triggering and flip-flop timings. It then deals with analysis and design of synchronous circuits and concludes with sequence detector circuits. *Chapter 7* deals with shift registers and counters. It introduces the basic idea of shift registers and then discusses various modes and application of shift registers. It then introduces the various types of modes of counters, and concludes with application. *Chapter 8* describes memory fundamentals and memory interfacing, and *Chapter 9* deals with asynchronous sequential circuits. It elaborates the analysis and design procedures with different considerations. *Chapter 10* describes the algorithmic state machine. It starts with basic concepts, design tools and concludes with design using multiplexers and PLA. Appendix A introduces the threshold logic and its capabilities to realize switching function.

All the topics are illustrated with clear diagram, and simple language is used throughout the text to facilitate easy understanding of the concepts. The author welcomes constructive suggestions and comments from the readers for the improvement of this book at aksingh_uptu@rediffmail.com.

AUTHOR

ACKNOWLEDGEMENT

This book is the result of the dedication and encouragement of many individuals. I would like to thank my family members especially wife Menka, daughter Omanshi and son Astitva for their patience and continuing support and parents for their blessings.

I am indebted to my friends and colleagues especially Manish Tiwari, and Arun Prakash co-authors of the book, Digital principles-switching theory .

I thankfully acknowledge the contribution of various authors, data manuals, journals, reference manuals, etc. from where materials have been collected to enrich the contents of the book.

Finally, I would like to thank the people at New Age International (P) Limited, Publishers, especially Mr. Saumya Gupta, Managing Director and Mr. L.N. Mishra, who continue to support and encourages in writing the present book.

Last but not the least. It is by the blessing of almighty and good fortune that I get such a supporting and cooperative people around me who in one way or other help me to complete the project in time.

ARUN KUMAR SINGH

SYLLABUS

ECS-301: Digital Logic Design

Unit-I

Digital system and binary numbers: Signed binary numbers, binary codes, cyclic codes, error detecting and correcting codes, hamming codes.

Floating point representation

Gate-level minimization: The map method up to five variable, don't care conditions, POS simplification, NAND and NOR implementation, Quine Mc-Clusky method (Tabular method).

Unit-II

Combinational logic: Combinational circuits, analysis procedure, design procedure, binary adder-subtractor, decimal adder, binary multiplier, magnitude comparator, decoders, encoders, multiplexers

Unit-III

Synchronous sequential logic: Sequential circuits, storage elements: latches, flip flops, analysis of clocked sequential circuits, state reduction and assignments, design procedure.

Registers and counters: Shift registers, ripple counter, synchronous counter, other counters.

Unit-IV

Memory and programmable logic: RAM, ROM, PLA, PAL.

Design at the register transfer level: ASMs, design example, design with multiplexers.

Unit-V

Asynchronous sequential logic: Analysis procedure, circuit with latches, design procedure, reduction of state and flow table, race free state assignment, hazards.

EEC-302: Digital Electronics

Unit-I

Digital system and binary numbers: Signed binary numbers, binary codes, cyclic codes, error detecting and correcting codes, hamming codes.

Floating point representation

Gate-level minimization: The map method up to five variable, don't care conditions, POS simplification, NAND and NOR implementation, Quine Mc-Clusky method (Tabular method).

Unit-II

Combinational logic: Combinational circuits, analysis procedure, design procedure, binary adder-subtractor, decimal adder, binary multiplier, magnitude comparator, decoders, encoders, multiplexers

Unit-III

Synchronous sequential logic: Sequential circuits, storage elements: latches, flip flops, analysis of clocked sequential circuits, state reduction and assignments, design procedure.

Registers and counters: Shift registers, ripple counter, synchronous counter, other counters.

Unit-IV

Memory and programmable logic: RAM, ROM, PLA, PAL.

Design at the register transfer level: ASMs, design example, design with multiplexers.

Unit-V

Asynchronous sequential logic: Analysis procedure, circuit with latches, design procedure, reduction of state and flow table, race free state assignment, hazards.

CONTENTS

1

NUMBER SYSTEMS AND CODES

1.0 INTRODUCTION

Inside today's computers, data is represented as 1's and 0's. These 1's and 0's might be stored magnetically on a disk, or as a state in a transistor, core, or vacuum tube. To perform useful operations on these 1's and 0's one have to organize them together into patterns that make up codes. Modern digital systems do not represent numeric values using the decimal system. Instead, they typically use a binary or two's complement numbering system. To understand the digital system arithmetic, one must understand how digital systems represent numbers.

This chapter discusses several important concepts including the binary, octal and hexadecimal numbering systems, binary data organization (bits, nibbles, bytes, words, and double words), signed and unsigned numbering systems. If one is already familiar with these terms he should at least skim this material.

1.1 A REVIEW OF THE DECIMAL SYSTEM

People have been using the decimal (base 10) numbering system for so long that they probably take it for granted. When one see a number like "123", he don't think about the value 123; rather, he generate a mental image of how many items this value represents. In reality, however, the number 123 represents:

$$1*10^2 + 2*10^1 + 3*10^0$$

or

$$100 + 20 + 3$$

Each digit appearing to the left of the decimal point represents a value between zero and nine times an increasing power of ten. Digits appearing to the right of the decimal point represent a value between zero and nine times an increasing negative power of ten. For example, the value 123.456 means:

$$1*10^2 + 2*10^1 + 3*10^0 + 4*10^{-1} + 5*10^{-2} + 6*10^{-3}$$

or

$$100 + 20 + 3 + 0.4 + 0.05 + 0.006$$

1.2 BINARY NUMBERING SYSTEM

Most modern digital systems operate using binary logic. The digital systems represent values using two voltage levels (usually 0 V and +5 V). With two such levels one can represent exactly two different values. These could be any two different values, but by convention we use the values zero and one. These two values, coincidentally, correspond to the two digits used by the binary numbering system.

1.2.1 Binary to Decimal Conversion

The binary numbering system works just like the decimal numbering system, with two exceptions: binary only allows the digits 0 and 1 (rather than 0–9), and binary uses powers of two rather than powers of ten. Therefore, it is very easy to convert a binary number to decimal. For each "1" in the binary string, add 2^n where "n" is the bit position in the binary string (0 to n–1 for n bit binary string).

For example, the binary value 1010_2 represents the decimal 10 which can be obtained through the procedure shown in the table 1.

Table 1

Binary No.	1	0	1	0
Bit Position (n)	3rd	2nd	1st	0th
Weight Factor (2^n)	2^3	2^2	2^1	2^0
bit * 2^n	$1*2^3$	$0*2^2$	$1*2^1$	$0*2^0$
Decimal Value	8	0	2	0
Decimal Number	8 + 0 + 2 + 0 = 10			

All the steps in above procedure can be summarized in short as

$$1*2^3 + 0*2^2 + 1*2^1 + 0*2^0 = 8 + 0 + 2 + 0 = 10_{10}$$

i.e.,

1. Multiply each digit of the binary number by its positional weight and then add up the result.
2. If any digit is 0, its positional weight is not to be taken into account.

1.2.2 Decimal to Binary Conversion

The inverse problem would be to find out the binary equivalent of given decimal number for instance let us find out binary of 19_{10} (decimal 19).

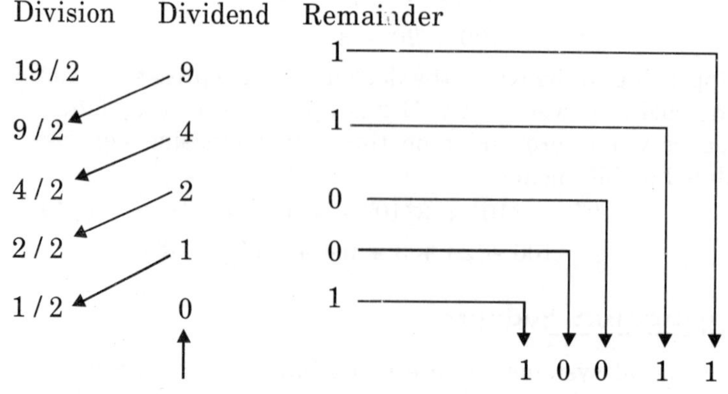

Dividend is 0, stop the procedure.

Our final number is $(10011)_2$.

i.e.,

1. Divide the decimal number by 2 producing a dividend and a remainder. This number is the LSB (least significant bit of the desired binary number).
2. Again divide the dividend obtained above by 2. This produces another dividend and remainder. The remainder is the next digit of the binary number.
3. Continue this process of division until the dividend becomes 0. The remainder obtained in the final division is the MSB (most significant bit of the binary number).

1.2.3 Binary Formats

In the purest sense, every binary number contains an infinite number of digits (or *bits* which is short for binary digits). Because any number of leading zero bits may precede the binary number without changing its value. For example, one can represent the number seven by:

111 00000111 ..0000000000111 000000000000111

Often several values are packed together into the same binary number. For convenience, a numeric value is assigned to each bit position. Each bit is numbered as follows:

1. The rightmost bit in a binary number is bit position zero.
2. Each bit to the left is given the next successive bit number.

An eight-bit binary value uses bits zero through seven:

$$X_7\ X_6\ X_5\ X_4\ X_3\ X_2\ X_1\ X_0$$

A 16-bit binary value uses bit positions zero through fifteen:

$$X_{15}\ X_{14}\ X_{13}\ X_{12}\ X_{11}\ X_{10}\ X_9\ X_8\ X_7\ X_6\ X_5\ X_4\ X_3\ X_2\ X_1\ X_0$$

Bit zero is usually referred to as the *low order* bit. The left-most bit is typically called the *high order* bit. The intermediate bits are referred by their respective bit numbers. The low order bit which is X_0 is called LEAST SIGNIFICANT BIT (LSB). The high order bit or left most bit *i.e.*, X_{15} is called MOST SIGNIFICANT BIT (MSB).

1.2.4 Data Organization

In pure mathematics a value may take an arbitrary number of bits. Digital systems, on the other hand, generally work with some specific number of bits. Common collections are single bits, groups of four bits (called *nibbles*), groups of eight bits (called *bytes*), groups of 16 bits (called *words*), and more. The sizes are not arbitrary. There is a good reason for these particular values.

Bits

The smallest "unit" of data on a binary computer or digital system is a single *bit*. **Bit**, an abbreviation for Binary Digit, can hold either a 0 or a 1. A bit is the smallest unit of information a computer can understand. Since a single bit is capable of representing only two different values (typically zero or one) one may get the impression that there are a very small number of items one can represent with a single bit. That's not true! There are an infinite number of items one can represent with a single bit.

With a single bit, one can represent any two distinct items. Examples include zero or one, true or false, on or off, male or female, and right or wrong. However, one are *not* limited to representing binary data types (that is, those objects which have only two distinct values). One could use a single bit to represent the numbers 321 and 1234. Or perhaps 6251 and 2. One could also use a single bit to represent the colours green and blue. One could even represent two unrelated objects with a single bit. For example, one could represent the colour red and the number 3256 with a single bit. One can represent *any* two different values with a single bit. However, one can represent *only* two different values with a single bit.

To confuse things even more, different bits can represent different things. For example, one bit might be used to represent the values zero and one, while an adjacent bit might be used to represent the values true and false. How can one tell by looking at the bits? The answer, of course, is that one can't. But this illustrates the whole idea behind computer data structures: *data is what one define it to be*. If one uses a bit to represent a boolean (true/false) value then that bit (by definition) represents true or false. For the bit to have any true meaning, one must be consistent. That is, if one is using a bit to represent true or false at one point in his program, he shouldn't use the true/false value stored in that bit to represent green or blue later.

Since most items one will be trying to model require more than two different values, single bit values aren't the most popular data type used. However, since everything else consists of groups of bits, bits will play an important role in programs. Of course, there are several data types that require two distinct values, so it would seem that bits are important by themselves. However, individual bits are difficult to manipulate, so other data types are often used to represent boolean values.

Nibbles

A *nibble* is a collection of four bits. It wouldn't be a particularly interesting data structure except for two items: BCD (*binary coded decimal*) numbers and hexadecimal numbers. It takes four bits to represent a single BCD or hexadecimal digit. With a nibble, one can represent up to 16 distinct values. In the case of hexadecimal numbers, the values 0, 1, 2, 3, 4, 5, 6, 7, 8, 9, A, B, C, D, E, and F are represented with four bits (see "The Hexadecimal Numbering System"). BCD uses ten different digits (0, 1, 2, 3, 4, 5, 6, 7, 8, 9) and requires four bits. In fact, any sixteen distinct values can be represented with a nibble, but hexadecimal and BCD digits are the primary items we can represent with a single nibble.

Bytes

Computer memory must be able to store letters, numbers, and symbols. A single bit by itself cannot be of much use. Bits are combined to represent some meaningful data. A group of eight bits is called a byte. It can represent a character and is the smallest addressable datum (data item) on the most of the digital systems (*e.g.* 80 × 86 microprocessor). The most important data type is the byte. Main memory and input/output addresses on the 80 × 86 are all byte addresses. This means that the smallest item that can be individually accessed by an 80 × 86 program is an eight-bit value. To access anything smaller requires that you read the byte containing the data and mask out the unwanted bits. The bits in a byte are normally numbered from zero to seven using the convention in Fig. 1.1.

Bit 0 is the *low order bit* or *least significant bit*, bit 7 is the *high order bit* or *most significant bit* of the byte. All other bits are referred by their number.

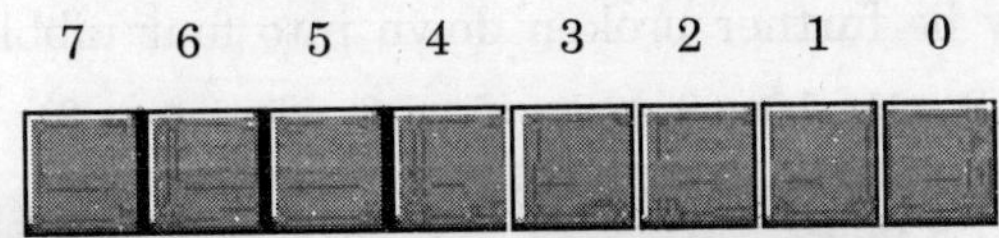

Fig. 1.1 Bit numbering in a byte

Note: That a byte also contains exactly two nibbles (see Fig. 1.2).

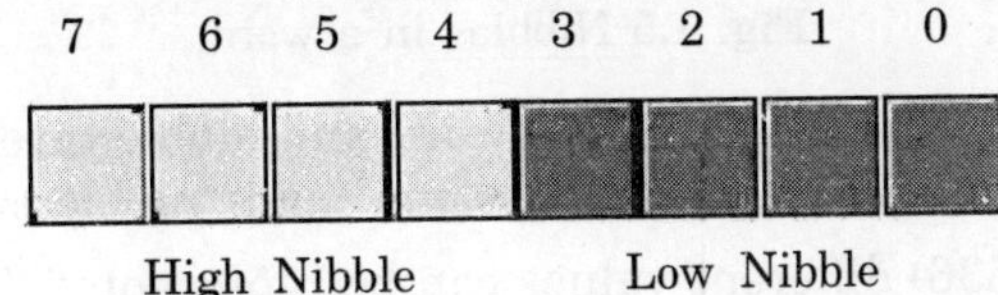

Fig. 1.2 The two nibbles in a byte

Bits 0–3 comprise the *low order nibble*, bits 4–7 form the *high order nibble*. Since a byte contains exactly two nibbles, byte values require two hexadecimal digits.

Since a byte contains eight bits, it can represent 2^8, or 256, different values. Generally, a byte is used to represent numeric values in the range 0.255, signed numbers in the range –128.. + 127 (refer "Signed binary representation"). Many data types have fewer than 256 items so eight bits is usually sufficient.

For a byte addressable machine, it turns out to be more efficient to manipulate a whole byte than an individual bit or nibble. For this reason, most programmers use a whole byte to represent data types that require no more than 256 items, even if fewer than eight bits would suffice. For example, we'll often represent the boolean values true and false by 00000001_2 and 00000000_2 (respectively).

Probably the most important use for a byte is holding a character code. Characters typed at the keyboard, displayed on the screen, and printed on the printer all have numeric values.

Words

A word is a group of 16 bits. Bits in a word are numbered starting from zero on up to fifteen. The bit numbering appears in Fig. 1.3.

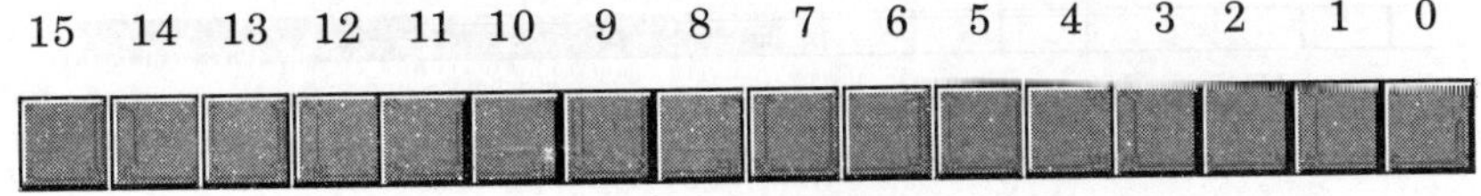

Fig. 1.3 Bit numbers in a word

Like the byte, bit 0 is the low order bit and bit 15 is the high order bit. When referencing the other bits in a word use their bit position number.

Notice that a word contains exactly two bytes. Bits 0 through 7 form the low order byte, bits 8 through 15 form the high order byte (see Fig. 1.4).

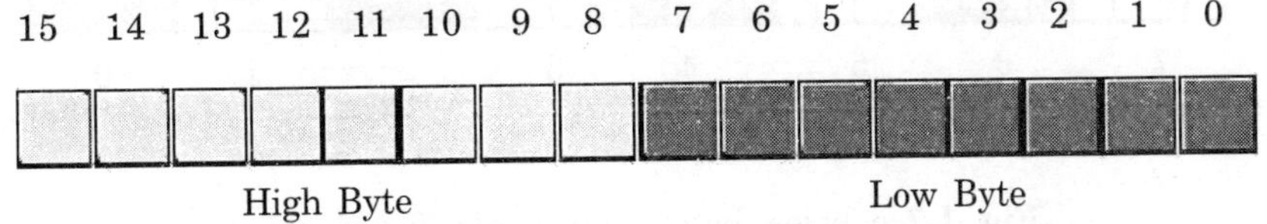

Fig. 1.4 The two bytes in a word

Naturally, a word may be further broken down into four nibbles as shown in Fig. 1.5.

Fig. 1.5 Nibbles in a word

Nibble zero is the low order nibble in the word and nibble three is the high order nibble of the word. The other two nibbles are "nibble one" and "nibble two".

With 16 bits, 2^{16} (65,536) different values can be represented. These could be the values in the range 0 to 65,535 (or –32,768 to +32,767) or any other data type with no more than 65,536 values.

Words can represent integer values in the range 0 to 65,535 or –32,768 to 32,767. Unsigned numeric values are represented by the binary value corresponding to the bits in the word. Signed numeric values use the two's complement form for numeric values (refer "Signed binary representation").

Double Words

A double word is exactly what its name implies, a pair of words. Therefore, a double word quantity is 32 bits long as shown in Fig. 1.6.

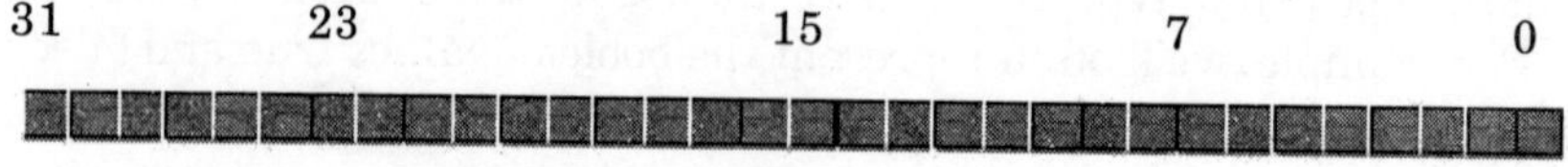

Fig. 1.6 Bit numbers in a double word

This double word can be divided into a high order word and a low order word, or four different bytes, or eight different nibbles (see Fig. 1.7).

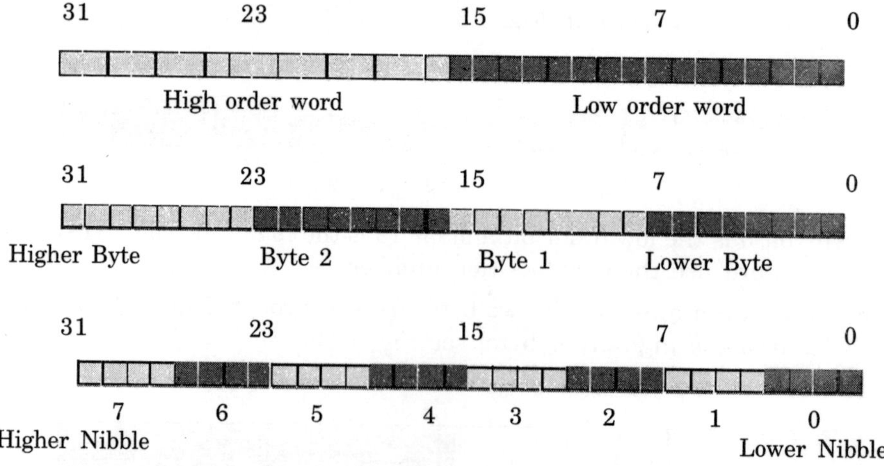

Fig. 1.7 Nibbles, bytes, and words in a double word

1.3 OCTAL NUMBERING SYSTEM

The octal number system uses base 8 instead of base 10 or base 2. This is sometimes convenient since many computer operations are based on bytes (8 bits). In octal, we have 8 digits at our disposal, 0–7.

Decimal	Octal
0	0
1	1
2	2
3	3
4	4
5	5
6	6
7	7
8	10
9	11
10	12
11	13
12	14
13	15
14	16
15	17
16	20

1.3.1 Octal to Decimal, Decimal to Octal Conversion

Converting octal to decimal is just like converting binary to decimal, except instead of powers of 2, we use powers of 8. That is, the LSB is 8^0, the next is 8^1, then 8^2, etc.

To convert 172 in octal to decimal:

$$\begin{array}{ccc} 1 & 7 & 2 \\ 8^2 & 8^1 & 8^0 \end{array}$$

$$\begin{aligned} \text{Weight} &= 1*8^2 + 7*8^1 + 2*8^0 \\ &= 1*64 + 7*8 + 2*1 \\ &= 122_{10} \end{aligned}$$

Converting decimal to octal is just like converting decimal to binary, except instead of dividing by 2, we divide by 8. To convert 122 to octal:

$$\begin{aligned} 122/8 &= 15 \text{ remainder } 2 \\ 15/8 &= 1 \text{ remainder } 7 \\ 1/8 &= 0 \text{ remainder } 1 \\ &= 172_8 \end{aligned}$$

If using a calculator to perform the divisions, the result will include a decimal fraction instead of a remainder. The remainder can be obtained by multiplying the decimal fraction by 8. For example, 122/8 = 15.25. Then multiply 0.25 * 8 to get a remainder of 2.

1.3.2 Octal to Binary, Binary to Octal Conversion

Octal becomes very useful in converting to binary, because it is quite simple. The conversion can be done by looking at 3 bit combinations, and then concatenating them together. Here is the equivalent for each individual octal digit and binary representation:

Octal	Binary
1	001
2	010
3	011
4	100
5	101
6	110
7	111

To convert back and forth between octal and binary, simply substitute the proper pattern for each octal digit with the corresponding three binary digits.

For example, 372 in octal becomes 010 111 010 or 010111010 in binary.

777 in octal becomes 111 111 111 or 111111111 in binary.

The same applies in the other direction:

100111010 in binary becomes 100 111 010 or 472 in octal.

Since it is so easy to convert back and forth between octal and binary, octal is sometimes used to represent binary codes. Octal is most useful if the binary code happens to be a multiple of 3 bits long. Sometimes it is quicker to convert decimal to binary by first converting decimal to octal, and then octal to binary.

1.4 HEXADECIMAL NUMBERING SYSTEM

The hexadecimal numbering system is the most common system seen today in representing raw computer data. This is because it is very convenient to represent groups of 4 bits. Consequently, one byte (8 bits) can be represented by two groups of four bits easily in hexadecimal.

Hexadecimal uses a base 16 numbering system. This means that we have 16 symbols to use for digits. Consequently, we must invent new digits beyond 9. The digits used in hex are the letters A, B, C, D, E, and F. If we start counting, we get the table below:

Decimal	Hexadecimal	Binary
0	0	0000
1	1	0001
2	2	0010
3	3	0011
4	4	0100

.....Contd

Decimal	Hexadecimal	Binary
5	5	0101
6	6	0110
7	7	0111
8	8	1000
9	9	1001
10	A	1010
11	B	1011
12	C	1100
13	D	1101
14	E	1110
15	F	1111
16	10	10000
17	11	10001
18 ...		

1.4.1 Hex to Decimal and Decimal to Hex Conversion

Converting hex to decimal is just like converting binary to decimal, except instead of powers of 2, we use powers of 16. That is, the LSB is 16^0, the next is 16^1, then 16^2, etc.

To convert 15E in hex to decimal:

$$\begin{array}{rcl} & & 1 \quad 5 \quad E \\ & & 16^2 \quad 16^1 \quad 16^0 \\ \text{Weight} & = & 1*16^2 + 5*16^1 + 14*16^0 \\ & = & 1*256 + 5*16 + 14*1 \\ & = & 350_{10} \end{array}$$

Converting decimal to hex is just like converting decimal to binary, except instead of dividing by 2, we divide by 16. To convert 350 to hex:

$$\begin{array}{rcl} 350/16 & = & 21 \text{ remainder } 14 = E \\ 21/16 & = & 1 \text{ remainder } 5 \\ 1/16 & = & 0 \text{ remainder } 1 \end{array}$$

So we get 15E for 350.

Again, note that if a calculator is being used, you may multiple the fraction remainder by 16 to produce the remainder. 350/16 = 21.875. Then to get the remainder, 0.875 * 16 = 14.

1.4.2 Hex to Binary and Binary to Hex Conversion

Going from hex to binary is similar to the process of converting from octal to binary. One must simply look up or compute the binary pattern of 4 bits for each hex code, and concatenate the codes together.

To convert AE to binary:

$$\begin{array}{rcl} A & = & 1010 \\ E & = & 1110 \end{array}$$

So AE in binary is 1010 1110

The same process applies in reverse by grouping together 4 bits at a time and then look up the hex digit for each group.

Binary 11000100101 broken up into groups of 4:

0110 0010 0101 (note the 0 added as padding on the MSB to get up to 4 bits)

6 2 5

$= 625_{16}$

1.4.3 Hex to Octal and Octal to Hex Conversion

These conversions are done through the binary conversion. Recall that, a group of 4-bits represent a hexadecimal digit and a group of 3-bits represent an octal digit.

Hex to Octal Conversion

1. Convert the given hexadecimal number into binary.
2. Starting from right make groups of 3-bits and designate each group an octal digit.

Example. *Convert $(1A3)_{16}$ into octal.*

Solution.

1. Converting hex to binary

$$(1\ A\ 3)_{16} = \underbrace{0001}_{1}\ \underbrace{1010}_{A}\ \underbrace{0011}_{3}$$

2. Grouping of 3-bits

$$(1A3)_{16} = \underset{0}{\underset{\downarrow}{000}}\ \underset{6}{\underset{\downarrow}{110}}\ \underset{4}{\underset{\downarrow}{100}}\ \underset{3}{\underset{\downarrow}{011}}$$

so $$(1A3)_{16} = (0643)_8 \equiv (643)_8$$

Octal to Hex Conversion

1. Convert the given octal number into binary.
2. Starting from right make groups of 4-bits and designate each group as a Hexadecimal digit.

Example. *Convert $(76)_8$ into hexadecimal.*

Solution. 1. Converting octal to binary

$$(76)_8 = \underbrace{111}_{7}\ \underbrace{110}_{6}$$

2. Grouping of 4-bits

$$(76)_8 = \underset{3}{\underset{\downarrow}{11}}\ \underset{E}{\underset{\downarrow}{1110}} = \underset{3}{\underset{\downarrow}{0011}}\ \underset{E}{\underset{\downarrow}{1110}}$$

$$\therefore \quad (76)_8 = (3E)_{16}$$

1.5 RANGE OF NUMBER REPRESENTATION

The range of numbers that can be represented is determined by the number of digits (or bits in binary) used to represent a number. Let us consider decimal number system to understand the idea.

Highest decimal number represented by 2 digits = 99

But $99 = 100 - 1 = 10^2 - 1$. The power of 10 (in $10^2 - 1$) indicates that it is 2 digit representation.

So, highest 2-digit decimal number $= 10^2 - 1$

and lowest 2-digit decimal number = 00

Thus, range of 2-digit decimal number = 00 to $10^2 - 1$

It is evident that a total of 100 or 10^2 numbers (00 to 99) can be represented by 2-digits.

So, we conclude that for n-digit representation

range of decimal numbers = 0 to $10^n - 1$

highest decimal number $= 10^n - 1$

total numbers that can be represented $= 10^n$

Note that highest *n*-digit decimal number can be represented by *n* 9s (*i.e.*, 10 – 1) *e.g.*, highest 2 digit decimal number is represented by 2 9s which is 99.

The above discussion can be generalized by taking base-*r* number system instead of base-10 (or decimal) number system. Thus, with *n*-digit representation–

Total distinct numbers that can be represented $= r^n$

Highest decimal Number $= r^{n-1}$

Range of Numbers = 0 to r^{n-1}

where, r = base or radix of Number system

n = Number of digits used for representation

It is worth noting that highest decimal number can be represented by *n* (*r* – 1)s in base-*r* system.

Let us consider the base-2 or binary number system. Thus 2^n distinct quantities, in the range 0 to $2^n - 1$, can be represented with *n*-bit. If *n* = 4-bits, total distinct quantities (*i.e.*, numbers) that can be represented

$$= N = 2^4 = 16$$

the range of numbers = 0 to $2^4 - 1$ = 0 to 15

and Highest decimal number $= 2^4 - 1 = 15$

The highest decimal number 15, is represented by four 1s *i.e.*, 1111. The range 0 to 15 corresponds to 0000 to 1111 in binary.

If we want to represent a decimal number M using n-bits, then the number M should lie in the range 0 to 2^n–1 *i.e.*,

$$0 \le M \le 2^n - 1$$

or

$$2^n - 1 \ge M$$

or $$2^n \geq M + 1$$

or $$n \geq \log_2 (M + 1)$$

where M and n are integers.

In the similar way, if we want to represent N distinct quantities in binary then N should not exceed 2^n.

$$2^n \geq N$$

or $$n \geq \log_2 N$$ Both n and N are integers

Example 1. *How many bits are required to represent*

(i) 16-distinct levels

(ii) 10 distinct levels

(iii) 32 distinct levels

Solution. (*i*) We have, $2^n \geq N$

or $2^n \geq 16 \Rightarrow 2^n \geq 2^4$

or $n \geq 4 \Rightarrow n = 4$

Thus, atleast 4-bits are required to represent 16 distinct levels, ranging from 0 to 15.

(*ii*) We have, $n \geq \log_2 N$

or $n \geq \log_2 10 \Rightarrow n \geq 3.32$

but n should be integer, so take next higher integer value

i.e., $n = 4$ bits

So, minimum 4-bits are required to represent 10 distinct levels, ranging from 0 to 9.

(*iii*) $n \geq \log_2 N$

or $n \geq \log_2 32 \Rightarrow n \geq \log_2 2^5$

or $n \geq 5 \Rightarrow n = 5$

So, minimum 5-bits are required to represent 32 levels, ranging from 0 to 31.

Example 2. *Calculate the minimum no. of bits required to represent decimal numbers*

(i) 16 *(ii) 63*

Solution. (*i*) We have, $n \geq \log_2(M + 1)$ where M = given number

so $n \geq \log_2(16 + 1) \Rightarrow n \geq \log_2(17)$

or $n \geq 4.09$

taking next higher integer *i.e.,* $n = 5$ bits.

Thus, atleast 5-bits are required to represent decimal number 16.

(*ii*) $n \geq \log_2 (M + 1)$

$n \geq \log_2 (63 + 1) \Rightarrow n \geq \log_2 64$

or $n \geq \log_2 2^6$ or $n \geq 6$ bits

So, minimum 6-bits are needed to represent decimal 63.

Example 3. *In a base-5 number system, 3 digit representation is used. Find out*

(i) Number of distinct quantities that can be represented.

(ii) Representation of highest decimal number in base-5.

Solution. Given radix of number system $r = 5$

digits of representation $n = 3$

digits in base-5 would be 0, 1, 2, 3, 4

(*i*) we have relation

$$\text{no of distinct quantities} = r^n = 5^3 = 125$$

So, 125 distinct levels (quantities) can be represented.

(*ii*) Highest decimal Number can be represented by $n(r - 1)$s *i.e.,* by three 4s.

So, highest decimal Number = 444.

1.6 BINARY ARITHMETIC

The binary arithmetic operations such as addition, subtraction, multiplication and division are similar to the decimal number system. Binary arithmetics are simpler than decimal because they involve only two digits (bits) 1 and 0.

Binary Addition

Rules for binary addition are summarized in the table shown in Fig. 1.8.

Augend	*Addend*	*Sum*	*Carry*	*Result*
0	0	0	0	0
0	1	1	0	1
1	0	1	0	1
1	1	0	1	10

Fig. 1.8 Rules for binary addition

As shown in 4th row adding 1 to 1 gives carry which, is given to next binary position, similar to decimal system. This is explained in examples below:

Example 1. (*i*) *Add 1010 and 0011* (*ii*) *Add 0101 and 1111*

Solution.

```
   1      ← Carry        1 1 1    ← Carry
   1 0 1 0               0 1 0 1
 + 0 0 1 1             + 1 1 1 1
   -------             ---------
   1 1 0 1             1 0 1 0 0
                       ↑
                       Carry
```

Binary Subtraction

The rules for binary subtraction is summarized in the table shown in Fig. 1.9.

Minuend	*Subtrahend*	*Difference*	*Borrow*
0	0	0	0
0	1	1	1
1	0	1	0
1	1	0	0

Fig. 1.9 Rules for binary subtraction

The process of subtraction is very similar to decimal system in which if a borrow is needed it is taken from next higher binary position, as shown in row 2.

Example 2. *Subtract 0100 from 1011*

Solution.

1				←Borrow
1	0	1	1	←Minuend
–0	1	0	0	←Subtrahend
0	1	1	1	←Difference
↑	↑	↑	↑	
C_3	C_2	C_1	C_0	

There is no problem in column C_0 and C_1. In column C_2 we made 0 –1, so result = 1 and borrow = 1. Then this borrow = 1 is marked in column C_3. So result in column C_2 is 1. Then in column C_3 we first made 1 – 0 to get result = 1 and then we subtracted borrow from result, thus we get 0 in column C_3.

"Thus in subtraction, first subtract the subtrahend bit from minuend and then subtract borrow from the result."

Watch out the next example to further clarify the concept.

Example 3. *Subtract 0110 from 1001*

Solution.

1	1			←Borrow
1	0	0	1	←Minuend
–0	1	1	0	←Subtrahend
0	0	1	1	←Difference
↑	↑	↑	↑	
C_3	C_2	C_1	C_0	

Here, in column C_1 we get difference = 1 and borrow = 1. This borrow is marked in column C_2, and difference = 1 is shown in the column C_1. We now come to column C_2. Here by 0–1 we get difference = 1 and borrow = 1. Now this borrow is marked in column C_3. But in column C_2 already we have borrow so this borrow = 1 is subtracted from difference = 1 which results in 0. Thus the difference = 0 is marked in column C_2.

In the similar way we process column C_3 and we get difference = 0 in column C_3.

Binary Multiplication

Binary multiplication is also similar to decimal multiplication. In binary multiplication if multiplier bit is 0 then partial product is all 0 and if multiplier bit is 1 then partial product is 1. The same is illustrated in example below:

Example 4.

```
    1 0 0 1  ←— MULTIPLICAND
      1 0 1  ←— MULTIPLIER
  -----------
    1 0 0 1  ←— Partial Product when multiplier bit = 1
  0 0 0 0 ×  ←— Partial Product when multiplier bit = 0
1 0 0 1 × ×  ←
  -----------
1 0 1 1 0 1  ←— FINAL PRODUCT
```

Binary Division

Binary division is also similar to decimal division as illustrated in example below:

Example 5.

```
                      1 0 1
Divisor → 1 0 0 1 ) 1 0 1 1 0 1  ← Dividend
                    1 0 0 1
                    × × 1 0 0 1
                        1 0 0 1
                        × × × ×
```

1.7 NEGATIVE NUMBERS AND THEIR ARITHMETIC

So far we have discussed straight forward number representation which are nothing but positive number. The negative numbers have got two representation

(*i*) complement representation.

(*ii*) sign magnitude representation.

We will discuss both the representation in following subsections.

1.7.1 1's and 2's Complement

These are the complements used for binary numbers. Their representation are very important as digital systems work on binary numbers only.

1's Complement

1's complement of a binary number is obtained simply by replacing each 1 by 0 and each 0 by 1. Alternately, 1's complement of a binary can be obtained by subtracting each bit from 1.

Example 1. Find 1's complement of (*i*) 011001 (*ii*) 00100111

Solution. (*i*) Replace each 1 by 0 and each 0 by 1

```
0 1 1 0 0 1
↓ ↓ ↓ ↓ ↓ ↓
1 0 0 1 1 0
```

So, 1's complement of 011001 is 100110.

(*ii*) Subtract each binary bit from 1.

```
  1 1 1 1 1 1 1 1
 −0 0 1 0 0 1 1 1
 ----------------
  1 1 0 1 1 0 0 0  ← 1's complement
```

one can see that both the method gives same result.

2's Complement

2's complement of a binary number can be obtained by adding 1 to its 1's complement.

Example 1. *Find 2's complement of (i) 011001 (ii) 0101100*

Solution. (*i*)

0	1	1	0	0	1	← Number
1	0	0	1	1	0	← 1's complement
				+	1	← Add 1 to 1's complement
1	0	0	1	1	1	← 2's complement

(*ii*)

0	1	0	1	1	0	0	← Number
1	0	1	0	0	1	1	← 1's complement
					+	1	← Add 1 to 1's complement
1	0	1	0	1	0	0	← 2's complement

There is an efficient method to find 2's complement based upon the observation made on the above 2 examples. Consider the number and its 2's complement of example (*ii*) as shown below:

0	1	0	1	1	0	0	← Number
1	0	1	0	1	0	0	← 2's Complement
1's complement				Same as number			

Fig. 1.10 Number and its 2's complement

The above figure clearly shows that to find 2's complement of a binary number start from right towards left till the first 1 appears in the number. Take these bits (including first 1) as it is and take 1's complement of rest of the bits. Workout below examples to enhance your understanding.

Example 2. *Find 2's complement of (i) 101100 (ii) 10010 (iii) 01001*

Solution. (*i*) Number = 101100

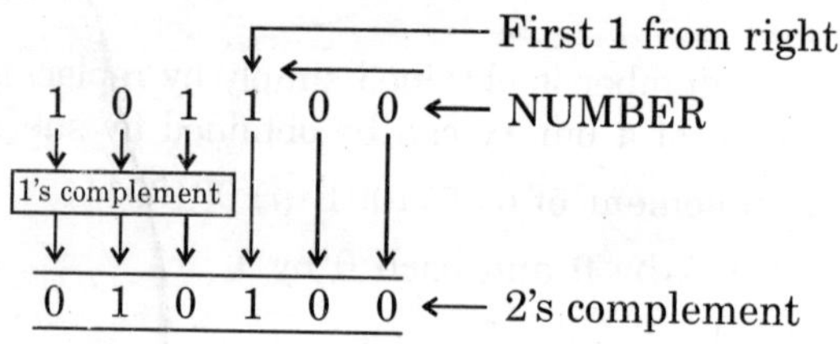

(*ii*) Number = 10010

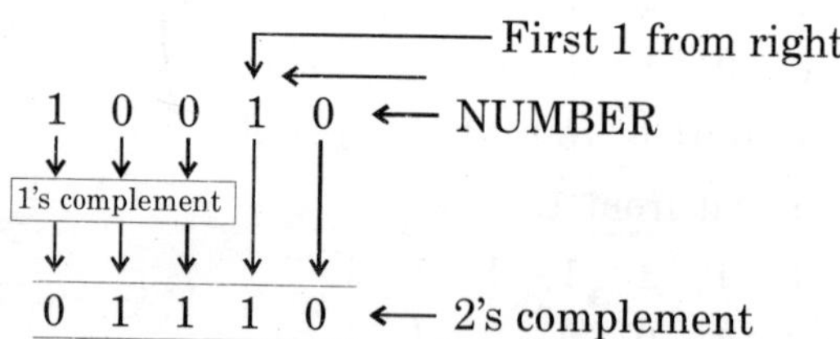

(*iii*) Number = 01001

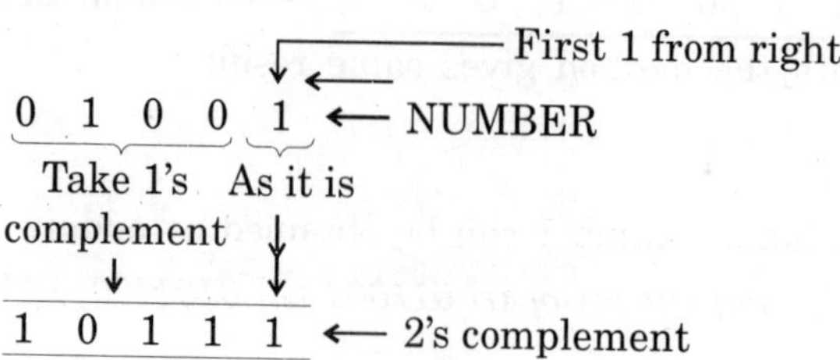

It is interesting to note that taking complement twice leaves the number as it is. This is illustrated in below Fig. 1.11.

Fig. 1.11 Effect of taking complement twice

To represent a negative number using complements the process involves two steps.

(1) obtain the binary representation of equivalent positive number for given negative number. *e.g.*, if given number is –2 then obtain binary representation of +2.

(2) Take the appropriate complement of representation obtained in step 1.

Example 3. *Obtain 1's and 2's complement representation of –5 and –7.*

Solution. (*i*) –5

1. binary of +5 $= (0101)_2$
2. 1's complement of $(0101)_2 = (1010)_2 \leftarrow$ Represents $(-5)_{10}$

   2's complement of $(0101)_2 = (1011)_2 \leftarrow$ Represents $(-5)_{10}$

(*ii*) –7

1. binary of +7 $= (0111)_2$
2. 1's complement of $(0111)_2 = (1000)_2$ Represents $(-7)_{10}$

   2's complement of $(0111)_2 = (1001)_2$ Represents $(-7)_{10}$

Note that in above two examples, for positive numbers we obtained such a binary representation in which MSB is 0. *e.g.*, for +7 we obtained $(0111)_2$ not just $(111)_2$. It is because for all positive numbers MSB must be 0 and for negative numbers MSB should be 1. This will be more clear in subsection 1.7.3.

1.7.2 Subtraction Using 1's and 2's Complement

Before using any complement method for subtraction equate the length of both minuend and subtrahend by introducing leading zeros.

1's complement subtraction following are the rules for subtraction using 1's complement.

1. Take 1's complement of subtrahend.
2. Add 1's complement of subtrahend to minuend.
3. If a carry is produced by addition then add this carry to the LSB of result. This is called as end around carry (EAC).
4. If carry is generated from MSB in step 2 then result is positive. If no carry generated result is negative, and is in 1's complement form.

Example 1. *Perform following subtraction using 1's complement.*

(*i*) 7 – 3 (*ii*) 3 – 7

Solution. (*i*) 7 – 3: binary of 7 $= (0111)_2$

binary of 3 $= (0011)_2$ both numbers have equal length.

Step 1. 1's complement of $(0011)_2 = (1100)_2$

Step 2. Perform addition of minuend and 1's complement of subtrahend

```
                 0 1 1 1  ← (7)
               + 1 1 0 0  ← (–3 or 1's complement of + 3)
Final          ---------
Carry        → 1 0 0 1 1
Step 3. EAC    └──────→ + 1   (EAC)
               ---------
                 0 1 0 0
```

Step 4. Since carry is generated in step 2 the result is positive.

since $(0100)_2 = (4)_{10}$

so, result = +4 which is correct answer

(*ii*) 3 – 7:

binary of 3 = 0011

binary of 7 = 0111

Step 1. 1's complement of 0111 = 1000

Step 2. Perform addition of minuend and 1's complement of subtrahend

```
          0 0 1 1  ← (3)
          1 0 0 0  ← (–7 or 1's complement of + 7)
         ---------
Result =  1 0 1 1
         ---------
```

Step 3. No carry produced so no EAC operation.

Step 4. Since no carry produced in step 2, result is negative and is in complemented form. So we must take 1's complement of result to find correct magnitude of result.

1's complement of result $(1011)_2 = (0100)_2$

so, final result $= -(0100)_2$ or $-(4)_{10}$

Note that when (in example (*ii*) the result was negative (step 2), MSB of the result was 1. When (in example (*i*)) the result was positive the MSB was 0. The same can be observed in 2's complement subtraction.

2's complement Subtraction Method of 2's complement is similar to 1's complement subtraction except the end around carry (EAC). The rules are listed below:

1. Take 2's complement of subtrahend.
2. Add 2's complement of subtrahend to minuend.
3. If a carry is produced, then discard the carry and the result is positive. If no carry is produced result is negative and is in 2's compliment form.

Example 2. *Perform following subtraction using 2's complement.*

(*i*) 7 – 5 (*ii*) 5 – 7

Solution. (*i*) 7 – 5: binary of 7 = $(0111)_2$, binary of 5 = $(0101)_2$ — both the numbers should have equal length.

Step 1. 2's complement of subtrahend $(=0101)_2 = (1011)_2$.

Step 2. Perform addition of minuend and 2's complement of subtrahend.

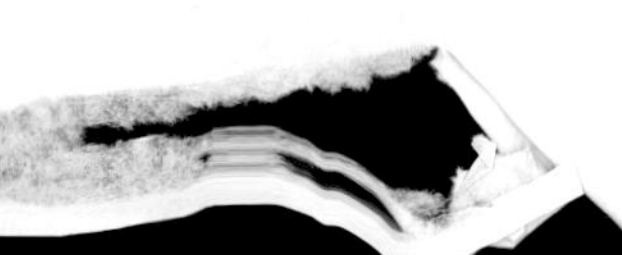

```
          0  1  1  1 ← (7)
        + 1  0  1  1 ← (–5 or 2's complement of +5)
Final   ──────────────
Carry ──→1  0  0  1  0
         ↑
Discard the final carry
```

Step 3. Since a final carry is produced in step 2 (which is discarded) the result is positive. So,

$$\text{result} = (0010)_2 = (2)_{10}$$

(*ii*) 5 – 7:

$$\text{binary of } 5 = (0101)_2$$

$$\text{binary of } 7 = (0111)_2$$

Step 1. 2's complement of subtrahend (= 0111) = 1001.

Step 2. Addition of minuend and 2's complement of subtrahend.

```
   0  1  0  1 ← 5
   1  0  0  1 ← (–7 or 2's complement of +7)
  ────────────
   1  1  1  0 ← Result
 ↑
No final carry
```

Step 3. Since final carry is not generated in step 2, the result is negative and is in 2's complement form. So we must take 2's complement of result obtained in step 2 to find correct magnitude of result.

$$\text{2's complement of result } (1110)_2 = (0010)_2$$

$$\text{so, final result} = -(0010)_2 = -(2)_{10}$$

1.7.3 Signed Binary Representation

Untill now we have discussed representation of unsigned (or positive) numbers, except one or two places. In computer systems sign (+ve or –ve) of a number should also be represented by binary bits.

The accepted convention is to use 1 for negative sign and 0 for positive sign. In signed representation MSB of the given binary string represents the sign of the number, in all types of representation. We have two types of signed representation:

1. Signed Magnitude Representation
2. Signed Complement Representation

In a **signed-magnitude** representation, the MSB represent the sign and rest of the bits represent the magnitude. *e.g.*,

```
+5 = (0  1 0 1)₂        –5 = (1  1 0 1)₂
      ↑   ↑                   ↑   ↑
      | Magnitude             | Magnitude
      |                       |
   + sign                  – sign
```

Note that positive number is represented similar to unsigned number. From the example it is also evident that out of 4-bits, only 3-bits are used to represent the magnitude. Thus in general, $n - 1$ bits are used to denote the magnitude. So, the range of signed representation becomes $-(2^{n-1} - 1)$ to $(2^{n-1} - 1)$.

In a **signed-complement** representation the positive numbers are represented in true binary form with MSB as 0. Whereas the negative numbers are represented by taking appropriate complement of equivalent positive number, including the sign bit. Both 1's and 2's complements can be used for this purpose *e.g.*,

$$+5 = (0101)_2$$
$$-5 = (1010)_2 \leftarrow \text{in 1's complement}$$
$$= (1011)_2 \leftarrow \text{in 2's complement}$$

Note that in signed complement representation the fact remains same that $n - 1$ bits are used for magnitude. The range of numbers

In 1's complement	0 to $(2^{n-1} - 1)$	Positive Numbers
	$-$ 0 to $-(2^{n-1} - 1)$	Negative Numbers
In 2's complement	0 to $(2^{n-1} - 1)$	Positive Numbers
	$-$ 1 to -2^{n-1}	Negative Numbers

To illustrate the effect of these 3 representations, we consider 4-bit binary representation and draw the below table. Carefully observe the differences in three methods.

Decimal	*Signed Magnitude*	*1's complement*	*2's complement*
+0	0 0 0 0	0 0 0 0	0 0 0 0
+1	0 0 0 1	0 0 0 1	0 0 0 1
+2	0 0 1 0	0 0 1 0	0 0 1 0
+3	0 0 1 1	0 0 1 1	0 0 1 1
+4	0 1 0 0	0 1 0 0	0 1 0 0
+5	0 1 0 1	0 1 0 1	0 1 0 1
+6	0 1 1 0	0 1 1 0	0 1 1 0
+7	0 1 1 1	0 1 1 1	0 1 1 1
–8	—	—	1 0 0 0
–7	1 1 1 1	1 0 0 0	1 0 0 1
–6	1 1 1 0	1 0 0 1	1 0 1 0
–5	1 1 0 1	1 0 1 0	1 0 1 1
–4	1 1 0 0	1 0 1 1	1 1 0 0
–3	1 0 1 1	1 1 0 0	1 1 0 1
–2	1 0 1 0	1 1 0 1	1 1 1 0
–1	1 0 0 1	1 1 1 0	1 1 1 1
–0	1 0 0 0	1 1 1 1	—

Fig. 1.12 Different signed representation

From the table, it is evident that both signed Magnitude and 1's complement methods introduce two zeros +0 and – 0 which is awkward. This is not the case with 2's complement. This is one among the reasons that why all the modern digital systems use 2's complement method for the purpose of signed representation. From the above table, it is also evident that in signed representation $\frac{2^n}{2}$ positive numbers and $\frac{2^n}{2}$ negative numbers can be represented with n-bits. Out of 2^n combinations of n-bits, first $\frac{2^n}{2}$ combinations are used to denote the positive numbers and next $\frac{2^n}{2}$ combinations represent the negative numbers.

Example 1. *In a signed representation given binary string is $(11101)_2$. What will be the sign and magnitude of the number represented by this string in signed magnitude, 1's complement and 2's complement representation.*

Solution.

The number N = $(11101)_2$

since MSB = 1 the given number is negative.

(*i*) In signed Magnitude MSB denotes sign and rest of the bits represent magnitude. So,

$$\left(\underset{\uparrow}{1}\ \underbrace{1\ 1\ 0\ 1}_{\text{Magnitude}}\right)_2 = -13$$

sign

(*ii*) In 1's complement if number is negative (*i.e.*, MSB = 1) then the magnitude is obtained by taking 1's complement of given number.

1's complement of $(11101)_2$ = $(00010)_2$

so, $(11101)_2$ = –2 in 1's complement.

(*iii*) In 2's complement if number is negative (*i.e.*, MSB = 1) then magnitude is obtained by taking 2's complement of given number.

2's complement of $(11101)_2$ = $(00011)_2$

= 3

so, $(11101)_2$ = –3 in 2's complement.

Example 2. *Obtain an 8-bit representation of –9 in signed Magnitude, 1's complement and 2's complement representation.*

Solution. We first find binary of 9 *i.e.*, $(9)_{10} = (1001)_2$

Next we represent 9 using 8-bits. So, N = $(00001001)_2$

= $(9)_{10}$

(*i*) In signed Magnitude, MSB shows sign and rest of the bits shows true magnitude. So,

$(-9)_{10} = (10001001)_2$

(*ii*) In 1's complement, negative number is represented by taking 1's complement of positive number. So,

$(-9)_{10}$ = 1's complement of $(00001001)_2$

= $(11110110)_2$

(*iii*) In 2's complement

$$\begin{aligned}(-9)_{10} &= \text{2's complement of } (00001001)_2 \\ &= (11110111)_2\end{aligned}$$

1.7.4 Arithmetic Overflow

When the result of an arithmetic operation requires n+1 bits, upon operating on n-bits number, an overflow occurs. Alternately, if result exceeds the range 0 to $2^n - 1$, an overflow occurs.

Let us consider the addition of two 4-bit numbers

$$\begin{array}{rr} 9 \longrightarrow & 1\ 0\ 0\ 1 \\ +8 \longrightarrow & 1\ 0\ 0\ 0 \\ \hline 17 \longrightarrow & 1\ 0\ 0\ 0\ 1 \\ \hline \end{array}$$

Thus, addition of two 4-bits numbers requires 5-bits (n+1 bits) to represent the sum. Alternately, the result of addition of 4-bits, falls outside the range 0 to 15 (*i.e.*, 0 to 2^4–1). Thus, overflow has occured.

In case of **signed arithmetic** the overflow causes the sign bit of the answer to change. In this case an overflow occurs if the result does not lie in the range -2^{n-1} to $2^{n-1} - 1$. In signed arithmetic overflow can occur only when two positive numbers or two negative numbers are added.

Let us consider 4-bit signed 2's complement representation.

1. Addition of two positive numbers +6 and +5

$$\begin{array}{rr} +6 \longrightarrow & 0\ 1\ 1\ 0 \\ + \; +5 \longrightarrow & 0\ 1\ 0\ 1 \\ \hline +11 & 1\ 0\ 1\ 1 \\ \hline \end{array}$$

Since, MSB of result is 1, if reflects a negative result which is incorrect. It happened because overflow has changed the sign of result.

2. Addition of two negative numbers –6 and –5

$$\begin{array}{rrl} -6 \longrightarrow & 1\ 0\ 1\ 0 & \longleftarrow \text{2's complement of 6} \\ -5 \longrightarrow & 1\ 0\ 1\ 1 & \longleftarrow \text{2's complement of 5} \\ \hline -11 & 1\ 0\ 1\ 0\ 1 & \end{array}$$

$\uparrow$ Carry

In 2's complement if a carry is generated after the addition then carry is discarded and result is declared positive. Thus, result = $(0101)_2$ = +5 which is wrong, because addition of two negative numbers should give a negative result. This happened due to overflow.

Note that overflow is a problem that occurs when result of an operation exceeds the capacity of storage device. In a computer system, the programmer must check the overflow after each arithmetic operation.

1.7.5 9's and 10's Complement

9's and 10's complements are the methods used for the representation of decimal numbers. They are identical to the 1's and 2's complements used for binary numbers.

9's complement: 9's complement of a decimal number is defined as $(10^n - 1) - N$, where n is no. of digits and N is given decimal numbers. Alternately, 9's complement of a decimal number can be obtained by subtracting each digit from 9.

$$\boxed{\textbf{9's complement of N} = (10^n-1) - N.}$$

Example 1. *Find out the 9's complement of following decimal numbers.*

(i) 459 *(ii) 36* *(iii) 1697*

Solution. (*i*) By using $(10^n-1) - N$; But, $n = 3$ in this case

So, $(10^n-1) - N = (10^3 - 1) - 459 = 540$

Thus, 9's complement of 459 = 540

(*ii*) By subtracting each digit from 9

$$\begin{array}{r} 9\ 9 \\ -3\ 6 \\ \hline 6\ 3 \\ \hline \end{array}$$

So, 9's complement of 36 is 63.

(*iii*) We have $N = 1697$, so $n = 4$

Thus, $10^n-1 = 10^4 - 1 = 9999$

So, $(10^n-1) - N = (10^4-1) - 1697 = 9999 - 1697$

$= 8302$

Thus, 9's complement of 1697 = 8302

10's complement: 10's complement of a decimal number is defined as $10^n - N$.

$$\boxed{\textbf{10's complement of N} = 10^n - N}$$

but $10^n - N = (10^n - 1) - N + 1$

$= 9\text{'s complement of } N + 1$

Thus, 10's complement of a decimal number can also be obtained by adding 1 to its 9's complement.

Example 2. *Find out the 10's complement of following decimal numbers. (i) 459 (ii) 36.*

Solution. (*i*) By using $10^n - N$; We have $N = 459$ so $n = 3$

So, $10^n - N = 10^3 - 459 = 541$

So, 10's complement of 459 = 541

(*ii*) By adding 1 to 9's complement

9's complement of 36 = 99 – 36

= 63

Hence, 10's complement of 36 = 63 + 1

= 64

1.7.6 r's Complement and (r – 1)'s Complement

The r's and $(r - 1)$'s complements are generalized representation of the complements, we have studied in previous subsections. r stands for radix or base of the number system, thus r's complement is referred as *radix complement* and $(r - 1)$'s complement is referred as *diminished radix complement*. Examples of r's complements are 2's complement and 10's complement. Examples of $(r - 1)$'s complement are 1's complement and 9's complement.

In a base-r system, the r's and $(r - 1)$'s complement of the number N having n digits, can be defined as:

$$\boxed{(r - 1)\text{'s complement of N} = (r^n - 1) - N}$$

and

$$\boxed{\begin{aligned} r\text{'s complement of N} &= r^n - N \\ &= (r - 1)\text{'s complement of N} + 1 \end{aligned}}$$

The $(r - 1)$'s complement can also be obtained by subtracting each digit of N from $r-1$. Using the above methodology we can also define the 7's and 8's complement for octal system and 15's and 16's complement for hexadecimal system.

1.7.7 Rules for Subtraction using (r–1)'s Complement

Let M (minuend) and S (subtrahend) be the two numbers to be used to evaluate the difference D = M – S by $(r - 1)$'s complement and either or both the numbers may be signed or unsigned.

Until and unless specified the given rules are equally applied for both signed and unsigned arithmetic. For the clarity of process, let us assume that two data sets are:

Unsigned data— $M_u = 1025$, $S_u = 50$ and $D_u = M_u - S_u$

Signed data— $M_s = -370$, $S_s = 4312$ and $D_s = M_s - S_s$

Step 1. Equate the Length

Find out the length of both the numbers (no. of digit) and see if both are equal. If not, then make both the numbers equal by placing leading zeroes.

$$M_u = 1025,\ S_u = 50 \rightarrow S_u = 0050$$

$$M_s = -370,\ S_s = 4312 \rightarrow M_s = -0370$$

Step 2. Represent Negative Operands (for Negative Numbers only)

If either or both the operands are negative then take the $(r - 1)$'s complement of the number as obtained in step 1.

$$M_s = -370 \quad (r - 1)\text{'s of } M_s = 9999 - 0370 = 9629$$

$$S_s = 4312$$

Step 3. Complement the Subtrahend

In order to evaluate difference take the $(r - 1)$'s complement of the representation obtained for the subtrahend S_u in step 1 and S_s in step 2.

$$S_u = 0050,\ (r - 1)\text{'s of } S_u = 9999 - 0050 = 9949 \text{ and } M_u = 1025$$

$$S_s = 4312,\ (r - 1)\text{'s of } S_s = 9999 - 4312 = 5687 \text{ and } M_s = 9629$$

Step 4. Addition and the Carry (CY)

Add the two numbers in the step 3 and check whether or not carry generated from MSB due to addition.

$$M_u = 1025,\ S_u = 9949 \rightarrow D_u = M_u - S_u = \underset{\text{CY}}{\underset{\downarrow}{1}}0974$$

$M_s = 9629$, $S_s = 5687 \rightarrow D_s = M_s - S_s = 15316$

$$\begin{array}{c} 15316 \\ \downarrow \\ \text{CY} \end{array}$$

Step 5. Process the Carry (CY)

In step 4, we obtained result as CY, D. The CY from MSB contains some useful information especially in some unsigned arithmetic. Processing of carry for $(r - 1)$'s complement is

- In this case if a carry is generated from MSB in step 4, add this carry to the LSB of the result. In step 4, we got CY = 1, $D_u = 0974$ also CY = 1, $D_s = 5316$. After adding carry to the LSB (generated in step 4) we get, $D_u = 0974 + 1 \rightarrow 0975$ and $D_s = 5316 + 1 \rightarrow 5317$. The carry in this case is called "end-around carry".

Step 6. Result Manipulation

The way result is manipulated is different for signed and unsigned arithmetic.

(*a*) UNSIGNED

1. If a carry is generated in step 4 then the result is positive (+) and the digits in the result shows the correct magnitude of result.
2. If there is no carry from MSB in step 4 then the result is negative (–) and the digits in result is not showing the correct magnitude. So, we must go for a post processing of result (Step 7) of result to determine the correct magnitude of the result.

(*b*) SIGNED

1. If the MSB of result obtained in step 5 is lesser than the half radix (*i.e.*, MSB $< r/2$) then the result is +ve and representing the correct magnitude. Thus, no post processing is required.
2. If the MSB of result obtained in step 5 is not lesser than the half radix (*i.e.*, MSB $\geq r/2$) then the result is –ve and correct magnitude of which must be obtained by post processing (Step 7).

Step 7. Post Processing and Result Declaration

By the step 6(*a*) – 1 and the step 6(*b*) – 1 we know that if the result is positive (+ve) it represents the correct magnitude whether it is signed or unsigned arithmetic. However, the negative results are not showing correct magnitudes *so post processing in principle is needed for declaration of negative results.*

(*a*) Declare positive results. As per the rules the result of the unsigned arithmetic is positive. Therefore,

$$D_u = +0975$$

(*b*) Process and declare negative results. As per the rules result of signed arithmetic is negative and is in complemented form. Take the $(r - 1)$'s complement to find the complement and declare the result.

$$(r - 1)\text{'s of } D_s = 9999 - 5317 = -4682$$

Therefore, $$D_s = -4682$$

1.7.8 Rules for Subtraction using r's Complement

For better understanding and clarity, let us assume the same data sets for (r – 1)'s complement method:

Unsigned data— $M_u = 1025$, $S_u = 50$ and $D_u = M_u - S_u$

Signed data— $M_s = -370$, $S_s = 4312$ and $D_s = M_s - S_s$

Step 1. Equate the Length

Same as for (r – 1)'s complement *i.e.*

$M_u = 1025, S_u = 50 \rightarrow S_u = 0050$

$M_s = -370, S_s = 4312 \rightarrow M_s = -0370$

Step 2. Represent Negative Operands

Take the r's complement of negative operands

$M_s = -370$, r's of $M_s = 9999 - 370 + 1 = 9630$

$S_s = 4312$

Step 3. Complement the Subtrahend

Take the r's complement of the representation obtained for the subtrahend S_u in step 1 and S_s in step 2 to evaluate the difference

$S_u = 0050$, r's of $S_u = 10000 - 0050 = 9950$ and $M_u = 1025$

$S_s = 4312$, r's of $S_s = 10000 - 4312 = 5688$ and $M_s = 9630$

Step 4. Addition and Carry (CY)

Add the two numbers in the step 3 and check whether or not carry generated from MSB due to addition. (Same as (r – 1)'s complement).

$M_u = 1025, S_u = 9950 \rightarrow D_u = 10975$

↓

CY

$M_s = 9630, S_s = 5688 \rightarrow D_s = 15318$

↓

CY

Step 5. Process the Carry (CY)

If there is carry from MSB in step 4 then simply discard it. In step 4, we got CY = 1, $D_u = 0975$ also CY = 1, $D_s = 5318$. After discarding the carry we get, $D_u = 0975$ and $D_s = 5318$.

Step 6. Result Manipulation

The way result is manipulated is different for signed and unsigned arithmetic.

(*a*) UNSIGNED

1. If a carry is generated in step 4 then the result is positive (+) and the digits in the result shows the correct magnitude of result.

2. If there is no carry from MSB in step 4 then the result is negative (–) and the digits in result is not showing the correct magnitude. So, we must go for a post processing of result (Step 7) to determine the correct magnitude of the result.

(*b*) SIGNED

1. If the MSB of result obtained in step 5 is lesser than the half radix (*i.e.*, MSB < $r/2$) then the result is +ve and representing the correct magnitude. Thus no post processing is required.
2. If the MSB of result obtained in step 5 is not lesser than the half radix (*i.e.*, MSB $\geq r/2$) then the result is –ve and correct magnitude of which must be obtained by post processing (Step 7).

Step 7. Post Processing and Result Declaration

(*a*) Declare positive results. As per the rules, the positive result shows the correct magnitude. Since, the result of the unsigned arithmetic is positive. Therefore,

$$D_u = +0975$$

(*b*) Process and declare negative results. As per the rule, the result obtained of signed arithmetic is negative and is in complemented form. Take the *r*'s complement to find the complement and declare the result.

$$r\text{'s of } D_s = 10000 - 5318 = -4682$$

Therefore,

$$D_s = -4682$$

1.7.9 Floating-Point Representation

Real numbers are presented in the computer in either fixed-point or floating-point form. The floating-point format is used to represent very large (or very small) numbers needed in scientific applications.

The floating-point representation of a number has *two* parts: the first part represents the signed or fixed-point number, called **mantissa** (or significand) and the second part represents the position of radix point in the number, called the **exponent.**

In general, a floating-point number *F*, is represented as,

$$F = \mp M \times r^E$$

where $M \Rightarrow$ mantissa (or significand), is a fixed point number

$E \Rightarrow$ exponent (also called the characteristics), is an integer

$r \Rightarrow$ base of the number.

For example, the decimal number +1632.789 is represented as floating point

mantissa	exponent
+0.6132	+04

The value of the exponent indicates that the actual position of the decimal point is four positions to the right of the indicated decimal point in the number. This representation is equivalent to the scientific notation $+ 0.6132789 \times 10^{+4}$.

Example 1. *Represent 1.23 in floating-point format, using the exponents of 0, 1 and 2.*

Solution. These representations are respectively

$$1.23 = 1.23 \times 10^0 = 0.123 \times 10^1 = 0.0123 \times 10^2$$

Binary Representation of Floating-Point

In binary representation of floating-point number, we have

$$F = (-1)^S M \times 2^E$$

Where M and E are as stated as previous but M representing a fractional part only, S is a single bit representing the sign of the number with $S = 0$ indicates a positive number and $S = 1$ indicates a negative number.

The range of floating point numbers depends on the number of bits allocated to each part of the mantissa and the exponent.

Example 2. *Give the 7-bit code word 1110101, decode the word assuming the code represents:*

(i) *An unsigned integer*

(ii) *A signed integer in 2's complement*

(iii) *A floating-point number of the form S, E, M with a 3-bit field.*

Solution. *(i)* As an unsigned integer, the code represents $2^6 + 2^5 + 2^4 + 2^2 + 2^0 = (117)_{10}$.

(ii) As a signed number in 2's complement, the code represents $-(128 - 117) = (-11)_{10}$.

(iii) As a floating-point number, we have

1	110	101
S : Sign	E : Exponent	M : Mantissa

∴ The code represents

$$(-1)^S M \times 2^E = -(0.101)_2 \times 2^{(110)_2}$$

$$= -0.625 \times 2^6$$

NOTE:

In the previous example, the same code word could be used to represent different data objects. Note that the number of maximum allowable objects that can be encoded does not exceed the range imposed by the size of the code word (2^7). As a result, if the code represents unsigned integers, then the range is the integers 0 through 127.

Similarly, if the code represents signed numbers in 2's complement, then this range is –64 to 63.

Finally, if the code represents a floating-point number then the largest possible number in the range corresponds to the code 0111111, which evaluates to $0.875 \times 128 = 112$. Similarly, the smallest number in the range would correspond to –112.

Since the floating-point rerpresentation is used to represent real number and since the maximum possible object that can be encoded is 128, there are many real numbers between –112 and + 112 that have no corresponding floating-point representation.

As a result, increasing the range size decreases the number of possible representations within a given sub-range.

Normalized and Biased Floating-Point Representation

Any floating-point number can be expressed in many ways. A floating-point number is said to be **normalized** if the most significant digit of the mantissa is non-zero. For binary numbers, a normalized number is one in which most significant bit of the mantissa is 1. The 8-bit binary number 00011010 is not normalized because of the three leading 0's. Any number that is not normalized may be normalized by shifting the radix point to the right of the leftmost 1 bit and adjust the exponent accordingly. This process is known as **normalization**.

In addition to represent a number in normalized form, the exponents of a floating-point number are generally represented as a biased signed (Excess-m) integer.

Here, bias = $2^{n-1} - 1$

where n is the number of bits in the exponent field. This bias is added to the exponent to produce a biased exponent.

Following figure shows a typical 32-bit floating-point format:

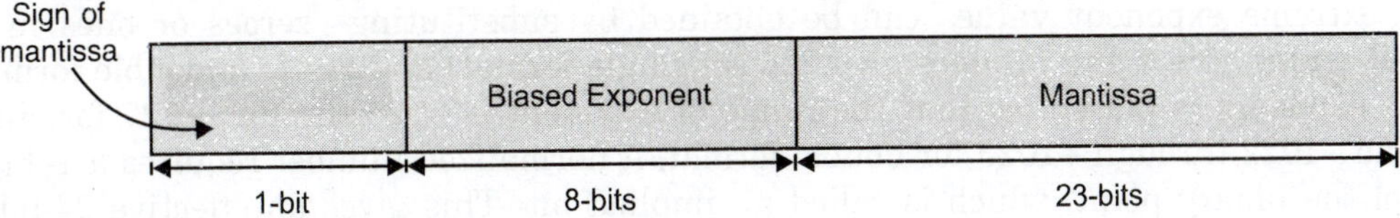

The leftmost bit stores the sign of the number (0 = positive, 1 = negative). The **exponent** value is stored in the next 8 bits.

This representation is known as **biased representation**. A fixed value (called bias) is *subtracted from the field to get the free exponent value*. The final portion of the word (i.e., 23-bits here) is the mantissa (or significand).

- Normalized floating-point numbers provide the maximum possible precision for the floating-point number.
- A zero cannot be normalized because it does not have a non-zero digit. It is represented by all 0's in the mantissa and exponent.
- The number represented in floating-point notation are not spaced evenly along the number line, as are real-numbers.

IEEE Standard for Binary Floating-Point Representation

IEEE Standard 754 defined by IEEE (*Institute of Electrical and Electronics Engineers*) for floating-point representation was developed to facilitate the portability of programs.

For base 2, IEEE standard defines following *two* formats:

1. a 32-bit single format with 8-bit exponent
2. a 64-bit double format with 16-bit exponent.

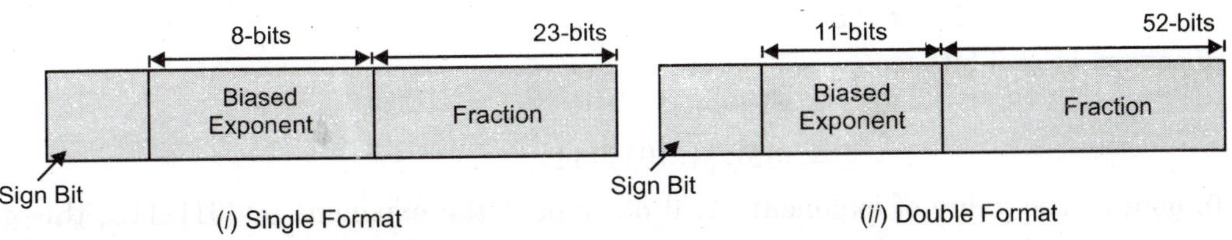

(*i*) Single Format (*ii*) Double Format

In addition, the standard defines two extended formats for single and double both named as single extended and double extended. These extended formats are implementation dependent.

All bit patterns in IEEE formats are not interpreted in the usual way. Some bit patterns represent special values.

- *An exponent of zero with a fraction of zero.* This represents positive or negative zero, depends upon the sign bit.
- *An exponent of all ones with a fraction of zero.* This represents positive or negative infinity depending on the sign bit.
- *An exponent of zero with a nonzero fraction.* This represents a denormalized number. In this case, the bit to the left of the binary point is zero and the true exponent is –126 (in single format) or –1022 (in double format). This number is positive or negative depends on the sign bit.
- *An exponent of all ones with a nonzero fraction.* This gives the value of NaN (Not a number). It is used to signal various exception conditions.

The extreme exponent values can be obtained by substituting zeroes or ones in all places. All zeroes give 0 and all ones give 255 (in single format) and 2047 (in double format). Here, the exponent is biased, so that the range of exponent is –126 through +127 for single format and –1022 through +1023 for double format. A normalized number requires a 1–bit to the left of the binary point, which is called an implied bit. This gives an effective 24-bit or 53-bit significand (called fraction in the standard).

Example 3. *What would be the bias value for*

(*a*) *A base-2 exponent in a 6-bit field?*

(*b*) *A base-8 exponent in a 7-bit field?*

Solution. Using, bias = $r^{n-1} - 1$, where n is the number of bits in the exponent field:

(*a*) For $r = 2$, $n = 6$

$$\text{bias} = 2^{6-1} - 1 = 31$$

(*b*) For $r = 8$, $n = 7$

$$\text{bias} = 8^{7-1} - 1 = 8^6 - 1 = 2,62,143$$

Example 4. *Given the normalized biased floating-point number shown below, find its decimal value:*

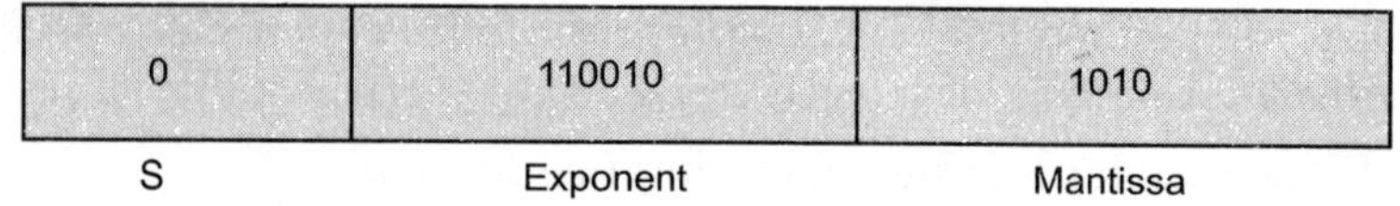

Solution. The sign bit is 0, so the number is positive.

Mantissa is 0.1010, which is

$$0.5 + 0.125 = 0.625$$

Exponent is of 6-bits, so

$$\text{bias} = 2^{6-1} - 1$$
$$= (31)_{10} = (011111)_2$$

To obtain true value of exponent, we'll decrement the exponent by $(011111)_2$, this gives

$$110010 - 011111 = 010011$$
$$= (19)_{10}$$

Hence, the decimal value of the number is 0.625×2^{19}

Example 5. *Consider a floating-point format with 8 bits for the biased exponent and 23 bits for the significand. Show the bit pattern for the following numbers in this format:*

(*a*) – 720 (*b*) 0.645

Solution. (*a*) $-(720)_{10} = -(1011010000)_2$

Normalized fraction is $0.1011010000 \times 2^{(10)_{10}}$

$$\text{Bias} = 2^{8-1} - 1$$
$$= 127$$

Bias exponent is $127 + 10 = (137)_{10}$
$$= (10001001)_2$$

∴ Floating-point representation is

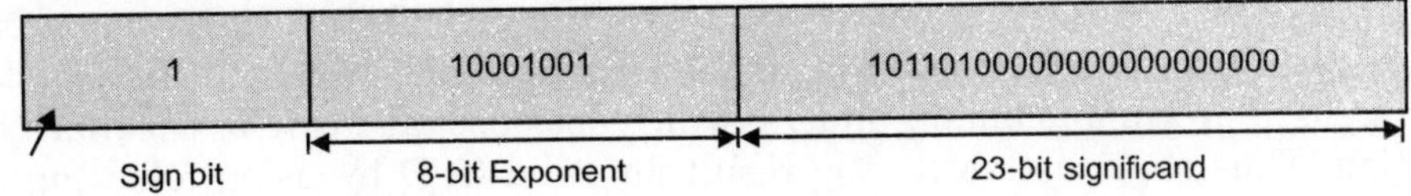

(*b*) $(0.645)_{10} = (0.101001)_2$

Here, exponent = 0

∴ Biased exponent = 127 + 0 = 127
$$= (01111111)_2$$

Thus, floating-point representation is

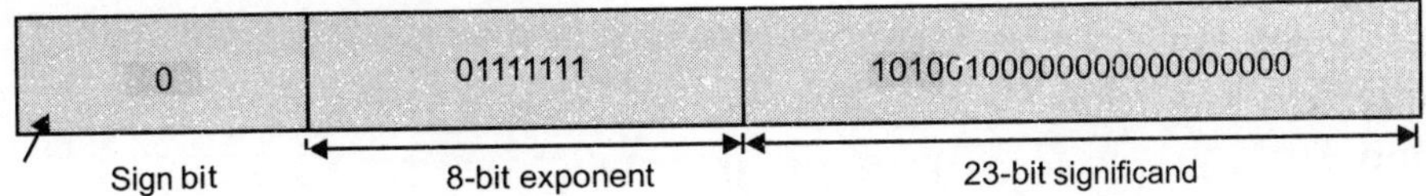

1.8 BINARY CODED DECIMAL (BCD) AND ITS ARITHMETIC

The BCD is a group of four binary bits that represent a decimal digit. In this representation each digit of a decimal number is replaced by a 4-bit binary number (*i.e.*, a nibble). Since a decimal digit is a number from 0 to 9, a nibble representing a number greater than 9 is invalid BCD. For example $(1010)_2$ is invalid BCD as it represents a number greater than 9. The table shown in Fig. 1.13 lists the binary and BCD representation of decimal numbers 0 to 15. Carefully observe the difference between binary and BCD representation.

Decimal Number	*Binary Representation*	*BCD Representation*
0	0 0 0 0	0 0 0 0
1	0 0 0 1	0 0 0 1
2	0 0 1 0	0 0 1 0
3	0 0 1 1	0 0 1 1
4	0 1 0 0	0 1 0 0
5	0 1 0 1	0 1 0 1
6	0 1 1 0	0 1 1 0
7	0 1 1 1	0 1 1 1
8	1 0 0 0	1 0 0 0
9	1 0 0 1	1 0 0 1

10	1010	0001 0000
11	1011	0001 0001
12	1100	0001 0010
13	1101	0001 0011
14	1110	0001 0100
15	1111	0001 0101

Fig. 1.13 Binary and BCD representation of decimal numbers

BCD Addition: In many applications it is required to add two BCD numbers. But the adder circuits used are simple binary adders, which does not take care of peculiarity of BCD representation. Thus one must verify the result for valid BCD by using following rules:

1. If Nibble (*i.e.,* group of 4-bits) is less than or equal to 9, it is a valid BCD number.
2. If Nibble is greater than 9, it is invalid. Add 6 (0110) to the nibble, to make it valid.

   Or

   If a carry was generated from the nibble during the addition, it is invalid. Add 6 (0110) to the nibble, to make it valid.
3. If a carry is generated when 6 is added, add this carry to next nibble.

Example 1. *Add the following BCD numbers. (i) 1000 and 0101 (ii) 00011001 and 00011000*

Solution. (*i*)

```
    1 0 0 0  →    8
  + 0 1 0 1  →  + 5
    -------      ---
    1 1 0 1      13
```

Since, $(1101)_2 > (9)_{10}$ add 6 (0110) to it

So,

```
      1 1 0 1
      0 1 1 0
    ---------
    1 0 0 1 1
    ⏟ ⏟⏟⏟⏟⏟⏟⏟
    1    3
```

So, result = 00010011

(*ii*)

```
               1            ← Carry generated from nibble
  0 0 0 1   1 0 0 1  →    19
  0 0 0 1   1 0 0 0  →   +18
  -----------------      ---
  0 0 1 1   0 0 0 1       37
```

Since, a carry is generated from right most nibble we must add 6 (0110) to it.

So,

```
  0 0 1 1   0 0 0 1
            0 1 1 0
  -----------------
  0 0 1 1   0 1 1 1  → (37)10
```

So, result = 00110111

BCD Subtraction. The best way to carry out the BCD subtraction is to use complements. The 9's and 10's complement, studied in subsection 1.7.5, are exclusively used for this purpose. Although any of the two complements can be used, we prefer 10's complement for subtraction. Following are the steps to be followed for BCD subtraction using 10's complement:

1. Add the 10's complement of subtrahend to minuend.
2. Apply the rules of BCD addition to verify that result of addition is valid BCD.
3. Apply the rules of 10's complement on the result obtained in step 2, to declare the final result *i.e.*, to declare the result of subtraction.

Example 2. *Subtract 61 from 68 using BCD.*

Solution. To illustrate the process first we perform the subtraction using 10's complement in decimal system. After that we go for BCD subtraction.

we have, D = 68 – 61

So, 10's complement of 61 = 99 – 61 + 1 = 39

So,

```
  6 8
+ 3 9
 1 0 7
 ↑
Carry
```

In 10's complement if an end carry is produced then it is discarded and result is declared positive. So,

D = +07

by using BCD

1.

```
                   1               ← Carry generated
                                     from nibble
BCD of 68 =    0 1 1 0 1 0 0 0
             +
BCD of 39 =    0 0 1 1 1 0 0 1
               ---------------
               1 0 1 0 0 0 0 1
```

2. Check for valid BCD– since a carry is generated from right most nibble, we must add 6 (0110) to it. Since the left most nibble is greater than 9, we must add 6 (0110) to it.

Thus,

```
     1 0 1 0  0 0 0 1
 +   0 1 1 0  0 1 1 0
   ------------------
   1 0 0 0 0  0 1 1 1
   ↑
 Carry
```

3. Declaration of result – We got end carry in step 2. In 10's complement arithmetic, end carry is discarded and result is declared positive. Hence,

$$D = (00000111)_2 = (7)_{10}$$

1.9 CODES

Coding and encoding is the process of assigning a group of binary digits, commonly referred to as 'bits', to represent, identify, or relate to a multivalued item of information. By assigning each item of information a unique combination of bits (1's and o's), we transform some given information into another form. In short, a code is a symbolic representation of an information transform. The bit combination are referred to as 'CODEWORDS'.

There are many different coding schemes, each having some particular advantages and characteristics. One of the main efforts in coding is to standardize a set of universal codes that can be used by all.

In a broad sense we can classify the codes into five groups:

(*i*) Weighted Binary codes
(*ii*) Non-weighted codes
(*iii*) Error–detecting codes
(*iv*) Error–correcting codes
(*v*) Alphanumeric codes.

1.9.1 Weighted Binary Codes

In weighted binary codes, each position of a number represents a specific weight. The bits are multiplied by the weights indicated; and the sum of these weighted bits gives the equivalent decimal digit. We have been familiar with the binary number system, so we shall start with straight binary codes.

(*a*) **Straight Binary coding** is a method of representing a decimal number by its binary equivalent. A straight binary code representing decimal 0 through 7 is given in Table 2.

Table 2

Decimal	*Three bit straight Binary Code*	*Weights MOI* 2^2	2^1	2^0	*Sum*
0	000	0	0	0	0
1	001	0	0	1	1
2	010	0	2	0	2
3	011	0	2	1	3
4	100	4	0	0	4
5	101	4	0	1	5
6	110	4	2	0	6
7	111	4	2	1	7

In this particular example, we have used three bits to represent 8 distinct elements of information *i.e.*, 0 through 7 in decimal form.

Now the question arises, if n elements of information are to be coded with binary (two valued bits), then how many bits are required to assign each element of information a unique code word (bit combination). Unique is important, otherwise the code would be ambiguous.

The best approach is to evaluate how many code words can be derived from a combination of n bits.

For example: Let n = no. of bits in the codeword and x = no. of unique words

Now, if

$$n = 1, \text{ then } x = 2 \ (0, 1)$$

$$n = 2, \text{ then } x = 4 \ (00, 01, 10, 11)$$

$$n = 3, \text{ then } x = 8 \ (000, 001, ..., 111)$$

and in general, $n = j$, then $x = 2^j$

that is, if we have available j no. of bits in the code word, we can uniquely encode max 2^j distinct elements of information.

Inversely, if we are given x elements of information to code into binary coded format, the following condition must hold:

$$x \leq 2^j$$

or $j \geq \log_2 x$

or $j \geq 3.32 \log_{10} x$

where j = number of bits in code word.

Example 1. *How many bits would be required to code the 26 alphabetic characters plus the 10 decimal digits.*

Solution. Here we have total 36 discrete elements of information.

i.e., $x = 36$

Now $j \geq \log_2 x$

therefore, $j \geq \log_2 36$ or $j \geq 3.32 \log_{10} 36$

or $j \geq 5.16$ bits

Since bits are not defined in fractional parts, we know $j \geq 6$.

In other words, a minimum of 6 bit code is required that leaves 28 unused code words out of the 64 which are possible ($2^6 = 64$ and $64 - 36 = 28$).

This system of straight binary coding has the disadvantage that the large numbers require a great deal of hardware to handle with. For example if we have to convert decimal 2869594 to straight binary code a regrous division of this number by 2 is required untill we get remainder 0 or 1.

The above difficulty is overcomed by using another coding scheme called as BCD codes.

(*b*) **Binary Codes Decimal Codes (BCD codes).** In BCD codes, individual decimal digits are coded in binary notation and are operated upon singly. Thus binary codes representing 0 to 9 decimal digits are allowed. Therefore, all BCD codes have at least four bits ($\because$ min. no. of bits required to encode to decimal digits = 4)

For example, decimal 364 in BCD

$$3 \rightarrow 0011$$
$$6 \rightarrow 0110$$
$$4 \rightarrow 0100$$
$$364 \rightarrow 0011\ 0110\ 0100$$

However, we should realize that with 4 bits, total 16 combinations are possible (0000, 0001, ..., 11 11) but only 10 are used (0 to 9). The remaining 6 combinations are unvalid and commonly referred to as 'UNUSED CODES'.

There are many binary coded decimal codes (BCD) all of which are used to represent decimal digits. Therefore, all BCD codes have atleast 4 bits and at least 6 unassigned or unused code words shown in Table 3.

Some example of BCD codes are:

(*a*) 8421 BCD code, sometimes referred to as the Natural Binary Coded Decimal Code (NBCD);

(*b*)* Excess-3 code (XS3);

(*c*)** 84 –2 –1 code (+8, +4, –2, –1);

(*d*) 2 4 2 1 code

Example 2. Lowest $[643]_{10}$ into XS3 code

Decimal	6 4 3
Add 3 to each	3 3 3
Sum	9 7 6

Converting the sum into BCD code we have

9	7	6
↓	↓	↓
1001	0111	0110

Hence, XS3 for $[643]_{10}$ = 1001 0111 0110

Table 3. BCD codes

Decimal Digit	*8421 (NBCD)*	*Excess-3 code (XS3)*	*84–2–1 code*	*2421 code*
0	0000	0011	0000	0000
1	0001	0100	0111	0001
2	0010	0101	0110	0010
3	0011	0110	0101	0011
4	0100	0111	0100	0100
5	0101	1000	1011	1011
6	0110	1001	1010	1100
7	0111	1010	1001	1101
8	1000	1011	1000	1110
9	1001	1100	1111	1111

* XS3 is an example of nonweighted code but is a type of BCD code. It is obtained by adding 3 to a decimal number. For example to encode the decimal number 7 into an excess 3 code. We must first add 3 to obtain 10. The 10 is then encoded in its equivalent 4 bit binary code 1010. Thus as the name indicates, the XS3 represents a decimal number in binary form, as a number greater than 3.

** Dashes (–) are minus signs.

There are many BCD codes that one can develop by assigning each column or bit position in the code, some weighting factor in such a manner that all of the decimal digits can be coded by simply adding the assigned weights of the 1 bits in the code word.

For example: 7 is coded 0111 in NBCD, which is interpreted as

$$0 \times 8 + 1 \times 4 + 1 \times 2 + 1 \times 1 = 7$$

The NBCD code is most widely used code for the representation of decimal quantities in a binary coded format.

For example: (26.98) would be represented in NBCD as

$$\begin{array}{cccccc} & & 2 & 6 & 9 & 8 \\ (26.98)_{10} & = & (0010 & 0110. & 1001 & 1000)\ \text{NBCD} \end{array}$$

It should be noted that on the per digit basis the NBCD code is the binary numeral equivalent of the decimal digit it represents.

Self complementing BCD codes

The excess-3, 84–2–1 and 2421 BCD codes are also known as self complementing codes.

Self complementing property– 9's complement of the decimal number is easily obtained by changing 1's to 0's and 0's to 1's in corresponding codeword or the 9's complement of self complementing code word is the same as its logical complement.

When arithmetic is to be performed, often an arithmetic "complement" of the numbers will be used in the computations. So these codes have a particular advantage in machines that use decimal arithmetic.

Example 3. *The decimal digit 3 in 8.4–2–1 code is coded as 0101. The 9's complement of 3 is 6. The decimal digit 6 is coded as 1010 that is 1's complement of the code for 3. This is termed as self complementing property.*

1.9.2 Non-Weighted Codes

These codes are not positionally weighted. This means that each position within a binary number is not assigned a fixed value. Excess-3 codes and Gray codes are examples of non-weighted codes.

We have already discussed XS3 code.

Gray code (Unit Distance code or Reflective code)

There are applications in which it is desirable to represent numerical as well as other information with a code that changes in only one bit position from one code word to the next adjacent word. This class of code is called a unit distance code (UDC). These are sometimes also called as 'cyclic', 'reflective' or 'gray' code. These codes finds great applications in Boolean function minimization using Karnaugh map.

The gray code shown in Table 4 is both reflective and unit distance.

Table 4. Gray codes*

Decimal Digit	*Three bit Gray code*	*Four bit Gray code*
0	0 0 0	0 0 0 0
1	0 0 1	0 0 0 1
2	0 1 1	0 0 1 1
3	0 1 0	0 0 1 0
4	1 1 0	0 1 1 0
5	1 1 1	0 1 1 1
6	1 0 1	0 1 0 1
7	1 0 0	0 1 0 0
8	–	1 1 0 0
9	–	1 1 0 1
10	–	1 1 1 1
11	–	1 1 1 0
12	–	1 0 1 0
13	–	1 0 1 1
14	–	1 0 0 1
15	–	1 0 0 0

Binary to Gray Conversion

(1) Place a leading zero before the most significant bit (MSB) in the binary number.

(2) Exclusive-OR (EXOR) adjacent bits together starting from the left of this number will result in the Gray code equivalent of the binary number.

Exclusive–OR– If the two bits EX–OR'd are identical, the result is 0; if the two bits differ, the result is 1.

---

*Gray codes are formed by reflection. The technique is as follows:

In binary we have two digits 0 and 1.

Step I. Write 0 and 1 and put a mirror, we first see 1 and then 0. Place 0's above mirror and 1's below mirror

We have got gray code for decimal digits 0 through 4.

Step II. Write those 4 codes and again put a mirror. The code will look in the order 10, 11, 01 and 00. Then place 0's above mirror and 1's below mirror.

Proceeding interactively in the same manner. We can form Gray code for any decimal digit.

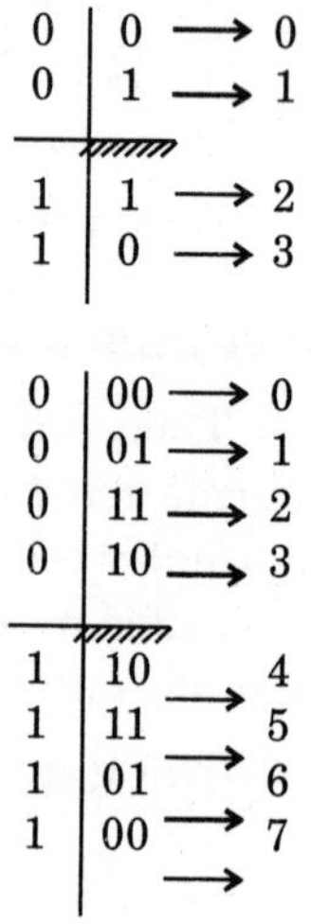

0 | 0 → 0
0 | 1 → 1
1 | 1 → 2
1 | 0 → 3

0 | 00 → 0
0 | 01 → 1
0 | 11 → 2
0 | 10 → 3
1 | 10 → 4
1 | 11 → 5
1 | 01 → 6
1 | 00 → 7

Example. *Convert binary 1010010 to Gray code word.*

```
 0  1  0  1  0  0  1  0
 /\ /\ /\ /\ /\ /\ /\ /
  1  1  1  1  0  1  1
```

$\Rightarrow (1010010)_2 = (1111011)_{Gray}$.

Gray to Binary Conversion

Scan the gray code word from left to right. The first 1 encountered is copied exactly as it stands. From then on, 1's will be written untill the next 1 is encountered, in which case a 0 is written. Then 0's are written untill the next 1 is encountered, in which case a 1 is written, and so on.

Example 1. *Convert Gray code word 1111011 into binary.*

1 1 1 1 0 1 1

↓ ↓ ↓ ↓ ↓ ↓ ↓ $\Rightarrow (1111011)_{Gray} = (1010010)_2$.

1 0 1 0 0 1 0

Example 2. *Convert Gray code word 10001011 into binary.*

1 0 0 0 1 0 1 1

↓ ↓ ↓ ↓ ↓ ↓ ↓ ↓

1 1 1 1 0 0 1 0

$(10001011)_{Gray} = (11110010)_2$.

1.9.3 Error Detecting Codes

Binary information is transmitted from one device to another by electric wires or other communication medium. A system that can not guarantee that the data received by one device are identical to the data transmitted by another device is essentially useless. Yet anytime data are transmitted from source to destination, they can become corrupted in passage. Many factors, including external noise, may change some of the bits from 0 to 1 or viceversa. Reliable systems must have a mechanism for detecting and correcting such errors.

Binary information or data is transmitted in the form of electromagnetic signal over a channel whenever an electromagnetic signal flows from one point to another, it is subject to unpredictable interference from heat, magnetism, and other forms of electricity. This interference can change the shape or timing of signal. If the signal is carrying encoded binary data, such changes can alter the meaning of data.

In a single bit error, a 0 is changed to a 1 or a 1 is changed to a 0.

In a burst error, multiple (two or more) bits are changed.

The purpose of error detection code is to detect such bit reversal errors. Error detection uses the concept of **redundancy** which means adding extra bits for detecting errors at the destination.

Four types of redundancy checks are used: Parity check (vertial redundancy check) (VRC), longitudinal redundancy check (LRC), cyclic redundancy check (CRC), and checksum.

For a single bit error detection, the most common way to achieve error detection is by means of a **parity bit.**

A parity bit is an extra bit (redundant bit) included with a message to make the total number of 1's transmitted either odd or even.

Table 5 shows a message of three bits and its corresponding odd and even parity bits.

If an odd parity is adopted, P bit is choosen such that the total no. of 1's is odd in four bit that constitute message bits and P.

If an even parity is adopted, the P bit is choosen such that the total number of 1's is even.

Table 5. Parity bit generation

Message			*Odd Parity bit (P)*	*Even Parity bit (P)*
x	*y*	*z*		
0	0	0	1	0
0	0	1	0	1
0	1	0	0	1
0	1	1	1	0
1	0	0	0	1
1	0	1	1	0
1	1	0	1	0
1	1	1	0	1

The message with the parity bit (either odd or even) is transmitted to its destination. The parity of the received data is checked at the receiving end. If the parity of the received data is changed (from that of transmitted parity), it means that at least one bit has changed their value during transmission. Though the parity code is meant for single error detection, it can detect any odd number of errors. However, in both the cases the original codeword can not be found.

If there is even combination of errors (means some bits are changed but parity remains same) it remains undetected.

Longitudinal Redundancy Check (LRC)

In LRC, a block of bits is organised in rows and columns (table). For example, a block of 16 bits can be organised in two rows and eight columns as shown in Fig. 1.14. We then calculate the parity bit (even parity/odd parity, here we are using even parity) for each column and create a new row of 8 bits, which are the parity bits for the whole block.

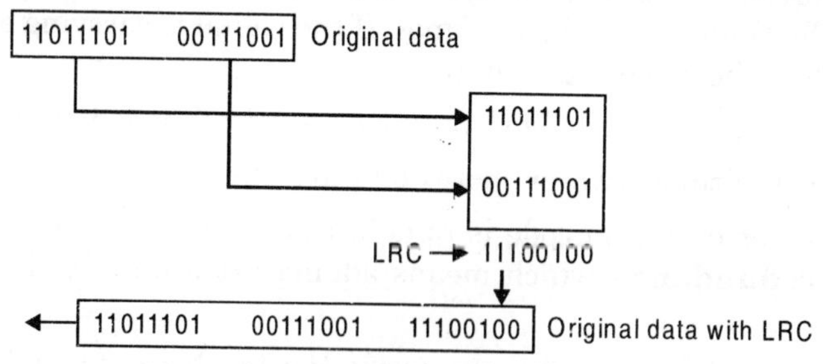

Fig. 1.14

Cyclic Redundancy Check

CRC is most powerful of the redundancy checking techniques. Cyclic redundancy check is based on binary division. A sequence of redundant bits called CRC or CRC remainder is appended to the end of a data unit. After adding CRC remainder, the resulting data unit becomes exactly divisible by another predetermined binary number. At the destination this data unit is divided by the same binary number. If there is no remainder, then there is no error. The question is how to obtain correct CRC? CRC is the remainder obtained by dividing the data unit by a predetermined divisor if it has exactly one bit less than divisor and by appending the CRC to the data string make it completely divisible by the divisior.

The CRC generator and CRC checker are shown in Fig. 1.15(*a*) and Fig. 1.15(*b*) respectively.

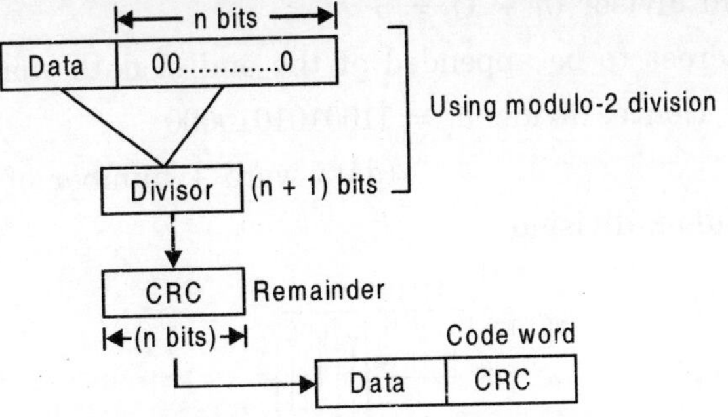

Fig. 1.15 (*a*) CRC generator

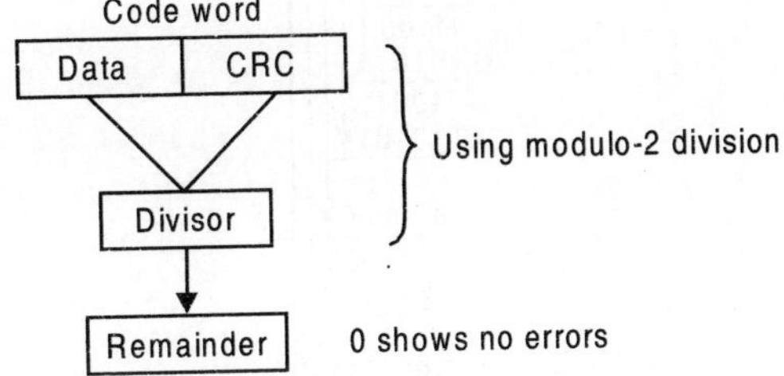

Fig. 1.15 (*b*) CRC checker

Example 1. For the divisor 10101, check whether there are errors in the received codeword 1100 1001 01011.

Solution. As shown in Fig. 1.15(*b*).

Code word: 1100 1001 01011.

Divisor : 10101

Using modulo 2 division, we obtain

```
                             111110001
             Divisor →10101 ) 1100100101011 ← Received code word
       Modulo-2 sum ← ⊕ 10101
                          11000
                        ⊕ 10101
                           11010
                         ⊕ 10101
                            11111
                          ⊕ 10101
                             10100
                           ⊕ 10101
                              00011011
                           ⊕     10101
                                  1110 ← Remainder
```

Since, remainder is non-zero shows that there is error in the received code word.

Example 2. Generate the CRC code for the data word of 1100 10101. The divisor is 10101.

Solution. As shown in Fig. 1.15(*a*).

Data word: 1100 10101.

Divisor : 10101

The no. of data bits (k) = 9

The no. of bits in divisor $(n + 1) = 5$

Therefore, no. of zeroes to be appended at the end of data word will be $n = 4$

Hence, dividend = 1100101010000

(data word + number of zeroes)

Carry out the modulo-2 division

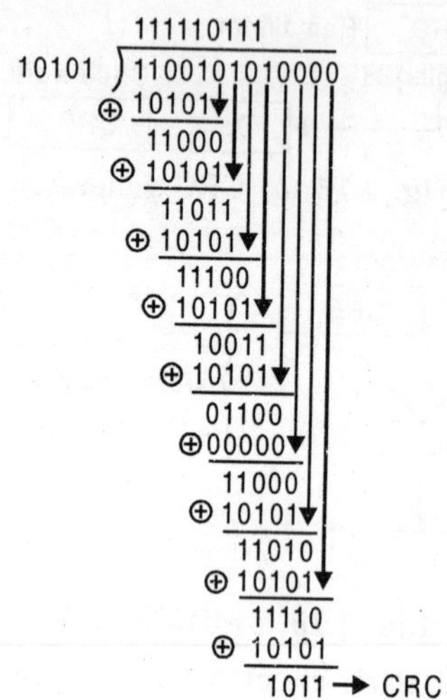

Therefore, the CRC will be 1011.

Hence, code word = Data + CRC

= 1100101011011

Checksums

The checksum method is used to detect double errors in bits. Since, the double error will not change the parity of the bits, the parity checker will not indicate any error.

In the checksums method, initially a word A (let 11001010) is transmitted, next word B (let 00101101) is transmitted. The sum of these two words is retained in the transmitter. Then a word C is transmitted and added to the previous sum; and the new sum is retained. Similarly, each word is added to the previous sum; after transmission of all the words, the final sum called the checksum is also transmitted. The same operation is done at the receiving end and the final sum obtained here is checked against the transmitted checksum. If the two sums are equal there is no error.

Burst Error Detection

So far we have considered detecting or correcting errors that occur independently or randomly in digit positions. But disturbances can wipe out an entire block of digits. For example, a stroke of lightening or a human made electrical disturbance can affect several transmitted digits. Burst errors are those errors that wipe out some or all of a sequential set of digits.

A burst of length b is defined as a sequence of digits in which the first digit and bth digit are in error, with the b-2 digits in between either in error or received correctly.

It can be shown that for detecting all burst errors of length b or less; b parity check bits are necessary and sufficient.

To construct such a code, lets group k data digits into segment of b digits in length as shown Fig. 1.16:

Fig. 1.16 Burst error detection

To this we add a last segment of b parity check digits, which are determined as follows:

"The modulo-2 sum* of the ith digit in each segment (including the parity check segment) must be zero."

It is easy to see that if a single sequence of length b or less is in error, parity will be violated and the error will be detected and the reciever can request retransmission of code.

1.9.4 Error Correcting Codes

The mechanism that we have covered upto this point detect errors but do not correct them. Error correction can be handled in two ways. In one, when an error is encountered the receiver can request the sender to retransmit entire data unit. In the other, a receiver can use an error correcting code, which automatically corrects certain errors.

In theory, it is possible to correct any binary code errors automatically using error correcting codes, however they require more reductant bits than error detecting codes. The number of bits required to correct a multiple-bit or burst error is so high that in most cases, it is inefficient to do so. For this reason, most error correction is limited to one, two, or three bit errors. However, we shall confine our discussion to only single bit error correction.

As we saw earlier, single bit errors can be detected by the addition of a redundant (parity) bit to the data (information) unit. This provides sufficient base to introduce a very popular error detection as well correction codes, known as Block codes.

Block codes: [(n, k) codes] In block codes, each block of k message bits is encoded into a larger block of n bits ($n > k$), as shown in Fig. 1.17. These are also known as (n, k) codes.

---

* Modulo–2 sum denoted by symbol ⊕ with the rules of addition as follows:

$$\begin{Bmatrix} 0 \oplus 0 = 0 \\ 0 \oplus 1 = 1 \\ 1 \oplus 0 = 1 \\ 1 \oplus 1 = 0 \end{Bmatrix}$$

The reductant* (parity) bits '*r*' are derived from message bits '*k*' and are added to them. The '*n*' bit block of encoder output is called a codeword.

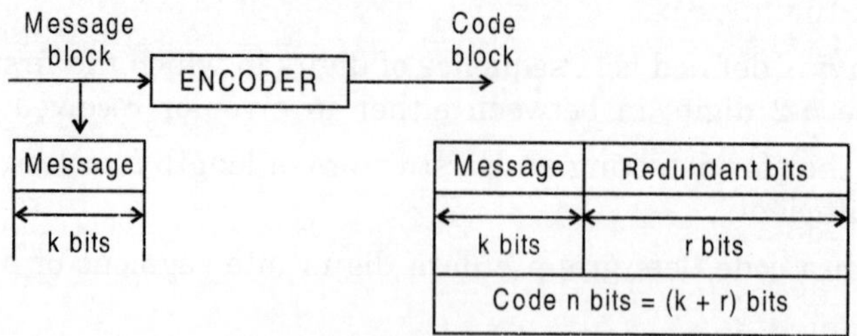

Fig. 1.17

The simplest possible block code is when the number of reductant or parity bits is one. This is known as parity check code. It is very clear that what we have studied in single bit error detection is nothing but a class of 'Block codes'.

R.W. Hamming developed a system that provides a methodical way to add one or more parity bit to data unit to detect and correct errors weight of a code.

Hamming distance and minimum distance

The weight of a code word is defined as the number of nonzero components in it. For example,

Code word	*Weight*
010110	3
101000	2
000000	0

The 'Hamming distance' between two code words is defined as the number of components in which they differ.

For example, Let

$$U = 1010$$
$$V = 0111$$
$$W = 1001$$

Then, D (U, V) = distance between U and V = 3

Similarly, D (V, W) = 3

and D (U, W) = 2

The 'minimum distance' (D_{min}) of a block code is defined is the smallest distance between any pair of codewords in the code.

From Hamming's analysis of code distances, the following important properties have been derived. If D_{min} is the minimum distance of a block code then

(*i*) '*t*' number of errors can be detected if

$$D_{min} = t + 1$$

(*ii*) '*t*' number of errors can be corrected if

$$D_{min} = 2t + 1$$

It means, we need a minimum distance (D_{min}) of at least 3 to correct single error and with this minimum distance we can detect upto 2 errors.

*The '*r*' bit are not necessarily appear after 'k' bits. They may appear at the starting, end or in between '*k*' data bits.

Now coming to our main objective *i.e.*, error correction, we can say that an error occurs when the receiver reads 1 bit as a 0 or a 0 bit as a 1. To correct the error, the reciever simply reverses the value of the altered bit. To do so, however, it must know which bit is in error. The secret of error correction, therefore, is to locate the invalid bit or bits.

For example, to correct a single bit error in a seven bit data unit, the error correction code must determine, which of the seven data bits has changed. In the case we have to distinguish between eight different states: no error, error in position 1, error in position 2, and so on, upto error in position 7. To do so requires enough redundant bits to show all eight states.

At first glance, it appears that a 3-bit redundant code should be adequate because three bits can show eight different states (000 to 111) and can thus indicate the locations of eight different possibilities. But what if an error occurs in the redundant bits themselves. Seven bits of data plus three bits of redundancy equals 10 bits. Three bits, however, can identify only eight possibilities. Additional bits are necessary to cover all possible error locations.

Redundant Bits

To calculate the number of redundant bits (r) required to correct a given no. of data bits (k), we must find a relationship between k and r. Figure 1.18 shows k bits of data with r bits of redundancy added to them. The length of the resulting code is thus $n = k + r$.

If the total no. of bits in code is $k + r$, then r must be able to indicate at least $k + r + 1$ different states. Of these, one state means no error and $k + r$ states indicate the location of an error in each of the $k + r$ positions.

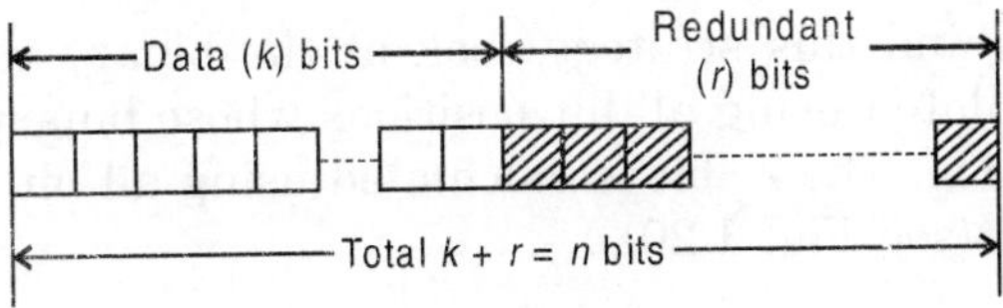

Fig. 1.18

Alternatively, we can say the $k + r + 1$ status must be discoverable by r bits; and r bits can indicate 2^r different states. Therefore, 2^r must be equal to or greater than $k + r + 1$:

$$2^r \geq k + r + 1$$

The value of r can be determined by plugging in the value of k (the length of data unit). For example, if the value of k is 7 ($\rightarrow$ seven bit data), the smallest r value that can satisfy this equation is 4:

$$2^4 \geq 7 + 4 + 1$$

and $$2^3 \not\geq 7 + 4 + 1$$

1.9.5 Hamming Code

So far, we have examined the number of bits required to cover all of the possible single bit error states in a transmission. But how do we manipulate those bits to discover which state has occured ? A technique developed by R.W. Hamming provides a practical solution, Hamming code is a class of block code (n, k) and we are going to discuss (11, 7) Hamming code.

Positioning the Redundant Bits

The 'Hamming code' can be applied to data units of any length and uses the relationship between data and redundant bits as discussed above. As we have seen, a 7 bit data unit ($k = 7$) requires 4 redundant bits ($r = 4$) that can be added to the end of data unit (or interspersed with data bits). Such that a code word of length 11 bits ($n = 11$) is formed.

In Fig. 1.19 these bits are placed in positions 1, 2, 4 and 8 (the positions in an 11-bit sequence that are powers of 2). We refer these bits as r_1, r_2, r_4 and r_8.

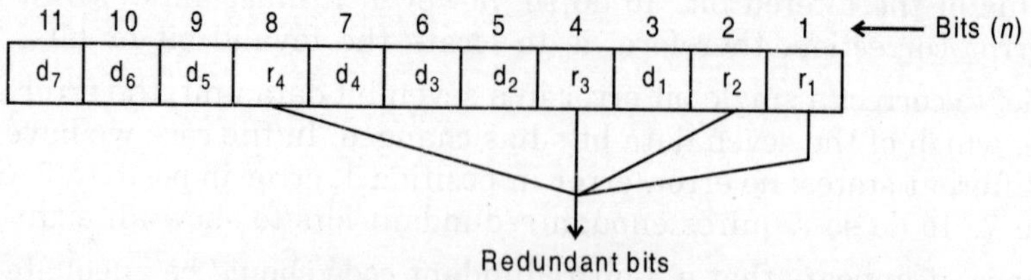

Fig. 1.19

In the Hamming code, each r bit is the redundant bit for one combination* of data bits. The combinations (modulo-2 additions) used to calculate each of four r values (viz, r_1, r_2, r_4 and r_8) for a 7 bit data sequence d_1 through d_7 are as follows:

r_1 : bits 1, 3, 5, 7, 9, 11

r_2 : bits 2, 3, 6, 7, 10, 11

r_4 : bits 4, 5, 6, 7

r_8 : bits 8, 9, 10, 11

Each data bit may be included in more than one redundant bit calculation. In the sequences above, for example, each of the original data bits is included in at least two sets, while the r bits are included in only one.

To see the pattern behind this strategy, look at the binary representation of each bit position. The r_1 bit is calculated using all bit positions whose binary representation includes a 1 in the right most position. The r_2 bit is calculated using all bit positions with a 1 in the second position, and so on (see Fig. 1.20).

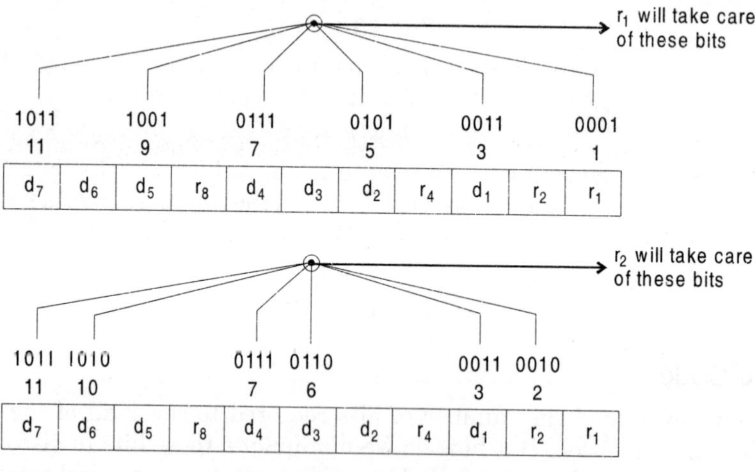

*In codes combination of bits means modulo-2 addition of data bits. Modulo-2 addition applies in binary field with following rules.

$0 \oplus 0 = 0$ Modulo-2 → $\oplus$

$0 \oplus 1 = 1$

$1 \oplus 0 = 1$

$1 \oplus 1 = 0$

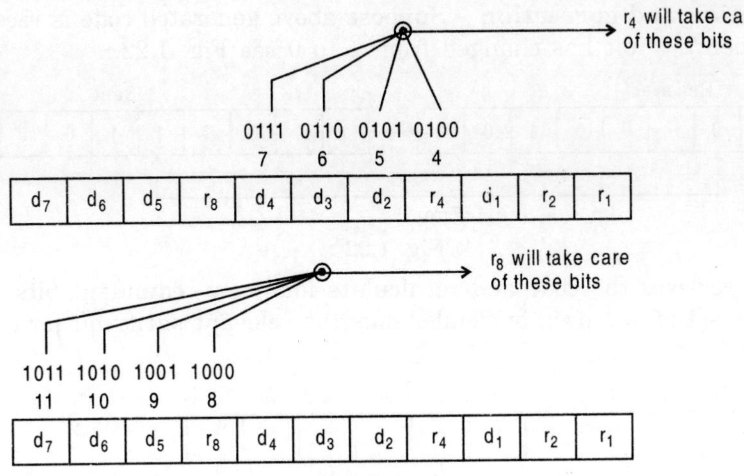

Fig. 1.20

Calculating the r values

Figure 1.21 shows a Hamming code implementation for a 7 bit data unit. In the first step; we place each bit of original data unit in its appropriate position in the 11-bit unit. For example, let the data unit be 1001101.

In the subsequent steps; we calculate the **EVEN** parities for the various bit combinations. The even parity value for each combination is the value of corresponding r bit. For example, the value of r_1 is calculated to provide even parity for a combination of bits 3, 5, 7, 9 and 11.

i.e., 1011 1001 0111 0101 0011 0001

Here the total no. of 1's are 3. Thus to provide even parity r_1 = 1.

Similarly the value of r_2 is calculated to provide even parity with bits 3, 6, 7, 10, 11, r_4 with bits 5, 6, 7 and r_8 with bits 9, 10, 11. The final 11-bit code is sent.

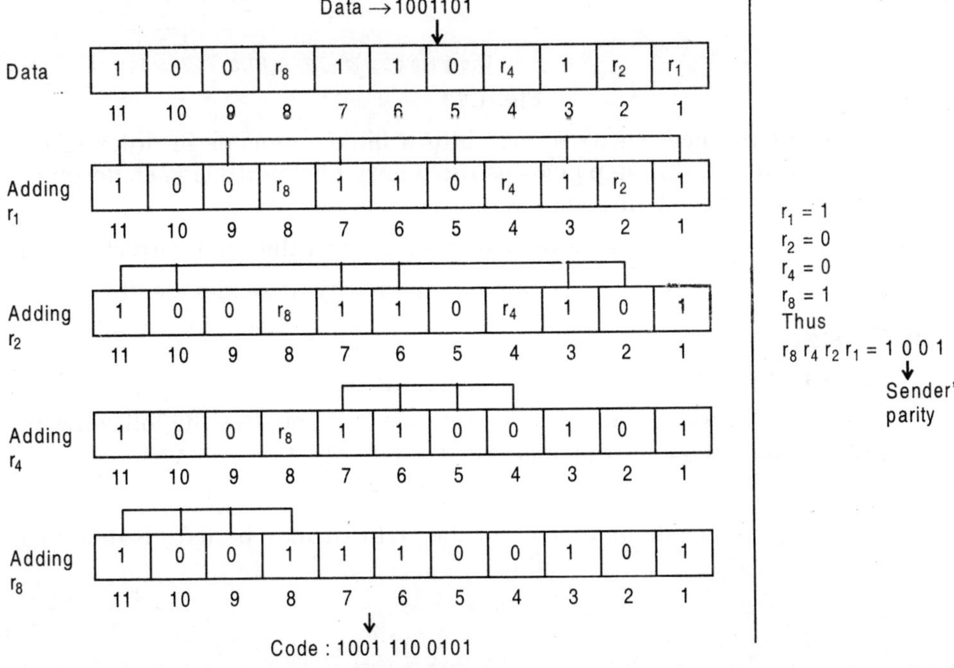

Fig. 1.21

Error detection and correction – Suppose above generated code is received with the error at bit number 7 ⇒ bit has changed from 1 to 0 see Fig. 1.22.

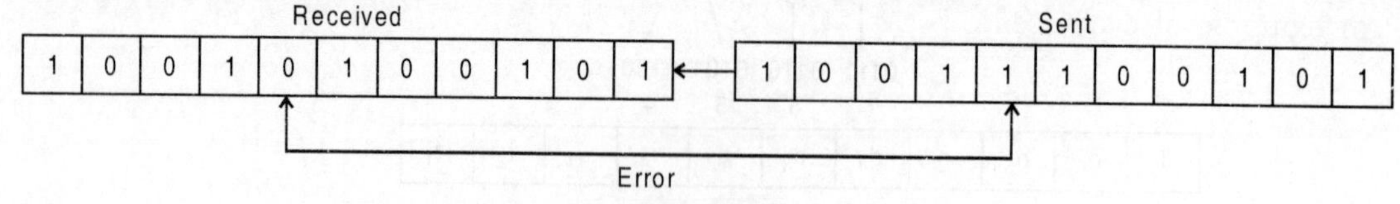

Fig. 1.22

The receiver receives the code and recalculate four new redundant bits (r_1, r_2, r_4 and r_8) using the same set of bits used by sender plus the relevant parity bit for each set shown in Fig. 1.23.

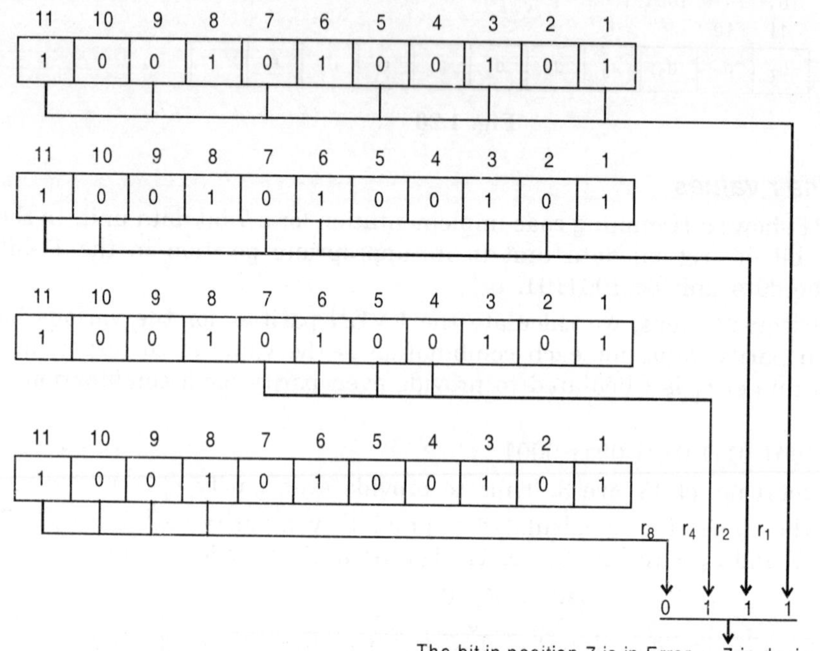

Fig. 1.23

Then it assembles the new parity values into a binary number in order of r position (r_8, r_4, r_2, r_1). In our example, this step gives us the binary number 0111 (7 in decimal), which is the precise location of the bit in error.

Once the bit is identified, the reciever can reverse its value and correct the error.

Note: If the new parity assembled is same as the parity at sender's end mean no error.

1.9.6 Cyclic Codes

Binary cyclic codes form a subclass of linear block codes.

An (n, k) linear block code is called the cyclic code if it satisfies the following property:

If an n tuple (a row vector of n elements), $V = (V_0, V_1, V_2, \ldots, V_{n-1})$

is a code vector or V, then the n tuple $(V_{n-1}, V_0, V_1, \ldots, V_{n-2})$

is obtained by shifting V cyclically one place to the right (it may be left also) is also a code vector of V.

$V^1 = (V_{n-1}, V_0, V_1, \ldots, V_{n-2})$. From the above definition it is clear that

$$V^{(i)} = (V_{n-i}, V_{n-i+1}, \ldots, V_0, V_1, \ldots V_{n-i-1}).$$

An example of cyclic code:

It can be seen that 1101, 1110, 0111, 1011 is obtained by a cyclic shift of n-tuple 1101 ($n = 4$). The code obtained by rearranging the four words is also a cyclic code. Thus 1110, 0111, 1011 are also cyclic codes.

This property of cyclic code allows to treat the codewords as a polynomial form. A procedure for generating an (n, k) cyclic code is as follows:

The bits of uncoded word (message), let D = $[d_0, d_1, d_2 \ldots d_{k-1}]$ are written as the coefficients of polynomial of degree $k - 1$.

$$D(x) = d_0 x^0 \oplus d_1 x^1 \oplus d_2 x^2 \oplus \ldots \oplus d_{k-1}\, x^{k-1}$$

Similarly, the coded word, let V = $[v_0, v_1, v_2, \ldots, v_{n-1}]$ are written as the coefficients of polynomial of degree $n - 1$.

$$V(x) = v_0 x^0 \oplus v_1 x^1 \oplus v_2 x^2 \oplus \ldots \oplus v_{n-1}\, x^{n-1}$$

The coefficients of the polynomials are 0's and 1's and they belong to the binary field with the modulo-2 rules for addition as described in Hamming codes.

Now, we will state a theorem* which is used for cyclic code generation.

Theorem. If $g(x)$ is a polynomial of degree $(n-k)$ and is a factor of x^n+1, *then* $g(x)$ generates an (n, k) cyclic code in which the code polynomial $V(x)$ for a data polynomial $D(x)$ is given by

$$V(x) = D(x).\, g(x)$$

where
$V(x)$ – Code word polynomial of degree $(n - 1)$
$D(x)$ – Data word polynomial of degree $(k - 1)$
$g(x)$ – Generator polynomial of degree $(n - k)$

Example. *Consider a (7, 4) cyclic code. The generator polynomial for this code is given as $g(x) = 1 + x + x^3$. Find all the code words of this code.*

Solution. It is a (7, 4) cyclic code

n = No. of bits in coded word = 7

and k = No. of bits in data word = 4.

$(n - k)$ = No. of redundant bits in code word = 3

It implies that, there are 16 different messages that are possible (0000, 0001, 0010 . . . 1110, 1111). Correspondingly, there will be 16 different codes (of 7 bits).

Now, according to above theorem, the generator polynomial $g(x)$ must be a factor of $(x^n + 1)$** and of degree $(n - k)$.

$$x^n + 1 = x^7 + 1$$

If we factorize this polynomial we get

$$x^7+1 = \underset{\text{I}}{(x + 1)}\ \underset{\text{II}}{(x^3 + x + 1)}\ \underset{\text{III}}{(x^3 + x^2 + 1)}$$

*Without giving proof that is beyond the scope of this book.

**+ means modulo-2 operation ⊕ in binary codes.

I Factor $\rightarrow x + 1$

II Factor $\rightarrow x^3 + x + 1$

III Factor $\rightarrow x^3 + x^2 + 1$

The I factor does not satisfy the requirement that it must be of degree $(n - k)$ but the II and III do satisfy.

Therefore, we can either choose II Factor or III Factor as generator polynomial $g(x)$. However, the set of codewords will naturally be different for these two polynomials.

In this example, we have taken $g(x)$ as $1 + x + x^3$.

i.e., we have to encode the 16 messages using generator polynomial.

$$g(x) = 1 + x + x^3.$$

Consider, for example, a data word 1010.

$$D = (d_0, d_1, d_2, d_3) = (1010)$$

Because the length is four, the data polynomial D(x) will be of the form $d_0 + d_1x + d_2x^2 + d_3x^3$

$$D(x) = 1 + 0.x + 1.x^2 + 0.x^3 = 1 + x^2$$

The code polynomial

$$V(x) = D(x) \,.\, g(x)$$

$$= (1 + x^2) \,.\, (1 + x + x^3)$$

$$= 1 + x + x^2 + \underbrace{x^3 + x^3}_{0} + x^5$$

i.e.,

$$V(x) = 1 + x + x^2 + x^5$$

Because if $x = 1$ then $x^3 = 1$

or if $x = 0$ then $x^3 = 0$

$$0 \oplus 0 = 1 \oplus 1 = 0$$

Because the length of codeword and (n) is 7.

So the standard polynomial will be of the form.

$$V(x) = V_0 + V_1x + V_2x^2 + V_3x^3 + V_4x^4 + V_5x^5 + V_6x^6$$

Comparing this standard polynomial with above poly. for V(x)

we get

$$V = [1110010]$$

In a similar way, all code vectors can be found out.

1.9.7 Alphanumeric Codes

For the inherently binary world of the computer, it is necessary to put all symbols, letters, numbers, etc. into binary form. The most commonly used alphanumeric code is the ASCII code, with other like the EBCDIC code being applied in some communication applications.

ASCII Alphanumeric Code

The American Standard Code for Information Interchange (ASCII) is the standard alphanumeric code for keyboards and a host of other data interchange tasks. Letters, numbers, and single keystroke commands are represented by a seven-bit word. Typically, a strobe bit or start bit is sent first, followed by the code with LSB first. Being a 7-bit code, it has 2^7 or 128 possible code groups.

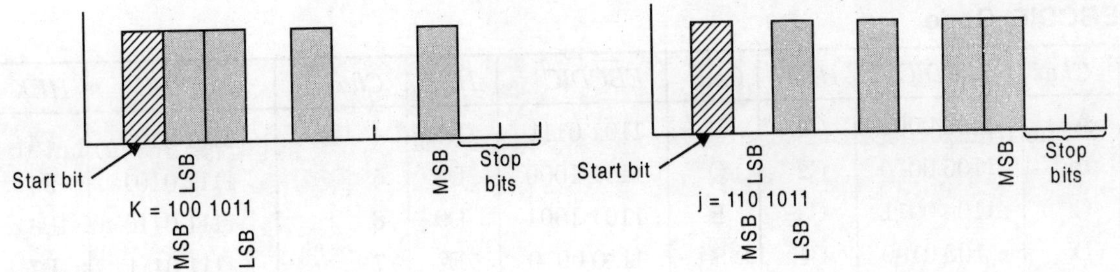

ASCII Alphanumeric Code

Char	7 bit ASCII	HEX	Char	7 bit ASCII	HEX	Char	7 bit ASCII	HEX
A	100 0001	41	a	1100001	61	0	0110000	30
B	100 0010	42	b	1100010	62	1	0110001	31
C	100 0011	43	c	1100011	63	2	0110010	32
D	100 0100	44	d	1100100	64	3	0110011	33
E	100 0101	45	e	1100101	65	4	0110100	34
F	100 0110	46	f	1100110	66	5	0110101	35
G	100 0111	47	g	1100111	67	6	0110110	36
H	100 1000	48	h	1101000	68	7	0110111	37
I	100 1001	49	i	1101001	69	8	011 1000	38
J	100 1010	4A	j	1101010	6A	9	011 1001	39
K	100 1011	4B	k	1101011	6B	blank	0100000	20
L	100 1100	4C	1	110 1100	6C	.	010 1110	2E
M	100 1101	4D	m	110 1101	6D	(	010 1000	28
N	100 1110	4E	n	110 1110	6E	+	010 1011	2B
O	100 1111	4F	o	110 1111	6F	$	010 0100	24
P	101 0000	50	p	111 0000	70	*	010 1010	2A
Q	101 0001	51	q	111 0001	71	)	010 1001	29
R	101 0010	52	r	111 0010	72	-	010 1101	2D
S	101 0011	53	s	111 0011	73	/	010 1111	2F
T	101 0100	54	t	111 0100	74	,	010 1100	2C
U	101 0101	55	u	111 0101	75	=	011 1101	3D
V	101 0110	56	v	111 0110	76	RETURN	000 1101	0D
W	101 0111	57	w	111 0111	77	LNFEED	000 1010	0A
X	101 1000	58	x	111 1000	78	0	011 0000	30
Y	101 1001	59	y	111 1001	79	0	011 0000	30
Z	101 1010	5A	z	111 1010	7A	0	011 0000	30

EBCDIC Alphanumeric Code

The Extended Binary Coded Decimal Interchange Code (EBCDIC) is an 8-bit alphanumeric code which has been extensively used by IBM in its mainframe applications.

EBCDIC Code

Char	*EBCDIC*	*HEX*	*Char*	*EBCDIC*	*HEX*	*Char*	*EBCDIC*	*HEX*
A	1100 0001	C1	P	1101 0111	D7	4	1111 0100	F4
B	1100 0010	C2	Q	1101 1000	D8	5	1111 0101	F5
C	1100 0011	C3	R	1101 1001	D9	6	1111 0110	F6
D	1100 0100	C4	S	1110 0010	E2	7	1111 0111	F7
E	1100 0101	C5	T	1110 0011	E3	8	1111 1000	F8
F	1100 0110	C6	U	1110 0100	E4	9	1111 10001	F9
G	1100 0111	C7	V	1110 0101	E5	blank	...	...
H	1100 1000	C8	W	1110 0110	E6	.	...	...
I	1100 1001	C9	X	1110 0111	E7	(	...	...
J	1101 0001	Dl	Y	1110 1000	E8	+	...	...
K	1101 0010	D2	Z	1110 1001	E9	$	...	...
L	1101 0011	D3	0	1111 0000	F0	*	...	...
M	1101 0100	D4	1	1111 0001	F1	)	...	...
N	1101 0101	D5	2	1111 0010	F2	-	...	...
O	1101 0110	D6	3	1111 0011	F3	/	...	...

1.10 SOLVED EXAMPLES

Example 1. *Convert each binary number to the decimal:*

(*a*) $(11)_2$, $(1011)_2$, $(10111)_2$, $(1111)_2$, $(11010111)_2$, $(1001)_2$

(*b*) $(.11)_2$, $(.111)_2$, $(.1011)_2$, $(.10101)_2$, $(.0101)_2$, $(.110)_2$

(*c*) $(11.11)_2$, $(1011.1011)_2$, $(1111.0101)_2$, $(11010111.110)_2$, $(1001.10101)_2$

Solution. (*a*)

$$(11)_2 = 1 \times 2^1 + 1 \times 2^0 = 2 + 1 = 3$$

$$(1011)_2 = 1 \times 2^3 + 0 \times 2^2 + 1 \times 2^1 + 1 \times 2^0 = 8 + 0 + 2 + 1 = 11$$

$$(10111)_2 = 1 \times 2^4 + 0 \times 2^3 + 1 \times 2^2 + 1 \times 2^1 + 1 \times 2^0 = 16 + 0 + 4 + 2 + 1 = 23$$

$$(1111)_2 = 1 \times 2^3 + 1 \times 2^2 + 1 \times 2^1 + 1 \times 2^0 = 8 + 4 + 2 + 1 = 15$$

$$(11010111)_2 = 1 \times 2^7 + 1 \times 2^6 + 0 \times 2^5 + 1 \times 2^4 + 0 \times 2^3 + 1 \times 2^2 + 1 \times 2^1 + 1 \times 2^0$$
$$= 128 + 64 + 0 + 16 + 0 + 4 + 2 + 1 = 215$$
$$(1001)_2 = 1 \times 2^3 + 0 \times 2^2 + 0 \times 2^1 + 1 \times 2^0$$
$$= 8 + 0 + 0 + 1 = 9$$

(*b*)
$$(.11)_2 = 1 \times 2^{-1} + 1 \times 2^{-2}$$
$$= .5 + .25 = (.75)_{10}$$
$$(.111)_2 = 1 \times 2^{-1} + 1 \times 2^{-2} + 1 \times 2^{-3}$$
$$= .5 + .25 + .125 = (.875)_{10}$$
$$(.1011)_2 = 1 \times 2^{-1} + 0 \times 2^{-2} + 1 \times 2^{-3} + 1 \times 2^{-4}$$
$$= .5 + 0 + .125 + .0625 = (.6875)_{10}$$
$$(.10101)_2 = 1 \times 2^{-1} + 0 \times 2^{-2} + 1 \times 2^{-3} + 0 \times 2^{-4} + 1 \times 2^{-5}$$
$$= .5 + 0 + .125 + 0 + .03125 = (.65625)_{10}$$
$$(.0101)_2 = 0 \times 2^{-1} + 1 \times 2^{-2} + 0 \times 2^{-3} + 1 \times 2^{-4}$$
$$= 0 + .25 + 0 + .0625 = (.3125)_{10}$$
$$(.110)_2 = 1 \times 2^{-1} + 1 \times 2^{-2} + 0 \times 2^{-3}$$
$$= .5 + .25 + 0 = (.75)_{10}$$

(*c*)
$$11.11 = ?$$

From part (*a*) and part (*b*), we see that

$$11 = 3$$
$$11 = .75$$

Therefore, $(11.11)_2 = (3.75)_{10}$

$$1011.1011 = ?$$
$$(1011)_2 = 11$$
$$(.1011)_2 = .6875$$

Therefore, $(1011.1011)_2 = (.6875)_{10}$

$$1111.0101 = ?$$
$$(1111)_2 = 15$$
$$(.0101)_2 = .3125$$

Therefore, $(1111.0101)_2 = (15.3125)_{10}$

$$11010111.110 = ?$$
$$11010111 = 215$$
$$.110 = .75$$
$$(11010111.110)_2 = (215.75)_{10}$$
$$1001.10101 = ?$$
$$1001 = 9$$
$$.10101 = .65625$$
$$(1001.10101)_2 = (9.65625)_{10}$$

Example 2. *How many bits are required to represent the following decimal numbers, represent them in binary.*

(i) $(17)_{10}$ *(ii)* $(27)_{10}$ *(iii)* $(81)_{10}$ *(iv)* $(112)_{10}$ *(v)* $(215)_{10}$

Solution. *(i)* Let n bits required

n should be such that

$$2^n \geq \text{Given Number (N)}$$

Therefore, $2^n \geq 17$

i.e., $n \geq 5$

Therefore, minimum number of bits required = 5.

		Remainder
2	17	1 ↑ LSB
2	8	0
2	4	0
2	2	0
2	1	1 MSB
	0	

$(17)_{10} = (10001)_2$

(ii) $(27)_{10}$

The minimum number of bits required is given by

$$2^n \geq \text{N (given number)}$$

$$2^n \geq 27$$

i.e., $n \geq 5$

2	27	1 ↑ LSB
2	13	1
2	6	0
2	3	1
2	1	1 MSB
	0	

$(27)_{10} = (11011)_2$

(iii) $(81)_{10}$

The minimum number of bits required is given by

$$2^n \geq \text{N}$$

$$2^n \geq 81$$

i.e., $n = 7$

		Remainder
2	81	1 ↑ LSB
2	40	0
2	20	0
2	10	0
2	5	1
2	2	0
2	1	1 MSB
	0	

$(81)_{10} = (1010001)_2$

(iv) $(112)_{10}$

The minimum number of bits required is given by

$$2^n \geq \text{N}$$

$$2^n \geq 112$$

i.e., $n = 7$

2	112	0 ↑ LSB
2	56	0
2	28	0
2	14	0
2	7	1
2	3	1
2	1	1 MSB
	0	

$(112)_{10} = (1110000)_2$

(v) $(215)_{10}$

The minimum number of bits required is given by

$$2^n \geq 215$$

i.e., $n = 8$

2	215	1 ↑ LSB
2	107	1
2	53	1
2	26	0
2	13	1
2	6	0
2	3	1
2	1	1 MSB
	0	

$(215)_{10} = (11010111)_2$

Example 3. *Convert the following numbers as indicated:*

(a) decimal 225.225 to binary, octal and hexadecimal.

(b) binary 11010111.110 to decimal, octal and hexadecimal.

(c) octal 623.77 to decimal, binary and hexadecimal.

(d) hexadecimal 2AC5.D to decimal, octal and binary.

Solution. (*a*) $225.225 = (?)_2$

		Remainder	
2	225	1	LSB ↑
2	112	0	
2	56	0	
2	28	0	
2	14	0	
2	7	1	
2	3	1	
2	1	1	MSB
	0		

Integer part = 11100001

			Carry
$(.225)_{10}$	=	$(?)_2$	
.225 × 2	=	0.450	0
.450 × 2	=	0.900	0
.900 × 2	=	1.800	1
.800 × 2	=	1.600	1
.600 × 2	=	1.200	1
.200 × 2	=	0.400	0
.400 × 2	=	0.800	0
.800 × 2	=	1.600	1
.600 × 2	=	1.200	1

Fraction part = 001110011

Therefore,

$$(225.225)_{10} = 11100001.001110011$$

$$(225.225)_{10} = (?)_8$$

From the previous step we know the binary equivalent of decimal no. as 11100001.001110011.

For octal number, binary number is partitioned into group of three digit each starting from right to left and replacing decimal equivalent of each group.

–11 100 001 . 001 110 011

011 100 001 . 001 110 011

↓ ↓ ↓

3 4 1 1 6 3

$$(225.225)_{10} = (341.163)_8$$

$$(225.225)_{10} = (?)_{16}$$

For hexadecimal number, instead of three four digits are grouped.

$$\underbrace{1110}\ \underbrace{0001}\ .\ \underbrace{0011}\ \underbrace{1001}\ \underbrace{1---}$$

$$\underbrace{1110}_{14 \equiv E}\ \underbrace{0001}_{1}\ .\ \underbrace{0011}_{3}\ \underbrace{1001}_{9}\ \underbrace{1000}_{8}$$

$$(225.225)_{10} = E1.398$$

(*b*) $$(11010111.110)_2 = (?)_{10}$$

From example 1.6.1

$$(11010111.110)_2 = (215.75)_{10}$$

$$(11010111.110)_2 = (?)_8$$

$$\underbrace{-11}\ \underbrace{010}\ \underbrace{111}$$

$$\underbrace{011}_{3}\ \underbrace{010}_{2}\ \underbrace{111}_{7} = 327$$

$$\underbrace{.110}_{6} = 6$$

$$(11010111.110)_2 = (327.6)_8$$

$$(11010111.110)_2 = (?)_{16}$$

$$\underbrace{1101}_{13 \equiv D}\ \underbrace{0111}_{7} = D7$$

$$\underbrace{.110-}$$

$$\underbrace{.1100}_{12 = C} = C$$

$$(11010111.110)_2 = (D7.C)_{16}$$

(*c*) $$(623.77)_8 = (?)_2$$

$$623 = 110010011$$

$$.77 = 111111$$

$$(623.77)_8 = (110010011.111111)_2$$

$$(623.77)_8 = (?)_{16}$$

$$\underbrace{---1}\ \underbrace{1001}\ \underbrace{0011}$$

$$\underbrace{0001}_{1}\ \underbrace{1001}_{9}\ \underbrace{0011}_{3} = 193$$

$$\underbrace{1111}\ \ \underbrace{11--}$$

$$\underbrace{1111}_{15 \equiv F}\ \ \underbrace{1100}_{12 = C} = FC$$

$$(623.77)_8 = (193.FC)_{16}$$

$(623.77)_8 = (?)_{10}$

$623 = 6 \times 8^2 + 2 \times 8^1 + 3 \times 8^0$

$= 384 + 16 + 3$

$= 403$

$.77 = 7 \times 8^{-1} + 7 \times 8^{-2}$

$= 7 \times .125 + 7 \times .015625$

$= .875 + .109375$

$= 0.9843$

$(623.77)_8 = (403.9843)_{10}$

(*d*) $(2AC5.D)_{16} = (?)_2$

$2AC5 = 0010101011000101$

$D = 1101$

$(2AC5.D)_{16} = (10101011000101.1101)_2$

$(2AC5.D)_{16} = (?)_8$

-10 101 011 000 101 . 110 1 - -

010 101 011 000 101 . 110 100

2 5 3 0 5 6 4

$(2AC5.D)_{16} = (25305.64)_8$

$(2AC5.D)_{16} = (?)_{10}$

$2AC5 = 2 \times 16^3 + 10 \times 16^2 + 12 \times 16^1 + 5 \times 16^0$

$= 2 \times 4096 + 10 \times 256 + 12 \times 16 + 5 \times 1$

$= 8192 + 2560 + 192 + 5$

$= 10949$

$D = 13 \times 16^{-1}$

$= 13 \times .0625$

$= .8125$

$(2AC5.D)_{16} = (10949.8125)_{10}$

Example 4. *Obtain the 9's and 10's complement of the following decimal numbers.*

(i) 10000, (ii) 00000, (iii) 13469, (iv) 90099, (v) 09900

Solution. 9's complement

(*i*)	(*ii*)	(*iii*)	(*iv*)	(*v*)
99999	99999	99999	99999	99999
10000	00000	13469	90099	09900
89999	99999	86530	09900	90099

10's complement = 9's complement + 1 (LSB)

(*i*)	(*ii*)	(*iii*)	(*iv*)	(*v*)
89999	99999	86530	09900	90099
+ 1	+ 1	+ 1	+ 1	+ 1
90000	100000	86531	09901	90100

Example 5. *Perform the subtraction with the decimal numbers given using*
(1) 10's complement.
(2) 9's complement.
Check the answer by straight subtraction
(a) 5249 – 320 (b) 3571 – 2101.

Solution. (*a*) **Using 10's complement.** 10's complement of 320

```
 9999
–0320
 9679
  + 1
 9680
```

Therefore, 5249 – 320 = 5249 + 9680

```
              5249
             +9680
CY        → (1)4929
discarded
```

Result = 4929

Using 9's complement. 9's complement of 320

```
 9999
–0320
 9679
```

Therefore, 5249 – 320 = 5249 + 9679

```
              5249
            + 9679
CY        → (1)4928
             └──→1
              4929
```

Result = 4929

By straight subtraction

```
 5249
– 320
 4929
```

Hence, results are same by each method.

(*b*) **Using 10's complement**

10's complement of 2101 =

```
 9999
–2101
 7898
  + 1
 7899
```

Therefore, 3571 – 2101 = 3571 + 7899

```
              3571
             +7899
CY        → (1)0470
discarded
```

Result = 470

Using 9's complement

	9's complement of 2101	=	7898
Therefore,	3571 – 2101	=	3571
			+7898
		CY →	①0469
			→1
			470

By straight subtraction

```
 3571
–2101
  470
```

Hence, results are same by each method.

Example 6. *Obtain the 1's and 2's complement of the following binary numbers (i) 11100101 (ii) 0111000 (iii) 1010101 (iv) 10000 (v) 00000.*

Solution.

(*i*)	1's complement	=	00011010
	2's complement	=	1's complement + 1 = 00011011
(*ii*)	1's complement	=	1000111
	2's complement	=	1000111
			1
			1001000
(*iii*)	1's complement	=	0101010
	2's complement	=	0101011
(*iv*)	1's complement	=	01111
	2's complement	=	10000
(*v*)	1's complement	=	11111
	2's complement	=	00000

1.11 EXERCISE

1. Write 9's and 10's complement of the following numbers:

   +9090

   –3578

   +136.8

   –136.28

2. (*a*) Convert the decimal integer's +21 and –21 into 10's complement and 9's complement.

   (*b*) Convert the above two numbers in binary and express them in six bit (total) signed magnitude and 2's complement.

3. (*a*) Find the decimal equivalent of the following binary numbers assuming signed magnitude representation of the binary number:

(*i*) 001000 (*ii*) 1111

(*b*) Write the procedure for the subtraction of two numbers with (r – 1)'s complement.

(*c*) Perform the subtraction with the following binary numbers using 2's complement and 1's complement respectively.

(*i*) 11010 – 1101 (*ii*) 10010 – 10011

(*d*) Perform the subtraction with following decimal numbers using 10's complement and 9's complement respectively.

(*i*) 5294 – 749 (*ii*) 27 – 289

4. Convert:

(*i*) $(225.225)_{12}$ to Hexadecimal number.

(*ii*) $(2AC5.15)_{16}$ to Octal number.

5. Perform the following using 6's complement:

(*i*) $(126)_7 + (42)_7$

(*ii*) $(126)_7 - (42)_7$

6. Represent the following decimal numbers in two's complement format:

(*i*) +5 (*ii*) +25 (*iii*) –5 (*iv*) –25 (*v*) –9

7. Represent the decimal numbers of question 6 in ones complement format.

8. Find the decimal equivalent of each of the following numbers assuming them to be in two's complement format.

(*a*) 1000 (*b*) 0110 (*c*) 10010 (*d*) 00110111

9. Convert the following octal numbers into equivalent decimal numbers:

(*a*) 237 (*b*) 0.75 (*c*) 237.75

10. Represent the following decimal numbers in sign-magnitude format:

(*a*) –11 (*b*) –7 (*c*) +12 (*d*) +25

2

DIGITAL DESIGN FUNDAMENTALS—BOOLEAN ALGEBRA AND LOGIC GATES

2.0 INTRODUCTORY CONCEPTS OF DIGITAL DESIGN

George Boole, in his work entitled 'An Investigation of the Laws of Thought', on which are founded the Mathematical Theories of Logic and Probability (1854), introduced the fundamental concepts of a two-values (binary) system called Boolean Algebra. This work was later organized and systemized by Claude Shannon in 'Symbolic Analysis of Relay and Switching Circuits (1938)'. Digital design since that time has been pretty much standard and advanced, following Boole's and Shannon's fundamentals, with added refinements here and there as new knowledge has been unearthed and more exotic logic devices have been developed.

Digital design is the field of study relating the adoptation of **Logic** concepts to the design of recognizable, realizable, and reliable digital hardware.

When we begin study of logic, digital logic, binary systems, switching circuits, or any other field of study that can be classified as being related to digital design, we must concern ourselves with learning some philosophical premises from which we must launch our studies. In order to reach a desirable theoretical, as well as conceptual, understanding of digital design, you must grasp some fundamental definitions and insight giving concepts.

Generally speaking, being involved in digital design is dealing in "LOGIC" a term that certainly needs some definition. LOGIC, by definition, *is a process of classifying information. Information is intelligence related to ideas, meanings, and actions which can be processed or transformed into other forms.* For example, NEWS is information by virtue of the fact that it is intelligence related to ACTIONS, be it good news or bad news. News can be heard, read, seen or even felt or any combination of all four, indicating the possibility of its transformation into different forms.

"BINARY LOGIC," or two-valued logic, *is a process of classifying information into two classes.* Traditionally, binary arguments, or that information which can be definitely classified as two valued, has been delivered either TRUE or FALSE. Thus, the Boolean variable is unlike the algebraic variables of the field of real numbers in that any Boolean variable can take on only two values, the TRUE or the FALSE. Traditionally, it is standard to use the shorthand symbols 1 for TRUE and 0 for the FALSE.

2.1 TRUTH TABLE

A Boolean variable can take on only two values, not an infinite number as, the variable of the real number system, can. This basic difference allows us to illustrate all possible logic conditions of a Boolean variable or a collection of Boolean variables using a finite tabular format called a 'truth-table'. Further, the nontrivial decisions in digital design are based on more than one-two valued variable. Thus, if an output is to be completely specified as a function of two inputs, there are four input combinations that must be considered. If there are three inputs, then eight combinations must be considered and from this we see that n inputs will require 2^n combinations to be considered.

A TRUTH-TABLE as suggested is a tabular or graphical technique for listing all possible combinations of input variables, arguments, or whatever they may be called, in a vertical order, listing each input combination one row at a time (Table 2.1). For example,

(*i*) Let we have a TV that operates with a switch. The TV, becomes on or off with the switch on or off respectively.

Table 2.1(*a*)

I/P (Switch)	O/P (TV)
Off 0	Off 0
On 1	On 1

(*ii*) Let we have a TV that operates with two switches. When both the switches are 'ON' then only TV becomes 'ON' and in all other cases TV is 'Off'.

Table 2.1(*b*)

S.1	*S.2*	*TV*	
0	0	0	OFF
0	1	0	OFF
1	0	0	OFF
1	1	1	ON

(*iii*) Let the TV operate with three switches. The condition now is that when at least two switches are 'ON' the TV becomes 'ON' and in all other conditions 'TV' is 'OFF'.

Table 2.1(*c*)

S.1	*S.2*	*S.3*	*TV*	
0	0	0	0	OFF
0	0	1	0	OFF
0	1	0	0	OFF
0	1	1	1	ON
1	0	0	0	OFF
1	0	1	1	ON
1	1	0	1	ON
1	1	1	1	ON

Table 2.1(*a*) illustrates the use of a one variable T.T. and how the output or combined interaction is manually listed to the right of each possible combination. Table 2.1(*b*) and

Table 2.1(*c*) show the standard form for two and three variable truth-tables. In review, what is suggested here is that after all the input variables have been identified, and all the possible combinations of these variables have been listed in the truth-table on the left, then each row should be studied to determine what output or combined interaction is desired for that input combination. Further, note that the input combinations are listed in ascending order, starting with the binary equivalent of zero. The TRUTH-TABLE also allows us to establish or prove Boolean identities without detailed mathematical proofs, as will be shown latter.

2.2 AXIOMATIC SYSTEMS AND BOOLEAN ALGEBRA

The AND, OR, and INVERTER functions make up a sufficient set to define a two valued Boolean Algebra. Now, we introduce some formal treatment to this two-valued Boolean algebra.

Axiomatic Systems

Axiomatic systems are founded on some fundamental statements referred to as 'axioms' and 'postulates.' *As you delve deeper into the origin of axioms and postualtes,* you find these to be predicted on a set of undefined objects that are accepted on faith.

Axioms or postulates are statements that make up the framework from which new systems can be developed. They are the basis from which theorems and the proofs of these theorems are derived. For example, *proofs are justified on the basis of a more primitive proof.* Thus, we use the statement—*'From this we justify this'*. Again, we find a process that is based on some point for which there exist no furhter primitive proofs. Hence, we need a starting point and that starting point is a set of axioms or postulates.

Axioms are formulated by combining intelligence and empirical evidence and should have some basic properties. These are:

1. They are statements about a set of undefined objects.
2. They must be consistent, that is, they must not be self-contradictory.
3. They should be simple but useful, that is, not lengthy or complex.
4. They should be independent, that is, these statements should not be interdependent.

The study of axiomatic systems related to logic motivated the creation of the set of postulates known as the 'HUNTINGTON POSTULATES'. E.V. Huntington (1904) formulated this set of postulates that have the basic properties described desirable, consistant, simple and independent. These postulates as set forth can be used to evaluate proposed systems and those systems that meet the criteria set forth by these posutlates become known as Huntington System. Further, once a proposed system meets the criteria set forth by the Huntington Postulates, automatically all theorems and properties related to other Huntington systems become immediately applicable to the new system.

Thus, we propose a Boolean algebra and test it with the Huntington postulates to determine its structure. We do this so that we can utilize the theorems and properties of other Huntington system for a new system that is defined over a set of voltge levels and hardware operators. Boolean algebra, like other axiomatic systems, is based on several operators defined over a set of undefined elements. A SET is any collection of elements having some common property; and these elements need not be defined. The set of elements we will be dealing with is {0, 1}. The 0 and 1, as far as we are concerned, are some special symbols. They are simply some objects we are going to make some statements about. An operation (., +) is defined as a rule defining the results of an operation on two elements of the set. Becuase these operators operate on two elements, they are commonly reflected to as "binary operators".

2.2.1 Huntington's Postulates

1. A set of elements S is closed with respect to an operator if for every pair of elements in S the operator specifies a unique result (element) which is also in the set S.

   Or

   For the operator + the result of A + B must be found in S if A and B are in S; and for the operator the result of A . B must also be found in S if A and B are elements in S.
2. (*a*) There exists an element 0 in S such that for every A in S, A + 0 = A.
   (*b*) There exists an element 1 in S such that for every A in S, A.1 = A.
3. (*a*) A + B = B + A
   (*b*) A . B = B . A } Commutative Law
4. (*a*) A . (B + C) = (A . B) + (A . C)
   (*b*) A + (B . C) = (A + B) . (A + C) } Distributive Law
5. For every element A in S, there exists an element A′ such that $A.\overline{A} = 0$ and $A + \overline{A} = 1$
6. There exist at least two elements A and B in S such that A is not equivalent to B.

Therefore, if we propose the following two values. Boolean algebra system, that is, if we define the set S = {0, 1} and prescribe the rules for ·, + and INVERTER as follows:

Rules for " . "

.	0	1
0	0	0
1	0	1

or

A	B	A.B
0	0	0
0	1	0
1	0	0
1	1	1

Rules for "+"

+	0	1
0	0	1
1	1	1

or

A	B	A + B
0	0	0
0	1	1
1	0	1
1	1	1

INVERT FUNCTION (COMPLEMENT)

A	A′
0	1
1	0

and test our system with postulates, we find

1. Closure is obvious—no results other than the 0 and 1 are defined.
2. From the tables (*a*) 0 + 0 = 0 0 + 1 = 1 + 0 = 1
   (*b*) 1.1 = 1 1.0 = 0.1 = 0
3. The commutative laws are obvious by the symmetry of the operator tables.
4. (*a*) The distributive law can be proven by a TRUTH-TABLE.

A	*B*	*C*	*B + C*	*A.(B + C)*	*A.B*	*A.C*	*(A.B) + (A.C)*
0	0	0	0	0	0	0	0
0	0	1	1	0	0	0	0
0	1	0	1	0	0	0	0
0	1	1	1	0	0	0	0
1	0	0	0	0	0	0	0
1	0	1	1	1	0	1	1
1	1	0	1	1	1	0	1
1	1	1	1	1	1	1	1

(*b*) Can be shown by a similar table.

5. From the INVERTER function table

(COMPLEMENT)

$$1.\overline{1} = 1.0 = 0, \qquad 0.\overline{0} = 0.1 = 0$$

$$1 + \overline{1} = 1 + 0 = 1, \qquad 0 + \overline{0} = 0 + 1 = 1$$

6. It is obvious that the set S = {0, 1} fulfills the minimum requirements of having at least two elements where $0 \neq 1$.

From this study the following postulates can be listed below:

Table 2.2

Postulate 2	(*a*) A + 0 = A	(*b*) A.1 = A	Intersection Law
Postulate 3	(*a*) A + B = B + A	(*b*) A.B = B.A	Commutating Law
Postulate 4	(*a*) A(B + C) = AB + AC	(*b*) A + BC = (A + B) (A + C)	Distributive Law
Postulate 5	(*a*) $A + \overline{A} = 1$	(*b*) A.A′ = 0	Complements Law

We have just established a two valued Boolean algebra having a set of two elements, 1 and 0, two binary operators with operation rules equivalent to the AND or OR operations, and a complement operator equivalent to the NOT operator. Thus, Boolean algebra has been defined in a formal mathematical manner and has been shown to be equivalent to the binary logic. The presentation is helpful in understanding the application of Boolean algebra in gate type circuits. The formal presentation is necessary for developing the theorems and properties of the algebraic system.

2.2.2 Basic Theorems and Properties of Boolean Algebra

Duality. The Huntington postulates have been listed in pairs and designated by part (*a*) and (*b*) in Table 2.3. *One part may be obtained from other if the binary operators* (+ and .) and identity elements (0 and 1) are interchanged. This important property of Boolean algebra is called the duality principle. It states that every algebraic expression deducible from the postulates of Boolean algebra remain valid if the operators and identity elements are interchanged. In a two valued Boolean algebra, the identity elements and the elements of the set are same: 1 and 0.

Basic Theorems. Table 2.3 lists six theorems of Boolean algebra. The theorems, like the postulates, are listed in pairs; each relation is the dual of the one paired with it. The postulates are basic axioms of the algebraic structure and need no proof. The theorems must be proven from the postulates.

Table 2.3 Theorems of boolean algebra

Theorem		
1. (*a*) A + A = A	(*b*) A.A = A	Tautology Law
2. (*a*) A + 1 = 1	(*b*) A.0 = 0	Union Law
3. (A′)′ = A		Involution Law
4. (*a*) A + (B + C) = (A + B) + C	(*b*) A.(B.C) = (A.B).C	Associative Law
5. (*a*) (A + B)′ = A′B′	(*b*) (A.B)′ = A′ + B′	De Morgan's Law

6. (*a*) A + AB = A (*b*) A(A + B) = A Absorption Law

7. (*a*) A + A′B = A + B (*b*) A(A′ + B) = AB

8. (*a*) AB + AB′ = A (*b*) (A + B) (A + B′) = A Logical adjancy

9. (*a*) AB + A′C + BC = AB + A′C (*b*) (A + B) (A′ + C) (B + C) = (A′ + B) (A′ + C)

Consensus Law

The proofs of the theorem are presented below. At the right is listed the number of postulate which justifies each step of proof.

Theorem 1(*a*) A + A = A

A + A = (A + A).1 by postulate 2(*b*)

= (A + A) (A + A′) 5(*a*)

= A + AA′ 4(*b*)

= A + 0 5(*b*)

= A 2(*a*)

Theorem 1(*b*) A.A = A.

A.A = A.A + 0 by postulate 2(*a*)

= A.A + A.A′ 5(*b*)

= A(A + A′) 4(*a*)

= A.1 5(*a*)

= A 2(*b*)

Note that theorem 1(*b*) is the dual of theorem 1(*a*) and that each step 7 the proof in part (*b*) is the dual of part (*a*). Any dual theorem can be similarly derived from the proof of its corresponding pair.

Theorem 2(*a*) A + A = 1

A + 1 = 1.(A + 1) by postulate 2(*b*)

= (A + A′) (A + 1) 5(*a*)

= A + A′.1 4(*b*)

= A + A′ 2(*b*)

= 1 5(*a*)

Theorem 2(*b*) A.0 = 0 by duality.

Theorem 3. (A′)′ = A From postulate 5, we have

A + A′ = 1 and A.A′ = 0, which defines the complement of A. The complement of A′ is A and is also (A′)′. Therefore, since the complement is unique, we have that (A′)′ = A.

Theorem 4(*a*) A + (B + C) = (A + B) + C

We can prove this by perfect induction method shown in table 2.4 below:

Table 2.4

A	B	C	(B + C)	A + (B + C)	(A + B)	(A + B) + C
0	0	0	0	0	0	0
0	0	1	1	1	0	1
0	1	0	1	1	1	1
0	1	1	1	1	1	1

(Contd.)...

A	B	C	$(B + C)$	$A + (B + C)$	$(A + B)$	$(A + B) + C$
1	0	0	0	1	1	1
1	0	1	1	1	1	1
1	1	0	1	1	1	1
1	1	1	1	1	1	1

We can observe that $A + (B + C) = (A + B) + C$

*Theorem 4(*b*)—can be proved in similar fashion.

*Theorem 5(*a*) and 5(*b*)—can also be proved by perfect induction method.

Theorem 6(*a*)

$$
\begin{aligned}
A + AB &= A \\
A + AB &= A\,(1 + B) \\
&= A\,(1) \\
&= A
\end{aligned}
$$

6(*b*) $A.(A + B) = A$ By duality.

Theorem 7(*a*)

$$
\begin{aligned}
A + A'B &= A + B \\
A + A'B &= A.1 + A'B \\
&= A\,(B + B') + A'B \\
&= AB + AB' + A'B \\
&= AB + AB + AB' + A'B \\
&= A(B + B') + B(A + A') \\
&= A + B.
\end{aligned}
$$

7(*b*) $A.(A' + B) = A.B$ By duality.

Theorem 8(*a*)

$$
\begin{aligned}
AB + AB' &= A \\
AB + AB' &= A(B + B') \\
&= A
\end{aligned}
$$

8(*b*) $(A + B)\,.\,(A + B') = A$ By duality.

Theorem 9(*a*)

$$
\begin{aligned}
AB + A'C + BC &= AB + A'C \\
AB + A'C + BC &= AB + A'C + BC(A + A') \\
&= AB + A'C + ABC + A'BC \\
&= AB\,(1 + C) + A'C\,(1 + B) \\
&= AB + A'C
\end{aligned}
$$

9(*b*) $(A + B)\,(A' + C)\,(B + C) = (A + B)\,(A' + C)$ By duality.

2.3 BOOLEAN FUNCTIONS

A binary variable can take the value of 0 or 1. A Boolean function is an expression formed with binary variable, the two binary operators OR and AND, the unary operator NOT, parantheses and an equal sign. For a given value of the variables, the function can be either 0 or 1. Consider, for example, the Boolean function

$$F_1 = xy'z$$

*The proofs of 4(*b*), 5(*a*) and 5(*b*) are left as an exercise for the reader.

The function F_1 is equal to 1 when $x = 1$, $y = 0$ and $z = 1$; otherwise $F_1 = 0$. This is an example of a Boolean function represented as an algebraic expression. A Boolean function may also be represented in a truth table. To represent a function in a truth table, we need a list of the 2^n combinations of 1's and 0's of n binary variables, and a column showing the combinations for which the function is equal to 1 or 0 as discussed previously. As shown in Table 2.5, there are eight possible distinct combination for assigning bits to three variables. The table 2.5 shows that the function F_1 is euqal to 1 only when $x = 1$, $y = 0$ and $z = 1$ and equal to 0 otherwise.

Consider now the function

$$F_2 = x'y'z + x'yz + xy'$$

$F_2 = 1$ if $x = 0, y = 0, z = 1$ or
$x = 0, y = 1, z = 1$ or
$x = 1, y = 0, z = 0$ or
$x = 1, y = 0, z = 1$

$F_2 = 0$, otherwise.

Table 2.5

x	y	z	F_1	F_2	F_3
0	0	0	0	0	0
0	0	1	0	1	1
0	1	0	0	0	0
0	1	1	0	1	1
1	0	0	0	1	1
1	0	1	1	1	1
1	1	0	0	0	0
1	1	1	0	0	0

The number of rows in the table is 2^n, where n is the number of binary variables in the function.

The question now arises, Is an algebraic expression of a given Boolean function unique? Or, is it possible to find two algebraic expressions that specify the same function? The answer is yes. Consider for example a third function.

$$F_3 = xy' + x'z$$

$F_3 = 1$ if $x = 1, y = 0, z = 0$ or
$x = 1, y = 0, z = 1$ or
$x = 0, y = 0, z = 1$ or
$x = 0, y = 1, z = 1$

$F_3 = 0$, otherwise.

From table, we find that F_3 is same as F_2 since both have identical 1's and 0's for each combination of values of the three binary variables. In general, two functions of n binary variables are said to be equal if they have the same value for all possible 2^n combinations of the n variables.

As a matter of fact, the manipulation of Boolean algebra is applied mostly to the problem of finding simpler expressions for the same function.

2.3.1 Transformation of Boolean Function into Logic Diagram

A Boolean function may be transformed from an algebraic expresion into a logic diagram composed of AND, OR and NOT gates. Now we shall implement the three functions discussed above as shown in Fig. 2.1.

- Here we are using inverters (NOT gates) for complementing a single variable. In general however, it is assumed that we have both the normal and complement forms available.
- There is an AND gate for each product term in the expression.
- An OR gate is used to combine two or more terms.

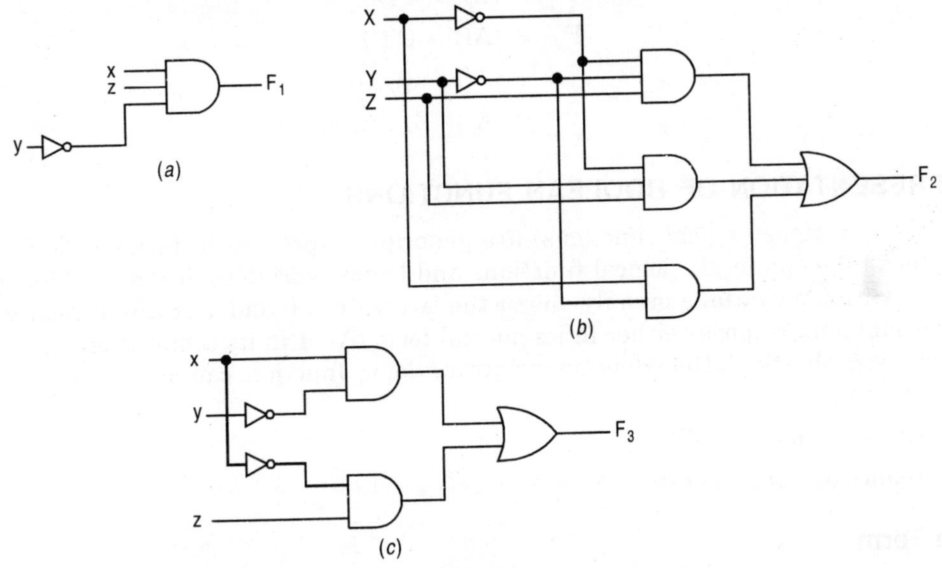

Fig. 2.1

From the Fig. 2.1, it is obvious that the implementation of F_3 requires fewer gates and fewer inputs than F_2. Since F_3 and F_2 are equal Boolean functions, it is more economical to implement F_3 form than the F_2 form. To find simpler circuits, we must know how to manipulate Boolean functions to obtain equal and simpler expression. These simplification (or minimization) techniques will be discuss in detail in next chapter.

2.3.2 Complement of a Function

The complement of a function F is F′ and is obtained from an interchange of 0's for 1's and 1's for 0's in the value of F. The complement of a function may be derived algebraically through De Morgan's theorem. De Morgan's theorem can be extended to three or more variables. The three-variable form of the De Morgan's theorem is derived below:

$$
\begin{aligned}
(A + B + C)' &= (A + X)' && \text{Let } B + C = X \\
&= A'X' && \text{by theorem } 5(a) \\
&= A'.(B + C)' && \text{substituting } B + C = X \\
&= A'.(B'C') && \text{theorem } 5(a) \\
&= A'B'C' && \text{theorem } 4(a)
\end{aligned}
$$

This theorem can be generalized as

$$(A + B + C + D + ... F)' = A'B'C'D'....F'$$

and its DUAL

$$(ABCD .... F)' = A' + B' + C' + D' + ... + F'$$

The generalized form of De Morgan's theorem states that the complement of a function is obtained by interchaning AND and OR operators and complementing each literal.

Example. *Determine the complements of the following function:*

$$F_1 = AB' + C'D$$

Solution.

$$F_1 = AB' + C'D$$

$$F_1' = (AB' + C'D)'$$

$$= (AB')' . (C'D)'$$

$$= (A' + B) . (C + D')$$

2.4 REPRESENTATION OF BOOLEAN FUNCTIONS

Boolean functions (logical functions) are generally expressed in terms of logical variables. Values taken on by the logical functions and logical variables are in the binary form. Any logical variable can take on only one of the two values 0 and 1 or any logical variable (binary variable) may appear either in its normal form (A) or in its complemented form (A'). As we will see shortly latter that an arbitrary logic function can be expressed in the following forms:

(*i*) Sum of Products (SOP)

(*ii*) Product of Sums (POS)

Product Term

The AND function is referred to as product. The logical product of several variables on which a function depends is considered to be a product term. The variables in a product term can appear either in complemented or uncomplemented (normal) form. For example AB′C is a product form.

Sum Term

The OR function is generally used to refer a sum. The logical sum of several variables on which a function depends is considered to be a sum term. Variables in a sum term also can appear either in normal or complemented form. For example A + B + C′, is a sum term.

Sum of Products (SOP)

The logic sum of two or more product terms is called a 'sum of product' expression. It is basically an OR operation of AND operated variables such as F = A′B + B′C + A′BC.

Product of Sums (POS)

The logical product of two or more sum terms is called a 'product of sum' expression. It is basically an AND operation of OR operated variables such as F = (A′ + B). (B′ + C) . (A′ + B + C).

2.4.1 Minterm and Maxterm Realization

Consider two binary variables A and B combined with an AND operation. Since each variable may appear in either form (normal or complemented), there are four combinations, that are possible—AB, A′B, AB′, A′B′.

Each of these four AND terms represent one of the four distinct combinations and is called a minterm, or a standard product or fundamental product.

Now consider three variables—A, B and C. For a three variable function there are 8 minterms as shown in Table 2.6(*a*). (Since there are 8 combinations possible). The binary numbers from 0 to 7 are listed under three varibles. Each minterm is obtained from an AND term of the three variables, with each variable being primed (complemented form) if the corresponding bit of the binary number is a 0 and unprimed (normal form) if a 1. The symbol is m_j, where j denotes the decimal equivalent of the binary number of the minterm designated.

In a similar manner, n variables can be combined to form 2^n minterms. The 2^n different minterms may be determined by a method similar to the one shown in table for three variables.

Similarly n variables forming an OR term, with each variable being primed or unprimed, provide 2^n possible combinations, called maxterms or standard sums.

Each maxterm is obtained from an OR term of the n variables, with each variable being unprimed if the corresponding bit is a 0 and primed if a 1.

It is interesting to note that each maxterm is the complement of its corresponding minterm and vice versa.

Now we have reached to a level where we are able to understand two very important properties of Boolean algebra through an example.

Table 2.6(*a*) Minterm and Maxterm for three binary variables

				MINTERMS		*MAXTERMS*	
Decimal Eqt.	A	B	C	*Term*	*Designation*	*Term*	*Designation*
0	0	0	0	A′B′C′	m_0	A + B + C	M_0
1	0	0	1	A′B′C	m_1	A + B + C′	M_1
2	0	1	0	A′BC′	m_2	A + B′ + C	M_2
3	0	1	1	A′BC	m_3	A + B′ + C′	M_3
4	1	0	0	AB′C′	m_4	A′ + B + C	M_4
5	1	0	1	AB′C	m_5	A′ + B + C′	M_5
6	1	1	0	ABC′	m_6	A′ + B′ + C	M_6
7	1	1	1	ABC	m_7	A′ + B′ + C′	M_7

Let we have a TV that is connected with three switches. TV becomes ‘ON’ only when at least two of the three switches are ‘ON’ (or high) and in all other conditions TV is ‘OFF’ (or low). The example is same as we have already discussed in section (2.1) TT.

Let the three switches are represented by three variables A, B and C. The output of TV is represented by F. Since there are three switches (three variables), there are 8 distinct combinations possible that is shown in TT (Table 2.6(*b*)).

Table 2.6(*b*)

SWITCHES			*TV (o/p)*
A	B	C	F
0	0	0	0
0	0	1	0
0	1	0	0
0	1	1	1
1	0	0	0
1	0	1	1
1	1	0	1
1	1	1	1

HIGH (ON) → 1
LOW (OFF) → 0.

The TV becomes 'ON' at four combinations. These are 011, 101, 110 and 111. We can say that F is determined by expressing the combinations A′BC, AB′C, ABC′ and ABC. Since each of these minterms result in F = 1, we should have

$$F = A'BC + AB'C + ABC' + ABC$$
$$= m_3 + m_5 + m_6 + m_7.$$

This demonstrates an important property of Boolean algebra that 'Any Boolean function can be expressed as sum of minterms or as 'Sum of product". However, there is no guarantee that this SOP expression will be a minimal expression. In other words, SOP expressions are likely to have redundancies that lead to systems which requires more hardware that is necessary. This is where the role of theorems and other reduction techniques come into play as will be shown in next chapter.

As mentioned, any TRUTH-TABLE INPUT/OUTPUT specifications can be expressed in a SOP expression. To facilitate this a shorthand symbology has been developed to specify such expressions. This is done by giving each row (MINTERM) in the TRUTH-TABLE a decimal number that is equivalent to the binary code of that row, and specifying the expression thus:

$$F = \Sigma(m_3, m_5, m_6, m_7)$$

which reads: F = the sum-of-products of MINTERMS 3, 5, 6 and 7. This shorthand notation can be further shortend by the following acceptable symbology:

$$F = \Sigma(3, 5, 6, 7)$$

Expression such as these serve as great aids to the simplification process, as shown in next chapter.

Now, continuing with the same example, consider the complement of Boolean function that can be read from Truth-table by forming a minterm for each combination that produces 0 in the function and by ORing

$$F' = A'B'C' + A'B'C + A'BC' + AB'C'$$

Now, if we take the complement of F′, we get F.

$$F = (A + B + C) . (A + B + C') . (A + B' + C) (A' + B + C)$$
$$= \quad M_0 \qquad M_1 \qquad M_2 \qquad M_4$$

This demonstrates a second important property of the Boolean algebra that 'Any Boolean function can be expressed as product-of-maxterms or as product of sums'. The procedure for obtaining the product of maxterms directly from Truth-table is as; Form a maxterm for each combination of the variables that produces a 0 in the function, and then form the AND of all those functions. Output will be equal to F because in case of maxterms 0 is unprimed.

The shortend symbology for POS expressions is as follows—

$$F = \Pi(M_0, M_1, M_2, M_4)$$

or

$$F = \Pi(0, 1, 2, 4)$$

Boolean functions expressed as sum of minterms (sum of product terms) SOP or product of maxterms, (Product of sum terms) POS are said to be in CANONICAL form or STANDARD form.

2.4.2 Standard Forms

We have seen that for n binary variables, we can obtain 2^n distinct minterms, and that any Boolean function can be expressed as a sum of minterms or product of maxterms. It is sometimes convenient to express the Boolean function in one of its standard form (SOP or POS). If not in this form, it can me made as follows:

1. Sum of Product. First we expand the expression into a sum of AND terms. Each term is then inspected to see if it contains all the variable. If it misses one or more variables, it is ANDed with an expression such as A + A′, where A is one of the missing variables.

Example 1. *Express the Boolean function $F = x + y'z$ in a sum of product (sum of minterms) form.*

Solution. The function has three variables x, y and z. The first term x is missing two variables; therefore

$$x = x\,(y + y') = xy + xy$$

This is still missing one variable:

$$\begin{aligned} x &= xy\,(z + z') + xy'\,(z + z') \\ &= xyz + xyz' + xy'z + xy'z' \end{aligned}$$

The second term $y'z$ is missing one variable:

$$y'z = y'z\,(x + x') = xy'z + x'y'z$$

Combining all terms, we have

$$F = x + y'z = xyz + xyz' + xy'z + xy'z' + xy'z + x'y'z$$

But $xy'z$ appears twice, and according to theorem 1 (A + A = A), it is possible to remove one of them. Rearranging the min terms in ascending order, we have:

$$\begin{aligned} F &= x'y'z + xy'z' + xy'z + xyz' + xyz \\ &= m_1 + m_4 + m_5 + m_6 + m_7. \\ F(x, y, z) &= \Sigma(1, 4, 5, 6, 7) \end{aligned}$$

An alternative method for product terms (minterms) is to make a TT directly from function. $F = x + y'z$. From TT, we can see directly five minterms where the value of function is equal to 1. Thus,

$$F(x, y, z) = \Sigma(1, 4, 5, 6, 7)$$

x	y	z	$F = x + y'z$
0	0	0	0
0	0	1	1
0	1	0	0
0	1	1	0
1	0	0	1
1	0	1	1
1	1	0	1
1	1	1	1

2. Product of Sums. To express the Boolean function as product of sums, it must first be brought into a form of OR terms. This may be done by using the distributive law A + BC = (A + B) . (A + C). Then any missing variable A in each OR term is ORed with AA′.

Example 2. *Express the Boolean function $F = AB + A'C$ in a product of sum (product of maxterm) form.*

Solution. First convert the function into OR terms using distributive law.

$$F = AB + A'C = (AB + A')(AB + C)$$
$$= (A + A')(B + A')(A + C)(B + C)$$
$$= (A' + B)(A + C)(B + C)$$

The function has three variables A, B and C. Each OR term is missing one variable therefore:

$$A' + B = A' + B + CC' = (A' + B + C)(A' + B + C')$$
$$A + C = A + C + BB' = (A + B + C)(A + B' + C)$$
$$B + C = B + C + AA' = (A + B + C)(A' + B + C)$$

Combining all these terms and removing those that appear more than once.

$$F = \underset{M_0}{(A + B + C)}\ \underset{M_2}{(A + B' + C)}\ \underset{M_4}{(A' + B + C)}\ \underset{M_5}{(A' + B + C')}$$

$$F(A, B, C) = \Pi(0, 2, 4, 5) = \Pi M(0, 2, 4, 5)$$

An alternative method for deriving sum terms (maxterms) again is to make a TT directly from function.

$$F = AB + A'C$$

From TT, we can see directly four maxterms where the value of function is equal to 0.

Thus, $F(A, B, C) = \Pi(0, 2, 4, 5)$

A	B	C	$F = AB + A'C$
0	0	0	0
0	0	1	1
0	1	0	0
0	1	1	1

(Contd.)...

A	B	C	F = AB + A′C
1	0	0	0
1	0	1	0
1	1	0	1
1	1	1	1

2.4.3 Conversion between Standard Forms

Consider the table 2.6(*b*) shown in section 2.4.1 to help you establish the relationship between the MAXTERM and MINTERM numbers.

From table we see that $m_j = \overline{M_j}$

An interesting point can be made in relationship between MAXTERM lists and MINTERMS lists. The subscript number of the terms in the MAXTERM list correspond to the same subscript numbers for MINTERMS that are not included in the MINTERM list. From this we can say the following:

Π (Set of MAXTERM numbers)

We know that the function derived from this list will yield precisely the same result as the following:

Σ(set of MINTERMS numbers that are not included in the MAXTERM list)

For example:

Given, F(A, B, C) = Π(0, 1, 4, 6)

We know immediately that

F(A, B, C) = Σ(2, 3, 5, 7)

2.5 LOGIC GATES

We have seen that the foundation of logic design is seated in a well defined axiomatic system called Boolean algebra, which was shown to be what is known as a "Huntington system". In this axiomatic system the definition of AND and OR operators or functions was set forth and these were found to be well defined operators having certain properties that allow us to extend their definition to Hardware applications. These AND and OR operators, sometimes referred to as connectives, actually suggest a function that can be emulated by some H/W logic device. The Hardware logic devices just mentioned are commonly referred to as "gates".

Keep in mind that the usage of "gate" refers to an actual piece of Hardware where "function" or "operation" refers to a logic operator AND. On the other hand, when we refer to a "gate" we are referring directly to a piece of hardware called a gate. The main point to remember is 'Don't confuse gates with logic operators'.

2.5.1 Positive and Negative Logic Designation

The binary signals at the inputs or outputs of any gate can have one of the two values except during transition. One signal levels represents logic 1 and the other logic 0. Since two signal values are assigned two logic values, there exist two different assignments of signals to logic.

Logics 1 and 0 are generally represented by different voltage levels. Consider the two values of a binary signal as shown in Fig. 2.2. One value must be higher than the other since the two values must be different in order to distinguish between them. We designate the higher voltage level by H and lower voltage level by L. There are two choices for logic values assignment. Choosing the high-level H to represent logic 1 as shown in (*a*) defines a positive logic system. Choosing the low level L to represent logic-1 as shown in (*b*), defines a negative logic system.

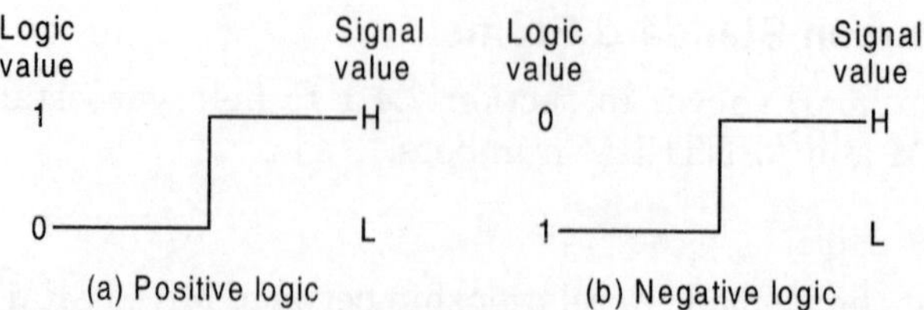

Fig. 2.2

The terms positive and negative are somewhat misleading since both signal values may be positive or both may be negative. Therefore, it is not signal polarity that determines the type of logic, but rather the assignment of logic values according to the relative amplitudes of the signals.

The effect of changing from one logic designation to the other equivalent to complementing the logic function because of the principle of duality of Boolean algebra.

2.5.2 Gate Definition

A 'gate' is defined as a multi-input (≥ 2) hardware device that has a two-level output. The output level (1–H/0–L) of the gate is a strict and repeatable function of the two-level (1–H/0–L) combinations applied to its inputs. Figure 2.3 shows a general model of a gate.

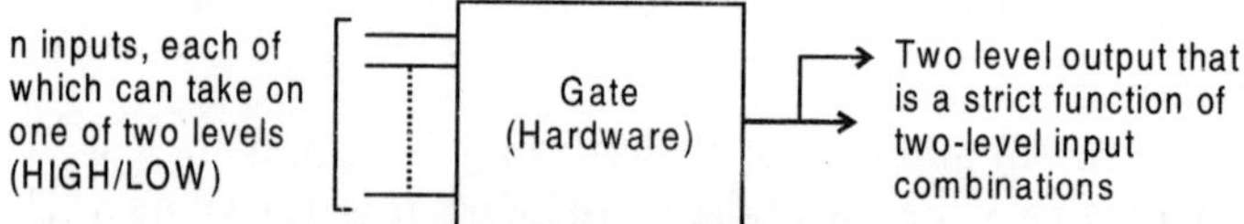

Fig. 2.3 The general model of a gate

The term "logic" is usually used to refer to a decision making process. A logic gate, then, is a circuit that can decide to say yes or no at the output based upon inputs.

We apply voltage as the input to any gate, therefore the Boolean (logic) 0 and 1 do not represent actual number but instead represent the state of a voltage variable or what is called its logic level. Sometimes logic 0 and logic 1 may be called as shown in table 2.7.

Table 2.7

Logic 0	*Logic 1*
False	True
Off	On
Low	High
No	Yes
Open switch	Close switch

2.5.3 The AND Gate

The AND gate is sometimes called the "all or nothing gate". To show the AND gate we use the logic symbol in Fig. 2.4(*a*). This is the standard symbol to memorize and use from now on for AND gates.

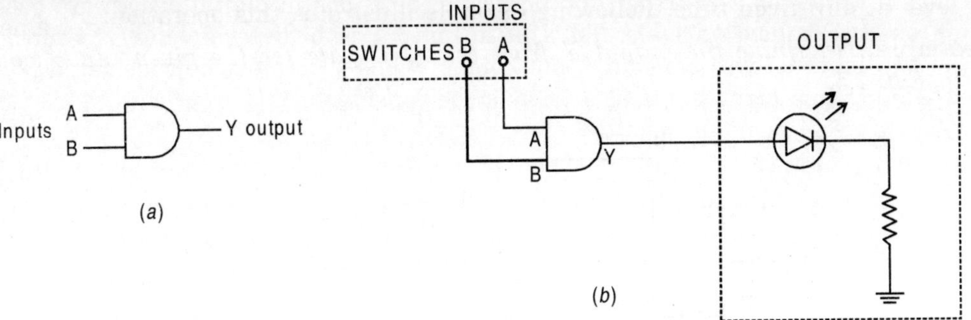

Fig. 2.4 (*a*) AND Gate logic symbol. (*b*) Practical AND gate circuit

Now, let us consider Fig. 2.4(*b*). The AND gate in this figure is connnected to input switches A and B. The output indicator is an LED. If a low voltage (Ground, GND) appears at inputs, A and B, then the output LED does not glow. This situation is illustrated in table 2.8. Line 1 indicates that if the inputs are binary 0 and 0, then the output will be binary 0. *Notice that only binary* 1s at both A and B will produce a binary 1 at the output.

Table 2.8 AND truth table

INPUTS				*OUTPUTS*	
A		*B*		*Y*	
Switch Voltage	*Binary*	*Switch Voltage*	*Binary*	*Light*	*Binary*
Low	0	Low	0	No	0
Low	0	High	1	No	0
High	1	Low	0	No	0
High	1	High	1	Yes	1

It is a +5V compared to GND appearing at A, B, or Y that is called a binary 1 or a HIGH voltage. A binary 0, or Low voltage, is defined as a GND voltage (near 0V compared to GND) appearing at A, B or Y. *We are using positive logic because it takes a positive +5V to produce what we call a binary 1.*

The truth table is said to describe the AND function. The unique output from the AND gate is a HIGH only when all inputs are HIGH.

Figure 2.4 (*c*) shows the ways to express that input A is ANDed with input B to produce output Y.

<table>
<tr><td>Boolean Expression</td><td>AND Symbol
↑
A . B = Y</td></tr>
<tr><td>Logic Symbol</td><td>A, B → Y</td></tr>
<tr><td>Truth Table</td><td>
<table>
<tr><td>A</td><td>B</td><td>Y</td></tr>
<tr><td>0</td><td>0</td><td>0</td></tr>
<tr><td>0</td><td>1</td><td>0</td></tr>
<tr><td>1</td><td>0</td><td>0</td></tr>
<tr><td>1</td><td>1</td><td>1</td></tr>
</table>
</td></tr>
</table>

Fig. 2.4 (*c*)

Pulsed Operation

In many applications, the inputs to a gate may be voltage that change with time between the two logic levels and are called as pulsed waveforms. In studying the pulsed operation of an AND gate, we consider the inputs with respect to each other in order to determine the output level at any given time. Following example illustrates this operation:

Example. *Determine the output Y from the AND gate for the given input waveform shown in Fig. 2.5.*

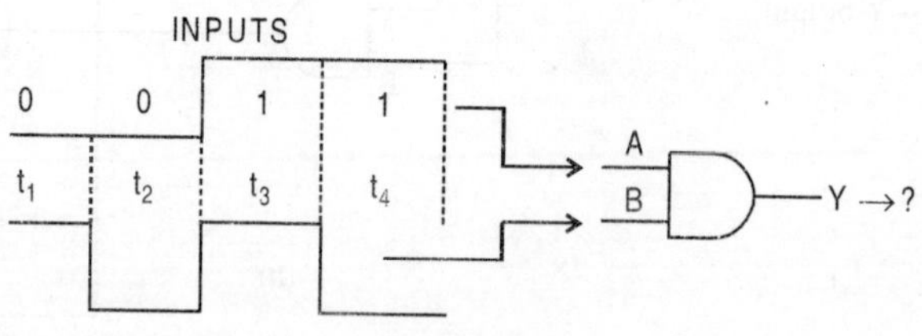

Fig. 2.5

Solution. The output of an AND gate is determined by realizing that it will be high only when both inputs are high at the same time. For the inputs the outputs is high only during t_3 period. In remaining times, the outputs is 0 as shown in Fig. 2.6.

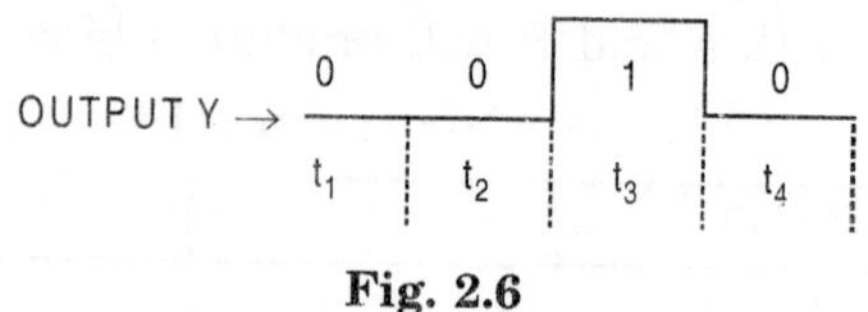

Fig. 2.6

2.5.4 The OR Gate

The OR gate is sometimes called the "any or all gate". To show the OR gate we use the logical symbol in Fig. 2.7(*a*).

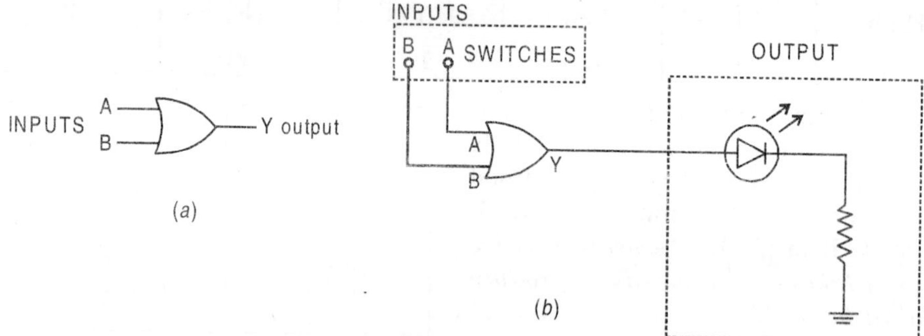

Fig. 2.7 (*a*) OR gate logic symbol. (*b*) Practical OR gate circuit

A truth-table for the 'OR' gate is shown below according to Fig. 2.7(*b*). The truth-table lists the switch and light conditions for the OR gate. The unique output from the OR gate is a LOW only when all inputs are low. The output column in Table (2.9) shows that only the first line generates a 0 while all others are 1.

Table 2.9

INPUTS				*OUTPUTS*	
A		*B*		*Y*	
Switch	*Binary*	*Switch*	*Binary*	*Light*	*Binary*
Low	0	Low	0	No	0
Low	0	High	1	Yes	1
High	1	Low	0	Yes	1
High	1	High	1	Yes	1

Figure 2.7(*c*) shows the ways to express that input A is ORed with input B to produce output Y.

Boolean Expression	OR Symbol ↑ A + B = Y
Logic Symbol	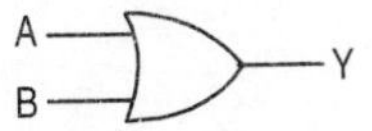 A, B → OR gate → Y
Truth Table	A B | Y 0 0 | 0 0 1 | 1 1 0 | 1 1 1 | 1

Fig. 2.7 (*c*)

Example 1. *Determine the output Y from the OR gate for the given input waveform shown in Fig. 2.8.*

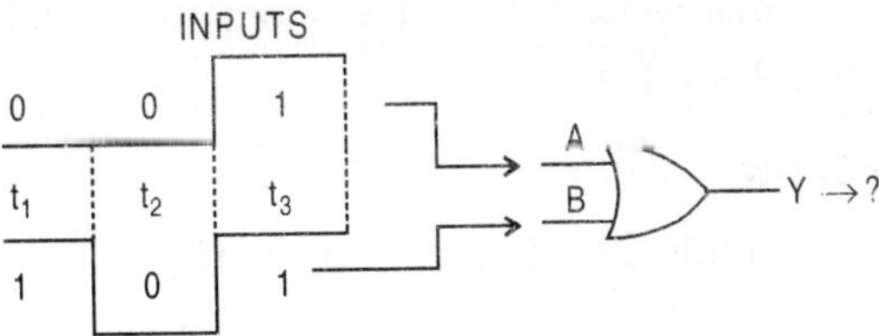

Fig. 2.8

Solution. The output of an OR gate is determined by realizing that it will be low only when both inputs are low at the same time. For the inputs the output is low only during period t_2. In remaining time output is 1 as shown in Fig. 2.9.

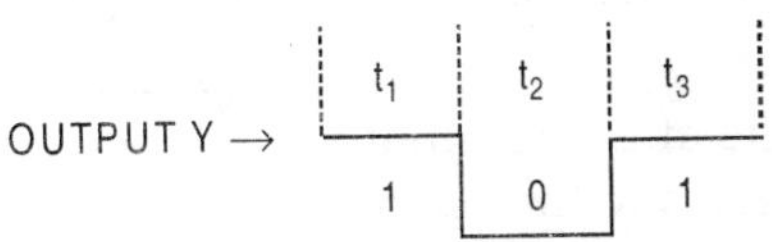

Fig. 2.9

We are now familiar with AND and OR gates. At this stage, to illustrate at least in part how a word statement can be formulated into a mathematical statement (Boolean expression) and then to hardware network, consider the following example:

Example 2. *Utkarsha will go to school if Anand and Sawan go to school, or Anand and Ayush go to school.*

Solution. → This statement can be symbolized as a Boolean expression as follows:

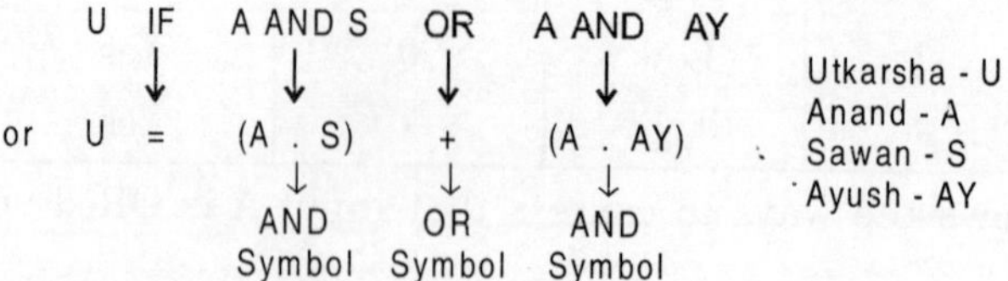

The next step is to transform this Boolean expression into a Hardware network and this is where AND and OR gates are used.

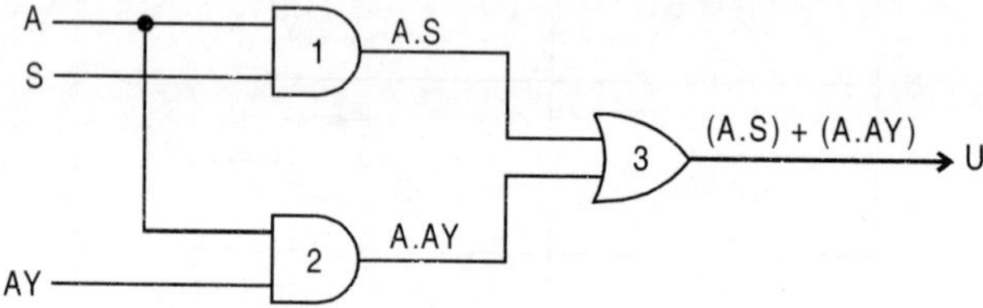

The output of gate 1 is high only if both the inputs A and S are high (mean both Anand and Sawan go to school). This is the first condition for Utkarsha to go to school.

The output of gate 2 is high only if both the inputs A and AY are high (means both Anand and Ayush go to school). This is the second condition for Utkarsha to go to school.

According to example at least one condition must be fullfilled in order that Utkarsha goes to school. The output of gate 3 is high when any of the input to gate 3 is high means at least one condition is fulfilled or both the inputs to gate 3 are high means both the conditions are fulfilled.

The example also demonstrates that Anand has to go to school in any condition otherwise Utkarsha will not go to school.

2.5.5 The Inverter and Buffer

Since an Inverter is a single input device, it performs no logic interaction function between two variables, and to say that merely changing a voltage level constitute a logic operation would be misleading. Therefore we define it as an Inverter function rather than a logic operator like the AND and OR operators. The NOT circuit performs the basic logical function called inversion or complementation. That is why, it is also known as Inverter. The NOT circuit has only one input and one ouput. The purpose of this gate is to give an output that is not the same as the input. When a HIGH level is applied to an inverter, a LOW level appears at its output and vice versa. The logic symbol for the inverter is shown in Fig. 2.10(*a*).

If we were to put in a logic at 1 and input A in Fig. 2.10(*a*), we would get out the opposite, or a logical 0, at output Y.

INPUT A ——— Y OUTPUT

$Y = A'$ or $\overline{A}$

Fig. 2.10 (*a*) A logic symbol and Boolean expression for an inverter

The truth-table for the inverter is shown in Fig. 2.10(*b*). If the voltage at the input of the inverter is LOW, then the output voltage is HIGH, if the input voltage is HIGH, then the output is LOW. We can say that output is always negated. The terms "negated", "complemented" and "inverted", then mean the same things.

INPUT		*OUTPUT*	
A		*B*	
Voltages	*Binary*	*Voltages*	*Binary*
LOW	0	HIGH	1
HIGH	1	LOW	0

Fig. 2.10 (*b*) Truth-table for an inverter

Now consider the logic diagram as shown in Fig. 2.10(*c*), that shows an arrangement where input A is run through two inverters. Input A is first inverted to produce a "not A" ($\overline{A}$) and then inverted a second time for "double not A" ($\overline{\overline{A}}$). In terms of binary digits, we find that when the input 1 is inverted twice, we end up with original digit. Therefore, we find $\overline{\overline{A}} = A$.

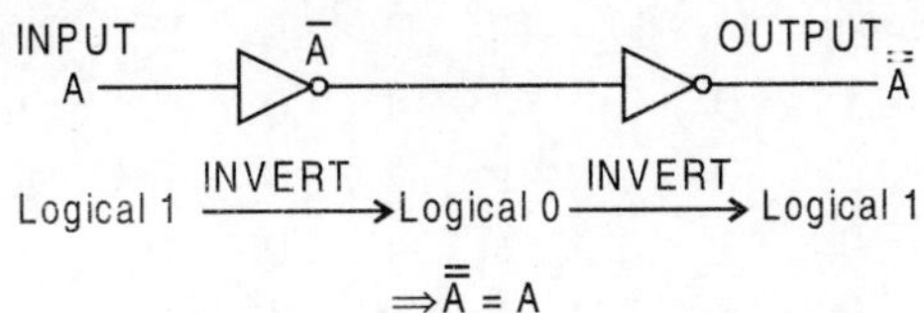

Fig. 2.10 (*c*) Effect of double inverting

The symbol shown in Fig. 2.11(*a*) is that of a non-inverting buffer/driver. A buffer produces the transfer function but does not produce any logical operation, since the binary value of the output is equal to the binary value of the input. The circuit is used merely for power amplification of the signal and is equivalent to two inverters connected in cascade. Figure 2.11(*b*) shows the TT for the buffer.

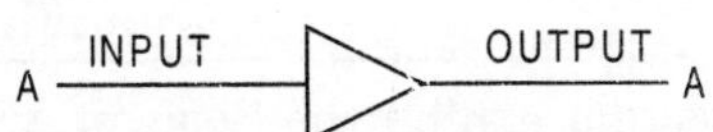

Fig. 2.11 (*a*) Non-inverting buffer/driver logic symbol.

INPUT		*OUTPUT*	
A		*B*	
Voltage	*Binary*	*Voltage*	*Binary*
HIGH	1	HIGH	1
LOW	0	LOW	0

Fig. 2.11 (*b*) Truth table for buffer

Example. *Determine the output Y from the inverter for the given input waveform shown in Fig. 2.12.*

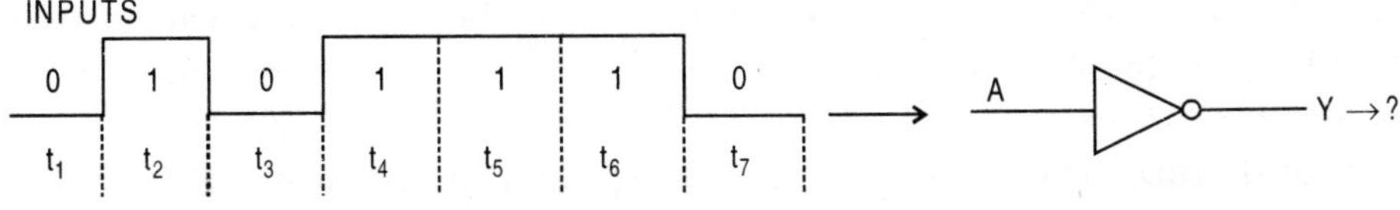

Fig. 2.12

Solution. The output of an Inverter is determined by realizing that it will be high when input is low and it will be low when input is high as shown in Fig. 2.13.

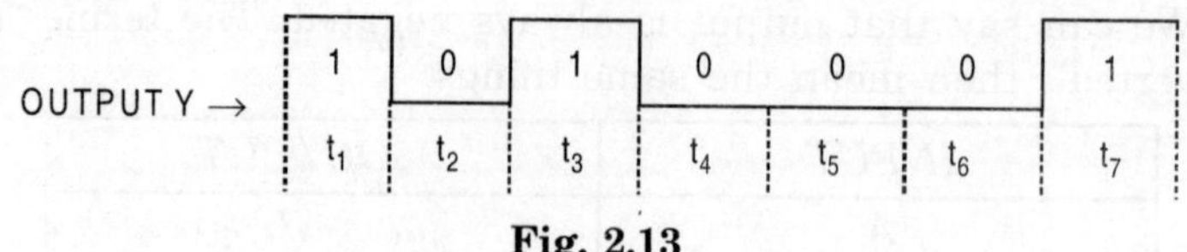

Fig. 2.13

2.5.6 Other Gates and their Functions

The AND, OR, and the inverter are the three basic circuits that make up all digital circuits. Now, it should prove interesting to examine the other 14 possible ouput specification (except AND and OR) for an arbitrary two-input gate.

Consider Table 2.10.

Table 2.10: Input/Output specifications for the 16 possible outputs derived from two-input gates

A	B	GND	F_1	F_2	F_3	F_4	F_5	F_6	F_7	F_8	F_9	F_{10}	F_{11}	F_{12}	F_{13}	F_{14}	+V
0	0	0	1	0	1	0	1	0	1	0	1	0	1	0	1	0	1
0	1	0	0	1	1	0	0	1	1	0	0	1	1	0	0	1	1
1	0	0	0	0	0	1	1	1	1	0	0	0	0	1	1	1	1
1	1	0	0	0	0	0	0	0	0	1	1	1	1	1	1	1	1
			↓ NOR					↓ EXOR	↓ NAND	↓ AND	↓ EXNOR					↓ OR	

Scanning the table for gates that exhibit the Boolean AND and OR operator, we see that F_1 (NOR), F_7 (NAND), F_8 (AND) and F_{14} (OR) are the only outputs from gates which manifest the AND and OR operators. *We shall see very shortly that both NAND and NOR gates can be used as AND and OR gates.* Because of this, they are found in integrated circuit form. All the rest are more complex and deemed unuseful for AND/OR implementation and are not normally found in gate form, with two exceptions. They are F_6 and F_9. F_6 is the Input/output specification for a gate called the EXCLUSIVE OR gate and F_9 is the specification for the COINCIDENCE, EQUIVALENCE, or EXNOR gate, also referred to as an EXCLUSIVE NOR.

2.5.7 Universal Gates

NAND and NOR gates. The NAND and NOR gates are widely used and are readily available in most integrated circuits. A major reason for the widespread use of these gates is that they are both *UNIVERSAL gates,* universal in the sense that both can be used for AND operators, OR operators, as well as Inverter. Thus, we see that a complex digital system can be completely synthesized using only NAND gates or NOR gates.

The NAND Gate. The NAND gate is a NOT AND, or an inverted AND function. The standard logic symbol for the NAND gate is shown in Fig. 2.14(*a*). The little invert bubble (small circle) on the right end of the symbol means to invert the output of AND.

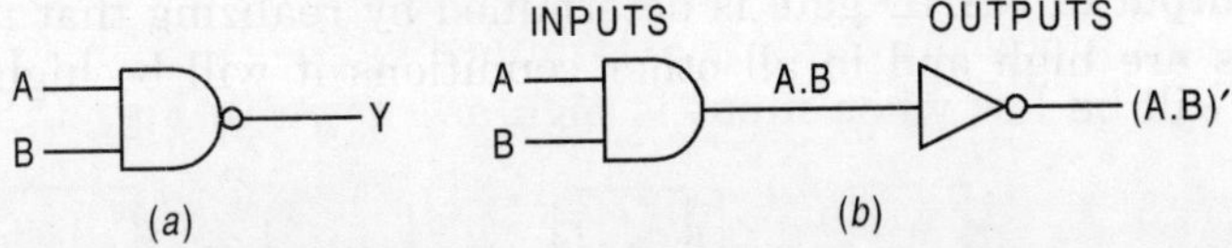

Fig. 2.14 (*a*) NAND gate logic symbol (*b*) A Boolean expression for the output of a NAND gate

Figure 2.14(*b*) shows a separate AND gate and inverter being used to produce the NAND logic function. Also notice that the Boolean expression for the AND gate, (A.B) and the NAND (A.B)′ are shown on the logic diagram of Fig. 2.14(*b*).

The truth-table for the NAND gate is shown in Fig. 2.14(*c*). The truth-table for the NAND gate is developed by inverting the output of the AND gate. 'The unique output from the NAND gate is a LOW only when all inputs are HIGH'.

INPUT		*OUTPUT*	
A	*B*	*AND*	*NAND*
0	0	0	1
0	1	0	1
1	0	0	1
1	1	1	0

Fig. 2.14 (*c*) Truth-table for AND and NAND gates

Figure 2.14 (*d*) shows the ways to express that input A is NANDed with input B yielding output Y.

Boolean Expression	NOT Symbol: $\overline{A \,.\, B} = Y$ (AND Symbol) or $\overline{AB} = Y$ or $(AB)' = Y$
Logic Symbol	A, B → NAND → Y
Truth Table	A B Y: 0 0 1; 0 1 1; 1 0 1; 1 1 0

Fig. 2.14 (*d*)

Example 1. *Determine the output Y from the NAND gate from the given input waveform shown in Fig. 2.15.*

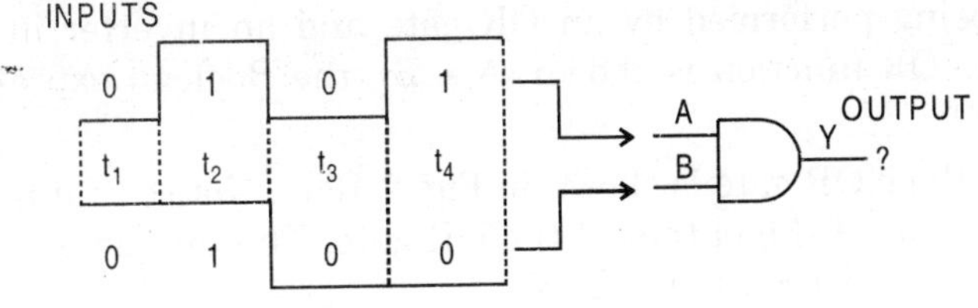

Fig. 2.15

Solution. The output of NAND gate is determined by realizing that it will be low only when both the inputs are high and in all other conditions it will be high. The ouput Y is shown in Fig. 2.16.

	t_1	t_2	t_3	t_4
OUTPUT Y →	1	0	1	1

Fig. 2.16

The NAND gate as a UNIVERSAL Gate

The chart in Fig. 2.17 shows how would you wire NAND gates to create any of the other basic logic functions. The logic function to be performed is listed in the left column of the table; the customary symbol for that function is listed in the center column. In the right column, is a symbol diagram of how NAND gates would be wired to perform the logic function.

Logic Function	*Symbol*	*Circuit using NAND gates only*
Inverter	A — A′	A — A′
AND	A, B — A.B	A, B — A.B
OR	A, B — A + B	A, B — A + B

Fig. 2.17

The NOR gate. The NOR gate is actually a NOT OR gate or an inverted OR function. The standard logic symbol for the NOR gate is shown in Fig. 2.18(*a*).

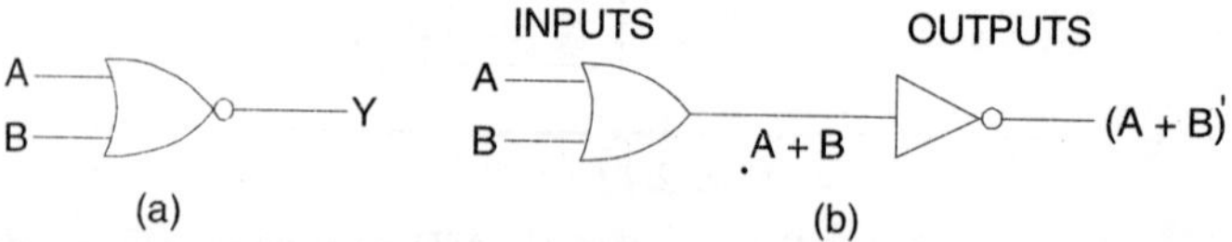

Fig. 2.18 (*a*) NOR gate logic symbol (*b*) Boolean expression for the output of NOR gate

Note that the NOR symbol is an OR symbol with a small invert bubble on the right side. The NOR function is being performed by an OR gate and an inverter in Fig. 2.18(*b*). The Boolean function for the OR function is shown (A + B), the Boolean expression for the final NOR function is (A + B)′.

The truth-table for the NOR gate is shown in Fig. 2.18(*c*). Notice that the NOR gate truth table is just the complement of the output of the OR gate. The unique output from the NOR gate is a HIGH only when all inputs are LOW.

INPUTS		*OUTPUTS*	
A	*B*	*OR*	*NOR*
0	0	0	1
0	1	1	0
1	0	1	0
1	1	1	0

Fig. 2.18 (*c*) Truth-table for OR and NOR gates

Figure 2.18(*d*) shows the ways to express that input A is ORed with input B yielding output Y.

Boolean Expression	NOT Symbol $\overline{A + B} = Y$ or $(A + B)' = Y$ OR Symbol
Logic Symbol	A, B → NOR gate → Y
Truth Table	A B Y 0 0 1 0 1 0 1 0 0 1 1 0

Fig. 2.18 (*d*)

Example 2. *Determine the output Y from the NOR gate from the given input waveform shown in Fig. 2.19.*

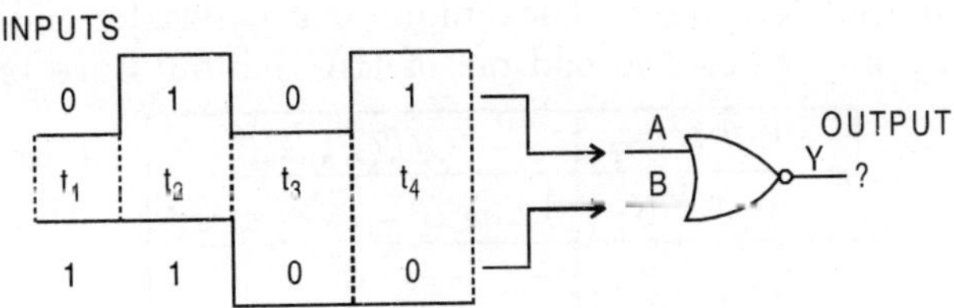

Fig. 2.19

Solution. The output of NOR gate is determined by realizing that it will be HIGH only when both the inputs are LOW and in all other conditions it will be high. The output Y is shown in Fig. 2.20.

	t_1	t_2	t_3	t_4
OUTPUT Y →	0	0	1	0

Fig. 2.20

The NOR gate as a UNIVERSAL gate.

The chart in Fig. 2.21 shows how would your wire NOR gates to create any of the other basic logic functions.

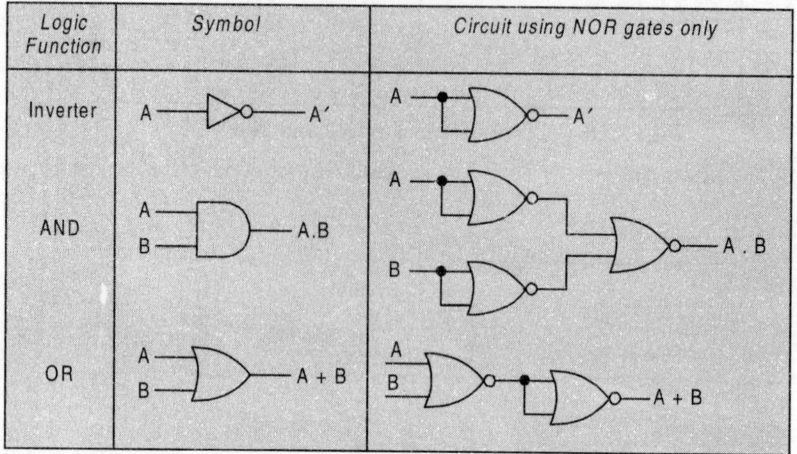

Fig. 2.21

2.5.8 The Exclusive OR Gate

The exclusive OR gate is sometimes referred to as the "Odd but not the even gate". It is often shortened as "XOR gate". The logic diagram is shown in Fig. 2.22 (*a*) with its Boolean expression. The symbol ⊕ means the terms are XORed together.

A
B
$Y = A \oplus B$
$= AB' + A'B$

Fig. 2.22 (*a*)

The truth table for XOR gate is shown in Fig. 2.22 (*b*). Notice that if any but not all the inputs are 1, then the output will be 1. 'The unique characteristic of the XOR gate that it produces a HIGH output only when the odd no. of HIGH inputs are present.'

INPUTS		*OUTPUTS*
A	*B*	$A \oplus B = AB' + A'B$
0	0	0
0	1	1
1	0	1
1	1	0

Fig. 2.22 (*b*)

To demonstrate this, Fig. 2.22 (*c*) shows a three input XOR gate logic symbol and the truth table of Fig. 2.22 (*d*). The Boolean expression for a three input XOR gate can be written as

$$Y = (A \oplus B) \oplus C$$

$$= (AB' + A'B) \oplus C$$

Now, Let $\quad X = AB' + A'B$

Therefore, $\quad X \oplus C = XC' + X'C$

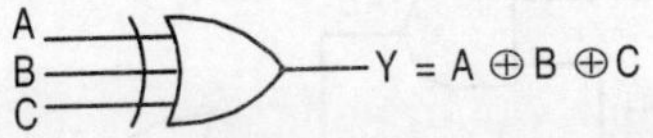

A
B
C
Y = A ⊕ B ⊕ C

Fig. 2.22 (*c*)

INPUTS			*OUTPUTS*
A	B	C	Y
0	0	0	0
0	0	1	1
0	1	0	1
0	1	1	0
1	0	0	1
1	0	1	0
1	1	0	0
1	1	1	1

Fig. 2.22 (*d*)

Putting the value of X, we get

$$Y = (AB' + A'B)C' + (A'B' + AB).C$$

$$Y = AB'C' + A'BC' + A'B'C + ABC$$

The HIGH outputs are generated only when odd number of HIGH inputs are present (see TT).

'This is why XOR function is also known as odd function'.

Figure 2.22 (*e*) shows the ways to express that input A is XORed with input B yielding output Y.

Boolean Expression	A ⊕ B = Y ↑XOR Symbol
Logic Symbol	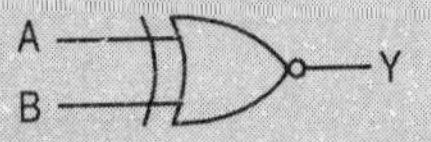 A, B → Y
Truth Table	A B Y 0 0 0 0 1 1 1 0 1 1 1 0

Fig. 2.22 (*e*)

The XOR gate using AND-OR-NOT gates:

we know $$A \oplus B = AB' + A'B$$

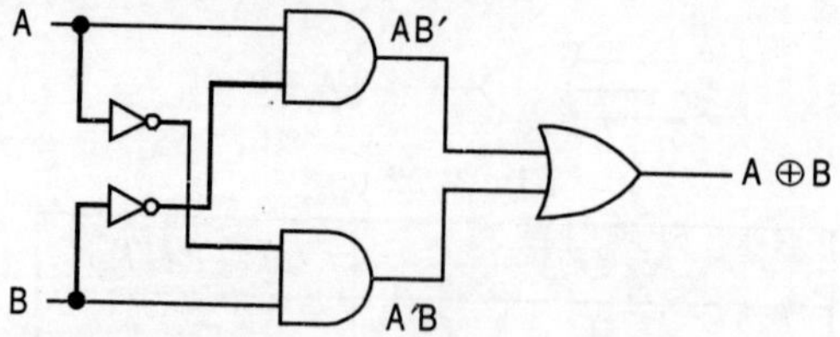

As we know that NAND and NOR are universal gates means any logic diagram can be made using only NAND gates or using only NOR gates.

XOR gate using NAND gates only.

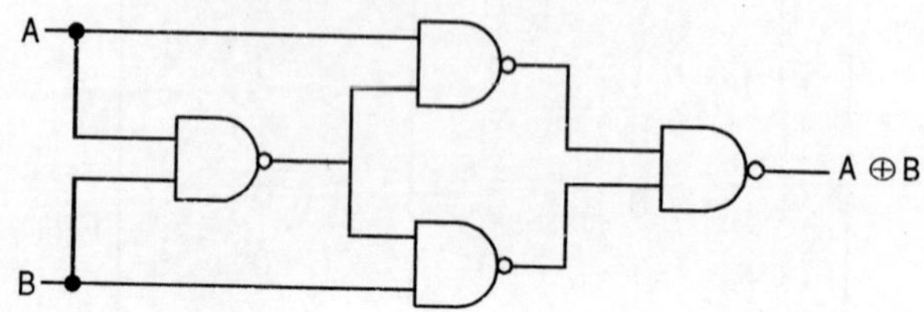

XOR using NOR gates only.

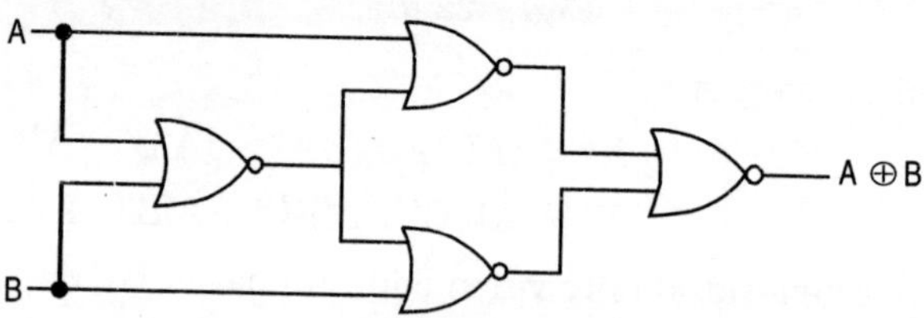

The procedure for implementing any logic function using only universal gate (only NAND gates or only NOR gates) will be treated in detail in section 2.6.

Example. *Determine the output Y from the XOR gate from the given input waveform shown in Fig. 2.23.*

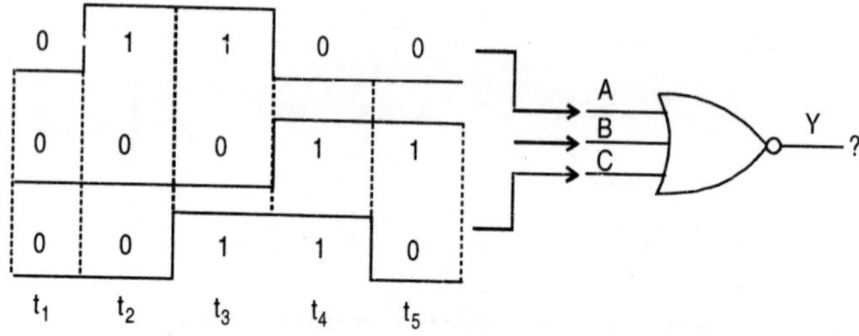

Fig. 2.23

Solution. The output XOR gate is determined by realizing that it will be HIGH only when the odd number of high inputs are present therefore the output Y is high for time period t_2 and t_5 as shown in Fig. 2.24.

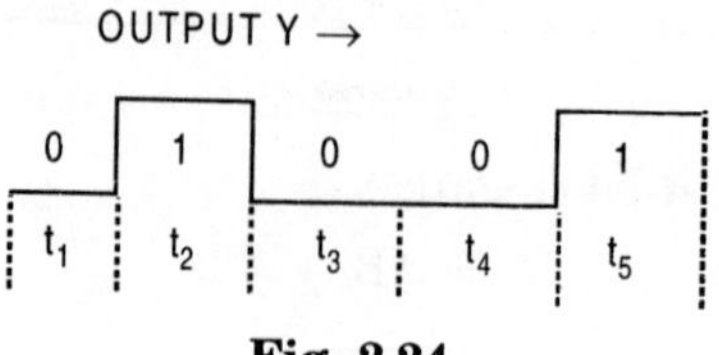

Fig. 2.24

2.5.9 The Exclusive NOR Gate

The Exclusive NOR gate is sometimes referred to as the 'COINCIDENCE' or 'EQUIVALENCE' gate. This is often shortened as 'XNOR' gate. The logic diagram is shown in Fig. 2.25(*a*).

A
B
Y = A ⊙ B
= AB + A'B'

Fig. 2.25 (*a*)

Observe that it is the XOR symbol with the added invert bubble on the output side. The Boolean expression for XNOR is therefore, the invert of XOR function denoted by symbol ⊙.

$$\begin{aligned} A \odot B &= (A \oplus B)' \\ &= (AB' + A'B)' \\ &= (A' + B) \,.\, (A + B') \\ &= AA' + A'B' + AB + BB' \\ &= AB + A'B'. \end{aligned}$$

The truth table for XNOR gate is shown in Fig. 2.25 (*b*).

INPUTS		*OUTPUTS*
A	*B*	*A⊙B = AB + A'B'*
0	0	1
0	1	0
1	0	0
1	1	1

Fig. 2.25 (*b*)

Notice that the output of XNOR gate is the complement of XOR truth table.

'The unique output of the XNOR gate is a LOW only when an odd number of input are HIGH'.

A
B
C
Y = A ⊙ B ⊙ C

Fig. 2.25 (*c*)

INPUTS			*OUTPUTS*
A	B	C	Y
0	0	0	1
0	0	1	0
0	1	0	0
0	1	1	1
1	0	0	0
1	0	1	1
1	1	0	1
1	1	1	0

Fig. 2.25 (*d*)

To demonstrate this, Fig. 2.25 (*c*) shows a three input XNOR gate logic symbol and the truth-table 2.25 (*d*).

Figure 2.25 (*e*) shows the ways to express that input A is XNORed with input B yielding output Y.

Boolean Expression	$A \odot B = Y$
Logic Symbol	A, B → XNOR → Y
Truth Table	A B Y: 0 0 1; 0 1 0; 1 0 0; 1 1 1

A	B	Y
0	0	1
0	1	0
1	0	0
1	1	1

Fig. 2.25 (*e*)

Now at this point, it is left as an exercise for the reader to make XNOR gate using AND-OR-NOT gates, using NAND gates only and using NOR gates only.

Example. *Determine the output Y from the XNOR gate from the given input waveform shown in Fig. 2.26.*

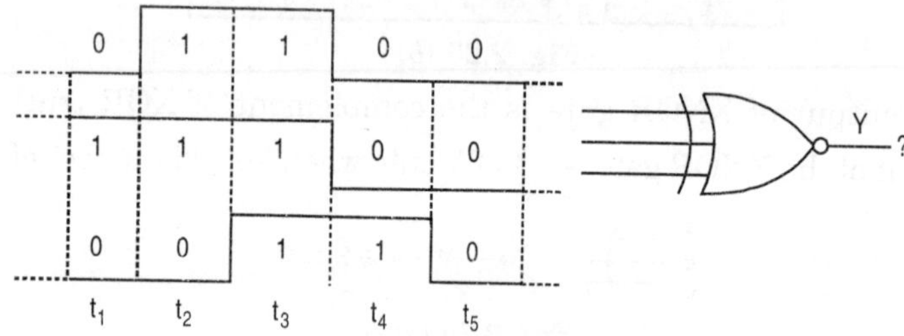

Fig. 2.26

Solution. The output of XNOR gate is determined by realizing that it will be HIGH only when the even-number of high inputs are present, therefore the output Y is high for time period t_2 and t_5 as shown in Fig. 2.27.

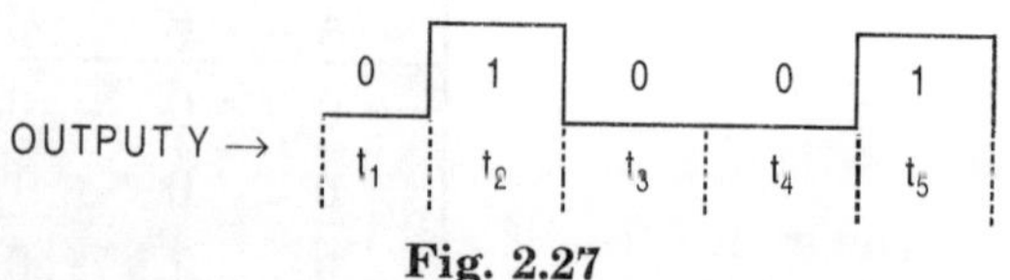

Fig. 2.27

2.5.10 Extension to Multiple Inputs in Logic Gates

The gates we have studied, except for the inverter and buffer can be extended to have more than two inputs.

A gate can be extended to have multiple inputs the binary operation it represents is commutative and associative.

The AND and OR operations defined in Boolean algebra, posseses these two properties.

For the NAD function, we have

$$x.y = y.x \text{ Commutative}$$

and

$$x.(yz) = (x.y).z \text{ Associative}$$

Similarly for the OR function,

$$x + y = y + x \text{ Commutative}$$

and

$$(x + y) + z = x + (y + z) \text{ Associative}$$

It means that the gate inputs can be interchanged and that the AND and OR function can be extended to one or more variables.

The NAND and NOR operations are commutative but not associative. For example

$$^*(x \uparrow y) = (y \uparrow x) \Rightarrow x' + y' = y' + x' \quad \text{commutative}$$

But

$$x \uparrow (y \uparrow z) \neq (x \uparrow y) \uparrow z$$

$$[x.(yz)']' \neq [(x.y)'.z)]'$$

$$x' + yz \neq xy + z'$$

Similarly,

$$^{**}(x \downarrow y) = (y \downarrow x) \Rightarrow x'y' = y'x' \rightarrow \text{commutative}$$

But

$$x\downarrow(y \downarrow z) \neq (x \downarrow y) \downarrow z$$

$$[x + (y + z)']' \neq [(x + y)' + z]'$$

$$x'y + x'z \neq xz' + yz'$$

This difficulty can be overcomed by slightly modifying the definition of operation. We define the multiple NAND (or NOR) gate as complemented AND (or OR) gate.

Thus by this new definition, we have

$$x \uparrow y \uparrow z = (xyz)'$$

$$x \downarrow y \downarrow z = (x + y + z)'$$

The graphic symbol for 3-input NAND and NOR gates is shown in Fig. 2.28(*a*) and (*b*).

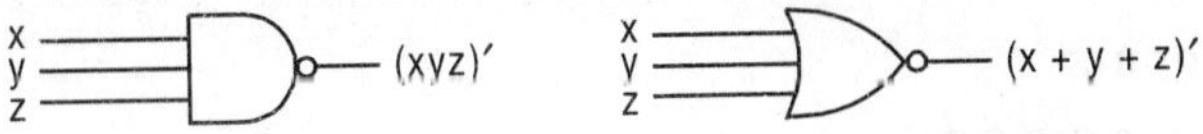

x y z — (xyz)′

(a) Three-input NAND gate

x y z — (x + y + z)′

(b) Three-input NOR gate

Fig. 2.28

The exclusive–OR and equivalence gates are both commutative and associative and can be extended to more than two inputs.

The reader is suggested to verify that, both X-OR and X-NOR possess commutative and associative property.

* NAND symbol

** NOR symbol

MORE EXAMPLES

Example 1. *Give the concluding statement of all the logic gates, we have studied in this chapter.*

Solution. AND: The output is HIGH only when all the inputs are HIGH, otherwise output is LOW.

OR: The output is LOW only when all the inputs are LOW, otherwise output is HIGH.

NOT: The output is complement of input.

NAND: The output is LOW only when all the inputs are HIGH, otherwise the output is HIGH.

NOR: The output is HIGH only when all the inputs are LOW, otherwise output is LOW.

EX-OR: The output is HIGH only when both the inputs are not same, otherwise output is LOW.

Or

The output is HIGH only when odd number of inputs are HIGH, otherwise output is LOW.

EX-NOR: The output is HIGH only when both the inputs are same, otherwise output is LOW.

Or

The output is HIGH only when even number of inputs are HIGH, otherwise output is LOW.

Example 2. *Show how an EX-OR gate can be used as NOT gate or inverter.*

Solution. The expression for NOT gate is

$$y = \overline{A} \text{ where } y = \text{output and A} = \text{input}$$

The expression for EX-OR gate is

$$y = \overline{A}B + A\overline{B}$$

where A and B are inputs.

In the expression of EX-OR we see that the first term $\overline{A}B$ can give the complement of input A, if B = 1 and second term $A\overline{B} = 0$. So we connect the B input to logic HIGH (*i.e.*, logic 1) to full fill the requirements above stated. *i.e.*,

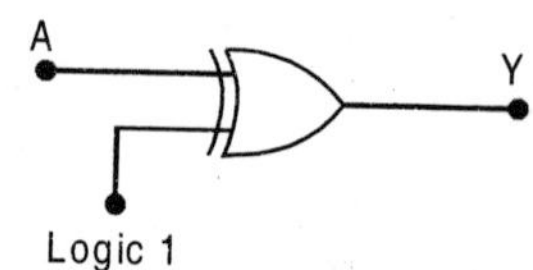

Fig. 2.29 EX-OR Gate connected as NOT Gate

From Fig. 2.29

$$y = \overline{A}.1 + A.0$$

or

$$y = \overline{A} \quad i.e., \text{ complement}$$

Thus, above connection acts as inverter or NOT gate.

Example 3. *Show, how an AND gate and an OR gate can be masked.*

Solution. Masking of gates means disabling the gate. It is the process of connecting a gate in such a way that the output is not affected by any change in inputs i.e., the output remains fixed to a logic state irrespective of inputs.

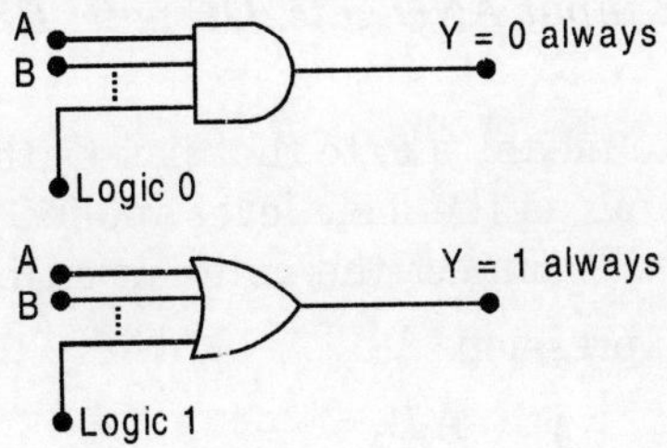

Fig. 2.30 Masking of AND gate and OR gate

AND gate can be masked by connecting one of its input to logic LOW (i.e. logic 0) and OR gate can be masked by connecting one of its input to logic HIGH (i.e. logic 1)

Example 4. *Below shown waveforms (Fig. 2.31) are applied at the input of 2-input logic gate.*

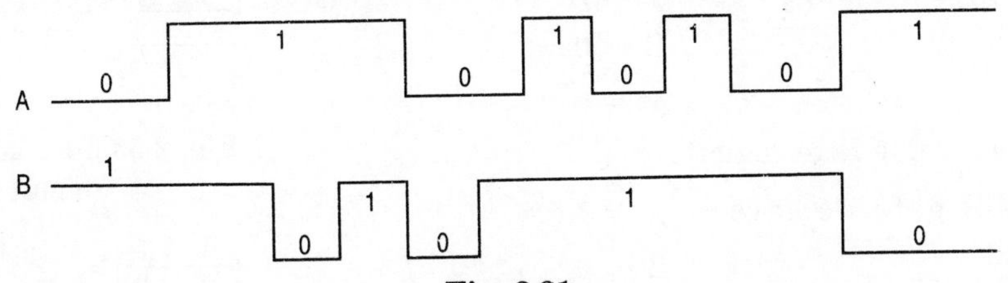

Fig. 2.31

Draw the output waveform if the gate is (a) AND gate (b) OR gate (c) EX-OR gate (d) NAND gate (e) NOR gate.

Solution. The waveforms can be drawn by recalling the concluding statements of logic gates given in example 1.

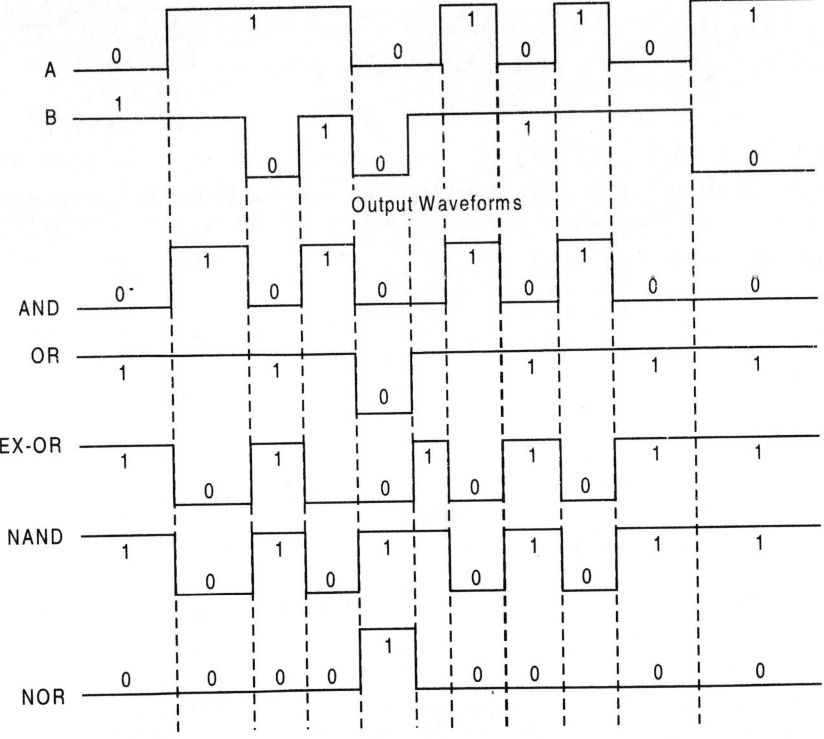

Fig. 2.32

Example 5. *Show how a 2 input AND gate, OR gate, EX-OR gate and EX-NOR gate can be made transparent.*

Solution. Transparent gate means a gate that passes the input as it is to the output i.e. the output logic level will be same as the logic level applied at input. Let the two inputs are A and B and output is y. We will connect the gates in such a way that it gives y = A.

For **AND gate** we have expression

$$y = A.B$$

if $B = 1$

$$y = A$$

So, connect input B to logic 1.

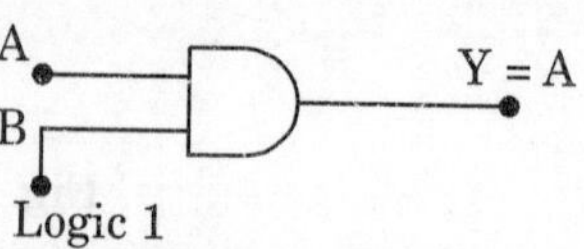

Fig. 2.33 *(a)* Transparent AND gate

For **OR gate** we have $y = A + B$

if $B = 0$

$$y = A$$

So connect input B to logic 0.

Fig. 2.33 *(b)* Transparent OR gate

For **EX-OR gate** we have $y = \overline{A}B + A\overline{B}$

if $B = 0,$

then $A\overline{B} = A$, and $\overline{A}B = 0$

and $y = A$

Hence, connect input B to logic 0.

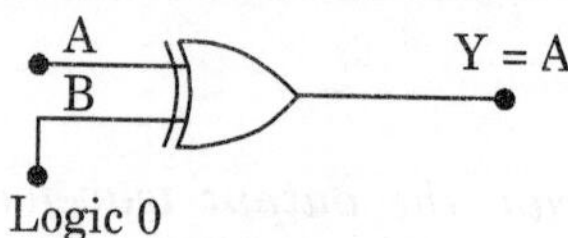

Fig. 2.33 *(c)* Transparent EX-OR gate

For **EX-NOR gate** we have $y = \overline{A}\overline{B} + AB$

if $B = 1,$

then $\overline{A}\overline{B} = 0$ and $AB = A$

so $y = A$

Hence, connect input B to logic 1

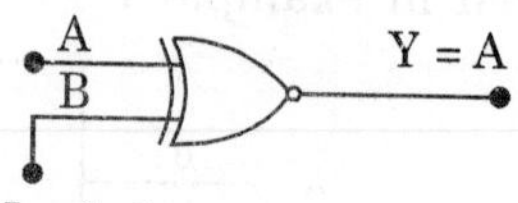

Fig. 2.33 *(d)* Transparent EX-NOR gate

It we take multiple input logic gates then connecting them as above is called *enabling* a gate.

Example 6. *Determine the purpose of digital circuit of Fig. 2.34.*

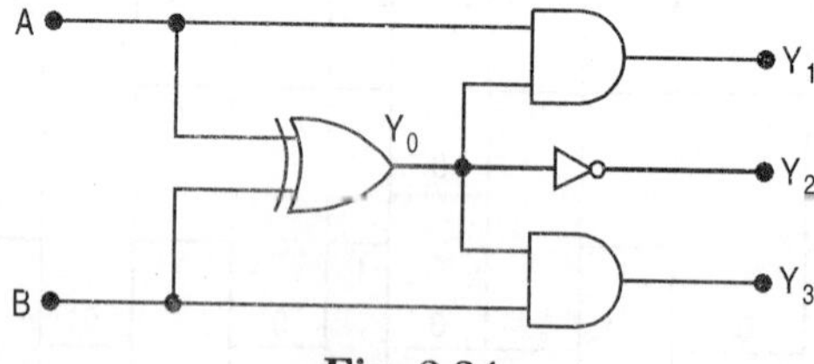

Fig. 2.34

Solution. From the Fig. 2.34 we see that

$$y_0 = A \oplus B = \overline{A}B + A\overline{B}$$

$$y_1 = A.y_0$$

$$y_2 = \overline{y_0}$$

$$y_3 = B.y_0$$

Now we draw three truth tables, one for each of the outputs y_1, y_2, and y_3 to determine the purpose of the circuit.

Table (*i*)

A	B	y_0	y_1
0	0	0	0
0	1	1	0
1	0	1	1
1	1	0	0

(*i*) From the table (*i*), it is evident that $y_1 = 1$, only when A = 1 and B = 0. It means that y_1 is HIGH only when A > B, as shown in third row of Table (*i*).

Table (*ii*)

A	B	y_0	y_2
0	0	0	1
0	1	1	0
1	0	1	0
1	1	0	1

(*ii*) It is evident from Table (*ii*) that $y_2 = 1$ if A = B = 0 and A = B = 1. Thus y_2 goes HIGH only when A = B, as shown by first and last row of Table (*ii*).

Table (*iii*)

A	B	y_0	y_3
0	0	0	0
0	1	1	1
1	0	1	0
1	1	0	0

(*iii*) It is evident from Table (*iii*) that $y_3 = 1$ if A = 0 and B = 1. Thus y_3 goes HIGH only when A < B (or B > A), as shown by the second row of table (*iii*).

Thus from the above discussion it can be concluded that the given circuit is 1-bit data comparator. In this circuit, y_1 indicates A > B, y_2 indicate the equality of two datas, and y_3 indicates A < B.

2.6 NAND AND NOR IMPLEMENTATION

In section 2.5.7 we have seen that NAND and NOR are universal gates. Any basic logic function (AND, OR and NOT) can be implemented using these gates only. Therefore digital circuits are more frequently constructed with NAND or NOR gates than with AND, OR and NOT gates. Moreover NAND and NOR gates are easier to fabricate with electronic components and are the basic gates used in all IC digital logic families. Because of this prominence, rules and procedures have been developed for implementation of Boolean functions by using either NAND gates only or NOR gates only.

2.6.1 Implementation of a Multistage (or Multilevel) Digital Circuit using NAND Gates only

To facilitate the implementation using NAND gates only, we first define two graphic symbols for these gates as shown in Fig. 2.35(*a*) and (*b*).

(*a*) The AND-invert symbol

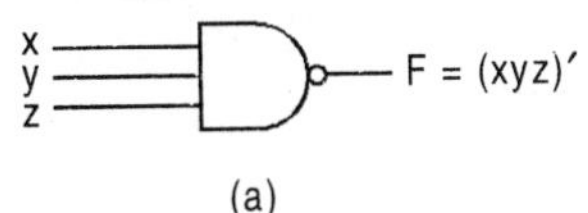

(a)

(*b*) The invert-OR symbol

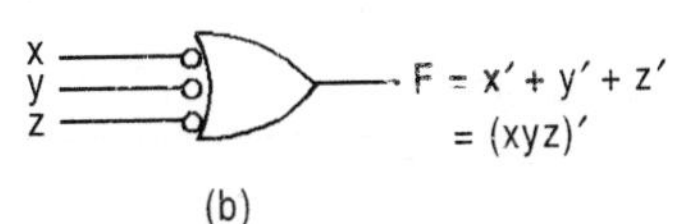

(b)

(*a*) This symbol we have been defined in section (2.5). It consists of an AND grpahic symbol followed by a small circle.

(*b*) This is an OR graphic symbol proceded by small circles in all the inputs.

Fig. 2.35

To obtain a multilevel NAND circuit from a given Boolean function, the procedure is as follows:

1. From the given Boolean function, draw the logic diagram using basic gates (AND, OR and NOT). In implementing digital circuit, it is assumed that both normal and inverted inputs are available. (*e.g.*, If x and x' both are given in function, we can apply them directly that is there is no need to use an inverter or NOT gate to obtain x' from x).
2. Convert all AND gates to NAND using AND-invert symbol.
3. Convert all OR gates to NAND using Invert-OR symbol.
4. Since each symbol used for NAND gives inverted output, therefore it is necessary to check that if we are getting the same value as it was at input. [For example if the input is in its normal form say x, the output must also be x, not x' (inverted or complemented value). Similarly if input is in its complemented form say x', the ouput must also be x', not x (normal value)].

   If it is not so, then for every small circle that is not compensated by another small circle in the same line, insert an inverter (*i.e.*, one input NAND gate) or complement the input variable.

Now consider a Boolean function to demonstrate the procedure:

$$Y = A + (B' + C)(D'E + F)$$

Step 1. We first draw the logic diagram using basic gates shown in Fig. 2.36. (It is assumed that both normal and complemented forms are available *i.e.*, no need of inverter).

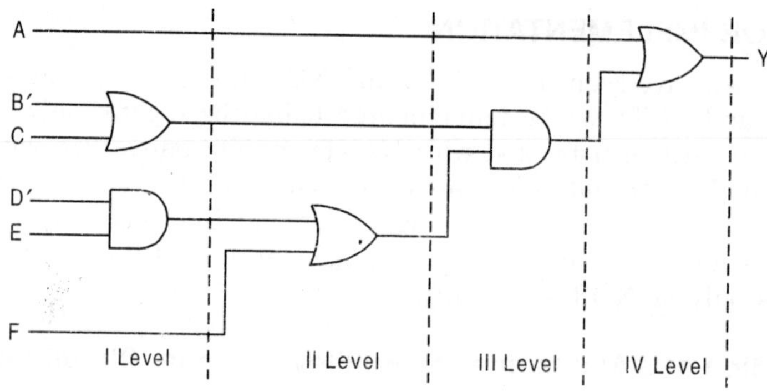

Fig. 2.36

There are four levels of gating in the circuits.

Step 2 and 3

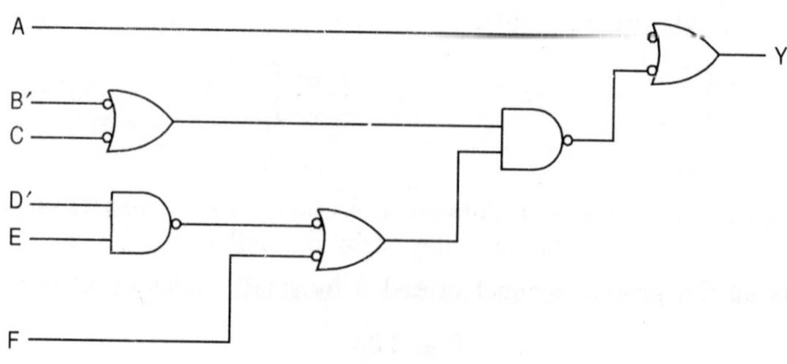

Fig. 2.37

Here, we have converted all AND gates to NAND using AND-invert and all OR gates to NAND using invert OR symbol shown in Fig. 2.37.

Step 4. From the Fig. 2.37 obtained from step 2 and 3, it is very clear that only two inputs D′ and E are emerging in the original forms at the output. Rest *i.e.,* A, B′, C and F are emerging as the complement of their original form. So we need an inverter after inputs A, B′, C and F or alternatively we can complement these variables as shown in Fig. 2.38.

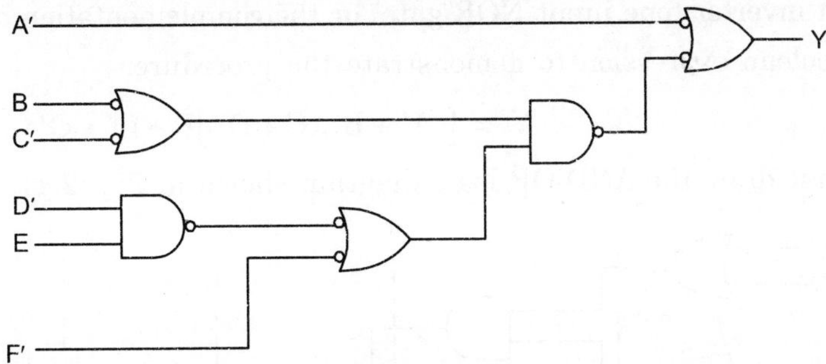

Fig. 2.38

Now because both the symbols AND-invert and invert-OR represent a NAND gate, Fig. 2.38 can be converted into one symbol diagram shown in Fig. 2.39. The two symbols were taken just for simplicity of implementation.

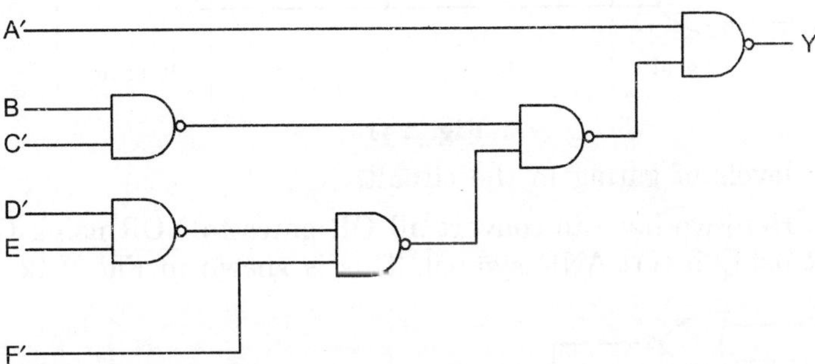

Fig. 2.39

After a sufficient practice, you can directly implement any Boolean function as shown in Fig. 2.39.

2.6.2 Implementation of a Multistage Digital Circuits using NOR Gates only

We first define two basic graphic symbols for NOR gates as shown in Fig. 2.40 (*a*) and (*b*).

(*a*) The OR-invert symbol

x
y
z
F = (x + y + z)′

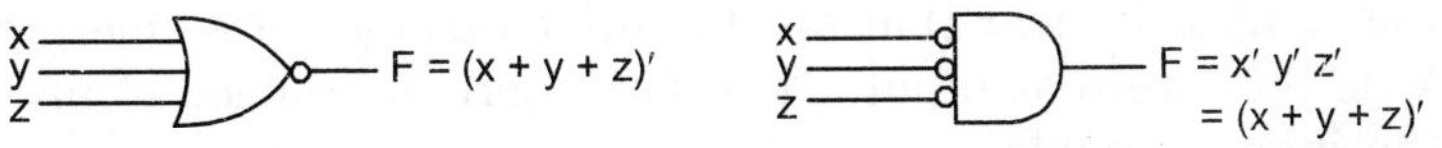

(a) This is an OR graphic symbol followed by a small circle.

(*b*) The invert-AND symbol

x
y
z
F = x′ y′ z′
= (x + y + z)′

(b) This is an AND graphic symbol proceded by small circles in all the inputs.

Fig. 2.40

Procedure to obtain a multilevel NOR circuit from a given Boolean function is as follows:

1. Draw the AND-OR logic diagram from the given expression. Assume that both normal and complemented forms are available.
2. Convert all OR gates to NOR gates using OR-invert symbol.
3. Convert all AND gates to NOR gates using invert-AND symbol.
4. Any small circle that is not complement by another small circle along the same line needs an inverter (one input NOR gate) or the complementation of input variable.

Consider a Boolean expression to demonstrate the procedure:

$$Y = \left[(A' + B).(C + D')\right]E + (F + G')$$

Step 1. We first draw the AND-OR logic diagram shown in Fig. 2.41.

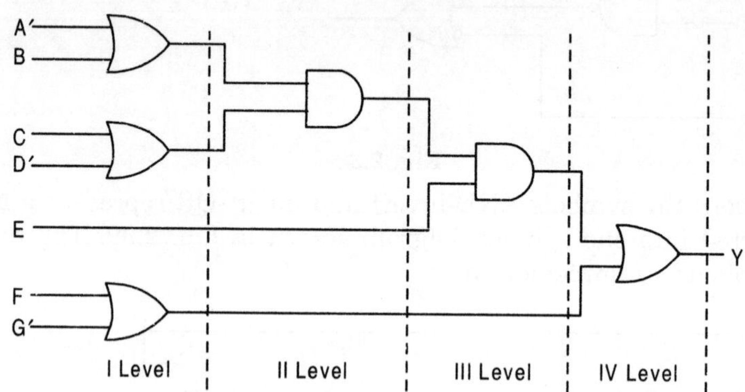

Fig. 2.41

There are four levels of gating in the circuit.

Step 2 and 3. Here we have to convert all OR gates to NOR using OR-invert and all AND gates to NOR using invert AND symbol. This is shown in Fig. 2.42.

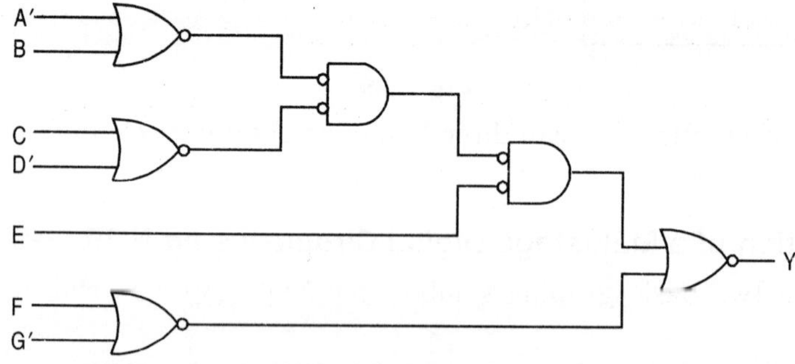

Fig. 2.42

Step 4. From Fig. 2.42, it is clear that all the input variables are emerging in the same form at the ouput Y as they were at input. Therefore there is no need of inverter at inputs or complementing the input variables.

Here again, both symbols OR-invert and invert-AND represent a NOR gate, so the diagram of Fig. 2.42 can be converted into one symbol shown in Fig. 2.43.

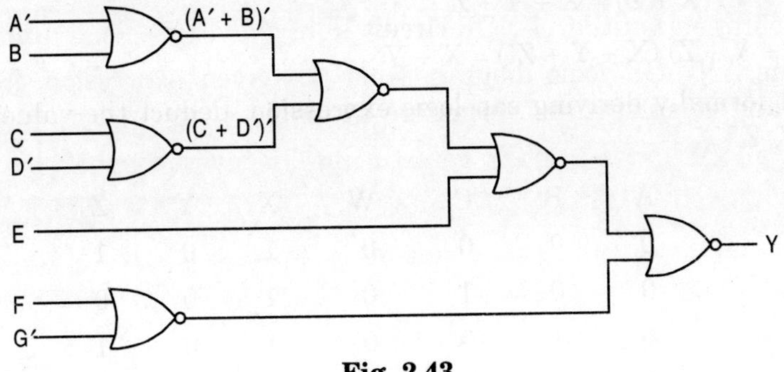

Fig. 2.43

2.7 EXERCISE

1. Write Boolean equations for F_1 and F_2.

A	B	C	F_1
0	0	0	1
0	0	1	0
0	1	0	0
0	1	1	0
1	0	0	0
1	0	1	1
1	1	0	0
1	1	1	1

A	B	C	F_2
0	0	0	0
0	0	1	1
0	1	0	1
0	1	1	1
1	0	0	0
1	0	1	1
1	1	0	1
1	1	1	0

2. Consider 2-bit binary numbers, say A, B and C, D. A function X is true only when the two numbers are different construct a truth table for X.

3. Draw truth table and write Boolean expression for the following:

   (*a*) X is a 1 only if A is a 1 and B is a 1 or if A is 0 and B is a 0.

   (*b*) X is a 0 if any of the three variables A, B and C are 1's. X is a 1 for all other conditions.

4. Prove the following identities by writing the truth tables for both sides:

   (*a*) A.(B + C) = (A.B) + (A.C)

   (*b*) (A.B.C)′ = A′ + B′ + C′

   (*c*) A.(A + B) = A

   (*d*) A + A′B = A + B

5. Prove the following:

   (*a*) (X + Y) (X + Y′) = X

   (*b*) XY + X′Z + YZ = XY + X′Z

   (*c*) (X + Y′) = X.Y

(*d*) $(X + Y)(X + Z) = X + Y + Z$

(*e*) $(X + Y + Z)(X + Y + Z') = X + Y$

6. Without formally deriving can logic expression, deduct the value of each function W, X, Y, Z.

A	B	C	W	X	Y	Z
0	0	0	0	1	0	1
0	0	1	0	1.	0	0
0	1	0	0	1	0	1
0	1	1	0	1	0	0
1	0	0	0	1	1	1
1	0	1	0	1	1	0
1	1	0	0	1	1	0
1	1	1	0	1	1	0

7. Use DeMorgan's theorm to complement the following Boolean expression.

(a) $Z = Z = x.(y + w.v)$

(*b*) $Z = x.y.w + y(w' + v')$

(*c*) $Z = x' + yY$

(*d*) $F = (AB + CD)''$

(*e*) $F = ((A + B')(C' + D)'$

(*f*) $F = ((A + B')(C + D))'$

(*g*) $F = \left[((ABC)'(EFG)')' + ((HIJ)'(KLM)')'\right]'$

(*h*) $F = \left[((A + B)'(C + D)'(E + F)'(G + H)')'\right]'$

8. A certain proposition is true when it is not true that the conditions A and B both hold. It is also true when conditions A and B both hold but condition C does not. Is the proposition true when it is not true that conditions B and C both hold? Use Boolean algebra to justify your answer.

9. Define the following terms:

(*a*) Canonical

(*b*) Minterm

(*c*) Maxterm

(*d*) Sum-of-products form

(*e*) Product-of-sum form

(*f*) Canonical sum-of-product

(*g*) Canonical product-of-sums

10. Write the standard SOP and the standard POS form for the truth tables:

(*a*)

x	y	z	F
0	0	0	1
0	0	1	1
0	1	0	1
0	1	1	0
1	0	0	1
1	0	1	0
1	1	0	0
1	1	1	1

(*b*)

x	y	z	F
0	0	0	1
0	0	1	0
0	1	0	0
0	1	1	1
1	0	0	1
1	0	1	1
1	1	0	0
1	1	1	1

11. Convert the following expressions into their respective canonical forms:

(*a*) AC + A′BD + ACD′

(*b*) (A + B + C′) (A + D)

12. Which of the following expressions is in sum of product form? Which is in product of sums form?

(*a*) A + (B.D)′

(*b*) C.D′.E + F′ + D

(*c*) (A + B).C

13. Find the canonical s-of-p form for the following logic expressions:

(*a*) W = ABC + BC′D

(*b*) F = VXW′Y + W′XYZ

14. Write down the canonical s-of-p form and the p-of-s form for the logic expression whose TT is each of the following.

(*a*)

x_1	y_2	z_3	Z
0	0	0	1
0	0	1	0
0	1	0	0
0	1	1	1
1	0	0	1
1	0	1	1
1	1	0	0
1	1	1	1

(*b*)

W	X	Y	Z	F
0	0	0	0	1
0	0	0	1	0
0	0	1	0	1
0	0	1	1	0
0	1	0	0	1
0	1	0	1	0
0	1	1	0	1
0	1	1	1	1
1	0	0	0	1
1	0	0	1	1
1	0	1	0	1
1	0	1	1	0
1	1	0	0	0
1	1	0	1	1
1	1	1	0	1
1	1	1	1	1

15. Convert the following expressions to sum-of-product forms:

(*a*) $AB + CD\,(AB' + CD)$

(*b*) $AB\,(B'C' + BC)$

(*c*) $A + B\,[AC + (B + C)'\,D]$

16. Given a Truth table

(*a*) Express W_1 and W_2 as a sum of minterms

(*b*) Express W_1 and W_2 as product of minterms

(*c*) Simplify the sum of minterm expressions for W_1 and W_2 using Boolean algebra.

17. Draw logic circuits using AND, OR and NOT gates to represent the following:

(*a*) $AB' + A'B$ (*b*) $AB + A'B' + A'BC$

(*c*) $A + B\,[C + D\,(B + C')]$ (*d*) $A + BC' + D\,(E' + F')$

(*e*) $[(AB)'\,(CD)']'$

18. Produce a graphical realization of Inverter, AND, OR and XOR using:

(*a*) NAND gates

(*b*) NOR gates

19. Implement the Boolean function

$F = AB'CD' + A'BCD' + AB'C'D + A'BC'D$ with exclusive-OR and AND gates.

20. Draw the logic circuits for the following functions.

21. A warning buzzer is to sound when the following conditions apply:

(*a*) Switches A, B, C are on.

(*b*) Switches A and B are on but switch C is off.

(*c*) Switches A and C are on but switch B is off.

(*d*) Switches B and C are on but switch A is off.

Draw a truth table and obtain a Boolean expression for all conditions. Also draw the logic diagram for it using (*i*) NAND gates (*ii*) NOR gates. If the delay of a NAND gate is 15 ns and that of a NOR gate is 12 ns, which implementation is faster.

22. Obtain the following operations using only NOR gates.

(*a*) NOT (*b*) OR (*c*) AND

23. Obtain the following operations using only NAND gates.

(*a*) NOT (*b*) AND (*c*) OR

24. Find the operation performed for each of the gets shown in figure below, with the help the truth table.

(*a*)

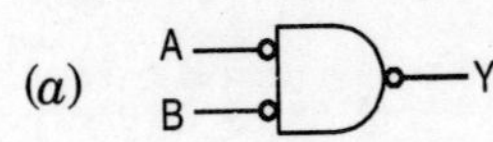

(*b*)

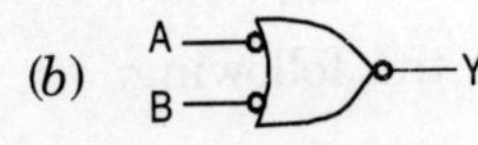

(*c*) A, B — Y

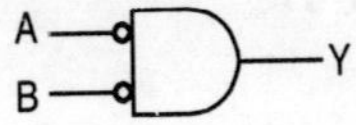

(*d*) A, B — Y

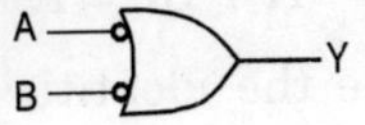

25. Write the expression for EX-OR operation and obtain

(*i*) The truth table

(*ii*) Realize this operation using AND, OR, NOT gates.

(*iii*) Realize this operation using only NOR gates.

26. Verify that the (*i*) NAND (*ii*) NOR operations are commutative but not associative.

27. Verify that the (*i*) AND (*ii*) OR (*iii*) EX-OR operations are commutative as well as associative.

28. Prove that

(*i*) A positive logic AND operation is equivalent to a negative logic OR operation and vice versa.

(*ii*) A positive logic NAND operation is equivalent to a negative logic NOR operation and vice versa.

29. Prove that the logic equations using the Boolean algebraic (switching algebraic) theorems.

(*i*) $A + A\overline{B} + \overline{A}B = A + B$

(*ii*) $AB + \overline{A}B + \overline{A}\overline{B} = \overline{A}B$

Verify these equations with truth table approach.

30. Prove De Morgan's theorems.

31. Using NAND gates produce a graphical realization of

(*a*) Inverter

(*b*) AND

(*c*) OR

(*d*) XOR

32. Using NOR gates also produce a graphical realization of

(*a*) Inverter

(*b*) AND

(*c*) OR

33. Prove (X + Y) (X + Y′) = X

XY + X′Z + YZ = XY + X′Z

34. Prove the following:

(*a*) (A + B)′ = A.B (*b*) (A + B) (A + C) = A + BC

35. Prove the identities:

(*i*) A = (A + B) (A + B)¢

(*ii*) A + B = (A + B + C) (A + B + C′)

36 Obtain the simplified expressions in s-of-p for the following Boolean functions:

(*a*) $xy + x'yz' + x'yz'$

(*b*) ABD + A′C′D′ + A′B + A′CD′ + AB′D′

(*c*) $x'z + w'xy' + w(x'y + xy')$

(*d*) F $(x, y, z) = \Sigma$ (2, 3, 6, 7)

(*e*) F (A, B, C′) (A + D)

37. Convert the following expressions into their respective Canonical forms

(*a*) AC + A'BD + ACD'

(*b*) (A + B + C') (A + D)

38. Write the standard SOP and the standard POS form for the truth tables

(*c*)

x	y	z	$F_{(x, y, z)}$
0	0	0	1
0	0	1	1
0	1	0	1
0	1	1	0
1	0	0	1
1	0	1	0
1	1	0	0
1	1	1	1

(*b*)

x	y	z	$F_{(x, y, z)}$
0	0	0	1
0	0	1	0
0	1	0	0
0	1	1	1
1	0	0	1
1	0	1	1
1	1	0	0
1	1	1	1

39. Consider 2-bit binary numbers, say A, B and C, D. A function X is true only when the two numbers are different.

(*a*) Construct a truth table for X

(*b*) Construct a four-variable K-Map for X, directly from the word definition of X

(*c*) Derive a simplified logic expression for X.

40. Show an implementation of the half-adder using NAND gates only.

41. A three-bit message is to be transmitted with an even-parity bit.

(*i*) Design the logic circuit for the parity generator.

(*ii*) Design the logic circuit for the parity checker.

42. Implement the Boolean function: F = AB′CD′ + A′BCD′ + AB′C′D + A′BC′D′ with exclusive-OR and AND gates.

43. Construct a 3-to-8 line decoder.

44. Construct a 3 × 8 multiplexer.

45. Construct a 8 × 3 ROM using a decoder and OR gates.

46. Construct a 16 × 4 ROM using a decoder and OR gates.

47. Determine which of the following diodes below are conducting and which are non conducting.

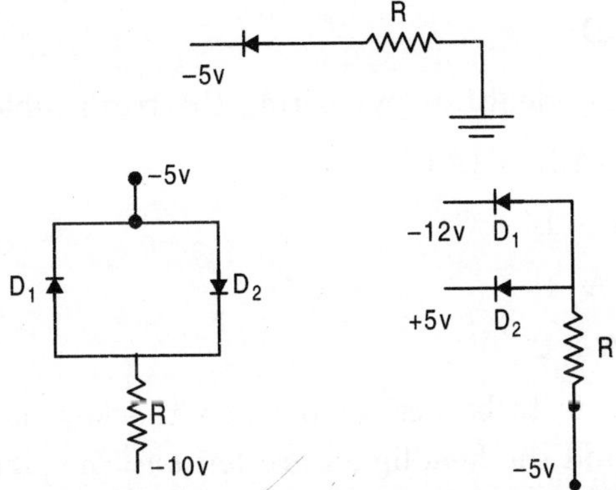

48. Determine which transistors are ON and which are OFF.

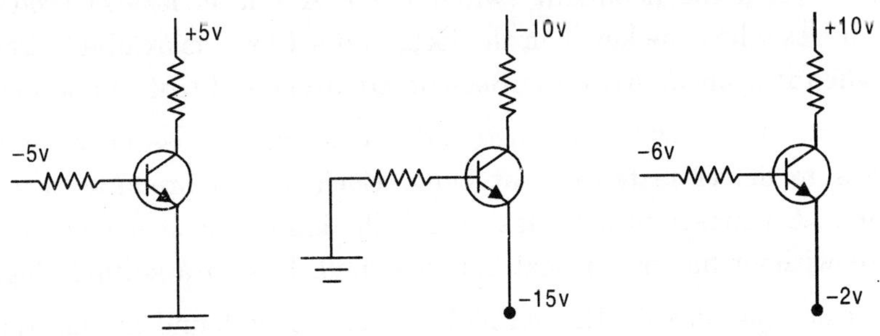

49. Construct voltage and logic truth tables for the circuits shown below. Determine

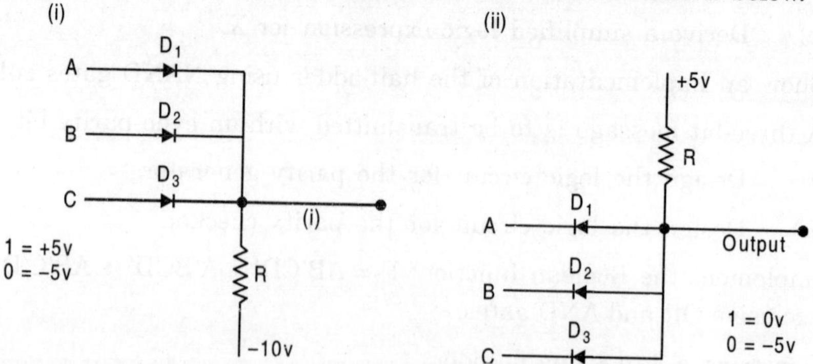

the logic operation performed by each. Assume ideal diodes i.e., –neglect the voltage drop across each diode when it is forward biased.

50. Draw the logic circuits for the following functions:

(*a*) B.(A.C′) + D + E′

(*b*) (A + B)′.C + D

(*c*) (A + B).(C′ + D)

51. Prove the following identities by writing the truth tables for both sides.

(*a*) A.(B + C) == (A.B) + (A.C)

(*b*) (A.B.C)′ == A′ + B′ + C′

(*c*) A.(A + B) == A

(*d*) A + A′.B == A + B

52. An audio (beeper) is to be activated when the key has been removed from the ignition of a car and the headlights are left *on.* The signal is also to be activated if the key is in the ignition lock and the driver's door is opened. A **1** level is produced when the headlight switch is *on.* A **1** level is also produced from the ignition lock when the key is in the lock, and a **1** level is available from the driver's door when it is open. Write the Boolean equation and truth table for this problem.

53. A car has two seat switches to detect the presence of a passenger and two seat belt switches to detect fastened seat belts. Each switch produces a *1* output when activated. A signal light is to flash when the ignition is switched on any passenger present without his or her seat belt fastened. Design a suitable logic circuit.

54. Draw logic diagrams for the simplified expressions found in Question 37 using.

- NAND gates only.
- NOR gates only.

Assume both A and A′, B and B′ .. etc. are available as inputs.

55. You are installing an alarm bell to help protect a room at a museum from unauthorized entry. Sensor devices provide the following logic signals:

ARMED = the control system is active

DOOR = the room door is closed

OPEN = the museum is open to the public

MOTION = There is motion in the room

Devise a sensible logic expression for ringing the alarm bell.

56. A large room has three doors, A, B and C, each with a light switch that can turn the room light ON or OFF. Flipping any switch will change the condition of the light. Assuming that the light switch is off when the switch variables have the values 0, 0, 0 write a truth table for a function LIGHT that can be used to direct the behaviour of the light. Derive a logic equation for LIGHT. Can you simplify this equation?

57. Design a combinational circuit that accepts a three-bit number and generates an output binary number equal to the square of the input number.

58. Design a combinational circuit whose input is a four-bit number and whose output is the 2's complement of the input number.

59. A combinational circuit has four inputs and one output. The output is equal to 1 when (I) all the inputs are equal to 1 or (II) none of the inputs are equal to 1 or (III) an odd number of inputs are equal to 1.

(*i*) Obtain the truth table.

(*ii*) Find the simplified output function in SOP

(*iii*) Find the simplified output function in POS

(*iv*) Draw the two logic diagrams.

60. Write down the canonical s-of-p form and the p-of-s form for the logic expression whose truth table is each of the following.

(*i*)

A	*B*	*C*	*W*
0	0	0	1
0	0	1	1
0	1	0	1
0	1	1	0
1	0	0	1
1	0	1	0
1	1	0	0
1	1	1	1

(ii)

W	X	Y	Z	F
0	0	0	0	1
0	0	0	1	0
0	0	1	0	1
0	0	1	1	0
0	1	0	0	1
0	1	0	0	1
0	1	1	1	1
1	0	0	0	1
1	0	0	1	1
1	0	1	1	0
1	0	1	1	0
1	1	0	0	0
1	1	0	1	1
1	1	1	0	1
1	1	1	1	1

61. Use Question 5.

(*a*) Using K-maps, obtain the simplified logic expressions for both s-of-p and p-of-s for each truth table in the question 5 above.

(*b*) From the p-of-s expressions, work backwards using a K-map to obtain the s-of-p canonical expression for each truth table.

62. Apply De-Morgan's theorems to each expression

(*a*) $\overline{A + \overline{B}}$ (*b*) $\overline{\overline{A}B}$ (*c*) $\overline{A + B + C}$ (*d*) $\overline{A\,B\,C}$

(*e*) $\overline{A\,(B + C)}$ (*f*) $\overline{AB} + \overline{CD}$ (*g*) $\overline{AB + CD}$ (*h*) $\overline{(A + \overline{B}) + (\overline{C} + D)}$

(*i*) $\overline{\overline{\overline{(ABC)}\,(EFG)} + \overline{\overline{(HIJ)}\,(KLM)}}$ (*j*) $\overline{A + \overline{B\overline{C}} + CD + \overline{\overline{BC}}}$

(*k*) $\overline{\overline{\overline{(A + B)}\,\overline{(C + D)}\,\overline{(E + F)}\,(G + H)}}$

63. Use Duality to derive new Boolean identities from the ones you obtained by simplification in the previous question.

64. Use De Morgan's Theorem to complement the following Boolean expressions

(*a*) $z = x.(y + w.v)$

(*b*) $z = x.y.w + y.(\overline{w} + \overline{v})$

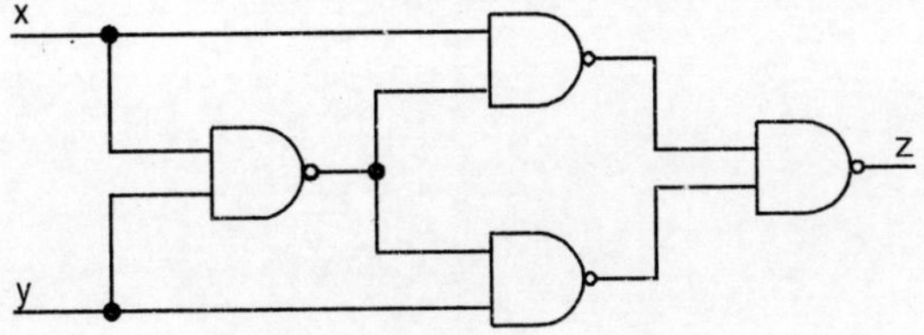

Prove this imlements XOR.

(*c*) $z = \overline{x} + \overline{y}$

(*d*) $z = x + y.\overline{w}$

(*e*) $z = (x + y).w$

(*f*) $z = x + \overline{y.w}$

65. Using Boolean Algebra verify that the circuit in figure 1 implements an exclusive OR (XOR) function

(*a*) Express z_1 and z_2 as a sum of minterms

(*b*) Express z_1 and z_2 as a product of maxterms

(*c*) Simplify the sum of the minterm for z_1 and z_2 using Boolean algebra.

3

BOOLEAN FUNCTION MINIMIZATION TECHNIQUES

3.0 INTRODUCTION

The minimization of combinational expression is considered to be one of the major steps in the digital design process. This emphasis on minimization steps from the time when logic gates were very expensive and required a considerable amount of physical space and power. However with the advent of Integrated circuits (SSI, MSI, LSI and VLSI) the traditional minimization process has lessened somewhat, there is still a reasonable degree of correlation between minimizing gate count and reduced package count.

It is very easy to understand that the complexity of logical implementation of a Boolean function is directly related to the complexity of algebraic expression from which it is implemented.

Although the truth table representation of a Boolean function is unique, but when expressed algebraically, it can appear in many different forms.

Because of the reason mentioned above, the objective of this chapter is to develop an understanding of how modern reduction techniques have evolved from the time consuming mathematical approach (Theorem reduction) to quick graphical techniques called 'mapping' and 'tabular method' for large number of variable in the combinational expression.

3.1 MINIMIZATION USING POSTULATES AND THEOREMS OF BOOLEAN ALGEBRA

The keys to Boolean minimization lie in the theorems introduced in chapter 2, section 2.3.2. The ones of major interest are theorem numbers 6, 7 and 8.

Then, 6 (*a*) A + AB = A   (*b*) A (A + B) = A   Absorption

7 (*a*) A + A′B = A + B   (*b*) A (A′ + B) = AB

8 (*a*) AB + AB′ = A   (*b*) (A + B) (A + B′) = A   Logic Adjacency

Theorem 6 and 7 have special significance when applied to expression in standard form, whereas theorem 8 is of particular importance in simplifying canonical form expression.

— Theorem 6 has a word statement—*If a smaller term or expression is formed entirely in a larger term, then the larger term is superfluous.*

Theorem 6 can be applied only when an expression is not in a 'standard form', that is, one that has at least one term which is not a MIN or MAX term.

Example 1 $F = CD + AB'C + ABC' + BCD$

Thm 6 $[\because A + AB = A]$

$= CD + AB'C + ABC'$

Select one of the smaller terms and examine the larger terms which contain this smaller term.

— for application of Theorem 7, the larger terms are scanned looking for of application the smaller in its complemented form.

Example 2

$$F = AB + BEF + A'CD + B'CD$$
$$= AB + BEF + CD\,(A' + B')$$
$$= AB + BEF + CD\,(AB)' \rightarrow \text{Using Demorgan's Theorem}$$

Thm 7

$$= AB + BEF + CD \qquad (\because A + A'B = A + B)$$

— Theorem 8 is the basis of our next minimization technique *i.e.*, Karnaugh map method. It has a word statement—*'If any two terms in a canonical* or *standard form expression vary in only one variable, and that variable in one term is the complement of the variable in the other term then of the variable is superfluous to both terms.*

Example 3 $F = A'B'C' + A'B'C + ABC + AB'C$

Thm 8 Thm 8

$F = A'B' + AC$

Example 4 $F = A'B'C' + AB'C' + ABC' + A'B'C$

Thm 8

Thm 8

$= A'B' + AC'$

By now it should become obvious that another techniques are needed because this technique backs specific rules to predict each succeeding step in manipulative process.

Therefore, if we could develop some graphical technique whereby the application of 'Theorem 8' the logical adjacency theorem is made obvious and where the desired grouping could be plainly displayed, we would be in mind better position to visualize the proper application of theorem.

3.2 MINIMIZATION USING KARNAUGH MAP (K-MAP) METHOD

In 1953 Maurice Karnaugh developed K-map in his paper titled 'The map method for synthesis of combinational logic circuits'.

The map method provides simple straight forward procedure for minimizing Boolean functions that may be regarded as pictorial form of truth table. K-map orders and displays the minterms in a geometrical pattern such that the application of the logic adjacency theorem becomes obvious.

— The K-map is a diagram made up of squares. Each square represents one minterm.

— Since any function can be expressed as a sum of minterms, it follows that a Boolean function can be recognized from a map by the area enclosed by those squares whose minterms are included in the operation.

— By various patterns, we can derive alternative algebraic expression for the same operation, from which we can select the simplest one. (One that has minimum member of literals).

Now, let us start with a two variable K-map.

3.2.1 Two and Three Variable K-Map

If we examine a two variable truth table, Fig. 3.1(*a*) we can make some general observations that support the geometric layout of K-Map shown in Fig. 3.1(*b*).

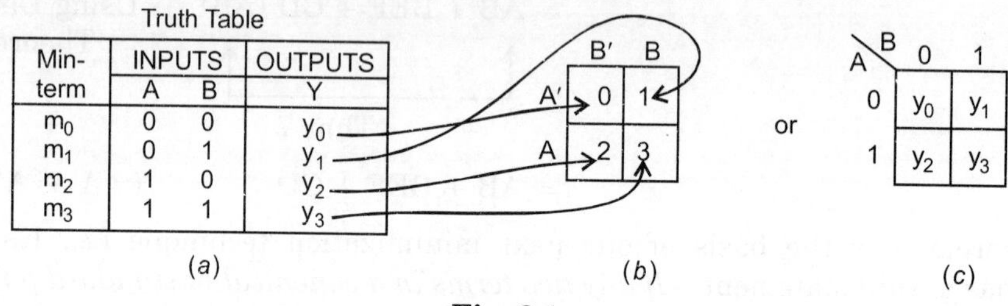

Fig. 3.1

The four squares (0, 1, 2, 3) represent the four possible combinations of A and B in a two variable truth table. Square 1 in the K-map; then, stands for A′B′, square 2 for A′B, and so forth. The map is redrawn in Fig. 3.1(*c*) to show the relation between the squares and the two variables. The 0's and 1's marked for each row and each column designate the values of variables A and B respectively. A appears primed in row 0 and unprimed in row 1. Similarly B appears primed in column 0 and unprimed in column 1.

Now let us map a Boolean function Y = A + B.

Method I—(*i*) Draw the truth table of given function. [Fig. 3.2(*a*)]

Min-term	*A*	*B*	*Y*
m_0	0	0	0
m_1	0	1	1
m_2	1	0	1
m_3	1	1	1

Fig. 3.2(*a*)

(*ii*) Draw a two variable K-map and fill those squares with a 1 for which the value of minterm in the function is equal to 1. [Fig. 3.2(*b*)]

A \ B	0	1
0	m_0	m_1
1	m_2	m_3

⇒

A \ B	0	1
0		1
1	1	1

Fig. 3.2(*b*)

The empty square in the map represent the value of minterms [$m0$ (or A′B′)] that is equal to zero in the function. Thus, actually this empty square represents zero.

Method II-(*i*) Find all the minterms of the function Y = A + B.

$$Y = A + B = A(B + B') + B(A + A')$$
$$= AB + AB' + AB + A'B$$
$$Y = AB + AB' + A'B$$

(*ii*) Draw a two variable K map using this sum of minterms expression.

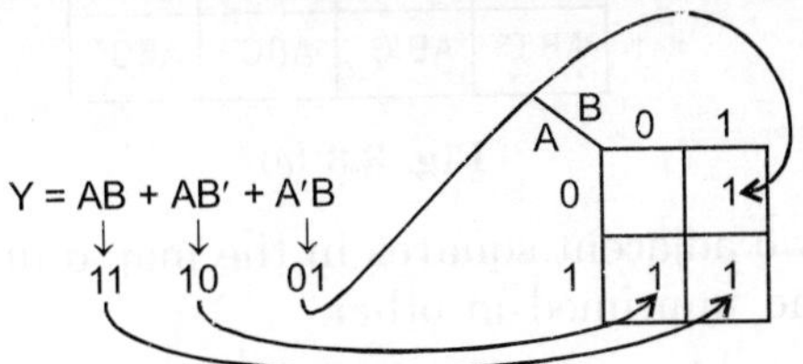

This K-map, is nothing more than an interesting looking Truth-Table, and it simply provide a graphical display of 'implicants' (minterms) involved in any SOP canonical or standard form expression.

Now examine a three variable truth table shown in Fig. 3.3 (*a*).

Truth Table

Min term	*INPUTS*			*OUTPUT*
	A	*B*	*C*	*Y*
m_0	0	0	0	y_0
m_1	0	0	1	y_1
m_2	0	1	0	y_2
m_3	0	1	1	y_3
m_4	1	0	0	y_4
m_5	1	0	1	y_5
m_6	1	1	0	y_6
m_7	1	1	1	y_7

Fig. 3.3 (*a*)

Here we need a K-map with 8 squares represent all the combination (Minterms) of input variables A, B and C distinctly. A three-variable map is shown in Fig. 3.3 (*b*).

A \ BC	00	01	11	10
0	m_0	m_1	m_3	m_2
1	m_4	m_5	m_7	m_6

Fig. 3.3 (*b*)

It is very important to realize that in the three variable K-map of Fig. 3.3 (*b*), the minterms are not arranged in a binary sequence, but similar to 'Gray Code' sequence.

The gray code sequence is a unit distance sequence that means only one bit changes in listing sequence.

Our basic objective in using K-map is to simplify the Boolean function to minimum number of literals. The gray code sequencing greatly helps in applying. 'Logic Adjacency theorem' to adjacent squares that reduces number of literals.

The map is Fig. 3.3 (*b*) is redrawn in Fig. 3.3 (*c*) that will be helpful to make the pictures clear.

A \ BC	00	01	11	10
0	A′B′C′	A′B′C	A′BC	A′BC′
1	AB′C′	AB′C	ABC	ABC′

Fig. 3.3 (*c*)

We can see that any two adjacent squares in the map differ by only one variable, which is primed in one square and unprimed in other.

For example m_3 (A′BC) and m_7 (ABC) are two adjacent squares. Variable A is primed in m_3 and unprimed in m_7, whereas the other two variables are same in both squares. Now applying 'logic adjacency' theorem, it is simplified to a single AND term of only two literals. To clarify this, consider the sum of m_3 and $m_7 \rightarrow m_3 + m_7 = A'BC + ABC = BC(A + A') = BC$.

3.2.2 Boolean Expression Minimization Using K-Map

1. Construct the K-map as discussed. Enter 1 in those squares corresponding to the minterms for which function value is 1. Leave empty the remaining squares. *Now in following steps the square means the square with a value 1.*
2. Examine the map for squares that can not be combined with any other squares and form group of such single squares.
3. Now, look for squares which are adjacent to only one other square and form groups containing only two squares and which are not part of any group of 4 or 8 squares. A group of two squares is called a *pair.*
4. Next, group the squares which result in groups of 4 squares but are not part of an 8-squares group. A group of 4 squares is called a *quad.*
5. Group the squares which result in groups of 8 squares. A group of 8 squares is called *octet.*
6. Form more pairs, quads and outlets to include those squares that have not yet been grouped, and use only a minimum no. of groups. There can be overlapping of groups if they include common squares.
7. Omit any redundant group.
8. Form the logical sum of all the terms generated by each group.

Using Logic Adjacency Theorem we can conclude that,

— a group of two squares eliminates one variable,

— a group of four squares eliminates two variable and a group of eight squares eliminates three variables.

Now let us do some examples to learn the procedure.

Example 1. *Simplify the boolean for F = AB + AB′ + A′B. Using two variable K-map. This function can also be written as*

$$F(A, B) = \Sigma(1, 2, 3)$$

Solution. Step 1. Make a two variable K-map and enter 1 in squares corresponding to minterms present in the expression and leave empty the remaining squares.

A \ B	0	1
0	m_0	m_1
1	m_2	m_3

Step 2. There are no 1's which are not adjacent to other 1's. So this step is discarded.

A \ B	0	1
0		1
1	1	1

Step 3. m_1 is adjacent to m_3, therefore, forms a group of two squares and is not part of any group of 4 squares. [A group of 8 squares is not possible in this case].

A \ B	0	1
0		1
1	1	1

Similarly m_2 is also adjacent to m_3 therefore forms another group of two squares and is not a part of any group of 4 squares.

Step 4 and 5. Discarded because there is no quad or octet.

Step 6. All the 1's have already been grouped.

There is an overlapping of groups because they include common minterm m_3.

Step 7. There is no redundant group.

Step 8. The terms generated by the two groups are 'OR' operated together to obtain the expression for F as follows:

$$F = A \quad + \quad B$$

A	B
↓	↓
From group m_2 m_3	From group m_1 m_3
↓	↓
This row is corresponding to the value of A is equal to 1.	This column is corresponding to the value of B is equal to 1.

Example 2. *Simplify the Boolean function F (A, B, C) = Σ (3, 4, 6, 7).*

Solution. Step 1. Construct K-map.

There are cases where two squares in the map are considered to be adjacent even though they do not touch each other. In a three variable K-map, m_0 is adjacent to m_2 and m_4 is adjacent to m_6.

Algebraically $m_0 + m_2 = A'B'C' + A'BC' = A'C'$

and $m_4 + m_6 = AB'C' + ABC' = AC'$

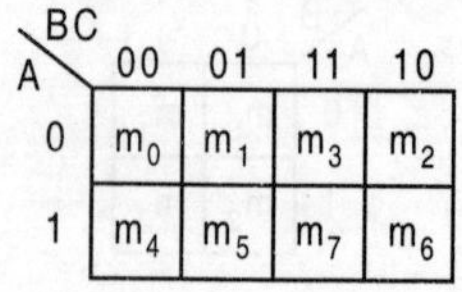

A \ BC	00	01	11	10
0	m_0	m_1	m_3	m_2
1	m_4	m_5	m_7	m_6

Consequently, we must modify the definition of adjacent squares to include this and other similar cases. This we can understand by considering that the map is drawn on a surface where the right and left edges touch each other to form adjacent squares.

Step 2. There are no 1's which are not adjacent to other 1's so this step is discarded.

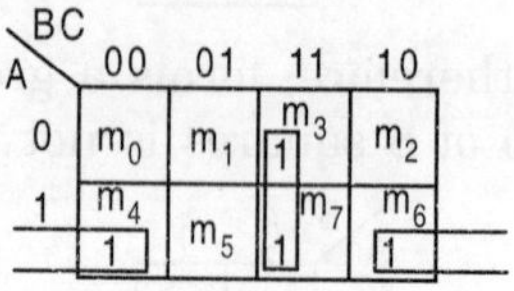

A \ BC	00	01	11	10
0	m_0	m_1	m_3 1	m_2
1	m_4 1	m_5	m_7 1	m_6 1

Step 3. m_3 is adjacent to m_7. It forms a group of two squares and is not a part of any group of 4 or 8 squares.

Similarly m_6 is adjacent to m_7. So this is second group (pair) that is not a part of any group of 4 or 8 squares.

Now according to new definition of adjacency m_4 and m_6 are also adjacent and form a pair. Moreover, this pair (group) is not a part of any group of 4 or 8 squares.

Step 4 and 5. Discarded, because no quad or octet is possible.

Step 6. All the 1's have already been grouped. There is an overlapping of groups because they include a common minterm m_7.

Step 7. The pair formed by m_6 m_7 is redundant because m_6 is already covered in pair m_4 m_6 and m_7 in pair m_3 m_7. Therefore, the pair m_6 m_7 is discarded.

Step 8. The terms generated by the remaining two groups are 'OR' operated together to obtain the expression for F as follows:

$$F = AC' + BC$$

AC' ↓ From group m_4 m_6. The row is corresponding to the value of A = 1 and in the two columns (00 → B′C′ and the 10 → BC′), the value C = 0 ⇒ C′ is common = AC′

BC ↓ From group m_3 m_7. Correspond to both rows (A = 0 and A = 1) so A is omitted, and single column (B = 1 and C = 1), *i.e.*, BC.

With a little more practice, you can make yourself comfortable in minimization using K-map technique. Then you have no used to write all the steps. You can directly minimize the given function by drawing simply the map.

Now let us do one more example of a three variable K-map.

Example 3. *Simply the following Boolean function by first expressing it in sum of minterms.*

$$F = A'B + BC' + B'C'$$

Solution. The given Boolean expression is a function of three variables A, B and C. The three product terms in the expression have two literals and are represented in a three variable map by two squares each.

The two squares corresponding to the first terms A′B are formed in map from the coincidence of A′ (A = 0, first row) and B (two last columns) to give squares 011 and 010.

A \ BC	00	01	11	10
0	1		1	1
1	1			1

Note that when marking 1's in the squares, it is possible to find a 1 already placed there from a preceding term. This happens with the second therm BC′, has 1's in the sqs. which 010 and 110, the sq. 010 is common with the first term A′B, so only one square (corresponding to 110) is marked 1.

Similarly, the third term B′C′ corresponds to column 00 that is squares 000 and 100.

The function has a total of five minterms, as indicated by five 1's in the map. These are 0, 2, 3, 4 and 6. So the function can be expressed in sum of minterms term:

$$F(A, B, C) = \Sigma(0, 2, 3, 4, 6)$$

Now, for the simplification purpose, let us redraw the map drawn above:

A \ BC	00	01	11	10
0	1		1	1
1	1			1

First we combine the four adjacent squares in the first and last columns to given the single literal term C′.

The remaining single square representing minterm 3 is combined with an adjacent square that has already been used once. This is not only permissible but rather desirable since the two adjacent squares give the two literal term A′B and the single sq. represent the three literal minterm A′BC. The simplified function is therefore,

$$\boxed{F = A'B + C'}$$

3.2.3 Minimization in Products of Sums Form

So far, we have seen in all previous examples that the minimized function were expressed in sum of products form. With a minor modification, product of sums (POS) form can be obtained. The process is as follows:

1. Draw map as for SOP; mark the 0 entries. The 1's places in the squares represents minterms of the function. The minterms not included in the function denote the complement of the function. Thus the complement of a function is represented on the map by the squares marked by 0's.
2. Group 0 entries as you group 1 entries for a SOP reading, to determine the simplified SOP expression for F′.

3. Use De Morgan's theorem on F′ to produce the simplified expression in POS form.

Example. *Given the following Boolean function:*

$$F = A'C + A'B + AB'C + BC$$

Find the simplified products of sum (POS) expression.

Solution. Step 1. We draw a three variable K-map. From the given function we observe that minterms 1, 2, 3, 5, and 7 are having the value 1 and remaining minterms *i.e.,* 0, 4 and 6 are 0. So we mark the 0 entries.

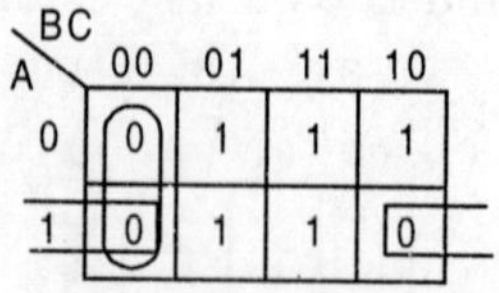

A \ BC	00	01	11	10
0	0	1	1	1
1	0	1	1	0

Step 2. Minterms 0 and 4 forms a pair, giving value = B′C′. Similarly minterms 4 and 6 forms a second pair giving value = AC′. Therefore we get F′ = AC′ + B′C′.

Step 3. Applying De Morgan's theorem automatically converts SOP expression into POS expression, giving the value of F

$$(F')' = [AC' + B'C']'$$

$$F = [(AC')' . (B'C')']$$

$$\boxed{F = (\bar{A} + C) . (B + C)}$$

3.2.4 Four Variable K-Map

Let us examine a four variable truth table shown in Fig. 3.4. We used a K-map with 16 squares to represent all the minterms of input variables A, B, C and D distinctly. A four-variable K-map is shown in Fig. 3.5.

Truth Table

Minterm	*Inputs*				*Output*
	A	*B*	*C*	*D*	*Y*
m_0	0	0	0	0	y_0
m_1	0	0	0	1	y_1
m_2	0	0	1	0	y_2
m_3	0	0	1	1	y_3
m_4	0	1	0	0	y_4
m_5	0	1	0	1	y_5
m_6	0	1	1	0	y_6
m_7	0	1	1	1	y_7
m_8	1	0	0	0	y_8
m_9	1	0	0	1	y_9
m_{10}	1	0	1	0	y_{10}
m_{11}	1	0	1	1	y_{11}
m_{12}	1	1	0	0	y_{12}

Minterm	*Inputs* A B C D	*Output* Y
m_{13}	1 1 0 1	y_{13}
m_{14}	1 1 1 0	y_{14}
m_{15}	1 1 1 1	y_{15}

Fig. 3.4

AB \ CD	00	01	11	10
00	m_0	m_1	m_3	m_2
01	m_4	m_5	m_7	m_6
11	m_{12}	m_{13}	m_{15}	m_{14}
10	m_8	m_9	m_{11}	m_{10}

Fig. 3.5

The rows and columns are numbered in a reflected-code sequence, with only one digit changing value between two adjacent rows or columns.

The minimization process is similar as well have done in a three variable K-Map. However the definition of adjacency can further be extended. Considering the map to be on a surface with the top and bottom edges, as well as right and left edges, touching each other to form adjacent squares.

For example m_0 is adjacent to m_2, m_4 as well as to m_8, similarly m_3 is adjacent to m_1, m_2, m_7 as well as to m_{11} and so on.

Example 1. *Simplify the given fraction.*

$$F = ABCD + AB'C'D' + AB'C + AB$$

Solution. Step 1. The given function is consisting of four variables A, B, C and D. We draw a four variable K-map. The first two terms in the function have four literals and are repeated in a four variable map by one square each. The square corresponding to first term ABCD is equivalent to minterm 1111. (m_{15}). Similarly the square for second term AB′C′D′ is equivalent to minterm 1000 (m_8) the third term in the function has three literals and is represented in a four var map by two adjacent squares. AB' corresponds to 4th row (*i.e.* 10) in the map and C corresponds to last two columns (*i.e.* 11 and 10) in the map. The last term AB has two (A term with only one literal is represented by 8 adjacent square in map. Finally a 1 in all 16 squares give F = 1. It means all the minterms are having the value equal to 1 in the function) literals and is represented by 4 adjacent squares. Here, AB simply corresponds to 3rd row (*i.e.*, AB = 11).

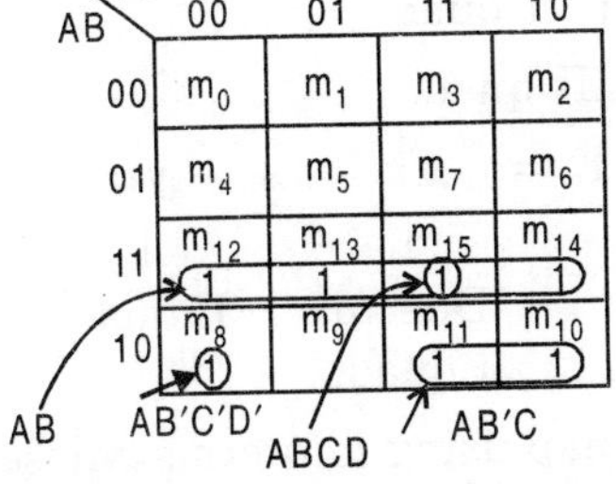

AB \ CD	00	01	11	10
00	m_0	m_1	m_3	m_2
01	m_4	m_5	m_7	m_6
11	m_{12} 1	m_{13} 1	m_{15} 1	m_{14} 1
10	m_8 1	m_9	m_{11} 1	m_{10} 1

AB AB′C′D′ ABCD AB′C

Now, let us redraw the map first for simplification purpose.

Step 2. Is discarded. (No 1's which are not adjacent to other 1's)

Step 3. Discard (No pairs which are not part of any larger group)

Step 4. There are three quads.

Minterms 8, 10, 12, 14 from first quad.

Minterms 12, 13, 14, 15 form second quad

and Minterms 10, 11, 14, 15 form third quad.

Step 5. Discarded. (No octetes)

Step 6. Discarded (All 1's have been grouped.)

Step 7. Discard (No redundant term)

AB \ CD	00	01	11	10
00				
01				
11	1	1	1	1
10	1		1	1

Step 8. The terms generated by three groups are 'OR' operated as follow

$$F = AD' + AB + AC.$$

AD' ↓ From First group; AB ↓ Second group; AC ↓ Third group.

Example 2. *Obtain (a) minimal sum of product (b) minimal product of sum expression for the function given below:*

$$F(w, x\ y, z) = \Sigma\ (0, 2, 3, 6, 7, 8, 10, 11, 12, 15).$$

Solution. The given function can also be written in product of minterm form as

$$F(w, x, y, z) = \Pi\ (1, 4, 5, 9, 13, 14).$$

Squares with 1's are grouped to obtain minimal sum of product; square with 0's are grouped to obtain minimal product of sum, as shown.

WX \ YZ	00	01	11	10
00	1		1	1
01			1	1
11	1		1	
10	1		1	1

(*a*) We draw a four variable map using minterms whose values in the function are equal to 1.

- Minterm 8 and 12. Form a pair.
- Minterms 0, 2, 8 and 10 form I quad.
- Minterms 3, 7, 11, 15 form II quad.
- Minterms 2, 3, 6, 7 form III quad.

Therefore,

$$F = x'z' + yz + w'y + wy'z'$$

$x'z'$ ↓ Due to I quad.; yz ↓ II quad; $w'y$ ↓ III quad; $wy'z'$ ↓ Due to pair

(*b*) We draw a four variable map using minterms whose values in the function are equal to zero. These minterms are 1, 4, 5, 9, 13 and 14.

WX \ YZ	00	01	11	10
00		0		
01	0	0		
11		0		0
10		0		

- Minterm 14 cannot be combined with any other square in the map 4 fours a group with single squares.
- Minterms 4 and 5 form a pair.
- Minterms 1, 5, 9 and 13 form a quad.

Therefore,
$$F' = wxyz' + w'xy' + y'z$$

Applying De Morgan's theorem

$$(F')' = [wxyz' + w'xy' + y'z]'$$
$$F = (wxyz')'.\ (w'xy')'.\ (y'z)'$$
$$F = (w' + x'y' + z).\ (w + x' + y).\ (y + z').$$

3.2.5 Prime and Essential Implicants

So far we have seen a method for drawing and minimising Karnaugh maps in such a way that unnecessary (redundant) groupings can be avoided. Now let us establish some important definitions that will be used to a systematic procedure for combining squares in the process of K-map minimization. To do this, consider a function defined as F (A, B, C, D) = Σ(0, 1, 2, 3, 5, 7, 8, 9, 10, 13, 15). Now we will analyze the grouping shown in the 4 variable map in Fig. 3.6.

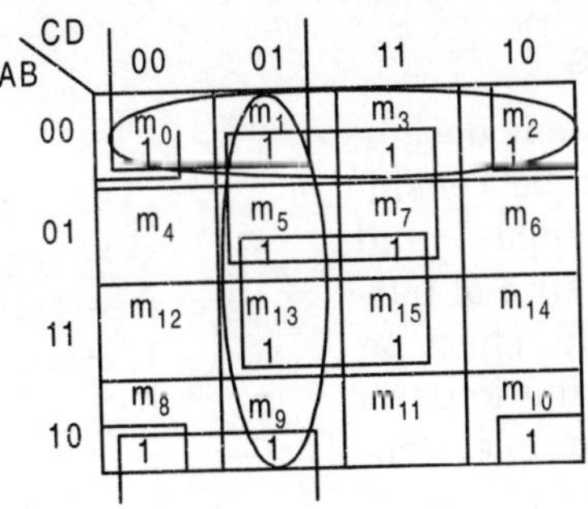

Fig. 3.6

Here, we see that all realistic groupings are shown. Note further that each group is sufficiently large that it can not be completely covered by any other simple grouping. Each of these five groupings is defined as a Prime implicant.

I group → covering minterms → 0, 2, 8, and 10, → B′D′

II group → covering minterms → 0, 1, 2, and 3 → A′B′

III group → covering minterms → 1, 5, 9, and 13 → C′D′

IV group → covering minterms → 1, 3, 5 and 7 → A′D′

V group → covering minterms → 5, 7, 13 and 15 → BD

VI group → covering minterms → 0, 1, 8, 9 → B′C′

Thus, '**a prime implicant is a product term (or minterm) obtained by combining the maximum possible number of adjacent squares in the map.**

As we examine the set of prime implicates that cover this map, it becomes obvious that some of the entries can be grouped in only one way. (Single way groupings). For example there is only one way to group m_{10} with 4 adjacent squares (I group only). Similarly there is only one way to group m_{15} with the adjacent square (V group only). The resultant terms from these groupings are defined as essential prime implicants.

Thus, **'if a minterm in a square is covered by only one prime implicant, the prime implicant is said to be essential'.** The two essential prime implicant *i.e.,* B′D′ and BD cover 8 minterms. The remaining three viz m_1, m_3 and m_9 must be considered next.

The prime implicant table shows that m_3 can be covered either with A'B' or with A′D. m_9 can be covered either with C′D or with B′C′.

m_1 can be covered with any of form prime implicates A'B', A′D, B′C' or C′D.

Now the simplified expression is obtained form the sum of two essentials prime implicates and two prime implicant that cover minterms m_1, m_3, and m_9. It means there are four possible ways to write the simplified expression.

(*a*) BD + B'D' + A'B' + C'D

(*b*) BD + B'D' + A'B' + B'C'

(*c*) BD + B'D' + A'D + C'D

(*d*) BD + B'D' + A'D + B'C'

The simplified expression is thus obtained from the logical sum of all the essential prime implicants plus the prime implicants that may be needed to cover any remaining minterms not covered by simplified prime implicants.

3.2.6 Don't care Map Entries

Many times in digital system design, some input combinations must be considered as cases that "Just don't happen", and there are cases when the occurrence of particular combinations will have no effect on the system, and if those combinations do occur, "you don't care". For example, consider the case where the outputs of a 4-bit binary counter; which happens to home a possible range from 0000 to 1111, is to be converted to a decimal display having the range of 0, 1, 2, ..., 9. The converter might be a digital system having 4-bit binary inputs and decimal upto as shown in Fig. 3.7. In this particular case, the input combinations 1001, 1010, 1011, 1100, 1101, 1110, 1111 are to be considered by combinations that just can not be accepted by the digital system if it functions properly.

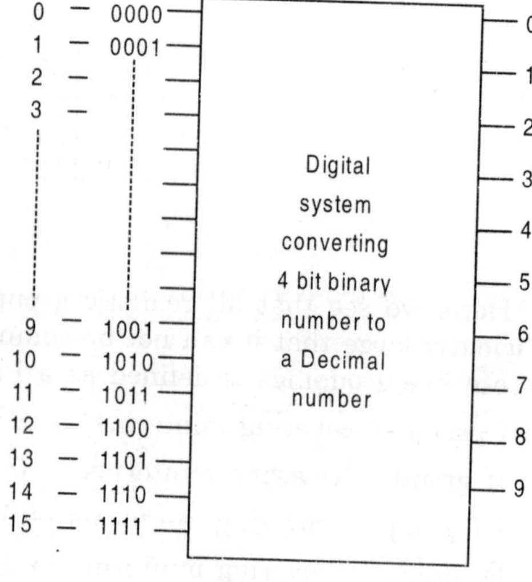

Fig. 3.7

Therefore, when this digital system is being designed, these minterms in the map are treated in a special way. That is a d or a × (cross) is entered into each square to signify "don't care" MIN/MAX terms.

Reading a map, or grouping a map with don't care entries is a simple process.

Group the don't care (d or ×) with a 1 grouping if and only if this grouping will result in greater, simplification; otherwise treat it as if it were a 0 entry.

Example. *Simplify the following Boolean function.*

$$F(A, B, C, D) = \Sigma(0, 1, 2, 10, 11, 14)$$
$$d(5, 8, 9)$$

Solution. The K-map for the given function is shown in Fig. 3.8 with entries X (don't care) in squares corresponding to combinations 5, 8 and 9.

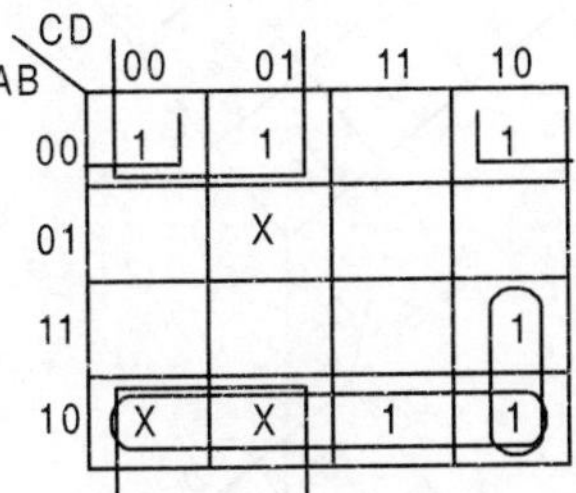

Fig. 3.8

As discussed above, the 1's and d's (Xs) are combined in order to enclose the maximum number of adjacent squares with 1. As shown in K-map in Fig. 3.8, by combining 1's and d's (Xs), three quads can be obtained. The X in square 5 is left free since it does not contribute in increasing the size of any group. Therefore, the

I Quad covers minterms 0, 2, 10 and d8

II Quad covers minterms 10, 11 and d8, d9.

III Quad covers minterms 0, 1 and d8, d9.

A pair covers minterms 10 and 14.

So,

F =	B'D'	+	AB'	+	B'C'	+	ACD'
	Due to I quad.		II quad		III Quad		Due to Pair.

3.2.7 Five Variable K-Map

The Karnaugh map becomes three dimensional when solving logic problems with more than four variables. A three dimensional K-map will be used in this section.

BC \ DE	00	01	11	10
00	0	1	3	2
01	4	5	7	6
11	12	13	15	14
10	8	9	11	10

A = 0

BC \ DE	00	01	11	10
00	16	17	19	18
01	20	21	23	22
11	28	29	31	30
10	24	25	27	26

A = 1

The 5 variable M-map contains $2^5 = 32$ squares. Instead of representing a single 32-square map, two 16-square K-maps are generally used. If the variable are A, B, C, D and E, the two identical 16-square maps contain B, C, D and E variable with one 16-sq. map for A = 1 and other 16-square map for A = 0 *i.e.*, ($\bar{A}$). This is shown in Fig. 3.9.

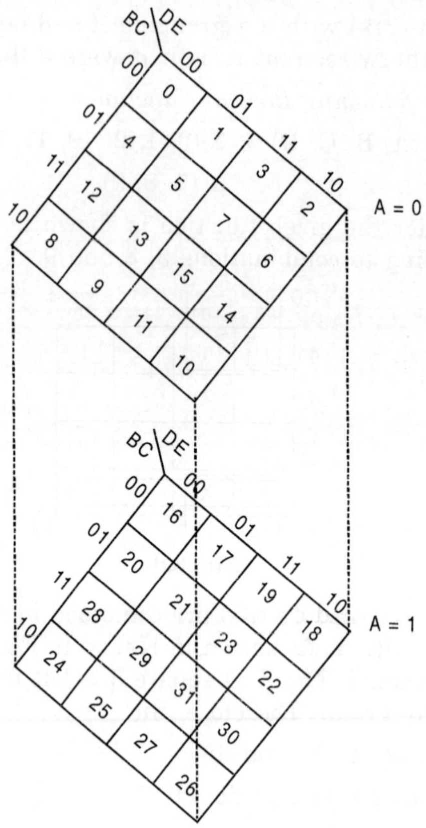

Fig. 3.9

The minimization procedure described so far with respect to functions of two, three or four variables can be extended to the case of five variables.

It is noted that in order to identify the adjacent grouping in the 5-variable maps, we must imagine that the two maps are superimposed on one another as shown in Fig. 3.9. Every square in one map is adjacent to the corresponding square in the other map, because only one variable, changes between such corresponding squares. Thus, rows and columns for one map is adjacent to the corresponding row and column on the other map and same rules are used for adjancencies within one 16 square map. This is illustrated in Fig. 3.10.

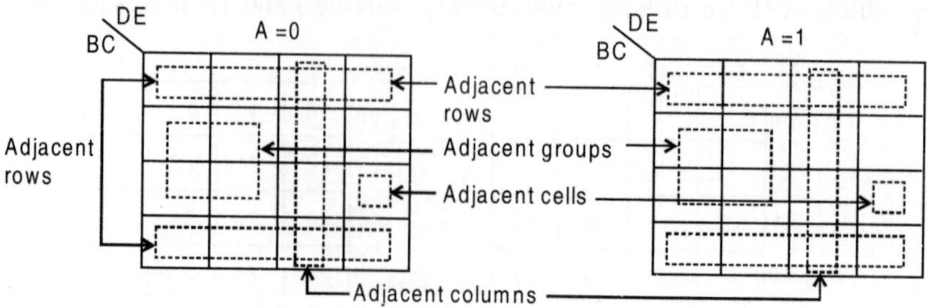

Fig. 3.10

Example. *Simplify the given function.*

$$F(A, B, C, D, E) = \Sigma (0, 4, 7, 8, 9, 10, 11, 16, 24, 25, 26, 27, 29, 31)$$

Solution. We make two 4-variable maps and fill minterms 0-15 in map corresponding to A = 0 and 16 to 31 corresponding to A = 1

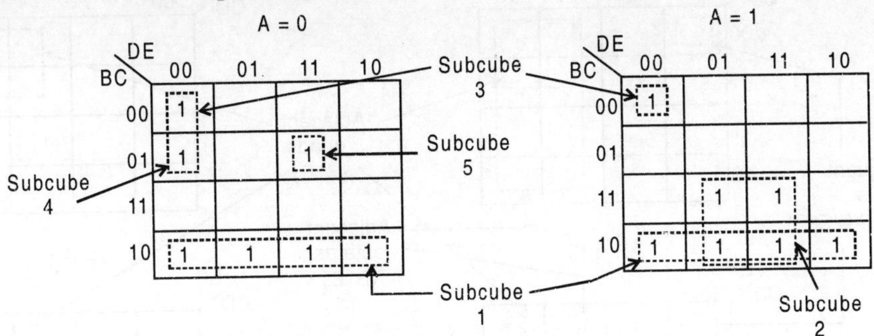

We have 5-subcubes after grouping adjacent squares.

Subcube 1 is an octate which gives BC'

Subcube 2 is a quad which gives ABE (In the map corresponding to A = 1)

Subcube 3 is a pair which gives B'C'D'E'

Subcube 4 is a pair which gives A'B'D'E' (In the map corresponding to A = 0)

Subcube 5 is a single square which gives A'B'CDE (In the map corresponding to A = 0).

Therefore, F (A, B, C, D, E) = BC' + ABE + B'C'D'B' + A'B'D'E' + A'B'CDE

3.2.8 Six Variable K-Map

A six variable K-map contains $2^6 = 64$ squares. These squares are divided into four identical 16-square maps as shown in Fig. 3.11. If the variables are A, B, C, D, E and F, then each 16 square map contains C, D, E, and F as variables along with anyone of 4 combinations of A and B.

AB = 00

CD \ EF	00	01	11	10
00	0	1	3	2
01	4	5	7	6
11	12	13	15	14
10	8	9	11	10

AB = 01

CD \ EF	00	01	11	10
00	16	17	19	18
01	20	21	23	22
11	28	29	31	30
10	24	25	27	26

AB = 10

CD \ EF	00	01	11	10
00	32	33	35	34
01	36	37	39	38
11	44	45	47	46
10	40	41	43	42

AB = 11

CD \ EF	00	01	11	10
00	48	49	51	50
01	52	53	55	54
11	60	61	63	62
10	56	57	59	58

Fig. 3.11

In order to identify the adjacent groupings in the 6-variable maps, we must imagine that the 4 maps are superimposed on one another. Figure 3.12 shows different possible adjacent squares:

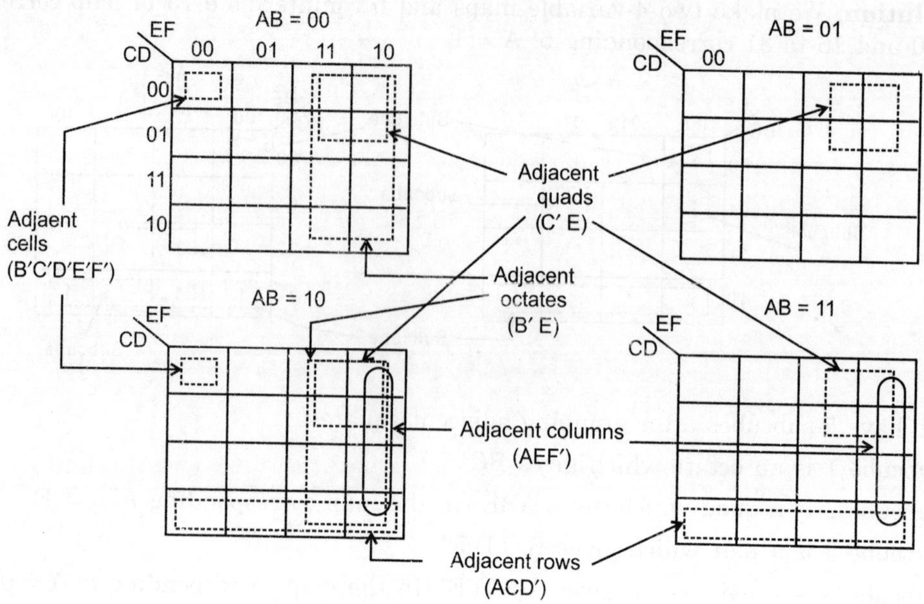

Fig. 3.12

Example. *Simplify the given Boolean function.*

$$F(A, B, C, D, E, F) = \Sigma (0, 4, 8, 12, 26, 27, 30, 31, 32, 36, 40, 44, 62, 63)$$

Solution. We make four 4-varibale maps and fill the mentions 0-15 in map corresponding to AB = 00, 16-31 corresponding to AB = 01, 32 – 47 corresponding to AB = 10 and 48 to 63 corresponding to A B = 11.

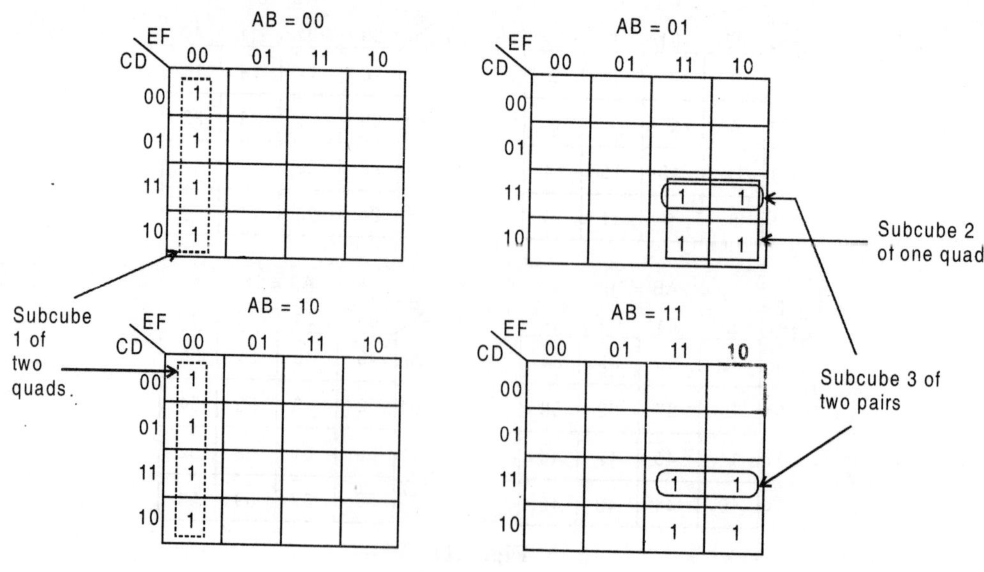

We have 3-subcubes after grouping adjacent square.

1. Subcube 1 contains two quads gives B'E'F'
2. Subcube 2 is form of one quad gives A'BCE
3. Subcube 3 is form of two pairs gives BCDE

Therefore, F(A, B, C, D, E, F) = B'EF' + A'BCE + BCDE.

Maps with seven or more variables needs too many squares and are impractical to use. The alternative is to employ computer programs specifically written to facilitate the simplification of Boolean functions with a large number of variables.

3.2.9 Multi Output Minimization

Finding the optimal cover for a system of output expressions all of which are a function of the some variables is somewhat tedious task. This task is basically one of identifying all possible prime implicants (PIs) that cover each implicated minterm in each O/P expression, then carrying out a search for the minimal cost cover by using 'shared' terms.

Suppose you were given the following system of expressions and asked to find the optimal cover for the complete system, implying that you must find how to optimally share terms between the expressions.

$$F_1\ (A, B, C) = \Sigma\ (0, 2, 3, 5, 6)$$

$$F_2\ (A, B, C) = \Sigma\ (1, 2, 3, 4, 7)$$

$$F_3\ (A, B, C) = \Sigma\ (2, 3, 4, 5, 6)$$

we first generate the maps for three expressions as shown in Fig. 3.13.

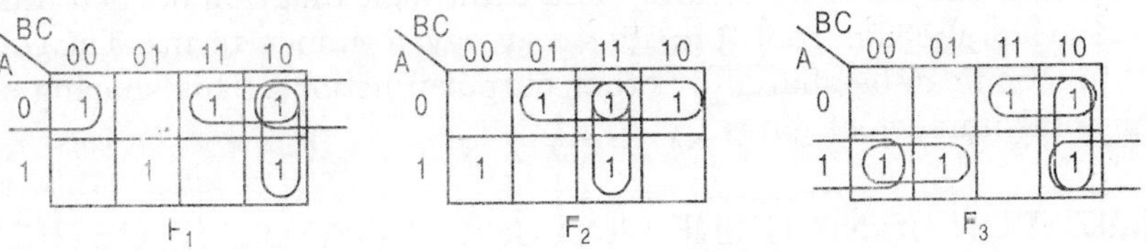

Fig. 3.13

Then we make up an implicant table as shown in Fig. 3.14, showing how each minterm can be covered:

Minterm	F_1	F_2	F_3
m_0	A'B'C'/(A'C')	–	–
m_1	–	A'B'C'/(A'C')	–
m_2	A'BC'/A'B/A'C/BC'	A'BC'/A'B	A'BC'/ A'B' /B'C'
m_3	A'BC/A'B	A'BC/A'B/A'C/BC	A'BC/A'B
m_4	–	AB'C'	AB'C'
m_5	AB'C	–	AB'C / AB'
m_6	ABC' /BC'	–	ABC' /BC'/AC'
m_7		ABC/BC	–

Fig. 3.14 Implicant table

We first scan the table for rows with only a single entry. These are related to essential implicants (m_0, m_1, m_7). We take the largest grouping and update the table (In table by making circle).

Next scan the rows for those which have two entries, selecting the functions that have only a single way grouping option (m_4 under F_2 and m_5 and m_6 under F_1). It means we have to find the common term. We take the common term and update the table (In table by making ovals).

Finally, scan the rows for those rows which have three entries, selecting the functions that have only a single way grouping option or we find the common term. We take the common term and update the table (In table by making rectangular boxes).

Now using implicant table, the three functions can be written as:

$$F_1 = A'C' + A'B + A'B + AB'C + BC'$$
$$= A'C' + A'B + BC' + AB'C$$
$$F_2 = A'C + A'B + A'B + AB'C' + BC$$
$$= A'C + A'B + AB'C' + BC$$
$$F_3 = A'B + A'B + AB'C' + AB'C + BC'$$
$$= A'B + BC' + AB'C' + AB'C$$

We see; F_3 is totally generated from shared terms from F_1 and F_2 with considerable saving over a combinational function by function reduction.

In summary, we can say that many times multiple outputs are derived from the same input variables. In this case, we simplify and draw logic diagram of each function separately. Sometimes, the simplified output functions may have common terms. The common term used by one O/P function can be shared by other output functions. This sharing of common terms reduces the total number of gates.

3.3 MINIMIZATION USING QUINE-McCLUSKEY (TABULAR) METHOD

The K-map method is suitable for simplification of Boolean functions up to 5 or 6 variables. As the number of variables increases beyond this, the visualization of adjacent squares is difficult as the geometry is more involved.

The 'Quine-McCluskey' or 'Tabular' method is employed in such cases. This is a systematic step by step procedure for minimizing a Boolean expression in standard form.

Procedure for Finding the Minimal Expression

1. Arrange all minterms in groups, such that all terms in the same group have same number of 1's in their binary representation. Start with the least number of 1's and continue with grouping of increasing number of 1's, the number of 1's in each term is called the index of that term *i.e.*, all the minterms of same index are placed in a same group. The lowest value of index is zero. Separate each group by a thick line. This constitutes the I stage.
2. Compare every term of the lowest index (say i) group with each term in the successive group of index (say, $i + 1$). If two minterms differ in only one variable, that variable should be removed and a dash (–) is placed at the position, thus a new term with one less literal is formed. If such a situation occurs, a check mark (✔) is

placed next to both minterms. After all pairs of terms with indices i and $(i + 1)$ have been considered, a thick line is drawn under the last terms.

When the above process has been repeated for all the groups of I stage, one stage of elimination have been completed. This constitutes the II stage.

3. The III stage of elimination should be repeated of the newly formed groups of second stage. In this stage, two terms can be compared only when they have dashes in same positions.

   The process continues to next higher stages until no further comparisons are possible. (*i.e.*, no further elimination of literals).
4. All terms which remain unchecked (No 4 sign) during the process are considered to be prime implicants (PIs). Thus, a set of all PIs of the function is obtained.
5. From the set of all prime implicants, a set of essential prime implicants (EPIs) must be determined by preparing prime implicant chart as follow.
   (*a*) The PIs should be represented in rows and each minterm of the function in a column.
   (*b*) Crosses should be placed in each row corresponding to minterms that makes the PIs.
   (*c*) A complete PIs chart should be inspected for columns containing only a single cross. PIs that cover minterms with a single cross in their column are called EPIs.
6. The minterms which are not covered by the EPIs are taken into consideration and a minimum cover is obtained from the remaining PIs.

Now to clarify the above procedure, lets do an example step by step.

Example 1. *Simplify the given function using tabular method.*

$$F(A, B, C, D) = \Sigma(0, 2, 3, 6, 7, 8, 10, 12, 13)$$

Solution. 1. The minterms of the function are represented in binary form. The binary represented are grouped into a number of sections in terms of the number of 1's index as shown in Table of Fig. 3.15.

Minterms	*Binary ABCD*	*No. of 1's*	*Minterms Group*	*Index*	*Binary ABCD*
m_0	0000	0	m_0	0	0000 4
m_2	0010	1	m_2	1	0010 4
m_3	0011	2	m_8		1000 4
m_6	0110	2	m_3	2	0011 4
m_7	0111	3	m_6		0110 4
m_8	1000	1	m_{10}		1010 4
m_{10}	1010	2	m_{12}		1100 4
m_{12}	1100	2	m_7	3	0111 4
m_{13}	1101	3	m_{13}		1101 4

Fig. 3.15

2. Compare each binary term with every term in the adjacent next higher category. If they differ only by one position put a check mark and copy the term into the next column with (–) in the place where the variable is unmatched, which is shown in next Table of Fig. 3.16.

Minterm Group	*Binary* A B C D
0, 2	0 0 – 0 ✔
0, 8	– 0 0 0 ✔
2, 3	0 0 1 – ✔
2, 6	0 – 1 0 ✔
2, 10	– 0 1 0 ✔
8, 10	1 0 – 0 ✔
8, 12	1 – 0 0 PI
3, 7	0 – 1 1 ✔
6, 7	0 1 1 – ✔
12, 13	1 1 0 – PI

Fig. 3.16

Minterm Group	*Binary* A B C D	
0, 2, 8, 10	– 0 – 0	PI
0, 8, 2, 10	– 0 – 0	PI eliminated
2, 3, 6, 7	0 – 1 –	PI
2, 6, 3, 7	0 – 1 –	PI eliminated.

Fig. 3.17

3. Apply same process to the resultant column of Table of Fig. 3.16 and continue until no further elimination of literals. This is shown in Table of Fig. 3.17 above.
4. All terms which remain unchecked are the PIs. However note that the minterms combination (0, 2) and (8, 10) form the same combination (0, 2, 8, 10) as the combination (0, 8) and (2, 10). The order in which these combinations are placed does not prove any effect. Moreover, as we know that $x + x = x$, thus, we can eliminate one of these combinations.

   The same occur with combination (2, 3) and (6, 7).
5. Now we prepare a PI chart to determine EPIs as follows shown in Table of Fig. 3.18.

Prime Implicants	*Minterms* 0	2	3	6	7	8	10	12	13
(8, 12)						×		×	
(12, 13) *								×	×
(0, 2, 8, 10) *	×	×				×	×		
(2, 3, 6, 7) *		×	×	×	×				
	✔		✔	✔	✔		✔		✔

Fig. 3.18

(*a*) All the PIs are represented in rows and each minterm of the function in a column.

(*b*) Crosses are placed in each row to show the composition of minterms that make PIs.

(*c*) The column that contains just a single cross, the PI corresponding to the row in which the cross appear is essential prime implicant. A tick mark is placed against each column which has only one cross mark. A star (*) mark is placed against each EPI.

6. All the minterms have been covered by EPIs.

Finally, the sum of all the EPIs gives the function in its minimal SOP form

EPIs.	*Binary representation*				*Variable Representation*
	A	*B*	*C*	*D*	
12, 13	1	1	0	–	ABC'
0, 2, 8, 10	–	0	–	0	B'D'
2, 3, 6, 7	0	–	1	–	A'C

Therefore, F = ABC' + B'D' + A'C.

If don't care conditions are given, they are also used to find the prime implicant, but it is not compulsory to include them in the final simplified expression.

Example 2. *Simplify the given function using tabular method.*

$$F(A, B, C, D) = \Sigma (0, 2, 3, 6,7)$$
$$d (5, 8, 10, 11, 15)$$

Solution. 1. Step 1 is shown in Table of Fig. 3.19. The don't care minterms are also included.

Minterms	*Binary ABCD*	*No. of 1's*	*Minterms Group*	*Index*	*Binary ABCD*
m_0	0000	0	m_0	0	0000✔
m_2	0010	1	m_2		0010✔
m_3	00 11	2	m_8	1	1000✔
m_5	0101	2	m_3		0011✔
m_6	0110	2	m_5	2	0101✔
m_7	0111	3	m_6		0110✔
m_8	1000	1	m_{10}		101 0✔
m_{10}	1010	2	m_7	3	0111✔
m_{11}	1011	3	m_{11}		1011✔
m_{15}	1111	4	m_{15}	4	1111✔

Fig. 3.19

2. Step 2 is shown in Table of Fig. 3.20.
3. Step 3 is shown in Table of Fig. 3.21.

Minterm Group	*Binary* A B C D
0, 2	0 0 – 0 ✔
0, 8	– 0 0 0 ✔
2, 3	0 0 1 – ✔
2, 6	0 – 1 0 ✔
2, 10	– 0 1 0 ✔
8, 10	1 0 – 0 ✔
3, 7	0 – 1 1 ✔
3, 11	– 0 1 1 ✔
5, 7	0 1 – 1 PI
6, 7	0 1 1 – ✔
10, 11	1 0 1 – ✔
7, 15	– 1 1 1 ✔
11, 15	1 – 1 1 ✔

Fig. 3.20

Minterm Group	*Binary* A B C D	
0, 2, 8, 10	– 0 – 0	PI
0, 8, 2, 10	– 0 – 0	PI Eliminated
2, 3, 6, 7	0 – 1 –	PI
2, 3 10, 11	– 0 1 –	PI
2, 6, 3, 7	0 – 1 –	PI Eliminated
2, 10, 3, 11	– 0 1 –	PI Eliminated
3, 7, 11, 15	– – 1 1	PI
3, 11, 7, 15	– – 1 1	PI Eliminated

Fig. 3.21

4. All the terms which remain unchecked are PIs. Moreover, one of two same combinations is eliminated.
5. Step 5 is to prepare a PI chart to determine EPIs as shown in Table of Fig. 3.22. Note, however, that don't care minterms will not be listed as column headings in the chart as they do not have to be covered by the minimal (simplified) expression.

Prime Implicants		*Minterms*				
		0	*2*	*3*	*6*	*7*
(5, 7)						×
(0, 2, 8, 10)	*	×	×			
(2, 3, 6, 7)	*		×	×	×	×
(2, 3, 10, 11)			×	×		
(3, 7, 11, 15)				×		×
		✔			✔	

Fig. 3.22

6. All the minterms have been covered by EPIs.

Therefore F (A, B, C, D) = B'D' + A'C

Example 3. *Simplify the given function using tabular method:*

$$F\ (A,\ B,\ C,\ D,\ E,\ F,\ G) = \Sigma\ (20,\ 28,\ 38,\ 39,\ 52,\ 60,\ 102,\ 103,\ 127)$$

Solution. Step 1 is shown in Table of Fig. 3.23.

Minterms	*Binary ABCDEFG*	*No. of 1's*	*Minterms Group*	*Index*	*Binary ABCDEFG*
m_{20}	0010100	2	m_{20}	2	0010100 ✔
m_{28}	0011100	3	m_{28}		0011100 ✔
m_{38}	0100110	3	m_{38}	3	0100110 ✔
m_{39}	0100111	4	m_{52}		0110100 ✔
m_{52}	0110100	3	m_{39}	4	0100111 ✔
m_{60}	0111100	4	m_{60}		0111100 ✔
m_{102}	1100110	4	m_{102}		1100110 ✔
m_{103}	1100111	5	m_{103}	5	1100111 ✔
m_{127}	1111111	7	m_{127}	7	1111111 PI

Fig. 3.23

2. Step 2 is shown in Table (Fig. 3.24).
3. Step 3 is shown in Table (Fig. 3.25).

Minterms Group	*Binary*							
	A	*B*	*C*	*D*	*E*	*F*	*G*	
20, 28	0	0	1	–	1	0	0	✔
20, 52	0	–	1	0	1	0	0	✔
28, 60	0	–	1	1	1	0	0	✔
38, 39	0	1	0	0	1	1	–	✔
38, 102	–	1	0	0	1	1	0	✔
52, 60	0	1	1	–	1	0	0	✔
39, 103	–	1	0	0	1	1	1	✔
102, 103	1	1	0	0	1	1	–	✔

Fig. 3.24

Minterms Group	*Binary* A B C D E F G	
20, 28, 52, 60	0 – 1 – 1 0 0	PI
20, 52, 28, 60	0 – 1 – 1 0 0	PI Eliminated
38, 39 102, 103	– 1 0 0 1 1 –	PI
38, 102, 39, 103	– 1 0 0 1 1 –	PI Eliminated

Fig. 3.25

4. All the terms which remain unchecked are PIs. Moreover one of two same combinations is eliminated.
5. PI chart to determine EPIs is shown in Table of Fig. 3.26.

Prime Implicants	*Minterms*								
	20	*28*	*38*	*39*	*52*	*60*	*102*	*103*	*127*
127 *									×
(20, 28, 52, 60) *	×	×			×	×			
(38, 39, 102, 103) *			×	×			×	×	
	✔	✔	✔	✔	✔	✔	✔	✔	✔

Fig. 3.26

6. All the minterms have been covered by EPIs.

Therefore, F (A, B, C, D, E, F, G) = ABCDEFG + A'CEF'G' + BC'D'EF

3.4 EXERCISE

1. Using Boolean algebra simplify each of the following logic expressions as much as possible:

   (*a*) Z = A(A + AB) (A + ABC) (A + ABCD)

   (*b*) C = $(X_1X_2' + X_2'X_3)'$

2. Draw the simplest possible logic diagram that implements the output of the logic diagram given below.

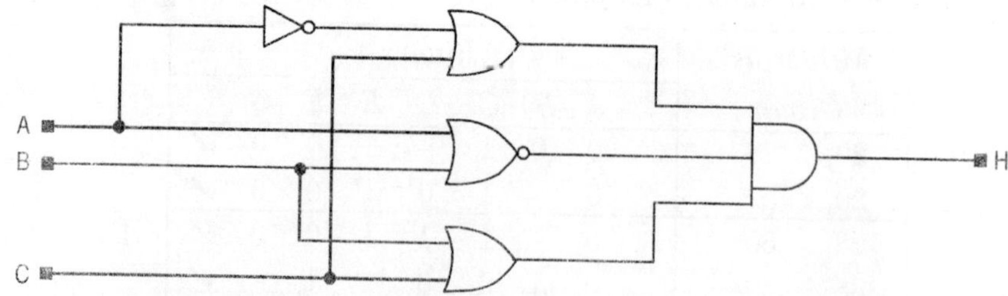

3. Write the logic expression and simplify it as much as possible and draw a logic diagram that implements the simplified expression.

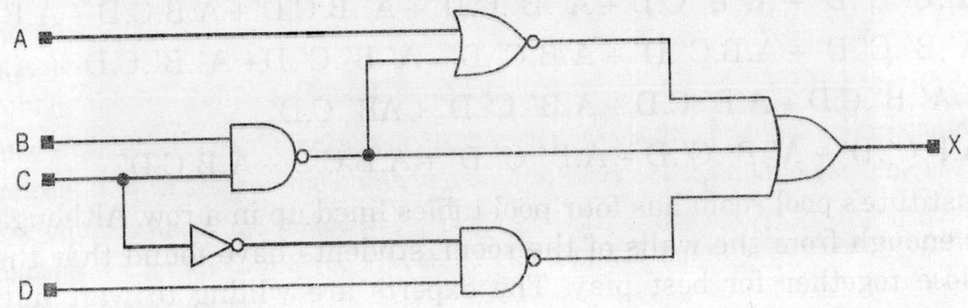

4. Obtain the simplified expression in s-of-p for the following Boolean functions:

   (*a*) $xy + x'y'z' + x'yz'$

   (*b*) ABD + A′C′D′ + A′B + ACD + AB′D′

   (*c*) $x'z + w'xy' + w(x'y + xy')$

   (*d*) F (*x*, *y*, *z*) = Σ (2, 3, 6, 7)

   (*e*) F (A, B, C, D) = Σ (7, 13, 14, 15)

5. Use a K-map to simplify each of the following logic expressions as much as possible:

   (*i*) F = AB′ + A′B + AB

   (*ii*) G = X′Y′Z′ + X′YZ′ + XY′Z′ + X′Y′Z′ + XYZ′

   (*iii*) H = A′B′CD + AB′C′D′ + A′B′C′D′ + ABC′D + A′B′C′D + AB′C′D + ABCD

   (*iv*) W = X′Y′Z + X′YZ + XYZ + XY′Z + X′YZ′

6. Simplify the following logic expressions using K-maps and tabular method.

   (*a*) F(A, B, C) = A′C + B′C+AB′C′

   (*b*) G(A, B, C, D) = B′CD + CD′ + A′B′C′D + A′B′C

7. Simplify the Boolean function $F_{(ABCDE)}$ = Σ(0, 1, 4, 5, 16, 17, 21, 25, 29)

8. Simplify the following Boolean expressions using K-maps and tabular method.

   (*i*) BDE + B′C′D + CDE + ABCE + ABC + BCDE

   (*ii*) ABCE + ABCD + BDE + BCD + CDE + BDE

   (*iii*) $F_{(ABCDEF)}$ – Σ(6, 9, 13, 18, 19, 27, 29, 41, 45, 57, 61)

9. Draw Karnaugh maps for the following expressions:

   F = A′.B′.C′ + A′.B′.C + A.B′.C + A.B.C

   F = A′.B.C′ + A.B.C′ + A′.B.C + A.B.C + A.B′.C′

   F = A.B.C′.D′ + A.B′.C′.D + A′.B.C.D + A′.B′.C′.D + A′.B′.C′.D
   + A′B′.C.D + A′.B′.C.D′ + A′B.C.D′

10. Simplify the following logic expressions using karnaugh maps. Draw logic diagrams for them using only (*a*) NAND, (b) NOR gates, assuming inputs A, B, C, and D only are available.

    Y = A′.B.C′.D′ + A.B.C′.D + A.B.C.D + A′.B′.C.D + A.B′.C′.D
    + A′.B′.C.D′ + A′.B.C.D′

$Y = A'.B'.C'.D' + A'.B'.C.D + A'.B'.C.D' + A'.B.C.D' + A.B.C.D + A.B'.C.D$

$Y = A'.B'.C'.D' + A.B.C'.D' + A.B.C'.D + A'.B'.C'.D + A'.B'.C.D' + A.B.C'.D + A'.B'.C.D + A.B'.C.D + A.B'.C'.D' + AB'.C.D'$

$Y = A.B.C'.D' + A'.B'.C'.D + A.B'.C'.D' + A.B.C.D + A.B.C.D'$

11. The institute's pool room has four pool tables lined up in a row. Although each table is far enough from the walls of the room, students have found that the tables are too close together for best play. The experts are willing to wait until they can reserve enough adjacent tables so that one game can proceed unencombered by nearby tables. A light board visible outside the pool room shows vacant tables. The manager has developed a digital circuit that will show an additional light whenever the experts' desired conditions arise. Give a logic equation for the assertion of the new light signal. Simplify the equation using a K-Map.

12. Simplify the Boolean functions using tabular method and verify result with K-map.

 (*a*) $F(w, x, y, z) = \Sigma(0, 1, 2, 4, 5, 6, 8, 9, 12, 13, 14)$

 (*b*) $F(w, x, y, z) = \Sigma(2, 3, 12, 13, 14, 15)$

 (*c*) $F(A, B, C, D) = \Sigma(4, 6, 7, 15)$

 (*d*) $F(A, B, C, D) = \Sigma(7, 13, 14, 15)$

 (*e*) $F(x, y, z) = \Sigma(7, 13, 14, 15)$

13. Simplify the Boolean function F using the don't care conditions d, in (*i*) SOP and (*ii*) POS:

 $F = A'B'D + A'CD + A'BC$ $\quad d = A'BC'D + ACD + AB'D'$

 $F = w'(x'y + x'y' + xyz) + x'z'(y + w)$ $\quad d = w'x(y'z + yz') + wyz$

 $F = ACE + A'CD'E' + A'C'DE$ $\quad d = DE' + A'D'E + AD'E'$

14. Use a Karnaugh map to simplify each of the following logic expressions as much as possible.

 (*a*) $F = A'B'CD + AB'C'D' + A'B'CD' + ABC'D + ABCD$

 Solution. $F = ABD + A'B'D + B'C'$

 (*b*) $G = A'C + B'C + AB'C' + A'B$

 Solution. $G = AB' + A'B + A'C$ or $AB' + A'B + B'C$

 (*c*) $H = B'CD + CD' + A'B'C'D' + A'B'C$

 Solution. $H = B'CD + CD' + A'B'C'D' + A'B'C$

 (*d*) $F = (A' + B + C')(A + B + C)(A + B + C')$

 Solution. $F = B + AC'$

15. Use a Karnaugh map to simplify each of the following logic expressions as much as possible.

 (*a*) $W = (AB'C')'(AB'C)(ABC)'$

(*b*) $M = X_2X_3 + X'_1X'_2X_3 + X'_3 + X_1X'_2X_3$

16. Using Boolean Algebra simplify

(*a*) $(A + \bar{B})(A + C)$

(*b*) $\bar{A}B + \bar{A}B\bar{C} + \bar{A}BCD + \bar{A}B\bar{C}\bar{D}E$

(*c*) $AB + \overline{ABC} + A$

(*d*) $(A + \bar{A})(AB + AB\bar{C})$

(*e*) $AB + (\bar{A} + \bar{B})C + AB$

17. Use a karnaugh map to simplify each function to a minimum sum-of-products form:

(*a*) $X = \bar{A}\bar{B}\bar{C} + A\bar{B}C + AB\bar{C}$

(*b*) $X = AC[\bar{B} + A(B + \bar{C})]$

(*c*) $X = DE\bar{F} + \bar{D}E\bar{F} + \bar{D}\bar{E}\bar{F}$

18.

A	*B*	*C*	F_1
0	0	0	1
0	0	1	0
0	1	0	0
0	1	1	0
1	0	0	0
1	0	1	1
1	1	0	0
1	1	1	1

A	*B*	*C*	F_2
0	0	0	0
0	0	1	1
0	1	0	1
0	1	1	1
1	0	0	0
1	0	1	1
1	1	0	1
1	1	1	0

Transfer the input-output specifications for F_1 and F_2 given above to 3 variable Karnaugh maps.

19. Using a Karnaugh map simplify the following equations

(*a*) $X = \bar{A}\bar{B} + \bar{A}C + BC + AB + A\bar{C} + \bar{A}\bar{B}\bar{C} + ABC$

(*b*) $X = \bar{A}\bar{B}\bar{C} + \bar{A}CD + \bar{A}B\bar{C} + B\bar{C}\bar{D} + A\bar{B}C + AB\bar{D} + A\bar{B}C\bar{D}$

(*c*) $X = \bar{D}(\bar{A}[\bar{C} + \bar{B}C] + A[\bar{C} + \bar{B}C]) + BC\bar{D}$

(*d*) $X = \bar{A}\bar{B}\bar{C} + \bar{B}C\bar{D} + \bar{A}BD + ABCD + A\bar{C}D + \bar{A}B\bar{C}\bar{D}$

20. Simplify the following using Boolean Algebra

(*a*) $z = w.x + w.\bar{x}.y$

(*b*) $z = \overline{(x + y).(\bar{x} + \bar{y})}$

(*c*) $z = x.y + w.\bar{y} + w.x + x.y.v$

(*d*) $z = (x + y).(x + \bar{w}).[y.(x + \bar{w}) + \bar{y}]$

21. Consider the function

$z = f(x, y, w, v) = (x.v + \bar{x}.w).[\bar{y}.(w + y.\bar{v})]$

(*a*) Draw a schematic diagram for a circuit which would implement this function.

22. Simplify the Boolean function by tabular method

$F(A, B, C, D, E) = \Sigma(0, 1, 4, 5, 16, 17, 21, 25, 29)$

23. Simplify the following function in (*a*) s–o–p and (*b*) p–o–s

$F(A, B, C, D) = \Pi(3, 4, 6, 7, 11, 12, 13, 14, 15)$

24. Simplify the Boolean function using tabular method.

$F(A, B, C, D, E) = \Sigma(0, 1, 4, 5, 16, 17, 21, 25, 29, 30)$

25. Simplify the Boolean function using tabular method

$F(A, B, C, D, E, F) = \Sigma(6, 9, 13, 18, 19, 27, 29, 41, 45, 57, 61, 63)$

4

COMBINATIONAL LOGIC

4.0 INTRODUCTION

Combinational logic circuits are circuits in which the output at any time depends upon the combination of input signals present at that instant only, and does not depend on any past conditions.

The block diagram of a combinational circuit with *m* inputs and *n* outputs is shown in Fig. 4.0.

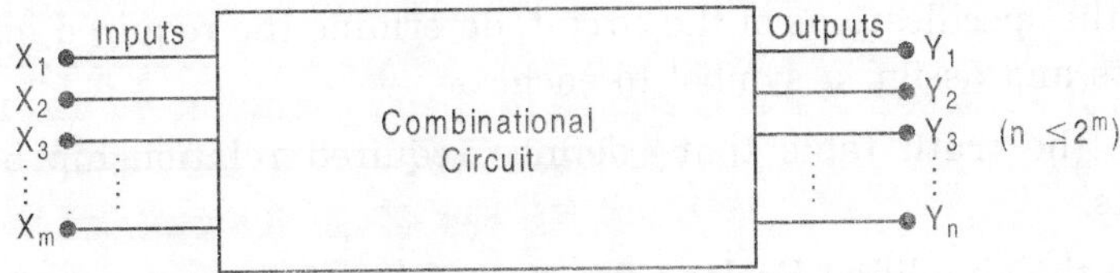

Fig. 4.0 Block diagram of combinational logic circuit

In particular, the output of particular circuit does not depend upon any past inputs or outputs i.e. the output signals of combinational circuits are not fedback to the input of the circuit. Moreover, in a combinational circuit, for a change in the input, the output appears immediately, except for the propagation delay through circuit gates.

The combinational circuit block can be considered as a network of logic gates that accept signals from inputs and generate signals to outputs. For m input variables, there are 2^m possible combinations of binary input values. Each input combination to the combinational circuit exhibits a distinct (unique) output. Thus a combinational circuit can be described by n boolean functions, one for each input combination, in terms of m input variables with n is always less than or equal to 2^m $[n \leq 2^m]$.

Thus, a combinational circuit performs a specific information processing operation which is specified by Boolean functions.

$n < 2^m$ represent the condition, that in a particular application there are some unused input combinations. For example, when we are using NBCD codes, the six combinations (1010, 1011, 1100, 1101, 1110 and 1111) are never used. So with four input variables (m = 4) we are using only 10 i/p combinations *i.e.*, 10 o/ps instead of 2^4 = 16.

The digital systems perform a number of information processing tasks. The basic arithmatic operations used by digital computers and calculators are implemented by combinational circuits using logic gates. We proceed with the implementation of these basic functions by first looking the simple design procedure.

Analysis and Design Procedure of Combinational Circuits

Analysis Procedure

The analysis of a combinational circuit requires that we determine the function that the circuit implements. The analysis starts with a logic diagram and terminates with a set of Boolean functions, a truth table or possibly, an explanation of the circuit operation.

Analysis procedure can be performed by using the following steps:

1. Firstly, make sure that the given circuit is combinational circuit. (The circuit with no feedback paths or memory elements).
2. After verification of the logic diagram, of the combinational circuit, proceed to obtain the output Boolean function.
3. After knowing the Boolean function, obtain the truth table for a circuit.

Design Procedure

A combinational circuit can be designed, by the following design procedure discussed in steps given below:

1. From the specifications of the circuit, determine the required number of inputs and outputs and assign a symbol to each.
2. Derive the truth table that defines required relationship between inputs and outputs.
3. Obtain the simplified Boolean functions for each output as a function of the input variables.
4. Draw the logic diagram and verify the correctness of the design.

1. How to obtain the output Boolean function from a logic diagram

- Label all the gate outputs that are functions of input variables with arbitrary symbols – but with meaningful names. Determine the Boolean function for each gate output.
- Label the gates that are a function of input variables and previously labelled gates with other arbitrary symbols. Find the Boolean functions for these gates.
- Repeat the process given in step 2 until the outputs of the circuit are obtained.
- By repeatedly substitution of previously defined functions, obtain the output Boolean function in terms of input variables.

2. Obtaining truth table directing from the logic diagram

- Determine the number of input variables in the circuit. For n inputs, form the 2^n possible input combinations and the binary numbers from 0 to $2^n - 1$ in a table.

- Label the outputs of selected gates with arbitrary symbols.
- Obtain the truth table for the outputs of those gates which are a function of the input variables only.
- Proceed to obtain the truth table for the outputs of those gates which are a function of previously defined values until the columns for all outputs are determined.

To explain the procedure, let us take an example that we have already been used in chapter 2.

Example: *A TV is connected through three switches. The TV becomes 'on' when at least two switches are in 'ON' position; In all other conditions, TV is 'OFF'.*

Solution. Step 1 : The TV is connected with 3 switches; thus there are three inputs to TV, represented by variables say A, B and C. The o/p of TV is represented by variable say, F.

The block diagram is shown in Fig. 4.1 :

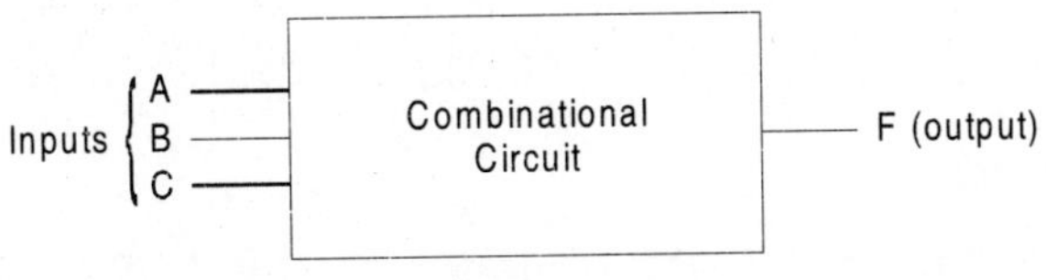

Fig. 4.1

Step 2.

Truth Tables

TV switches ← INPUTS			*OUTPUTS*
A	*B*	*C*	*F*
0	0	0	0
0	0	1	0
0	1	0	0
0	1	1	1
1	0	0	0
1	0	1	1
1	1	0	1
1	1	1	1

0 → switch off

1 → switch on

It means for the input combinations in which there are two or more 1's, the output F = 1 (TV is ON) and for rest combinations, output F = 0 (TV is OFF).

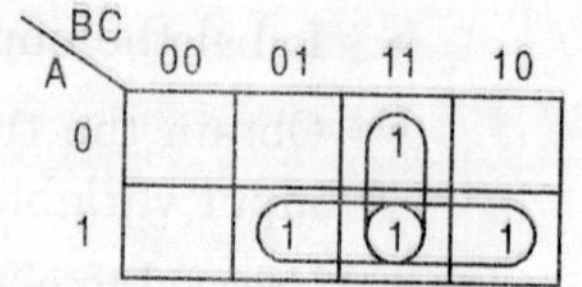

Step 3 : In general, in simplifying boolean functions upto four variables, the best method is K-map technique. Thus, using a 3 variable K-map, we can simplify the function obtained in step 2.

We get
$$F = AB + AC + BC$$

We can observe that if the value of any two variables is equal to 1, the output is equal to 1.

Step 4. For implementation we need three 'AND' gates and one 'OR' gate as shown in Fig. 4.2.

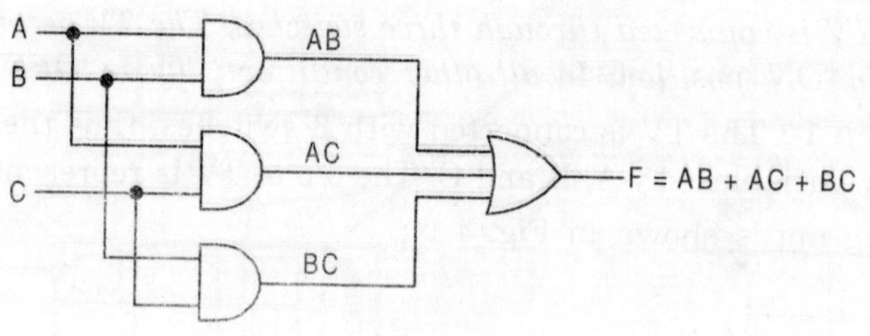

Fig. 4.2

4.1 ARITHMETIC CIRCUITS

The logic circuits which are used for performing the digital arithmetic operations such as addition, subtraction, multiplication and division are called 'arithmetic circuits'.

4.1.1 Adders

The most common arithmetic operation in digital systems is the addition of two binary digits. The combinational circuit that performs this operation is called a half-adder.

Half Adder

1. Figure 4.3 shows a half adder (HA).

It has two inputs A and B that are two 1-bit members, and two output sum (S) and carry (C) produced by addition of two bits.

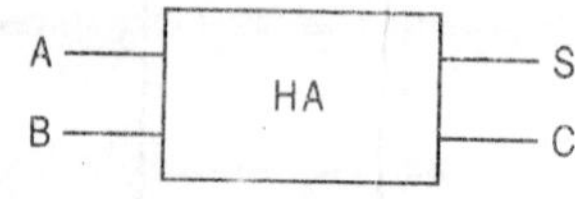

Fig. 4.3 Half adder

2. **Truth Table :**

Inputs		*Outputs*	
A	*B*	*S*	*C*
0	0	0	0
0	1	1	0
1	0	1	0
1	1	0	1

The sum output is 1 when any of inputs (A and B) is 1 and the carry output is 1 when both the inputs are 1.

3. Using a two variable *k*-map, separately for both outputs S and C.

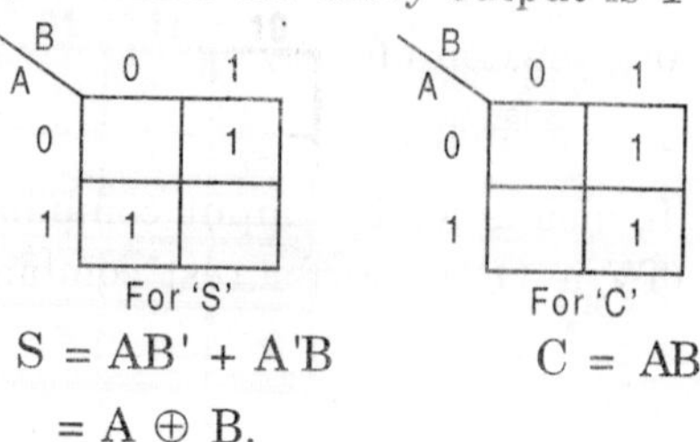

$S = AB' + A'B$
$= A \oplus B.$

$C = AB$

4. Logical Implementation.

(*i*) Using Basic gates (as shown in Fig. 4.4(*a*)).

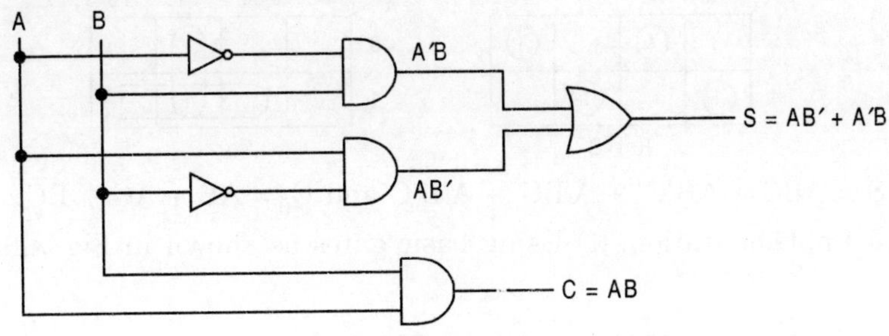

Fig. 4.4 (*a*)

(*ii*) Using XOR gate as shown in Fig. 4.4 (*b*).

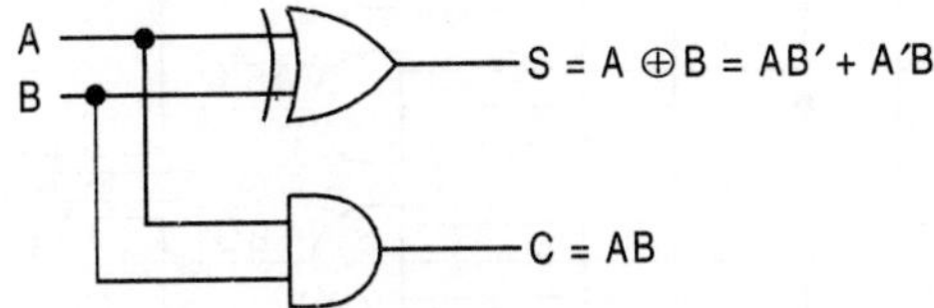

Fig. 4.4 (*b*)

Implementation using only NAND or only NOR gates is left as an exercise.

Full Adder

Full adder is a combinational circuit that performs the addition of three binary digits.

1. Figure 4.5 shows a full adder (FA). It has three inputs A, B and C'and two outputs S and C_0 produced by addition of three input bits. Carry output is designated C_0 just to avoid confusion between with i/p variable C.

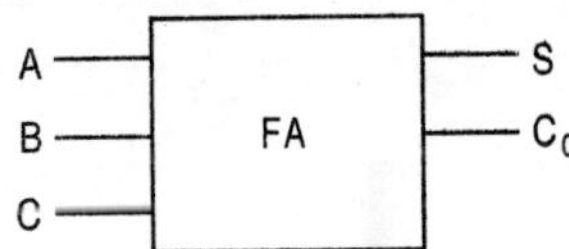

Fig. 4.5 Full adder

2. **Truth Table :** The eight possible combinations of three input variables with their respective outputs is shown. We observe that when all the three inputs are 1, the sum and carry both outputs, are 1.

Inputs			*Output*	
A	*B*	*C*	*S*	C_0
0	0	0	0	0
0	0	1	1	0
0	1	0	1	0
0	1	1	0	1
1	0	0	1	0
1	0	1	0	1
1	1	0	0	1
1	1	1	1	1

3. Using a three variable map for both outputs.

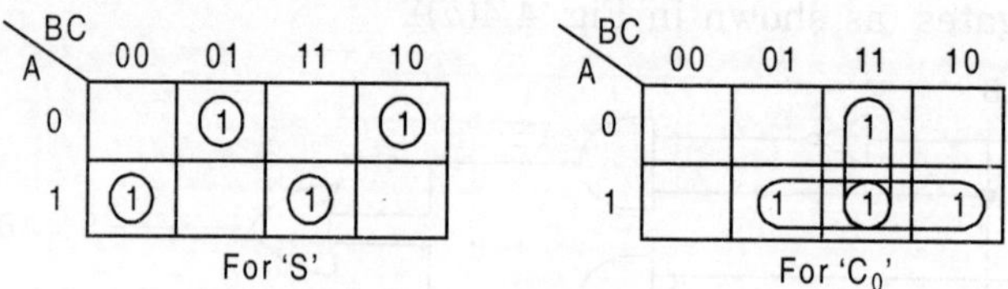

S = ABC + AB'C' + A'BC' + A'B'C and C_0 = AB + AC + BC.

4. Logical Implementation. (*i*) Using basic gates as shown in Fig. 4.6.

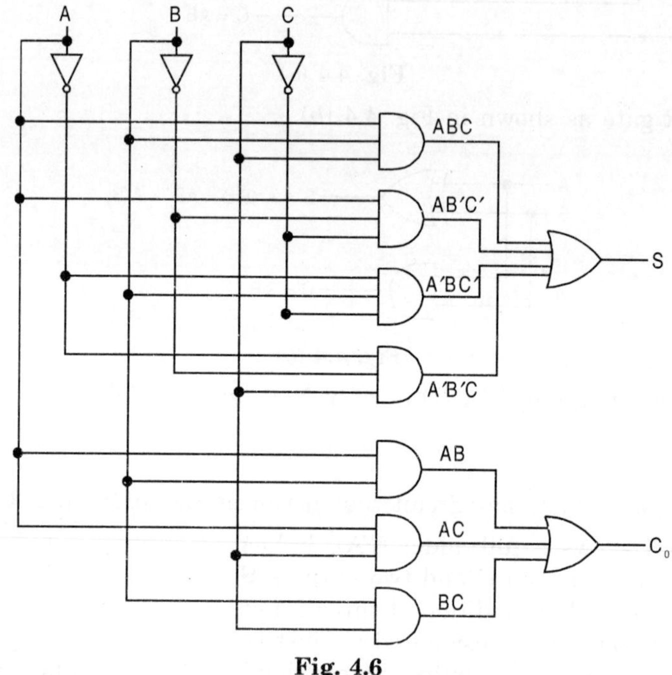

Fig. 4.6

(*ii*) A 'Full Adder' can also be implemented using two half adders and an 'OR' Gate as shown in Fig. 4.7

The Sum S = ABC + AB'C' + A'BC' + A'B'C

= ABC + A'B'C + AB'C' + A'BC'

= C (AB + A'B') + C' (AB' + A'B)

= C (AB' + A'B)' + C' (AB' + A'B)

= (A ⊕ B) ⊕ C

and the carry C_0 = AB + AC + BC

= AB + C (A + B)

= AB + C (A + B) (A + A') (B + B')

= AB + C [AB + AB' + A'B]

= AB + ABC + C (AB' + A'B)

= AB (1 + C) + C (A ⊕ B)

= AB + C (A ⊕ B)

Therefore, $S = (A \oplus B) \oplus C$ and $C_0 = AB + C\,(A \oplus B)$

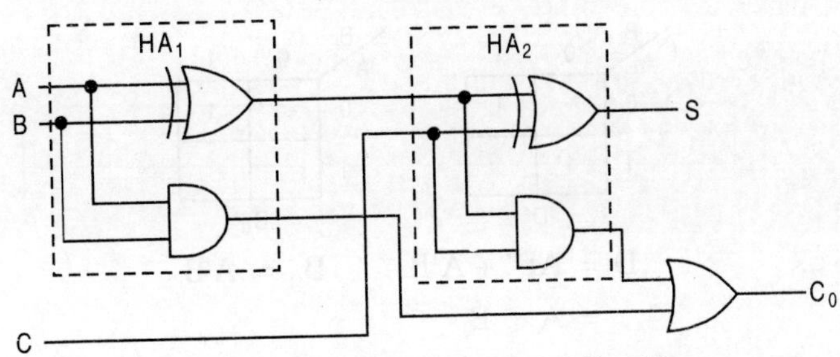

Fig. 4.7 Implementation of full adder

Block Diagram representation of a full adder using two half adders :

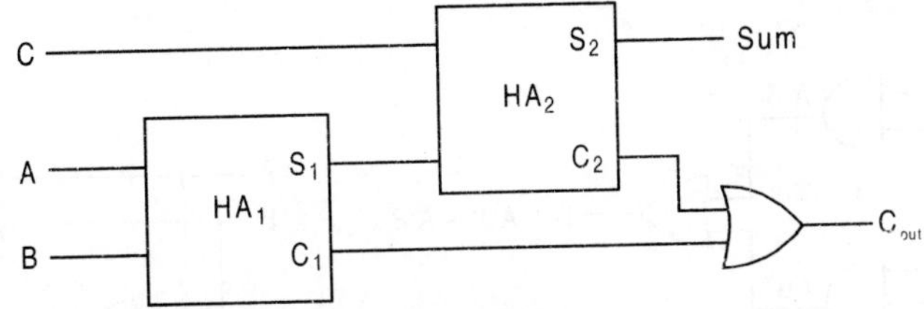

S_1 and C_1 are outputs of first half adder (HA_1)

S_2 and C_2 are outputs of second half adder (HA_2)

A, B and C are inputs of Full adder.

Sum and C_{out} are outputs of full adder.

4.1.2 Subtractors

The logic circuits used for binary subtraction, are known as 'binary subtractors'.

Half Subtractor : The half subtractor is a combinational circuit which is used to perform the subtraction of two bits.

1. Figure 4.8 shows a half subtractor (HS).

   It has two inputs, A (minuend) and B (subtrahend) and two outputs D (difference) and B_0 (Borrow). [The symbol for borrow (B_0) is taken to avoid confusion with input variable B] produced by subtractor of two bits.

Fig. 4.8 Half subtractor

2. **Truth Table**

   The difference output is 0 if A = B and 1 if A ≠ B; the borrow output is 1 whenever A < B. If A < B, the subtraction is done by borrowing 1 from the next higher order bit.

Inputs		*Outputs*	
A	*B*	*D*	B_0
0	0	0	0
0	1	1	1
1	0	1	0
1	1	0	0
1	1	0	0

3. Using a two variable map, for outputs D and B_0.

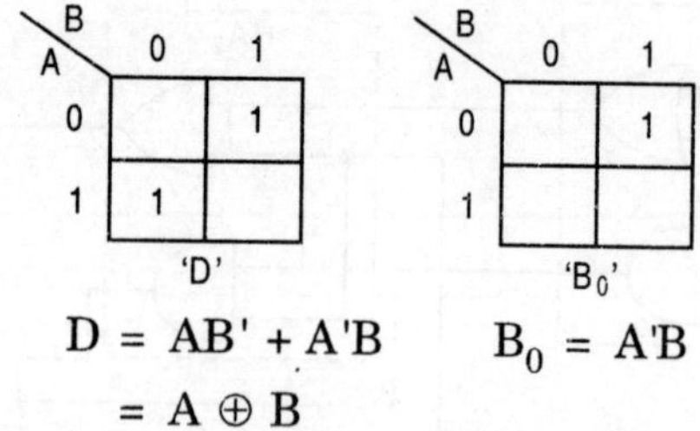

$D = AB' + A'B$ $\quad B_0 = A'B$

$= A \oplus B$

4. Logical Implementation shwon in Fig. 4.9

(*a*) using basic gates (*b*) using XOR gate

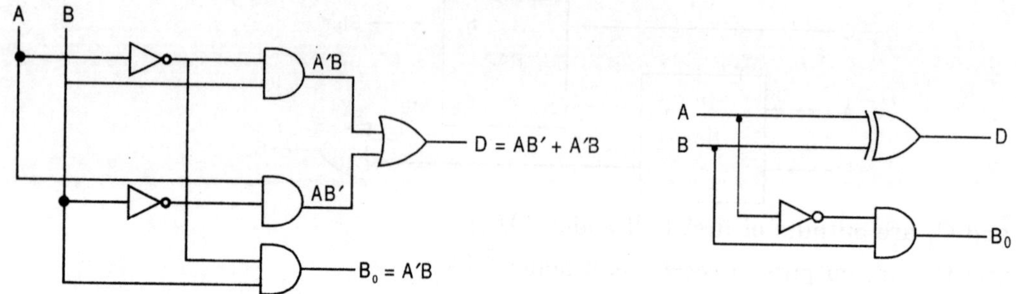

Fig. 4.9 (*a*) Basic gate implementation of half subtractor.

Fig. 4.9 (*b*) X-OR gate implementation of half subtractor

Full subtractor: Full subtractor is a combinational circuit that performs the subtraction of three binary digits.

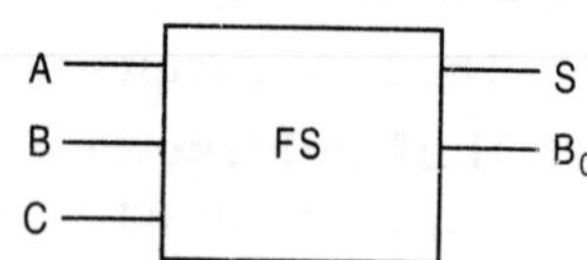

Fig. 4.10 Full subtractor

1. Figure 4.10 shows a full subtractor (FS).

   It has three inputs A, B and C and two outputs D and B_0 produced by subtraction of three input bits.

2. **Truth Table**

   The eight possible combinations of three input variables with their respective outputs is shown. We observe that when all the three inputs are 1, the difference and borrow both outputs are 1.

Inputs			*Output*	
A	*B*	*C*	B_0	*D*
0	0	0	0	0
0	0	1	1	1
0	1	0	1	1
0	1	1	0	1
1	0	0	1	0
1	0	1	0	0
1	1	0	0	0
1	1	1	1	1

3. Using a three variable map for both outputs.

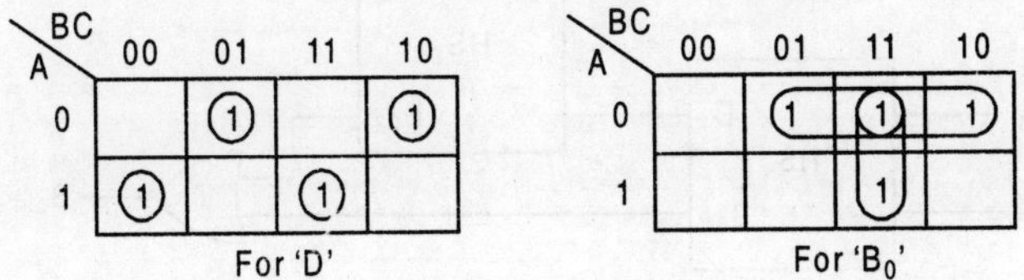

$D = ABC + AB'C' + A'BC' + A'B'C$, $B_0 = A'B + A'C + BC$

4. Logical implementation—

(*i*) Using basic gates : Left as an exercise.

(*ii*) A 'full subtractor' can also be implemented using two 'half subtractors' and an 'OR' gate as shwon in Fig. 4.11.

The difference

$$\begin{aligned}
'D' &= ABC + AB'C' + A'BC' + A'B'C \\
&= ABC + A'B'C + AB'C' + A'BC' \\
&= C\,(AB + A'B') + C'\,(AB' + A'B) \\
&= C\,(AB' + A'B)' + C'\,(AB' + A'B) \\
&= C\,(A \oplus B)' + C'\,(A \oplus B) \\
&= (A \oplus B) \oplus C
\end{aligned}$$

and the borrow

$$\begin{aligned}
B_0 &= A'B + A'C + BC \\
&= A'B + C\,(A' + B) \\
&= A'B + C\,(A' + B)\,(A + A')\,(B + B') \\
&= A'B + C\,[A'B + AB + A'B'] \\
&= A'B + A'BC + C\,(AB + A'B') \\
&= A'B\,(C + 1) + C\,(A \oplus B)' \\
&= A'B + C\,(A \oplus B)'
\end{aligned}$$

$D = (A \oplus B) \oplus C$ and $B_0 = A'B + C\,(A \oplus B)'$

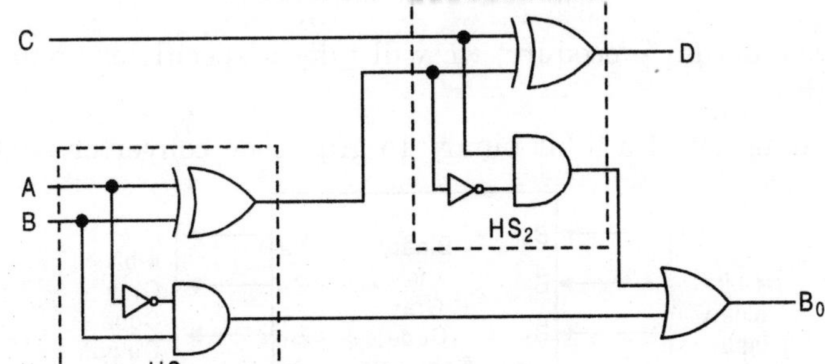

Fig. 4.11 (*a*)

Block Diagram Representation of a full subtractor using two half subtractors :

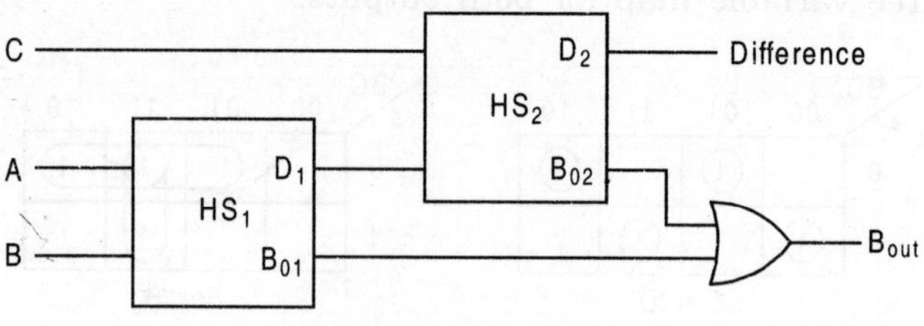

Fig. 4.11 *(b)*

D_1 and B_{01} are outputs of first half subtractor (HS_1)

D_2 and B_{02} are outputs of second half subtractor (HS_2)

A, B and C are inputs of full subtractor.

Difference and B_{out} are outputs of full subtractor.

4.1.3 Code Converters

In the previous study of codes, coding was defined as the use of groups of bits to represent items of information that are multivalued. Assigning each item of information a unique combination of bits makes a transformation of the original information. This we recognize as information being processed into another form. Moreover, we have seen that there are many coding schemes exist. Different digital systems may use different coding schemes. It is sometimes necessary to use the output of one system as the input to other. Therefore a sort of code conversion is necessary between the two systems to make them compatible for the same information.

'A code converter is a combinational logic circuit that changes data presented in one type of binary code to another type of binary code.' A general block diagram of a code converter is shown in Fig. 4.12.

Fig. 4.12 Code converter

To understand the design procedure; we will take a specific example of 4-bit Binary to Gray code conversion.

1. The block diagram of a 4-bit binary to gray code converter is shown in Fig. 4.13.

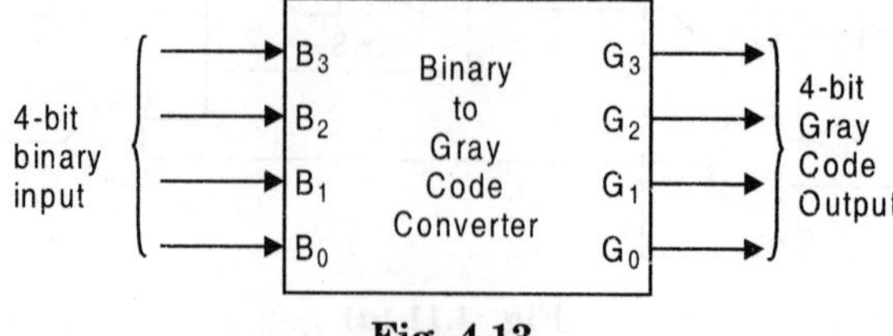

Fig. 4.13

If has four inputs (B_3 B_2 B_1 B_0) representing 4-bit binary numbers and four outputs (G_3 G_2 G_1 G_0) representing 4-bit gray code.

2. Truth table for binary to gray code converters.

Binary Inputs				*Gray code Outputs*			
B_3	B_2	B_1	B_0	G_3	G_2	G_1	G_0
0	0	0	0	0	0	0	0
0	0	0	1	0	0	0	1
0	0	1	0	0	0	1	1
0	0	1	1	0	0	1	0
0	1	0	0	0	1	1	0
0	1	0	1	0	1	1	1
0	1	1	0	0	1	0	1
0	1	1	1	0	1	0	0
1	0	0	0	1	1	0	0
1	0	0	1	1	1	0	1
1	0	1	0	1	1	1	1
1	0	1	1	1	1	1	0
1	1	0	0	1	0	1	0
1	1	0	1	1	0	1	1
1	1	1	0	1	0	0	1
1	1	1	1	1	0	0	0

3. Now we solve all the gray outputs distantly with respect to binary inputs. From the truth table; the logic expressions for the gray code outputs can be written as

$$G_3 = \Sigma\ (8, 9, 10, 11, 12, 13, 14, 15)$$

$$G_2 = \Sigma\ (4, 5, 6, 7, 8, 9, 10, 11)$$

$$G_1 = \Sigma\ (2, 3, 4, 5, 10, 11, 12, 13)$$

$$G_0 = \Sigma\ (1, 2, 5, 6, 9, 10, 13, 14).$$

The above expressions can be simplified using K-map

Map for G_3:

From the octet, we get

$G_3 = B_3$

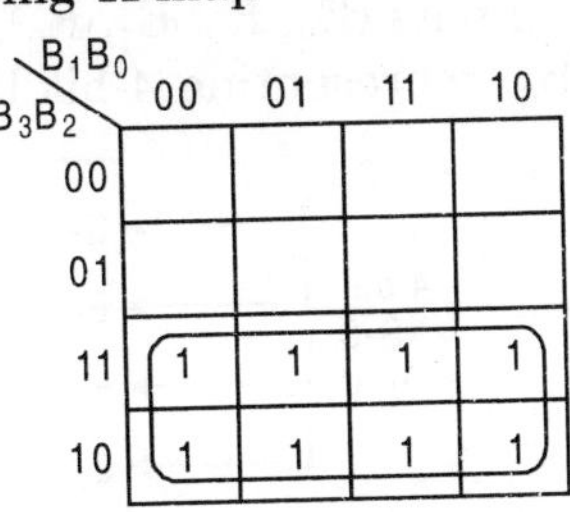

Map for G_2:

From the two quads, we get

$$G_2 = B_3' B_2 + B_3 B_2'$$
$$= B_3 \oplus B_2.$$

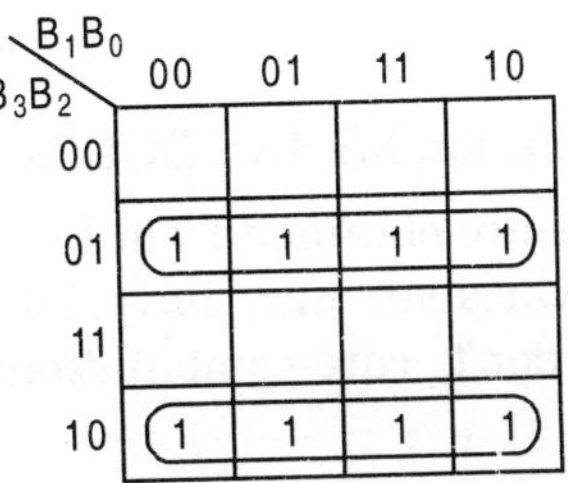

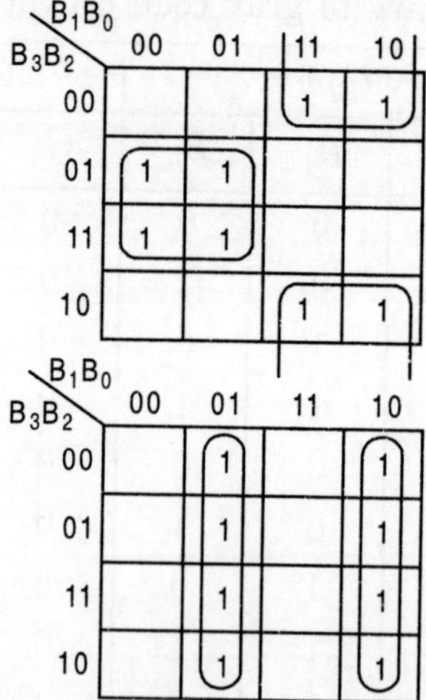

Map for G_1:

From the two quads, we get

$$G_1 = B_2\,B_1' + B_2'\,B_1$$
$$= B_2 \oplus B_1$$

Map for G_0:

From the two quads. we get

$$G_0 = B_1'\,B_0 + B_1B_0'$$
$$= B_1 \oplus B_0.$$

4. Now the above expressions can be implemented using X-OR gates to yield the desired code converter circuit shown in Fig. 4.14.

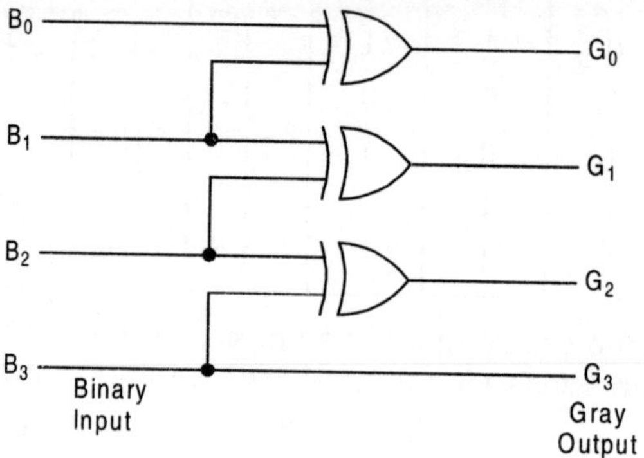

Fig. 4.14

Let us see one more example of XS-3 to BCD code converter.

1. The block diagram of an XS-3 to BCD code converter is shown in Fig. 4.15.

   It has four inputs (E_3, E_2, E_1, E_0) representing 4 bit XS-3 number and four outputs (B_3B_2 B_1 B_0) representing 4-bit BCD code.

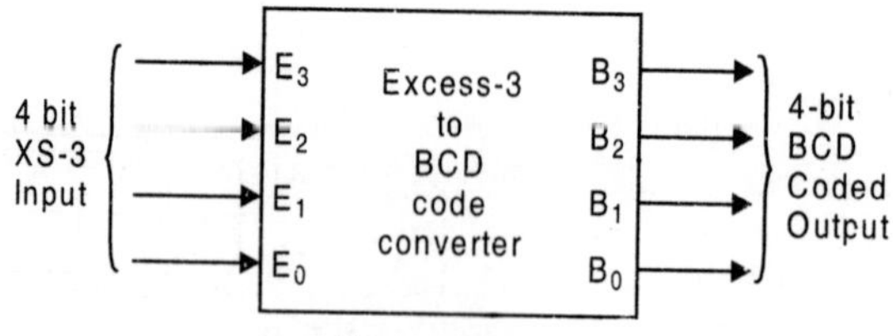

Fig. 4.15

2. Truth Table for XS-3 to BCD code converter.

   XS-3 codes are obtained from BCD code by adding 3 to each coded number. Moreover 4 binary variables may have 16 combinations, but only 10 are listed. The six not listed are don't care-combinations. Since they will never occur, we are at liberty to assign to the output variable either a 1 or a 0, whichever gives a simpler circuit. In this particular example, the unused i/o combinations are listed below the truth table.

Min Terms	*Excess-3 Inputs*				*BCD Outputs*				*Decimal Equivalent*
	E_3	E_2	E_1	E_0	B_3	B_2	B_1	B_0	
m_3	0	0	1	1	0	0	0	0	0
m_4	0	1	0	0	0	0	0	1	1
m_5	0	1	0	1	0	0	1	0	2
m_6	0	1	1	0	0	0	1	1	3
m_7	0	1	1	1	0	1	0	0	4
m_8	1	0	0	0	0	1	0	1	5
m_9	1	0	0	1	0	1	1	0	6
m_{10}	1	0	1	0	0	1	1	1	7
m_{11}	1	0	1	1	1	0	0	0	8
m_{12}	1	1	0	0	1	0	0	1	9
	Unused I/Ps				Outputs				
m_0	0	0	0	0	x	x	x	x	
m_1	0	0	0	1	x	x	x	x	
m_2	0	0	1	0	x	x	x	x	
m_{13}	1	1	0	1	x	x	x	x	
m_{14}	1	1	1	0	x	x	x	x	
m_{15}	1	1	1	1	x	x	x	x	

* XS-3 is also a class of BCD codes.

3. Now we solve all the BCD outputs. From the truth table, the logic expressions for the BCD coded outputs can be written as :

$B_3 = \Sigma\ (m_{11}, m_{12}),\ d\ (m_0, m_1, m_2, m_{13}, m_{14}, m_{15})$

$B_2 = \Sigma\ (m_7, m_8, m_9, m_{10}),\ d\ (m_0, m_1, m_2, m_{13}, m_{14}, m_{15})$

$B_1 = \Sigma\ (m_5, m_6, m_9, m_{10}),\ d\ (m_0, m_1, m_2, m_{13}, m_{14}, m_{15})$

$B_0 = \Sigma\ (m_4, m_6, m_8, m_{10}, m_{12}),\ d\ (m_0, m_1, m_2, m_{13}, m_{14}, m_{15})$.

These expressions can be simplified using k-map

Map for B_3

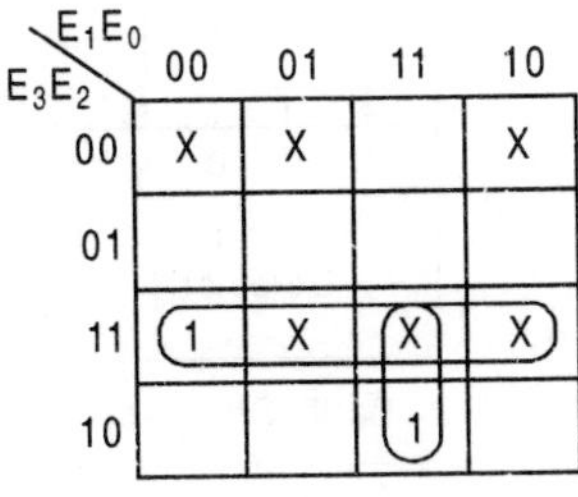

$$B_3 = E_3 E_2 + E_3 E_1 E_0$$

Map for B_2

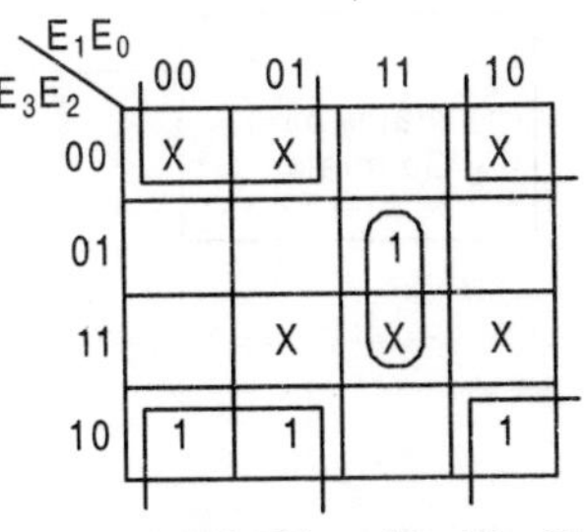

$$B_2 = E_2' E_0' + E_2 E_1 E_0 + E_2' E_1'$$

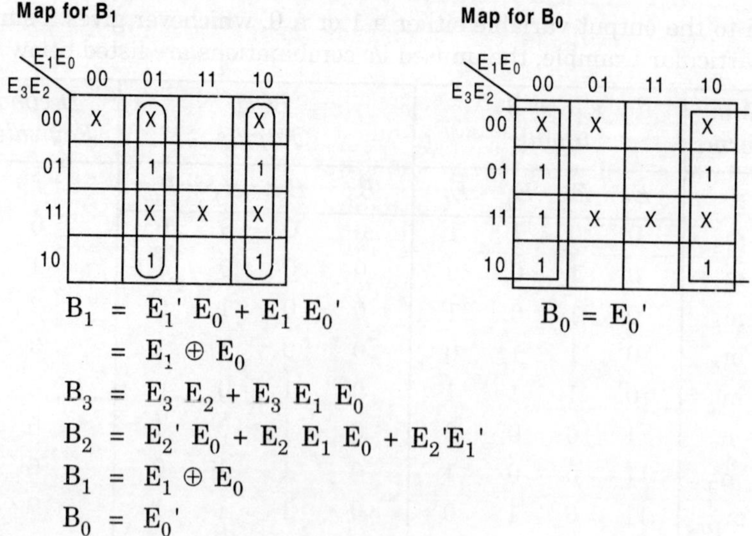

$$B_1 = E_1' E_0 + E_1 E_0'$$
$$= E_1 \oplus E_0$$

$$B_0 = E_0'$$

$$B_3 = E_3 E_2 + E_3 E_1 E_0$$
$$B_2 = E_2' E_0 + E_2 E_1 E_0 + E_2' E_1'$$
$$B_1 = E_1 \oplus E_0$$
$$B_0 = E_0'$$

4. The expressions for BCD outputs (B_3 B_2 B_1 B_0) can be implemented for terms of inputs (E_3 E_2 E_1 E_0) to form a XS-3 to BCD code converter circuit.

   The implementation is left as an exercise.

4.1.4 Parity Generators and Checkers

When digital data is transmitted from one location to another, it is necessary to know at the receiving end, whether the received data is free of error. To help make the transmission accurate, special error detection methods are used.

To detect errors, we must keep a constant check on the data being transmitted. To check accuracy we can generate and transmit an extra bit along with the message (data). This extra bit is known as the parity bit and it decides whether the data transmitted is error free or not. There are two types of parity bits, namely, even parity and odd parity that we have discussed in chapter 1 under error detecting codes.

Figure 4.16 shows an error detecting circuit using a parity bit.

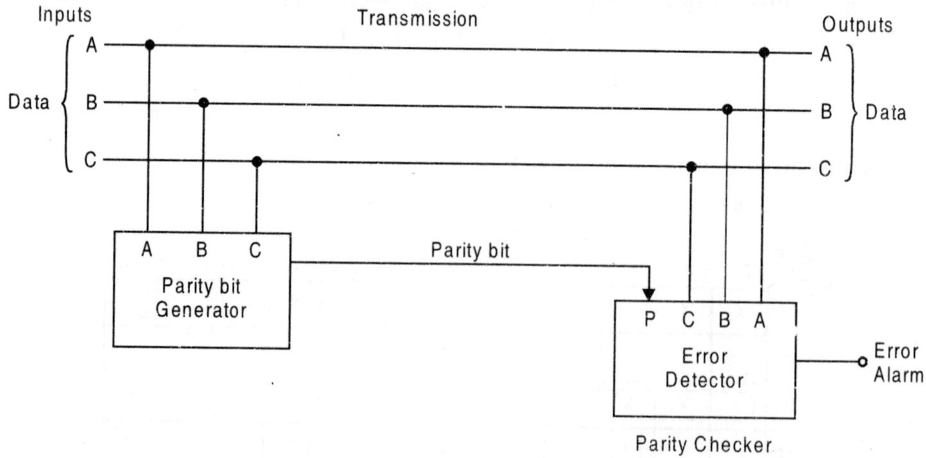

Fig. 4.16

In this system three parallel bits A, B and C are being transmitted over a long distance. Near the input they are fed into a parity bit generator circuit. This circuit generates what is called a parity bit. It may be either even or odd. For example, if it is a 3-bit even parity generator, the parity bit generated is such that it makes total number of 1s even. We can make a truth table of a 3-bit even parity generator circuit.

Truth Table for a 3-bit even parity generator.

Inputs Data			*Output* *Even parity bit*
A	*B*	*C*	*P*
0	0	0	0
0	0	1	1
0	1	0	1
0	1	1	0
1	0	0	1
1	0	1	0
1	1	0	0
1	1	1	1

Next, the truth table is converted to a logic circuit shown in Fig. 4.17.

$$
\begin{aligned}
P &= A'B'C + A'BC' + AB'C' + ABC \\
&= A'(B'C + BC') + A(B'C' + BC) \\
&= A'(B \oplus C) + A(BC' + B'C)' \\
&= A'(B \oplus C) + A(B \oplus C)' \\
&= A \oplus (B \oplus C) = (A \oplus B) \oplus C = A \oplus B \oplus C.
\end{aligned}
$$

A
B
C
Parity bit
P

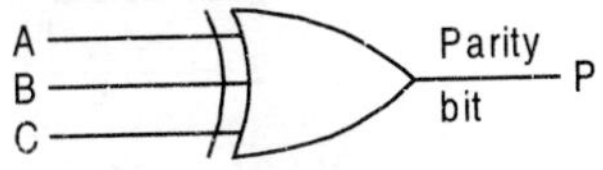

3-bit even parity generator circuit

Fig. 4.17

The generated parity bit is transmitted with the data and near the output it is fed to the error detector (parity checker) circuit. The detector circuit checks for the parity of transmitted data. As soon as the total number of 1's in the transmitted data are found 'odd' it sounds an alarm, indicating an error. If total number of 1's are even, no alarm sounds, indicating no error.

In above example we are transmitting 4 bits (3 bits of message plus 1 even parity bit). So, it is easy to understand that error detector is nothing but a 4 bit even-parity checker circuit. Figure 4.18 (*a*) shows a truth table of a 4 bit even parity checker circuit.

Inputs *Transmitted Data* *with parity bit*				*Outputs* *Parity error check*
A	*B*	*C*	*P*	*E*
0	0	0	0	0
0	0	0	1	1
0	0	1	0	1
0	0	1	1	0
0	1	0	0	1
0	1	0	1	0

(Contd.)

0	1	1	0	0
0	1	1	1	1
1	0	0	0	1
1	0	0	1	0
1	0	1	0	0
1	0	1	1	1
1	1	0	0	0
1	1	0	1	1
1	1	1	0	1
1	1	1	1	0

Fig. 4.18 (*a*)

Now, we convert this truth table into logic circuit shown in Fig. 4.18(*b*).

$$E = A'B'C'P + A'B'CP' + A'BC'P' + A'BCP + AB'C'P' + AB'CP + ABC'P + ABCP'$$

$$= A'B'\,(C \oplus P) + A'B\,(C\oplus P)' + AB'\,(C\oplus P)' + AB\,(C\oplus P)$$

$$= (C \oplus P)\,(A \oplus B)' + (C \oplus P)'\,(A \oplus B)$$

$$= (A \oplus B) \oplus (C \oplus P)$$

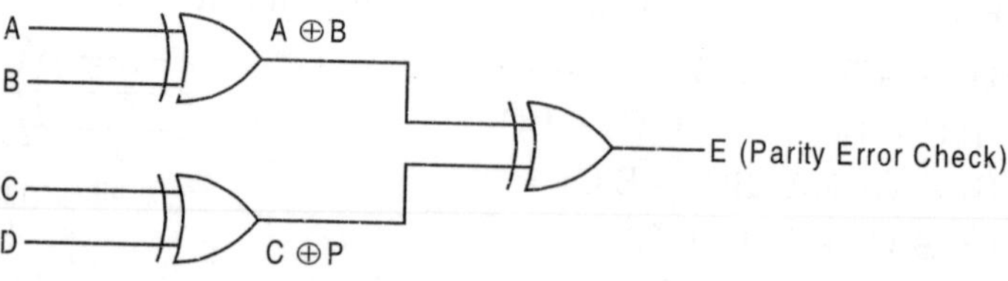

Fig. 4.18 (*b*) 4-bit even parity checker

If E = 1, Alarm sounds means error.

If E = 0, No alarm sounds means no error.

Now, it is possible to implement the parity generator with the circuit of parity checker. If the input P is connected to logic-0, causing the value of C to pass through the gate unchanged (because $C \oplus 0 = C$). The advantage of this is that the same circuit can be used for both parity generation and checking.

4.2 MSI AND LSI CIRCUITS

When designing logic circuits, the "discrete logic gates"; *i.e.*, individual AND, OR, NOT *etc.* gates, are often neither the simplest nor the most economical devices we could use. There are many standard MSI (medium scale integrated) and LSI (large scale integrated) circuits, or functions available, which can do many of the things commonly required in logic circuits. Often these MSI and LSI circuits do not fit our requirements exactly, and it is often necessary to use discrete logic to adapt these circuits for our application.

However, the number and type of these LSI and VLSI (very large scale integrated) circuits are steadily increasing, and it is difficult to always be aware of the best possible circuits available for a given problem. Also, systematic design methods are difficult to devise when the types of logic device available keeps increasing. **In general the "best" design procedure is to attempt to find a LSI device which can perform the required function, or which can be modified using other devices to perform the required function.** If nothing is available, then the function should be implemented with several MSI devices. Only as a last option should the entire function be implemented with discrete logic gates. In fact, with present technology, it is becoming increasingly cost-effective to implement a design as one or more dedicated VLSI devices.

When designing all but the simplest logic devices, a "top-down" approach should be adopted. The device should be specified in block form, and attempt to implement each block with a small number of LSI or MSI functions. Each block which cannot be implemented directly can be then broken into smaller blocks, and the process repeated, until each block is fully implemented.

Of course, **a good knowledge of what LSI and MSI functions are available in the appropriate technology makes this process simpler.**

4.2.1 The Digital Multiplexers

One MSI function which has been available for a long time is the digital selector, or multiplexer. It is the digital equivalent of the rotary switch or selector switch (*e.g.*, the channel selector on a TV set). Its function is to accept a binary number as a "selector input," and present the logic level connected to that input line as output from the data selector.

A digital multiplexer (MUX) is a combinational circuit that selects one input out of several inputs and direct it to a single output. The particular input selection is controlled by a set of select inputs. Figure 4.19 shows block diagram of a digital multiplexer with n inputs lines and single output line.

For selecting one out of n input, a set of m select inputs is required where

$$n = 2^m$$

On the basis of digital (binary) code applied at the select inputs, one output of n data sources is selected. Usually, an enable (or strobe) input (E) is built-in for cascading purpose. Enable input is generally active-low, *i.e.*, it performs its intended operation when it is low (logic).

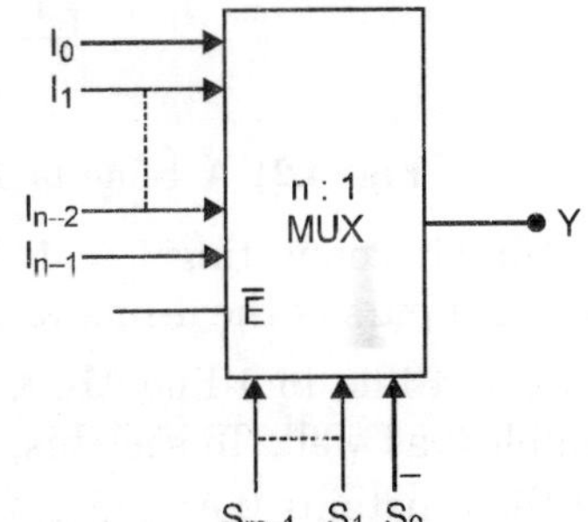

Fig. 4.19 Block diagram of the digital multiplexer

Note. 16:1 are the largest available ICs, therefore for larger input requirements there should be provision for expansion. This is achieved through enable/strobe input (multiplexer stacks or trees are designed).

A circuit diagram for a possible 4-line to 1-line data selector/multiplexer (abbreviated as MUX for multiplexer) is shown in Fig. 4.20. Here, the output Y is equal to the input I_0, I_1, I_2, I_3 depending on whether the select lines S_1 and S_0 have values 00, 01, 10, 11 for S_1 and S_0 respectively. That is, the output Y is *selected* to be equal to the input of the line given by the binary value of the select lines S_1S_0.

The logic equation for the circuit shown in Fig. 4.20 is:

$$Y = I_0.\overline{S}_1.\overline{S}_0 + I_1.\overline{S}_1.S_0 + I_2.S_1.\overline{S}_0 + I_3.S_1.S_0$$

This device can be used simply as a data selector/multiplexer, or it can be used to perform logic functions. Its simplest application is to implement a truth table directly, *e.g.*, with a 4 line to 1 line MUX, it is possible to implement any 2-variable function directly, simply by connecting I_0, I_1, I_2, I_3 to logic 1 in logic 0, as dictated by a truth table. In this way, a MUX can be used as a simple look-up table for switching functions. This facility makes the MUX a very general purpose logic device.

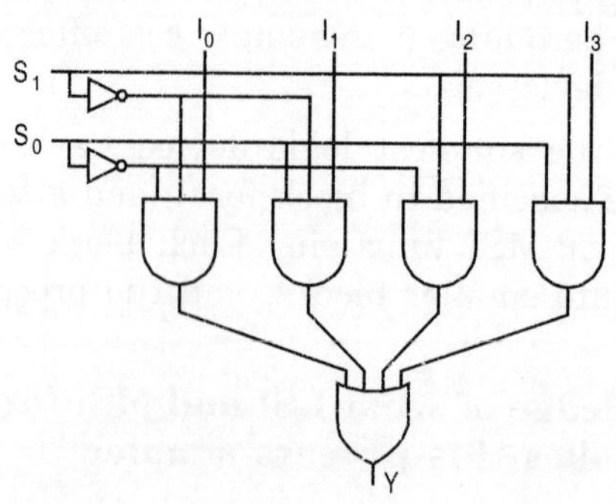

Fig 4.20 A 4-line to 1-line multiplexer

Example 1. Use a 4 line to 1 line MUX to implement the function shown in the following truth table $(Y = \overline{A}.\overline{B} + A.B)$.

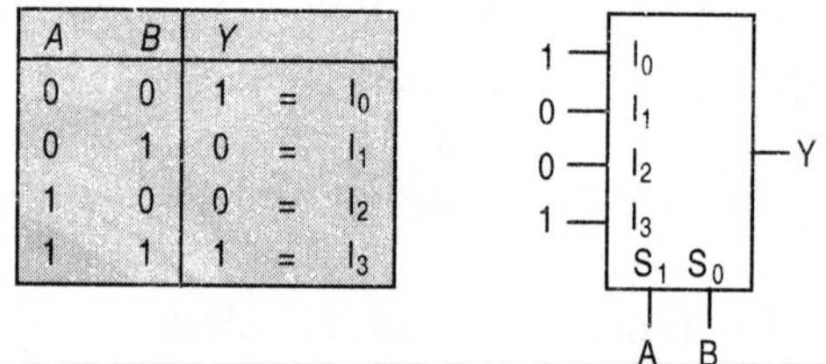

A	B	Y		
0	0	1	=	I_0
0	1	0	=	I_1
1	0	0	=	I_2
1	1	1	=	I_3

Fig. 4.21 A 4-line to 1-line MUX implementation of a function of 2 variables

Simply connecting $I_0 = 1$, $I_1 = 0$, $I_2 = 0$, $I_3 = 1$, and the inputs A and B to the S_1 and S_0 selector inputs of the 4-line to 1-line MUX implement this truth table, as shown in Fig. 4.21.

The 4-line to 1-line MUX can also be used to implement any function of three logical variables, as well. To see this, we need only possible functions of one variable C, are C, $\overline{C}$, and the constants 0 or 1 (*i.e.*, C, $\overline{C}$, $C + \overline{C} = 1$, and 0). We need only connect the appropriate value, C, $\overline{C}$, 0 or 1, to I_0, I_1, I_2, I_3 to obtain a function of 3 variables. The MUX still behaves as a table lookup device; it is now simply looking up values of another variable.

Example 2. Implement the function

$$Y(A, B, C) = \overline{A}.\overline{B}.C + \overline{A}.B.\overline{C} + A.B.C + A.B.C$$

Using a 4-line to 1-line MUX.

Here, again, we use the A and B variables as data select inputs. We can use the above equation to construct the table shown in Fig. 4.22. The residues are what is "left over" in each minterm when the "address" variables are taken away. To implement this circuit, we connect I_0 and I_3 to C, and I_1 and I_2 to $\overline{C}$, as shown in Fig. 4.22.

Input	"Address"	Other variables (residues)
I_0	$\bar{A}.\bar{B}$	C
I_1	$\bar{A}.B$	$\bar{C}$
I_2	$A.\bar{B}$	$\bar{C}$
I_3	$A.B$	C

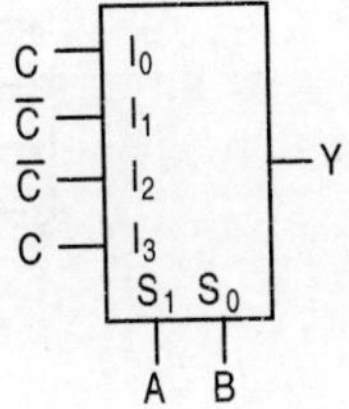

Fig. 4.22 A 4-line to 1-line MUX implementation of a function of 3 variables

In general a 4 input MUX can give any function of 3 inputs, an 8 input MUX can give any function of 4 variables, and a 16 input MUX, any function of 5 variables.

Example 3. Use an 8 input MUX to implement the following equation:

$$Y = \bar{A}.\bar{B}.\bar{C}.\bar{D} + \bar{A}.\bar{B}.C.D + \bar{A}.B.\bar{C}.D + \bar{A}.B.\bar{C}.\bar{D} + A.\bar{B}.\bar{C}.D + A.\bar{B}.C.\bar{D} + A.B.\bar{C}.\bar{D} + A.B.\bar{C}.D$$

Again, we will use A, B, C as data select inputs, or address inputs, connected to S_2, S_1 and S_0, respectively.

Input	"Address"	Residues
I_0	$\bar{A}.\bar{B}.\bar{C}$	$\bar{D}$
I_1	$\bar{A}.\bar{B}.C$	D
I_2	$\bar{A}.B.\bar{C}$	$D + \bar{D} = 1$
I_3	$\bar{A}.B.C$	
I_4	$A.\bar{B}.\bar{C}$	D
I_5	$A.\bar{B}.C$	$\bar{D}$
I_6	$A.B.\bar{C}$	$\bar{D} + D = 1$
I_7	$A.B.C$	

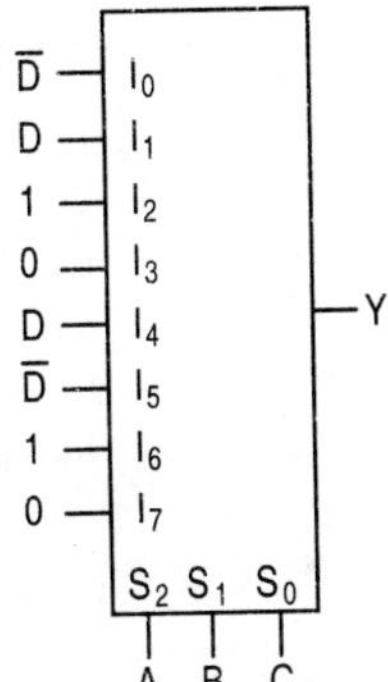

Fig. 4.23 An 8-line to 1-line MUX implementation of a function of 4 variables

Values of the address set A, B, C with no residues corresponding to the address in the above table must have logic value 0 connected to the corresponding data input. The select variables A, B, C must be connected to S_2, S_1 and S_0 respectively. A circuit which implements this function is shown in Fig. 4.23.

This use of a MUX as a "table look up" device can be extended to functions of a larger number of variables; the MUX effectively removes the terms involving the variables assigned to its select inputs from the logic expression. This can sometimes be an effective way to reduce the complexity of implementation of a function. For complex functions, however, there are often better implementations, as we use PLDs (see chapter 5).

Although it is obvious how the function shown in Fig. 4.20 can be extended a 2^n line to 1 line MUX, for any n, in practice, about the largest devices available are only to 16 line to 1 line functions. It is possible to use a "tree" of smaller MUX's to make arbitrarily large MUX's. Figure 4.24 shows an implementation of a 16 line to 1 line MUX using five 4 line to 1 line MUX's.

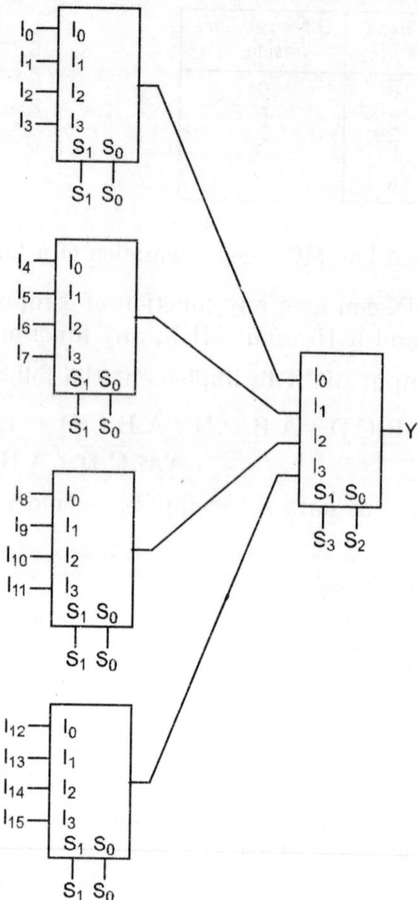

Fig. 4.24 A 16-line to 1-line MUX made from five 4-line to 1-line MUX's

4.2.2 Decoders (Demultiplexers)

Another commonly used MSI device is the decoder. Decoders, in general, transform a set of inputs into a different set of outputs, which are coded in a particular manner; *e.g.*, certain decoders are designed to decode binary or BCD coded numbers and produce the correct output to display a digit on a 7 segment (calculator type) display. Decoders are also available to convert numbers from binary to BCD, from binary to hexadecimal, etc.

Normally, however, the term "decoder" implies a device which performs, in a sense, the inverse operation of a multiplexer. A decoder accepts an n digit number as its n "select" inputs and produces an output (usually a logic 0) at one of its 2^n possible outputs. Decoders are usually referred to as n line to 2^n line decoders; *e.g.* a 3 line to 8 line decoder. This type of decoder is really a binary to unary number system decoder. Most decoders have inverted outputs, so the selected output is set to logic 0, while all the other outputs remain at logic 1. As well, most decoders have an "enable" input $\overline{E}$, which "enables" the operation of the decoder—when the $\overline{E}$ input is set to 0, the device behaves as a decoder and selects the output determined by the select inputs; when the $\overline{E}$ input is set to 1, the outputs of the decoder are *all* set to 1. (The bar over the E indicates that it is an "active low" input; that is, a logic 0 enables the function).

The enable input also allows decoders to be connected together in a tree like fashion, much as we saw for MUX's, so large decoders can be easily constructed from smaller devices. The enable input also allows the decoder to perform the inverse operation of a MUX; a MUX selects as output one of 2^n inputs, the decoder can be used to present an input to one of 2^n outputs, simply by connecting the input signal to the $\overline{E}$ input; the signal at the selected output will then be the same as the input at $\overline{E}$—this application is called "demultiplexing." The demulti-plexer (DEMUX) performs the reverse operation of a multiplexer. A demultiplexer is a circuit that accepts single input and transmit it over several (one of 2^n possible) outputs.

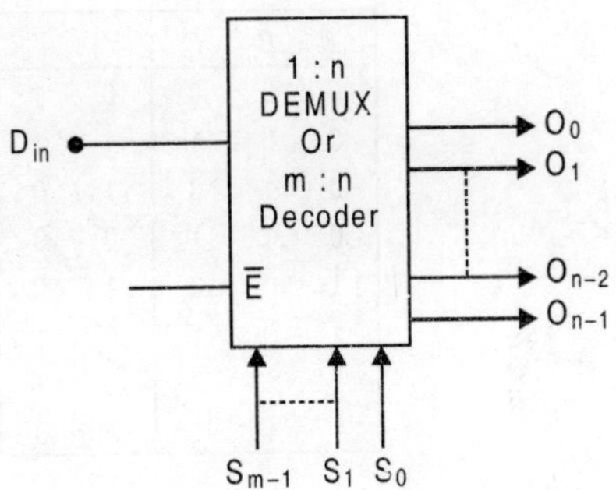

Fig. 4.25 Block diagram of the demultiplexer/decoder

In the block diagram (Fig. 4.25) a demultiplexer, the number of output lines is n and the number of selected lines is m, where $n = 2^m$.

On the basis of select input code, to which the output data will be transmitted is determined. There is an active-low (low-logic) enable/data input. The output for these devices are also active-low.

Note. 4-line to 16-line decoders are the largest available circuits in ICs.

A typical 3 line to 8 line decoder with an enable input behaves according to the following truth table, and has a circuit symbol as shown in Fig. 4.26.

E	S_2	S_1	S_0	O_0	O_1	O_2	O_3	O_4	O_5	O_6	O_7
1	x	x	x	1	1	1	1	1	1	1	1
0	0	0	0	0	1	1	1	1	1	1	1
0	0	0	1	1	0	1	1	1	1	1	1
0	0	1	0	1	1	0	1	1	1	1	1
0	0	1	1	1	1	1	0	1	1	1	1
0	1	0	0	1	1	1	1	0	1	1	1
0	1	0	1	1	1	1	1	1	0	1	1
0	1	1	0	1	1	1	1	1	1	0	1
0	1	1	1	1	1	1	1	1	1	1	0

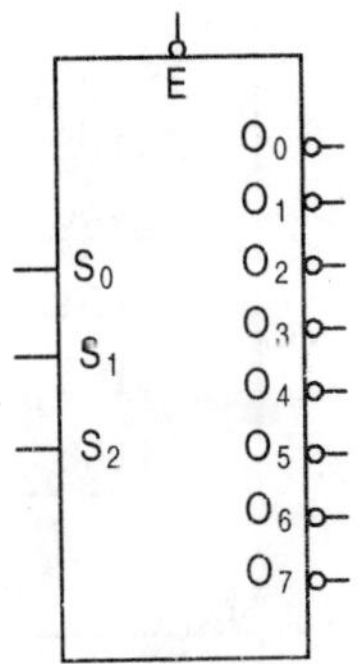

Fig. 4.26 A 3-line to 8-line decoder

Note that, when the $\overline{E}$ input is enabled, an output of 0 is produced corresponding to each minterm of S_2, S_1, S_0. These minterm can be combined together using other logic gates to form any required logic function of the input variables. In fact, several functions can be produced at the same time. If the selected output was a logic 1, then the required minterms could simply be ORed together to implement a switching function directly from its minterm form. Using de Morgans theorem, we can see that when the outputs are inverted, as is normally the case, then the minterm form of the function can be obtained by NANDing the required terms together.

Example 1. An implementation of the functions defined by the following truth table using a decoder and NAND gates is shown in Fig. 4.27.

A	B	C	Y_1	Y_2
0	0	0	0	1
0	0	1	1	1
0	1	0	1	0
0	1	1	0	0
1	0	0	1	0
1	0	1	0	1
1	1	0	0	1
1	1	1	0	0

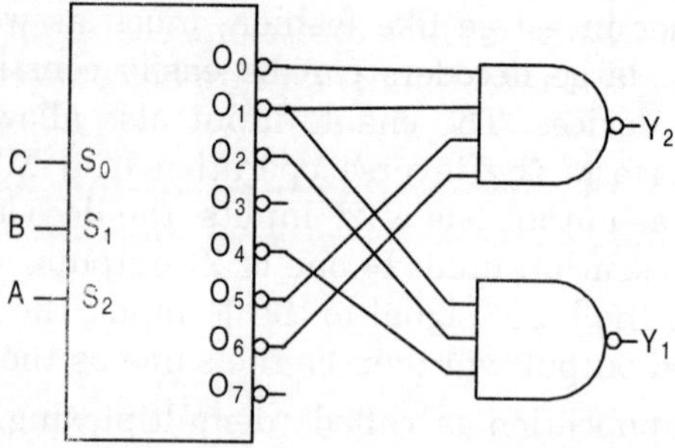

Fig. 4.27

IMPLEMENTATION EXAMPLES OF COMBINATIONAL LOGIC DESIGN USING MUX/DEMUX

We have already seen how to implement combinational circuits using MUX/DEMUX. The standard ICs available for multiplexers are 2:1, 4:1, 8:1 and 16:1. The different digital ICs are given in appendix B, but for sake of convenience some of the MUX/DEMUX ICs are given here in Tables A and B.

Table A: Standard Multiplexer ICs

IC No.	*Description*	*Output*
74157	Quad. 2:1 Multiplexer	Same as input
74158	Quad 2:1 MUX	Inverted input
74153	Dual 4:1 MUX	Same as input
74352	Dual 4:1 MUX	Inverted input
74151A	8:1 MUX	Complementary outputs
74152	8:1 MUX	Inverted input
74150	16:1 MUX	Inverted input

Table B: Standard Demultiplexer/Decoder ICs

IC No.	*Description*	*Output*
74139	Dual 1:4 Demultiplexer (2-line-to-4-line decoder)	Inverted input
74155	Dual 1:4 Demultiplexer (2-line-to-4-line decoder)	1Y-Inverted I/P 2Y-Same as I/P
74138	1:8 Demultiplexer (3-line-to-8-line decoder)	Inverted I/P
74154	1:16 Demultiplexer (4-line-to-16-line decoder)	Same as input

When using the multiplexer as a logic element either the truth table or one of the standard forms of logic expression must be available. The design procedure for combinational circuits using MUX are as follows:

STEP 1: Identify the decimal number corresponding to each minterm in the expression. The input lines corresponding to these numbers are to be connected to logic 1 (high).

STEP 2 : All other input lines except that used in step 1 are to be connected to logic 0 (low).

STEP 3 : The control inputs are to be applied to select inputs.

Example 2. *Implement the following function with multiplexer.*

$$Y = F(A, B, C, D) = \Sigma m\ (0, 1, 3, 4, 8, 9, 15)$$

Solution. STEP 1 : The input lines corresponding to each minterms (decimal number) are to be connected to logic 1.

Therefore input lines 0, 1, 3, 4, 8, 9, 15 have to be connected to logic 1.

STEP 2 : All other input lines except 0, 1, 3, 4, 8, 9, 15 are to be connected to logic 0.

STEP 3 : The control inputs A, B, C, D are to be applied to select inputs.

Note: Although the given procedure is simple to implement but the 16 to 1 multiplexers are the largest available ICs, therefore to meet the larger input needs there should be provision for expansion. This is achieved with the help of enable/stroke inputs and multiplexer stacks or trees are designed.

Example 3. *Implement the following function with a 4×1 multiplexer.*

$$Y = F(A, B, C) = \Sigma m\ (1, 3, 5, 6)$$

Solution. Given

$$Y = F(A, B, C) = \Sigma m\ (1, 3, 5, 6)$$

$$= \overline{A}\,\overline{B}C + \overline{A}BC + A\overline{B}C + AB\overline{C}$$

We use the A and B variables as data select inputs. We can use the above equation to construct the table shown in Fig. 4.28. The residues are what is "left over" in each minterm when the "address" variables are taken away.

Input	*"Address"*	*Other variables (residues)*
I_0	$\overline{A}\,\overline{B}$	C
I_1	$\overline{A}B$	C
I_2	$A\overline{B}$	C
I_3	AB	$\overline{C}$

Fig. 4.28 A 4-line to 1-line MUX implementation of a function of 3 variables

To implement this circuit, we connect I_0, I_1 and I_2 to C and I_3 to $\overline{C}$ as shown in Fig. 4.28.

Example 4. *Using four-input multiplexer, implement the following function*

$$Y = F(A, B, C) = \Sigma m\ (0, 2, 3, 5, 7)$$

Control variables A, B.

Solution. Given

$$Y = F(A, B, C) = \Sigma m\,(0, 2, 3, 5, 7)$$
$$= \overline{A}\,\overline{B}\,\overline{C} + \overline{A}B\overline{C} + \overline{A}BC + A\overline{B}C + ABC$$

We can use the above equation to construct the table shown in Fig. 4.29. The residues are what is "left over" in each minterm when the "address/control" variables are taken away.

Input	*"Address"*	*Other variables (residues)*
I_0	$\overline{A}\,\overline{B}$	$\overline{C}$
I_1	$\overline{A}B$	$C + \overline{C} = 1$
I_2	$A\overline{B}$	C
I_3	AB	C

Fig. 4.29 A 4-line to 1-line MUX implementation of a function of 3 variables

To implement this function, we connect I_0 to $\overline{C}$, I_1 to 1 and I_2 and I_3 to C, as shown in Fig. 4.29.

Example 5. *Design a full adder using 8:1 multiplexer.*

Solution. The truth table of a full adder is given as

A	*B*	*C*	*S*	C_F
0	0	0	0	0
0	0	1	1	0
0	1	0	1	0
0	1	1	0	1
1	0	0	1	0
1	0	1	0	1
1	1	0	0	1
1	1	1	1	1

$$S(A, B, C) = \overline{A}\,\overline{B}C + \overline{A}B\overline{C} + A\overline{B}\,\overline{C} + ABC = \Sigma m\,(1, 2, 4, 7)$$

$$C_F(A, B, C) = \overline{A}BC + A\overline{B}C + AB\overline{C} + ABC = \Sigma m\,(3, 5, 6, 7)$$

The implementation for summation expression is

Step 1: The input lines corresponding to 1, 2, 4, 7 are to be connected to logic 1.

Step 2: Other input lines are to be connected to logic 0.

Step 3: Control inputs A, B, C are to be applied to select inputs. Fig. 4.30 A.

Similarly for carry expression.

Step 1: The input lines corresponding to 3, 5, 6, 7 are to be connected to logic 1.

Step 2: Other input lines are to be connected to logic 0.

Step 3: Control inputs A, B, C are to be applied to select inputs. Fig. 4.30 B.

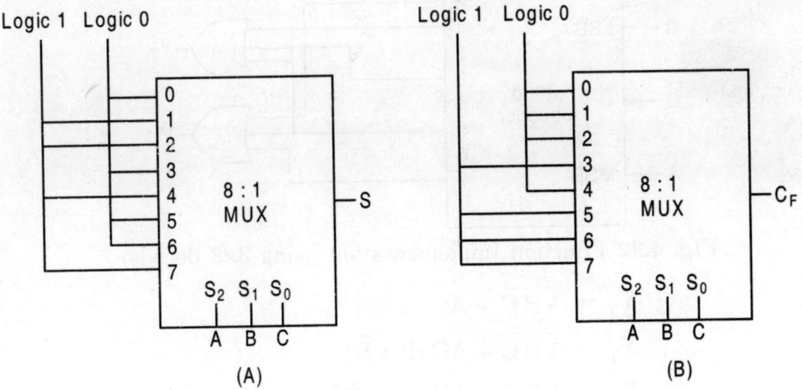

Fig. 4.30 Full adder implementation using 8:1 multiplexer

Example 6. *Implement a full adder with a decoder and two OR-gates.*

Solution. From the previous example we note that expression for summation is given by

$$S\ (A, B, C) = \Sigma m\ (1, 2, 4, 7)$$

and expression for carry is given by

$$C_F\ (A, B, C) = \Sigma m\ (3, 5, 6, 7)$$

The combinational logic of full adder can be implemented with due help of 3-line to 8-line decoder/1:8 demultiplexer as shown in Fig. 4.31.

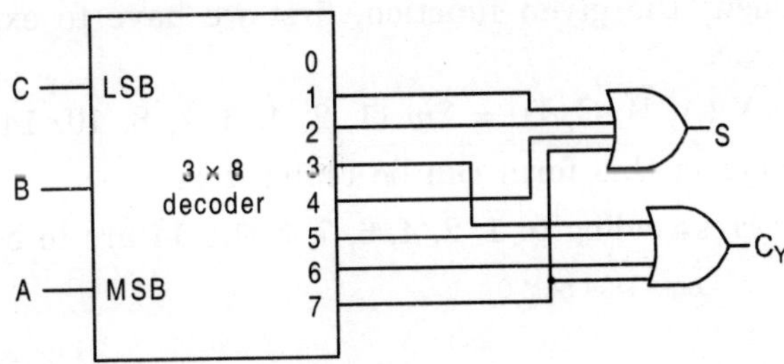

Fig. 4.31 Full adder implementation using 3 × 8 decoder

Example 7. *A combinational circuit is defined by the following Boolean functions. Design circuit with a decoder and external gates.*

Solution. $Y_1 = F_1(A, B, C) = \overline{A}\,\overline{B}\,\overline{C} + AC$

$Y_2 = F_2(A, B, C) = A\overline{B}C + \overline{A}C$

Given $Y_1 = \overline{A}\,\overline{B}\,\overline{C} + AC$

First we have to write the expression in minterms, if the expression is not in the form of minterms by using $(x + \overline{x} = 1)$

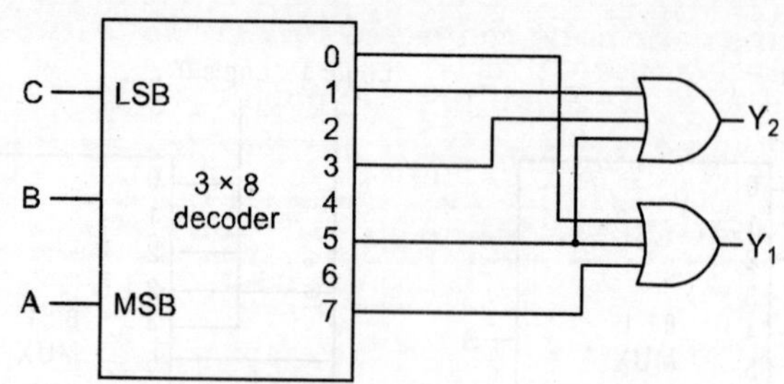

Fig. 4.32 Function implementation using 3×8 decoder

Therefore,

$$Y_1 = \overline{A}\,\overline{B}\,\overline{C} + AC$$

$$Y_1 = \overline{A}\,\overline{B}\,\overline{C} + AC\,(B+\overline{B})$$

$$Y_1 = \overline{A}\,\overline{B}\,\overline{C} + ABC + A\overline{B}C$$

$$Y_1 = \Sigma m\ (0, 5, 7)$$

$$Y_2 = A\overline{B}C + \overline{A}C$$

$$Y_2 = A\overline{B}C + \overline{A}C\,(B+\overline{B})$$

$$Y_2 = A\overline{B}C + \overline{A}BC + \overline{A}\,\overline{B}C$$

$$Y_2 = \Sigma m\ (1, 3, 5)$$

The combinational logic for the boolean function can be implemented with the help of 3-line to 8-line decoder as shown in Fig. 4.32.

Example 8. *Realise the given function using a multiplexer*

$$Y(A, B, C, D) = \Pi M\ (0, 3, 5, 9, 11, 12, 13, 15)$$

Solution. To implement the given function, first we have to express the function in terms of sum of product. *i.e.,*

$$Y\ (A, B, C, D) = \Sigma m\ (1, 2, 4, 6, 7, 8, 10, 14)$$

Now the given function in this form can be realized as

Step 1: Input lines corresponding to 1, 2, 4, 6, 7, 8, 10, 14 are to be connected to logic 1.

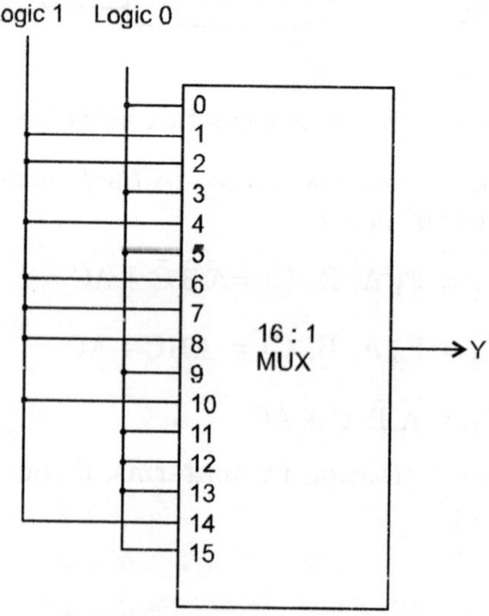

Fig. 4.33 A 16-line to 1-line MUX implementation

Step 2: Other input lines are to be connected to logic 0.

Step 3: Control inputs A, B, C, D are to be applied to select inputs.

Example 9. *Realize the following boolean expression using 4:1 MUX(S) only.*

$$Z = \overline{A}\,\overline{B}\,\overline{C}\,\overline{D} + \overline{A}BC\overline{D} + A\overline{B}C\overline{D} + A\overline{B}C\overline{D} + A\overline{B}\,\overline{C}\,\overline{D} + ABCD$$

Solution. Given $Z = \Sigma m\ (0, 6, 8, 10, 11, 15)$

To implement the given boolean expression we must have 16 input and 4 selection inputs.

Since 4:1 mux has 4 input lines and two selection lines. Therefore we can use 4, 4:1 MUX with their select lines connected together. This is followed by a 4:1 MUX to select one of the four outputs. The select lines of the 4:1 MUX (final) are driven from inputs A, B. The complete circuit is shown in Fig. 4.34.

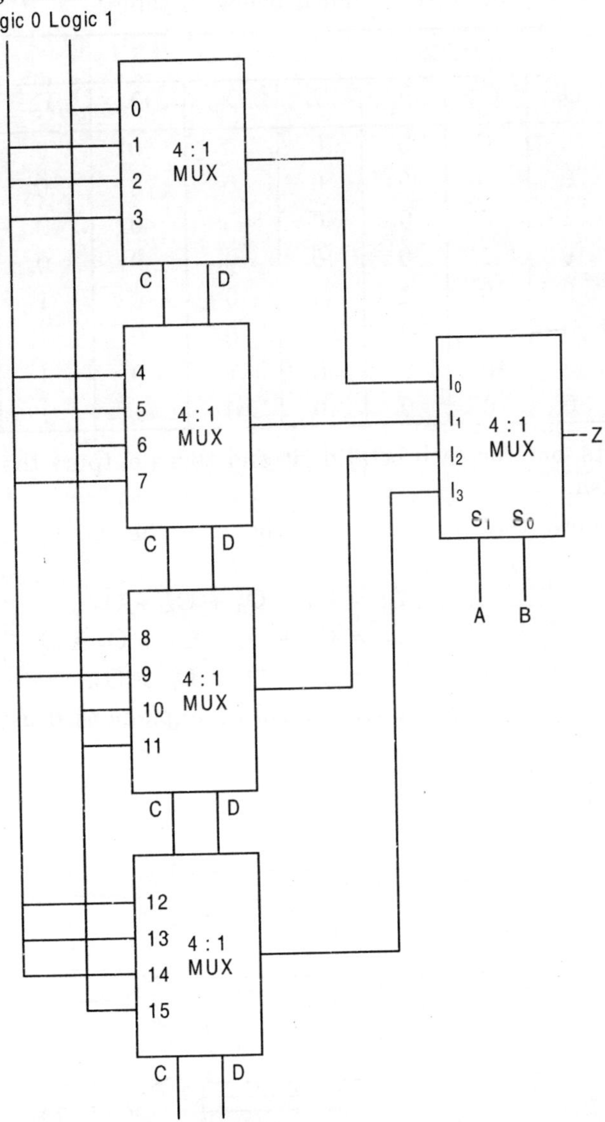

Fig. 4.34 A 4-line to 1-line MUX implementation of a function of 4 variable

4.2.3 Encoders

The encoder is another example of combinational circuit that performs the inverse operation of a decoder. It is designed to generate a different output code for each input which becomes active. In general, the encoder is a circuit with m input lines $(m \leq 2^n)$* (* $m < 2^n \rightarrow$ If unused input combinations occur) and n output lines that converts an active input signal into a coded output signal. In an encoder, the number of outputs is less than the number of inputs. The block diagram of an encoder is shown in Fig. 4.35.

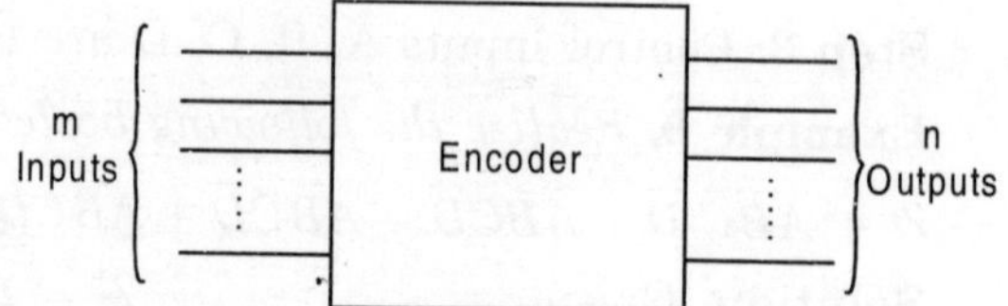

Fig. 4.35 Block diagram of an encoder

An example of an encoder is an octal to binary encoder. An octal to binary encoder accept eight inputs and produces a 3-bit output code corresponding to the activated input. The truth table for the octal to binary encoder is shown below in table.

Inputs								*Outputs*		
O_0	O_1	O_2	O_3	O_4	O_5	O_6	O_7	Y_2	Y_1	Y_0
1	0	0	0	0	0	0	0	0	0	0
0	1	0	0	0	0	0	0	0	0	1
0	0	1	0	0	0	0	0	0	1	0
0	0	0	1	0	0	0	0	0	1	1
0	0	0	0	1	0	0	0	1	0	0
0	0	0	0	0	1	0	0	1	0	1
0	0	0	0	0	0	1	0	1	1	0
0	0	0	0	0	0	0	1	1	1	1

It has eight inputs, one for each octal digit and three outputs that generate the corresponding binary number.

The truth table shows that Y_0 must be 1 whenever the input O_1 or O_3 or O_5 or O_7 is high. Thus,

$$Y_0 = O_1 + O_3 + O_5 + O_7$$

Similarly

$$Y_1 = O_2 + O_3 + O_6 + O_7 \text{ and}$$

$$Y_2 = O_4 + O_5 + O_6 + O_7.$$

Using these three expressions, the circuit can be implemented using three 4-input OR gates as shown in Fig. 4.36.

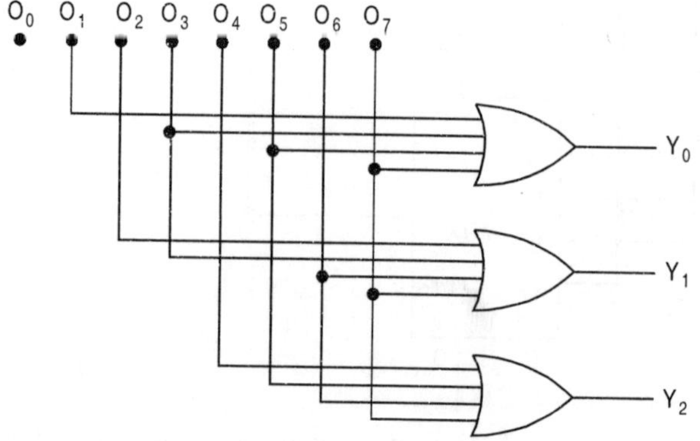

Fig. 4.36 Octal to binary encoder

The encoder has two limitations:

1. Only one input can be active at any given time. If two or more inputs are equal to 1 at the same time, the O/P is undefined. For example if O_2 and O_5 are active simultaneously, the o/p of encoder will be 111 that is equal to binary 7. This does not represent binary 2 or 5.
2. The output with all 0's is generated when all inputs are '0', and is also true when O_0 = '1'.

The first problem is taken care by a circuit, called as 'priority encoder'. It establishes a priority to ensure that only one input is active (High) at a given time.

The second problem is taken care by an extra line in the encoder output, called 'valid output indicator' that specifies the condition that none of the inputs are active.

Priority Encoder

A priority encoder is an encoder that includes priority function. If two or more inputs are equal to 1 at the same time, the input having the highest priority will take precedence. To understand priority encoder, consider a 4 to 2 line encoder which gives priority to higher subscript number input than lower subscript number. The truth table is given below.

Truth Table of 4 to 2 line priority encoder:

Inputs				*Outputs*		
D_0	D_1	D_2	D_3	Y_1	Y_2	V
0	0	0	0	x	x	0
1	0	0	0	0	0	1
x	1	0	0	0	1	1
x	x	1	0	1	0	1
x	x	x	1	1	1	1

The X's are don't care conditions. Input D_3 has the highest priority, so regardless of values of other inputs, when this input is 1, the output Y_1 Y_2 = 11. D_2 has next priority level. The o/p is 10 if D_2 is 1, provided D_3 = 0, irrespective of the values of the other two lower-priority inputs. The o/p is 01 if D_1 is 1, provided both D_2 and D_3 are 0, irrespective of the value of lower-priority input D_0. The o/p is 00 if D_0 = 1, provided all other inputs are 0.

A valid output indicator, V is set to 1, only when one or more of the inputs are equal to 1. If all the inputs are 0, V is equal to 0 and the other two outputs of the circuit are not used.

Now, simplifying using k-map the outputs can be written as :

$$Y_1 = D_2 + D_3$$

$$Y_2 = D_3 + D_1 D_2'$$

$$V = D_0 + D_1 + D_2 + D_3.$$

The logic diagram for a 4 to 2 line priority encoder with 'valid output indicator' is shown below in Fig. 4.37.

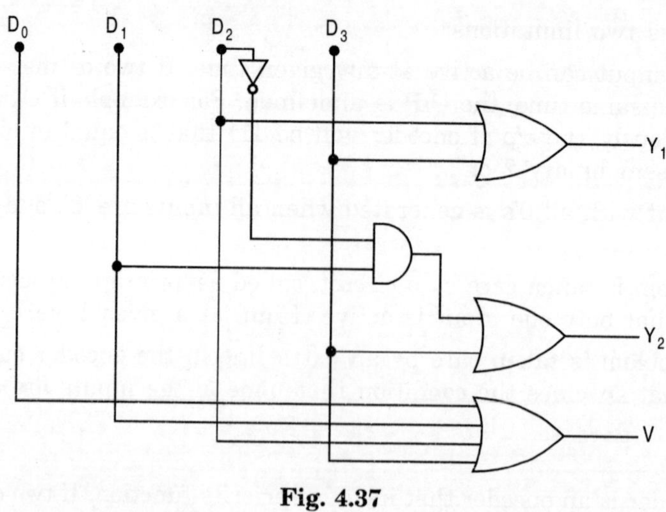

Fig. 4.37

4.2.4 Serial and Parallel Adders

In section 4.1.1 we have discussed the full-adder circuit. Full adder is a combinational circuit that adds three binary digits. When we add two numbers of any length, the terms we have to deal with are :

Input carry, Augend, Addend, sum and output carry. We simply start adding two binary digits from LSB (rightmost positioned bits). At this position, the input carry is always equal to zero. After addition, we get sum and output carry. This output carry works as the input carry to the next higher positioned augend and addend bits. Next we add augend and addend bits alongwith the input carry that again produces sum and output carry. The process repeats upto MSB position (leftmost positioned bits).

We observe that in the process of addition we are actually adding three digits – the input carry, the augend bit and the addend bit. And, we are getting two outputs the sum and the output carry.

This can be illustrated by the following example. Let the 4-bits words to be added be represented by:

$A_3\ A_2\ A_1\ A_0 = 1\ 1\ 0\ 1$ and $B_3\ B_2\ B_1\ B_0 = 0\ 0\ 1\ 1$.

Significant Place	3	2	1	0	
Input Carry	1	1	1	0	← Carry-In
Augend Word	1	1	0	1	
Addend Word	0	0	1	1	
Sum	0	0	0	0	
Output carry	1	1	1	1	

Carry-out

Now, if we compare this with the full adder circuit, we can easily observe that the two inputs (A and B) are augend and addend bits with the third input (c) as the input carry. Similarly two outputs are sum (s) and output carry (C_0).

In general, the sum of two *n*-bit numbers can be generated by using either of the two methods : the serial addition and the parallel addition.

Serial Adder

In serial addition, the addition operation is carried out bit by bit. The serial adder uses one full adder circuit and some storage device (memory element) to hold generated output carry. The diagram of a 4 bit serial adder is shown in Fig. 4.38.

The two bits at the same positions in augend and addend word are applied serially to A and B inputs of the full adder respectively. The single full adder is used to add one pair of bits at a time along with the carry C_{in}. The memory element is used to store the carry output of the full adder circuit so that it can be added to the next significant position of the numbers in the augend and addend word. This produces a string of output bits for sum as S_0, S_1, S_2 and S_3 respectively.

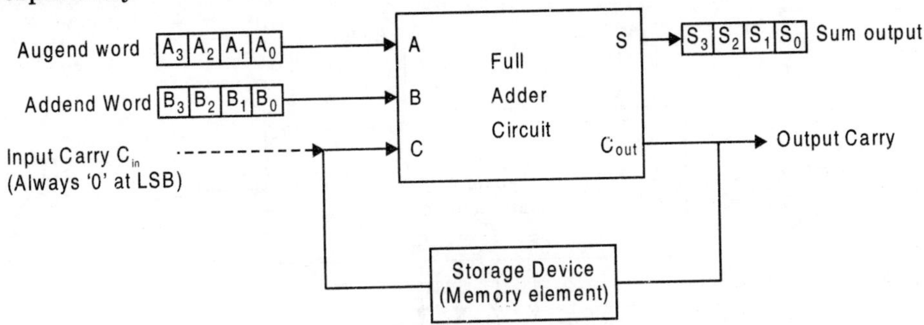

Fig. 4.38 4-bit serial adder

Parallel Adder

To add two n-bit numbers, the parallel method uses n full adder circuits and all bits of addend and augend are applied simultaneously. The output carry from one full adder is connected to the input carry of the full adder one position to its left.

The 4-bit adder using full adder circuit is capable of adding two 4-bit numbers resulting in a 4-bit sum and a carry output as shown in Fig. 4.39.

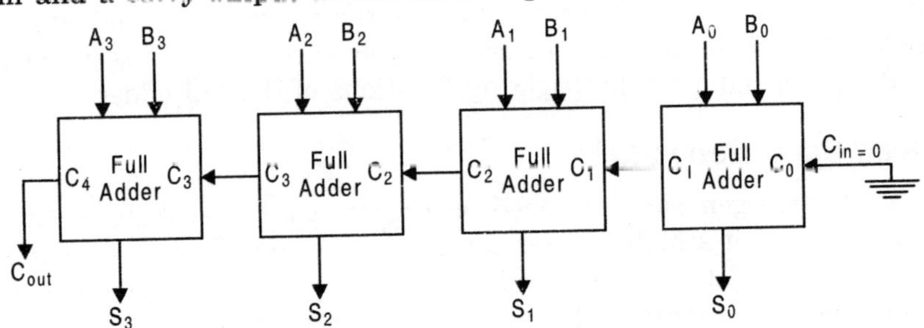

Fig. 4.39 4-bit binary parallel adder

The addition operation is illustrated in the following example. Let the 4-bit words to be added be represented by $A_3\ A_2\ A_1\ A_0$ = 1 0 1 0 and $B_3\ B_2\ B_1\ B_0$ = 0 0 1 1.

Subscript i	*3*	*2*	*1*	*0*
Input carry C_i	0	1	0	0
Augend A_i	1	0	1	0
Addend B_i	0	0	1	1
Sum S_i	1	1	0	1
Output carry C_{i+1}	0	0	1	0

← Significant place.

In a 4-bit parallel adder, the input to each full adder will be A_i, B_i and C_i, and the outputs will be S_i and C_{i+1}, where i varies from 0 to 3.

In the least significant stage, A_0, B_0 and C_0 (which is 0) are added resulting in sum S_0 and carry C_1. This carry C_1 becomes the carry input to the second stage. Similarly, in the second stage, A_1 B_1 and C_1 are added resulting in S_1 and C_2; in the third stage, A_2 B_2 and C_2 are added resulting in S_2 and C_3; in the fourth stage A_3, B_3 and C_3 are added resulting in S_3 and C_4 which is the output carry. Thus the circuit results in a sum (S_3, S_2, S_1, S_0) and a carry output (C_{out}).

An alternative block diagram representation of a 4 bit binary parallel adder is shown in Fig. 4.40.

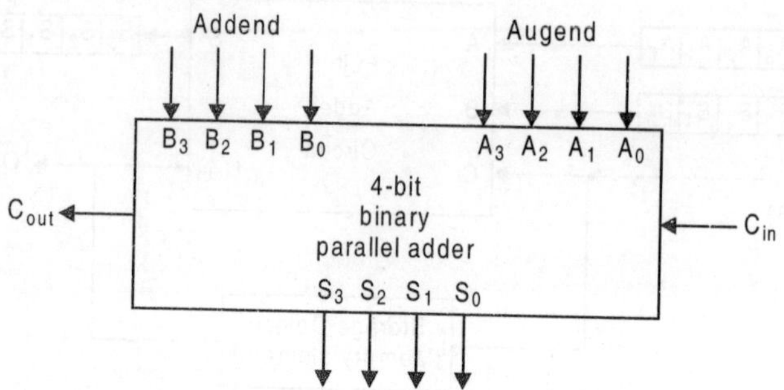

Fig. 4.40 4-bit binary parallel adder

Propagation Delay: Though the parallel binary adder is said to generate its output immediately after the inputs are applied, its speed of operation is limited by the carry propagation delay through all stages. In the parallel binary adder, the carry generated by the adder is fed as carry input to $(i + 1)$th adder. This is also called as 'ripple carry adder'. In such adders, the output (C_{out}, S_3, S_2, S_1, S_0) is available only after the carry is propagated through each of the adders, *i.e.*, from LSB to MSB adder through intermediate adders. Hence, the addition process can be considered to be complete only after the carry propagation delay through adders, which is proportional to number of stages in it; one of the methods of speeding up this process is look-ahead carry addition, which eliminates the ripple carry delay. This method is based on the carry generating and the carry propagating functions of the full adder.

4-bit Look-ahead Carry Generator

The look-ahead carry generator is based on the principle of looking at the lower order bits of addend and augend if a higher order carry is generated. This reduces the carry delay by reducing the number of gates through which a carry signal must propagate. To explain its operation consider the logic diagram of a full addor circuit Fig. 4.41.

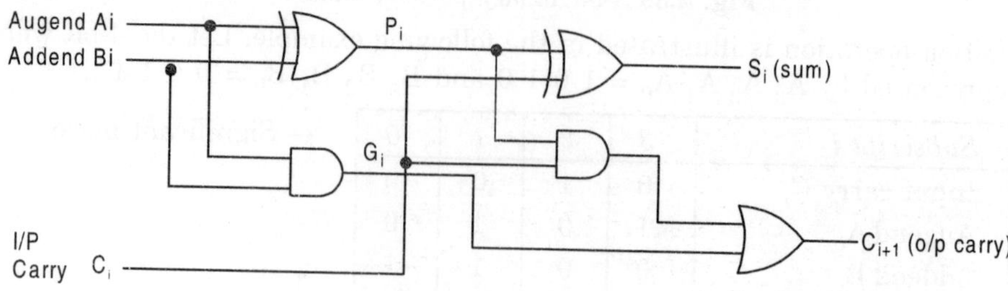

Fig. 4.41 Full adder

We define two new intermediate output variable P_i and G_i.

$P_i = A_i \oplus B_i$; called carry propagate, and

$G_i = A_i \cdot B_i$, called carry generate.

We now write the Boolean function for the carry output of each stage and substitute for each C_i its value from the previous equations :

$$C_1 = G_0 + P_0 C_0$$
$$C_2 = G_1 + P_1 C_1 = G_1 + P_1 (G_0 + P_0 C_0) = G_1 + P_1 G_0 + P_1 P_0 C_0.$$
$$C_3 = G_2 + P_2 C_2 = G_2 + P_2 (G_1 + P_1 G_0 + P_1 P_0 C_0)$$
$$= G_2 + P_2 G_1 + P_2 P_1 G_0 + P_2 P_1 P_0 C_0.$$

Note that C_3 does not have to wait for C_2 and C_1 to propagate; in fact C_3 is propagated at the same time as C_1 and C_2.

Next we draw the logic diagram of this 4 bit look-ahead carry generator as shown in Fig. 4.42.

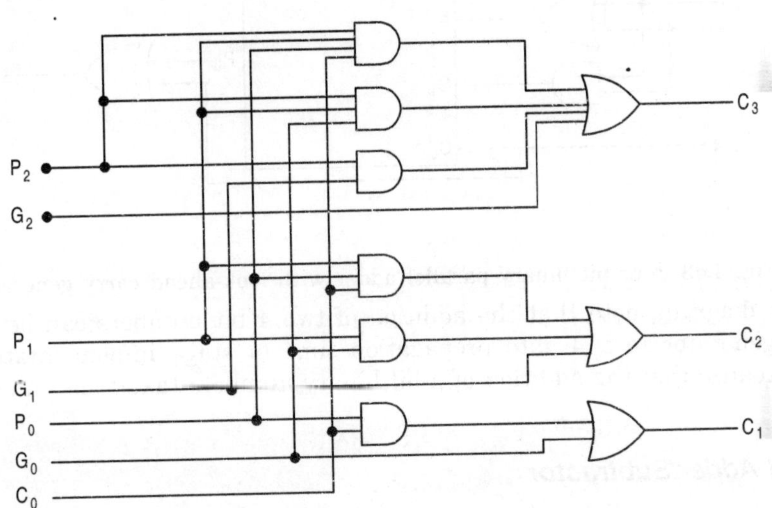

Fig. 4.42 4-bit look-ahead carry generator

4-bit binary parallel adder with a look-ahead carry generator (FAST ADDER)

In the 4-bit look ahead carry generator. We have seen that all the carry outputs are generated simultaneously with the application of Augend word, addend word and the input carry. What is remaining are the sum outputs. From the newly defined full adder circuit of Fig. 4.41, we notice that the sum output $S_i = P_i \oplus C_i$.

$$S_0 = P_0 \oplus C_0 = A_0 \oplus B_0 \oplus C_0$$
$$S_1 = P_1 \oplus C_1 = A_1 \oplus B_1 \oplus C_1$$
$$S_2 = P_2 \oplus C_2 = A_2 \oplus B_2 \oplus C_2$$
$$S_3 = P_3 \oplus C_3 = A_3 \oplus B_3 \oplus C_3.$$

Similarly carry output $C_{i+1} = G_i + P_i C_i$

Therefore, Final o/p carry $C_4 = G_3 + P_3 C_3.$

Using the above equations, the 4-bit binary parallel adder with a look ahead carry generator can be realized as shown in Fig. 4.43.

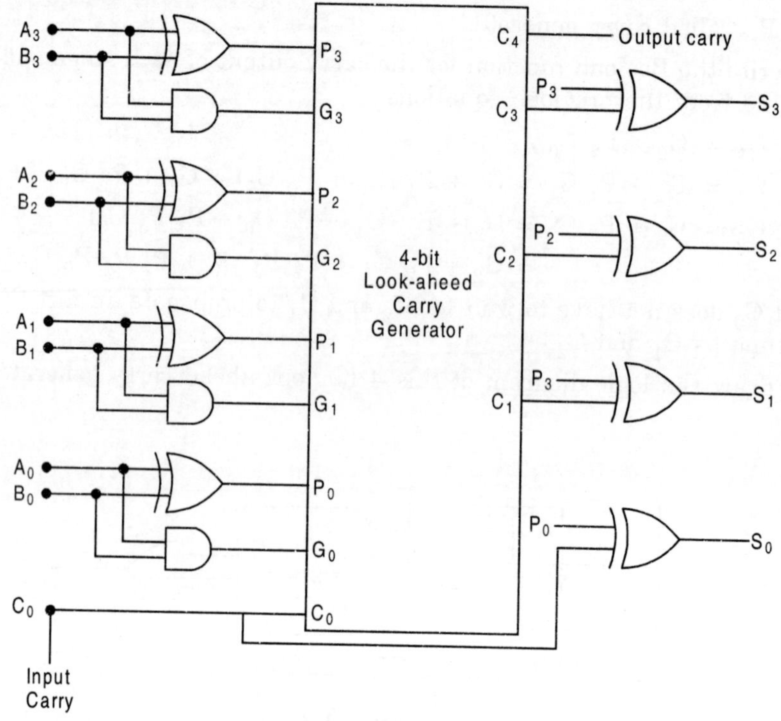

Fig. 4.43 Four bit binary parallel adder with look-ahead carry generator

From the diagram, note that the addition of two 4 bit numbers can be done by a look ahead carry generator in a 4 gate propagation time (4 stage implementation). Also, it is important to realize that the addition of n-bit binary numbers takes the same 4-stage propagation delay.

4-bit Parallel Adder/Subtractor

The 4-bit binary parallel adder/subtractor can be realized with the same circuit taking into consideration the 2's complement method of subtraction and the controlled inversion property of the Exclusive OR gate.

The subtraction of two binary number by taking 2's complement of the subtrahend and adding to the minuend. The 2's complement of the subtrahend can be obtained by adding 1 to the 1's complement of the subtrahend.

From the Exclusive OR operation, we know when one of the input is low the output is same as the other input and when one of the input is high the output is the complement of the other input.

$$Y = C X' + C'X$$

Naturally, if $C = 0, Y = X$

$C = 1, Y = X'$

The 4-bit binary parallel adder/subtractor circuit is shown in Fig. 4.44. that perform the operation of both addition and subtraction. It has two four bit inputs A_3 A_2 A_1 A_0 and B_3 B_2 B_1 B_0. The control input line C, connected with the input carry of the LSB of the full adder, is used to perform both operations.

To perform subtraction, the C (control input) is kept high. The controlled inverter produces the 1's complement of the adder (B_3' B_2' B_1' B_0'). Since 1 is given to input carry of the LSB of the adder, it is added to the complemented addend producing 2's complement of the addend before addition.

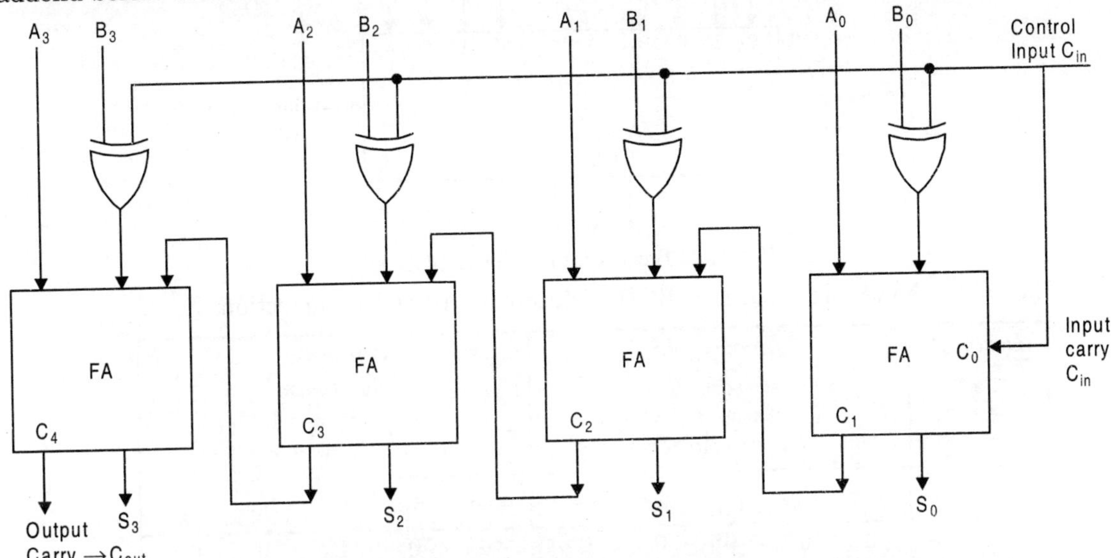

Fig. 4.44 Bit binary parallel adder/subtractor

Now the angend ($A_3A_2A_1A_0$) will be added to the 2's complement of addend ($B_3B_2B_1B_0$) to produce the sum, i.e., the diffrence between the addend and angend, and C_{out} (output carry), i.e. the borrow output of the 4-bit subtractor.

When the control input 'C' is kept low, the controlled inverter allows the addend (B_3 B_2 B_1 B_0) without any change to the input of full adder, and the input carry C_{in} of LSB of full adder, becomes zero, Now the augend (A_3 A_2 A_1 A_0) and addend (B_3 B_2 B_1 B_0) are added with C_{in} = 0. Hence, the circuit functions as 4-bit adder resulting in sum S_3 S_2 S_1 S_0 and carry output C_{out}.

4.2.5 Decimal Adder

A BCD adder is a combinational circuit that adds two BCD digits in parallel and produces a sum digit which is also in BCD. The block diagram for the BCD adder is shown in Fig. 4.45. This adder has two 4-bit BCD inputs A_8 A_4 A_2 A_1 and B_8 B_4 B_2 B_1 and a carry input C_{in}. It also has a 4-bit sum output S_8 S_4 S_2 S_1 and a carry output C_{out}. Obviously the sum output is in BCD form. (This is why subscripts 8, 4, 2, 1 are used).

If we consider the arithmetic addition of two decimal digits in BCD, the sum output can not be greater than 9 + 9 + 1 = 19. (Since each input digit does not exceed 9 and 1 being the possible carry from previous stage).

Suppose, we apply two BCD digits to a 4-bit binary parallel adder. The adder will form the sum in binary and produce a sum that will range from 0 to 19. But if we wish to design a BCD adder, it must be able to do the following.

1. Add two 4-bit BCD numbers using straight binary addition.
2. If the four bit sum is equal to or less than 9, the sum is in proper BCD form.
3. If the four bit sum is greater than 9 or if a carry is generated from the sum, the sum is not in BCD form. In this case a correction is required that is obtained by adding the digit 6 (0110) to the sum produced by binary adder.

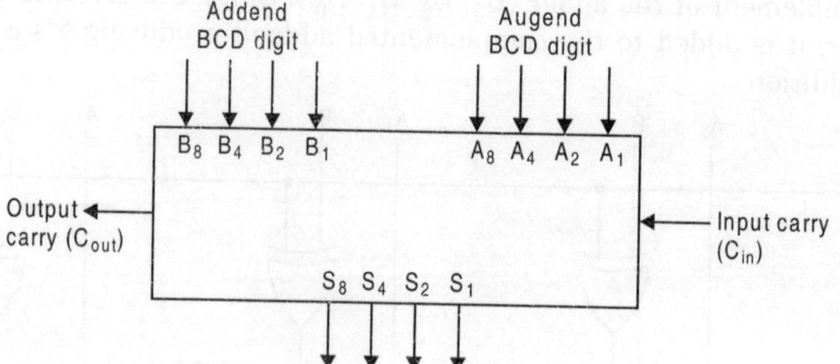

Fig. 4.45 Block diagram of a BCD adder

The table shows the results of BCD addition with needed correction.

Decimal Digit	*Uncorrected BCD sum produced by Binary Adder.*					*Corrected BCD sum produced by BCD Adder.*					
	K	Z_8	Z_4	Z_2	Z_1	C	S_8	S_4	S_2	S_1	
0	0	0	0	0	0	0	0	0	0	0	No Correction Required
1	0	0	0	0	1	0	0	0	0	1	
2	0	0	0	1	0	0	0	0	1	0	
3	0	0	0	1	1	0	0	0	1	1	
4	0	0	1	0	0	0	0	1	0	0	
5	0	0	1	0	1	0	0	1	0	1	
6	0	0	1	1	0	0	0	1	1	0	
7	0	0	1	1	1	0	0	1	1	1	
8	0	1	0	0	0	0	1	0	0	0	
9	0	1	0	0	1	0	1	0	0	1	
10	0	1	0	1	0	1	0	0	0	0	Correction Required
11	0	1	0	1	1	1	0	0	0	1	
12	0	1	1	0	0	1	0	0	1	0	
13	0	1	1	0	1	1	0	0	1	1	
14	0	1	1	1	0	1	0	1	0	0	
15	0	1	1	1	1	1	0	1	0	1	
16	1	0	0	0	0	1	0	1	1	0	
17	1	0	0	0	1	1	0	1	1	1	
18	1	0	0	1	0	1	1	0	0	0	
19	1	0	0	1	1	1	1	0	0	1	

The binary numbers are listed, labled as K Z_8 Z_4 Z_2 Z_1. K is the output carry and subscript under Z represent the weight 8, 4, 2 and 1. The first table lists the binary sums as produced by a 4-bit binary adder. For representing them in BCD, they must appear as second table.

From the two tables it is clear that upto 9, the binary sum is same as the BCD sum, so no correction is required. When the sum is greater than 9, the binary sum is diffrent from BCD sum, means correction is required. Therefore, a BCD adder must include the correction logic in its internal construction. Moreover, the circuit is required to develop a logic in its internal construction that indicates for needed correction.

This later logic can be developed by observing the two table entries. From tables it is clear that the correction is required when K = 1 or $Z_8 Z_4 = 1$ or $Z_8 Z_2 = 1$.
or when k = 1 or $Z_8 (Z_4 + Z_2) = 1$.

K = 1, means the result is 16 or above,

$Z_8 Z_4 = 1$, means the result is 12 or above and

$Z_8 Z_2 = 1$, means the result is 10 or above.

Therefore, the condition for correction can be written as :

$$C = K + Z_8 (Z_4 + Z_2)$$

i.e., whenever C = 1, we need correction ⇒ Add binary 0110 (decimal 6) to the sum produced by 4 bit binary adder. It also produce an output carry for the next stage. The BCD adder can be implemented using two 4-bit binary parallel adders as shown in Fig. 4.46.

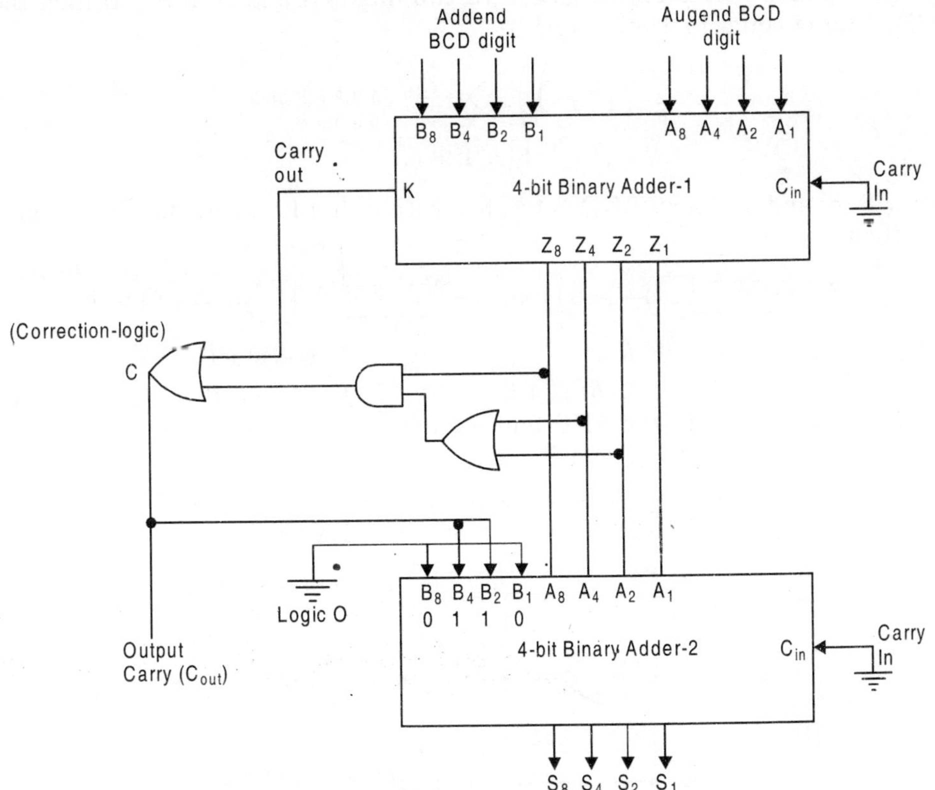

Fig. 4.46 BCD adder using two 4-bit binary adders along with the correction logic C

Here A_8 A_4 A_2 A_1 and B_8 B_4 B_2 B_1 are the BCD inputs. The two BCD inputs with input carry C_{in} are first added in the 4-bit binary adder-1 to produce the binary sum Z_8, Z_4, Z_2, Z_1 and output carry K. The outputs of adder-1 are checked to ascertain wheather the output is greater than 9 by AND-OR logic circuitry. If correction is required, then a 0110 is added with the output of adder-1. Now the 4-bit binary adder-2 forms the BCD result (S_8 S_4 S_2 S_1) with carry out C. The output carry generated from binary adder-2 can be ignored, since it supplies information already available at output carry terminal C.

4.2.6. Magnitude Comparator

A magnitude comparator is a combinational circuit designed primarily to compare the relative magnitude of the two binary numbers A and B. Naturally, the result of this comparison is specified by three binary variables that indicate, wheather $A > B$, $A = B$ or $A < B$.

The block diagram of a single bit magnitude comparator is shown in Fig. 4.47.

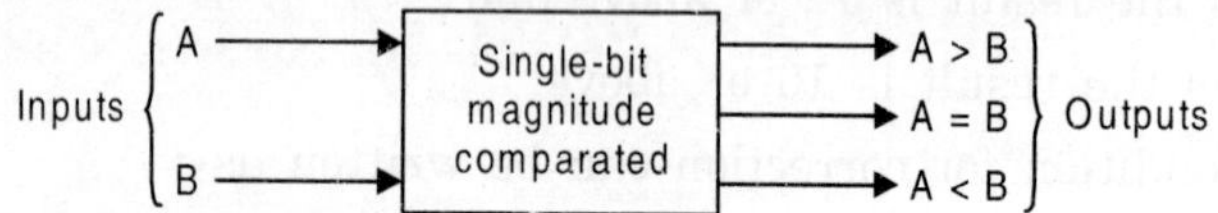

Fig. 4.47 Block diagram of single bit magnitude comparator

To implement the magnitude comparator the properties of Ex-NOR gate and AND gate is used.

Fig. 4.48(*a*) shows an EX-NOR gate with two inputs A and B. If A = B then the output of Ex-NOR gate is equal to 1 otherwise 0.

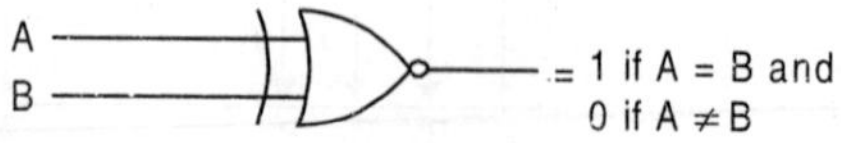

A
B
= 1 if A = B and
0 if A ≠ B

Fig. 4.48 (*a*)

Fig. 4.48 (*b*) and (*c*) shows AND gates, one with A and B' as inputs and another with A' and B as their inputs.

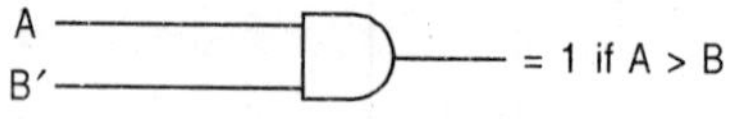

A
B'
= 1 if A > B

Fig. 4.48 (*b*)

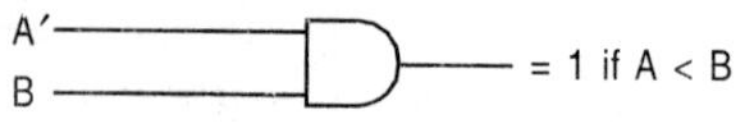

A'
B
= 1 if A < B

Fig. 4.48 (*c*)

The AND gate output of 4.48(*b*) is 1 if $A > B$ (i.e. $A = 1$ and $B = 0$) and 0 if $A < B$ (i.e. $A = 0$ and $B = 1$). Similarly the AND gate output of 4.48(*c*) is 1 if $A < B$ (i.e. $A = 0$ and $B = 1$) and 0 if $A > B$ (i.e. $A = 1$ and $B = 0$).

If the EX-NOR gate and two AND gates are combined as shown in Fig. 4.49(*a*), the circuit with function as single bit magnitude comparator. For EX-NOR implementation.

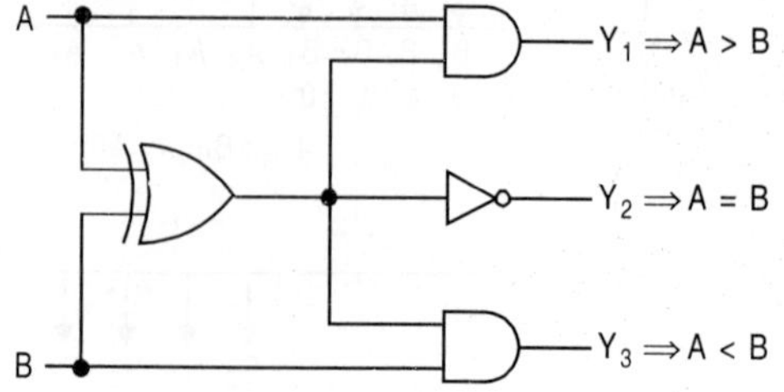

Fig. 4.49 (*a*) Single bit magnitude comparator

We have used EX-OR followed by an inverter.

Truth table of a single bit magnitude comparator.

Inputs		*Output*		
A	*B*	Y_1	Y_2	Y_3
0	0	0	1	0
0	1	0	0	1
1	0	1	0	0
1	1	0	1	0

It clearly shows Y_1 is high when A > B.
Y_2 is high when A = B
Y_3 is high when A < B.

The same principle can be extended to an n-bit magnitude comparator.

4-bit Magnitude Comparator

Consider two numbers A and B, with four digits each.

$$A = A_3\, A_2\, A_1\, A_0$$
$$B = B_3\, B_2\, B_1\, B_0.$$

(*a*) The two numbers are equal if all pairs of significant digits are equal i.e. if $A_3 = B_3$, $A_2 = B_2$, $A_1 = B_1$ and $A_0 = B_0$. We have seen that equality relation is generaed by EX-NOR gate. Thus

$x_i = A_i \,.\, B_i = A_i\, B_i + A_i'\, B_i'$, $i = 0, 1, 2, 3.$

Where x_i represents the equality of two numbers

$x_i = 1$, if A = B.

$x_i = 0$, otherwise.

If follows an AND operation of all variables.

Therefore, $(A = B) = x_3\, x_2\, x_1\, x_0 = 1$ only if all pairs are equal.

(*b*) To determine if A > B or A < B, we check the relative magnitude of pairs of significant digits starting from MSB. If the two digits are equal, we compare the next lower significant pair of digits. The comparison follows until a pair of unequal digits are reached. If the corresponding digit of A is 1 and that of B is 0, we say that A > B. If the corresponding digit A is 0 and that of B is 1 ⇒ A < B.

This discussion can be expressed logically as :

$$(A > B) = A_3\, B_3' + x_3\, A_2\, B_2' + x_3\, x_2\, A_1\, B_1' + x_3\, x_2\, x_1\, A_0\, B_0'$$
$$(A < B) = A_3'\, B_3 + x_3\, A_2'\, B_2 + x_3\, x_2\, A_1'\, B_1 + x_3\, x_2\, x_1\, A_0'\, B_0.$$

The logical implementation is shown in Fig. 4.49(*b*)

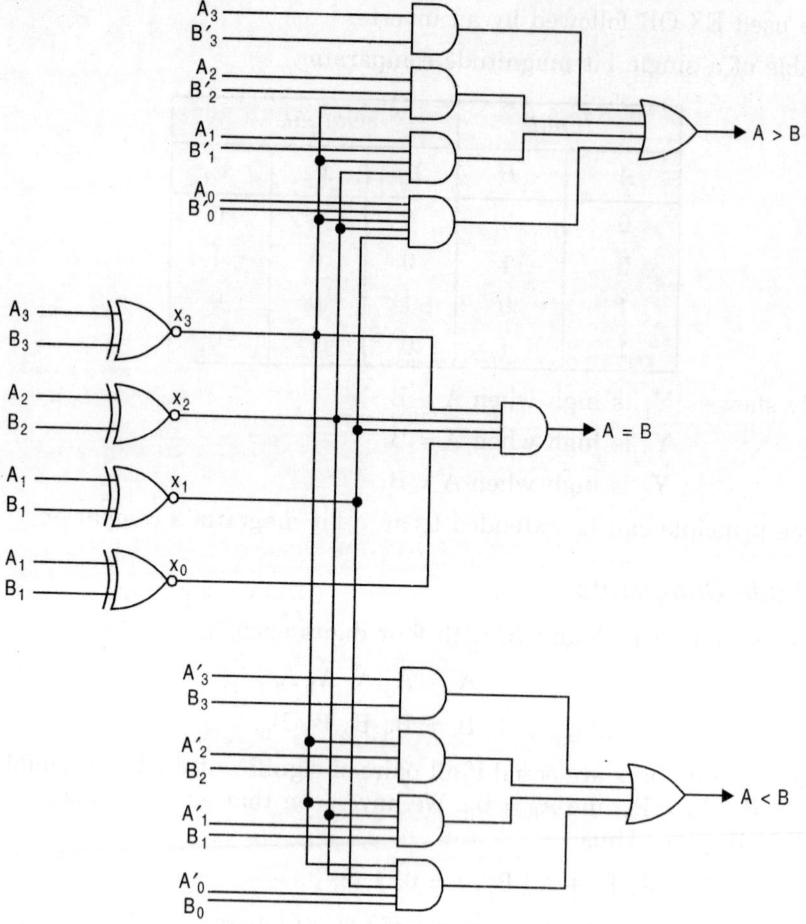

Fig. 4.49 (*b*) Logical implementation of 4-bit magnitude comparator

4.2.7 Binary Multiplier

Multiplication of binary number is similar, as the multiplication of decimal number. The multiplicand is multiplied by each bit of the multiplier. The process starts from the least significant bit. Each such multiplication forms a partial product. Successive partial products are shifted one position to the left. The final product is obtained from the sum of the partial products. Binary multiplier can be implemented with a combinational circuit. The procedure is as follows:

Procedure

Let us consider two 2-bit numbers $A(A_1\ A_0)$ and $B(B_1\ B_0)$. B is the multiplicand and A is the multiplier and the product is $C_3\ C_2\ C_1\ C_0$.

- The first partial product is formed by multiplying $B_1\ B_0$ by A_0.
  The multiplication of two bits such as A_0 and B_0 produces 1 if both bits are 1, otherwise 0 will be produced. This is identical to an AND operation.
  Hence, partial product can be implemented using NAND gate.

- The second partial product is formed by multiplying $B_1\ B_0$ by A_1 and shifting one position to the left.
- The two partial products are added with two half adders.

NOTE

The least significant bit of the product does not have to go through an adder, since it is formed by the output of the first AND gate.

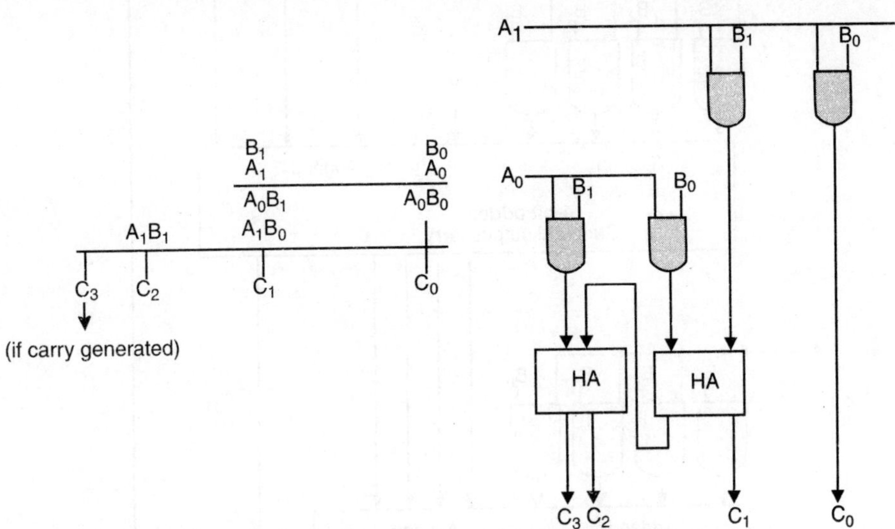

Fig. 4.49 (*c*) 2-bit binary multiplier

A combinational circuit binary multiplier with more bits can be constructed in a similar manner.

The generalised concept is given below.

Generalized way to Implement Binary Multiplier

- For X multiplier bits and Y multiplicand bits, we need (XY) AND gates.
- A bit of multiplier is ANDed with each bit of the multiplicand in as many levels, as there are bits in the multiplier.
- We required (X – 1), Y bit adder to produce a product of (X + Y) bits.

The binary output in each level of AND gates is added with the partial product of the previous level to form a new partial product. The least level produces the product.

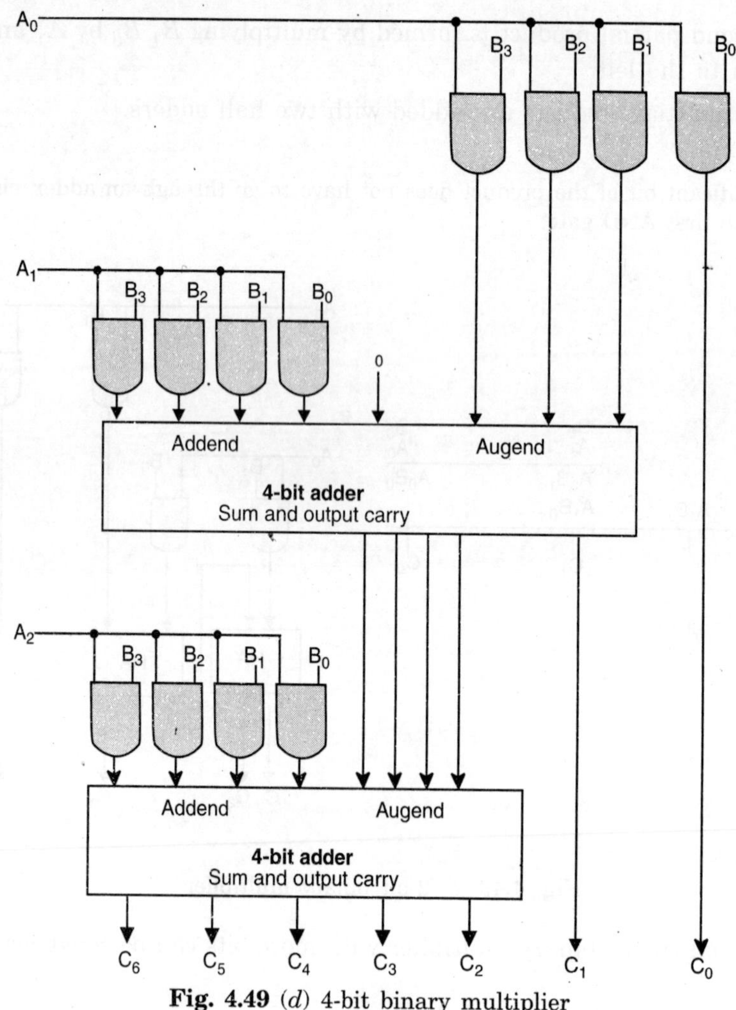

Fig. 4.49 (*d*) 4-bit binary multiplier

4.3 HAZARDS

In digital circuits it is important that undesirable glitches (spikes) on signal should not occur. Therefore in circuit design one must be aware of the possible sources of glitches (spikes) and ensure that the transitions in a circuit will be glitch free. The glitches (spikes) caused by the structure of a given circuit and propagation delays in the circuit are referred to as *hazards*. Hazards occur in combinational circuits, where they may cause a temporary false-output value.

4.3.1 Hazards in Combinational Circuits

Hazard is **unwanted** switching transients appearing in the output while the input to a combinational circuit (network) changes. The reason of hazard is that the different paths from input to output have different propagation delays, since there is a finite propagation delay through all gates. Figure 4.50 depicts the propagation delay in NOT gate.

In the circuit analysis, dynamic behaviour is an important consideration. The propagation delay of circuit varies and depends upon two factors.

- Path of change through circuit.
- Direction of change within gates.

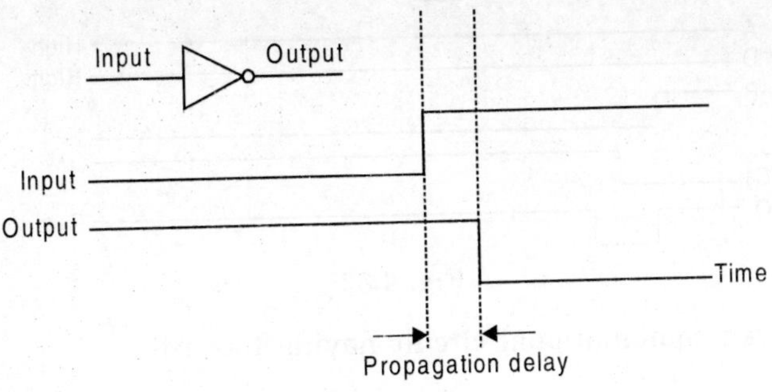

Fig. 4.50

The glitches (spikes) are momentary changes in output signal and are property of circuit, not function as depicted in Fig. 4.51.

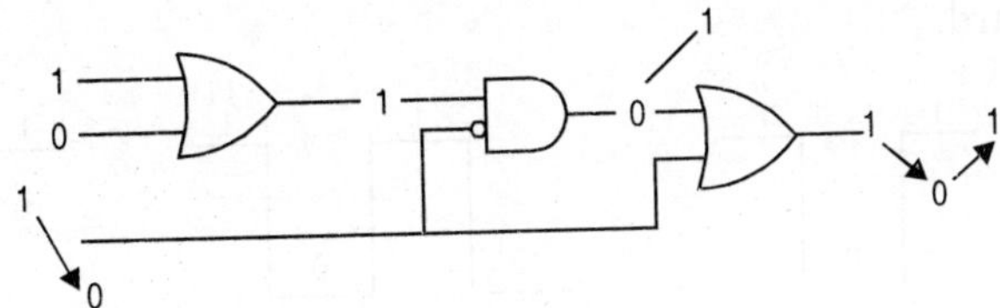

Fig. 4.51

Hazards/Glitches (spikes) are dangerous if

- Output sampled before signal stabilizes
- Output feeds asynchronous input (immediate response)

The usual solutions are :

- Use synchronous circuits with clocks of sufficient length
- Minimize use of circuits with asynchronous inputs
- Design hazard free circuits.

Example. *Show that the combinational circuit* $Q = A\overline{B} + BD$ *having hazards.*

Solution. For $Q = A\overline{B} + BD$; if B and D are 1 then Q should be 1 but because of propagation delays, if B changes stage then Q will become unstable for a short time, as follows :

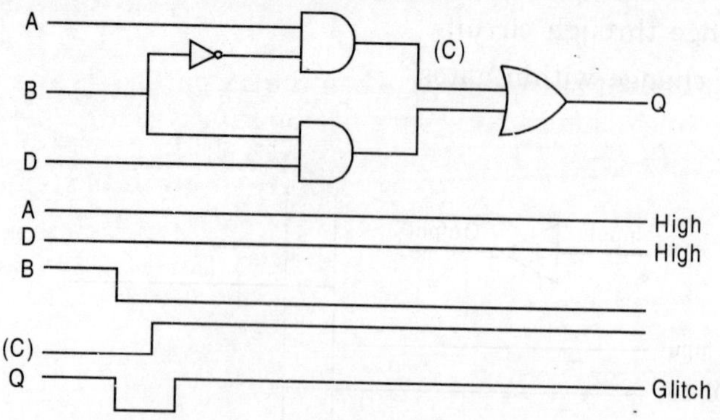

Fig. 4.52

Therefore, the given combinational circuit having hazards.

4.3.2 Types of Hazards

Two types of hazards are illustrated in Fig. 4.53.

- Static hazard
- Dynamic hazard.

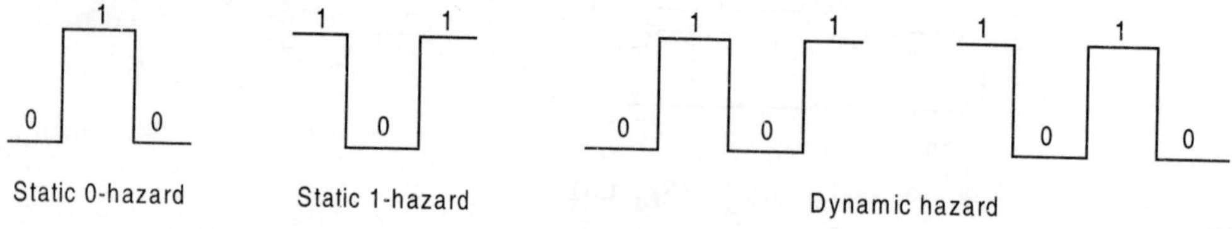

Fig. 4.53

1. Static 1 (0) hazard

A static hazard exists if, in response to an output change and for some combination of propagation delays, a network output may momentarily go to 0 (1) when it should remain a constant 1 (0), we say the network has a static 1 (0) hazard. Example of static hazard is shown in Fig. 4.54 (*a*).

Example.

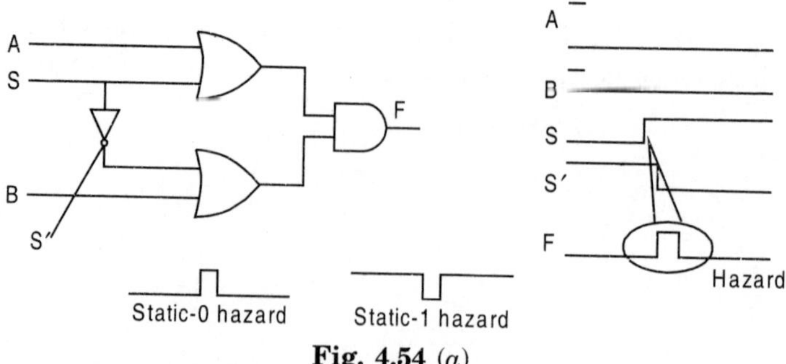

Fig. 4.54 (*a*)

2. Dynamic hazard

A different type of hazard may occur if, when the output is suppose to change from 0 to 1 (or 1 to 0), the output may change three or more times, we say the network has a dynamic hazard. Example of dynamic hazard is shown in Fig. 4.54 (*b*).

Example:

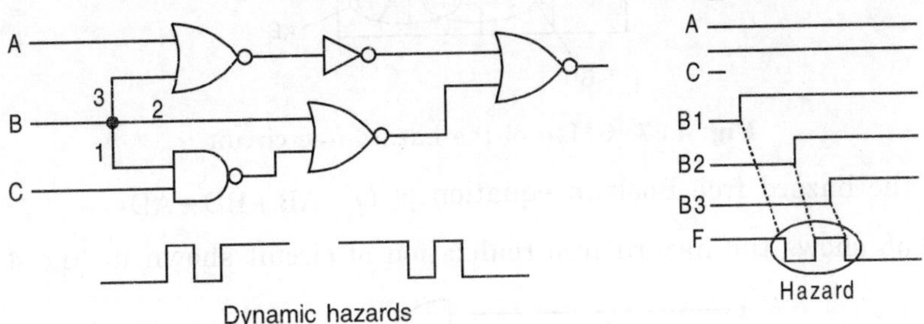

Fig. 4.54 (*b*)

4.3.3 Hazard Free Realizations

The occurrence of the hazard can be detected by inspecting the Karnaugh Map of the required function. Hazards like example (Fig. 4.52) are best eliminated logically. The Fig. 4.52 is redrawn here (Fig. 4.55).

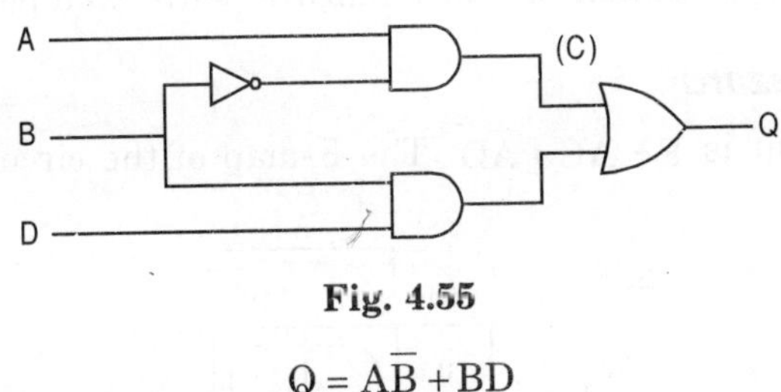

Fig. 4.55

$$Q = A\overline{B} + BD$$

The Karnaugh Map of the required function is given in Fig. 4.56.

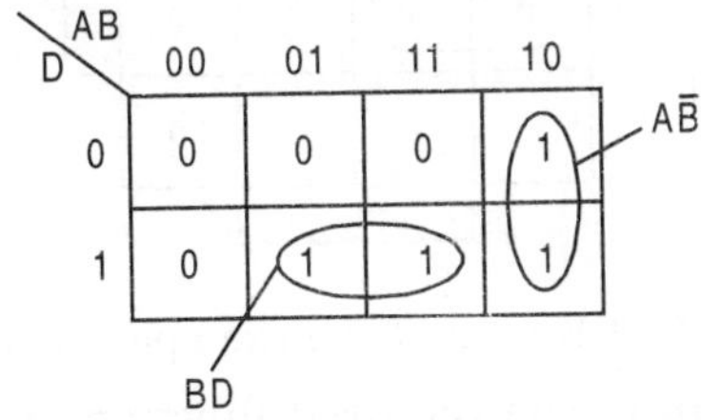

Fig. 4.56 K-Map of the given circuit

Whenever the circuit move from one product term to another there is a possibility of momentary interval when neither term is equal to 1, giving rise to an undesirable output. The remedy for eliminating hazard is to enclose the two minterms with another product term that overlaps both groupings.

The covering of hazard causing the transition with a redundant product term (AD) will eliminate the hazard. The K-Map of the hazard-free circuit will be as shown in

Fig. 4.57.

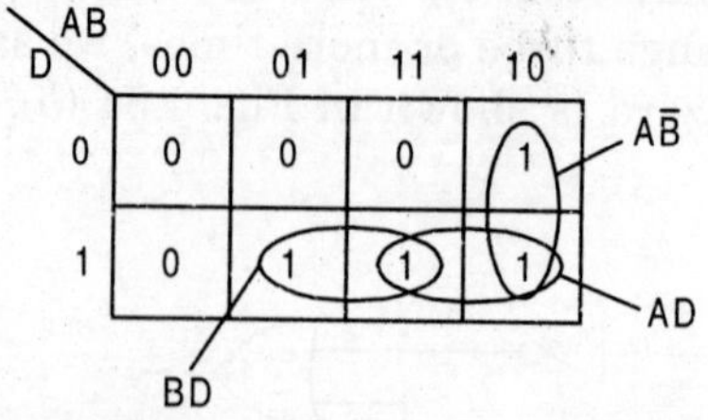

Fig. 4.57 K-Map of the hazard-free circuit

Therefore, the hazard free Boolean equation is $Q = A\overline{B} + BD + AD$.

The Fig. 4.58 shows the hazard free realization of circuit shown in Fig. 4.55.

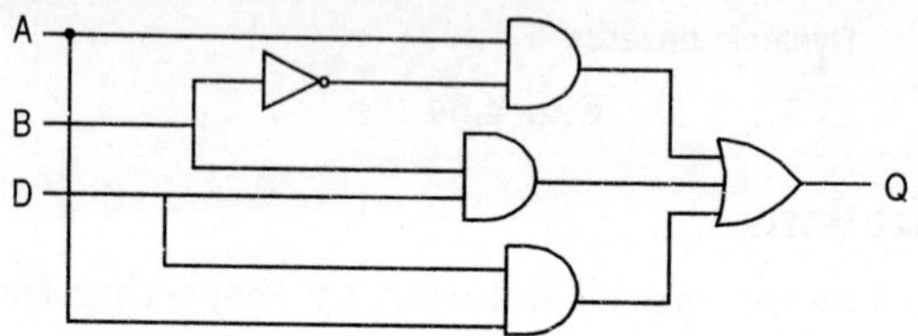

Fig. 4.58 Hazard free circuit

Now, we will discuss elimination of static hazards with examples.

Eliminating a static-1 hazard

Let the example circuit is $F = A\overline{C} + \overline{A}D$. The K-map of the circuit is given in Fig. 4.59.

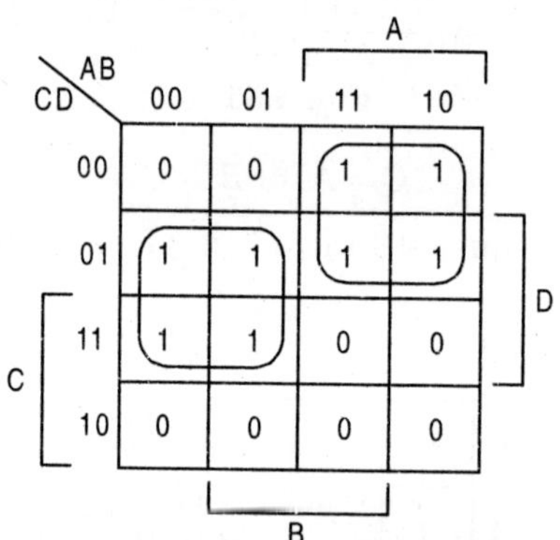

Fig. 4.59 K-Map of the example circuit

By inspecting Karnaugh Map of the required function, we notice the following points.

- Input change within product term (ABCD = 1100 to 1101)
- Input change that spans product terms (ABCD = 1101 to 0101)
- Glitch only possible when move between product terms.

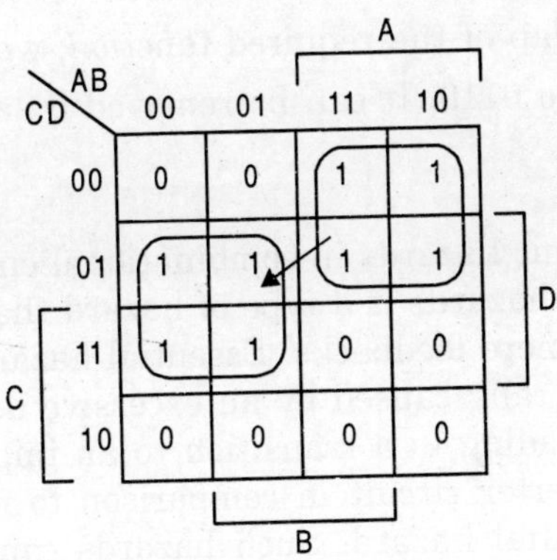

From the above three points it is clear that addition of redundant prime implicants so that all movements between adjacent on-squares remains in a prime implicant will remove hazard.

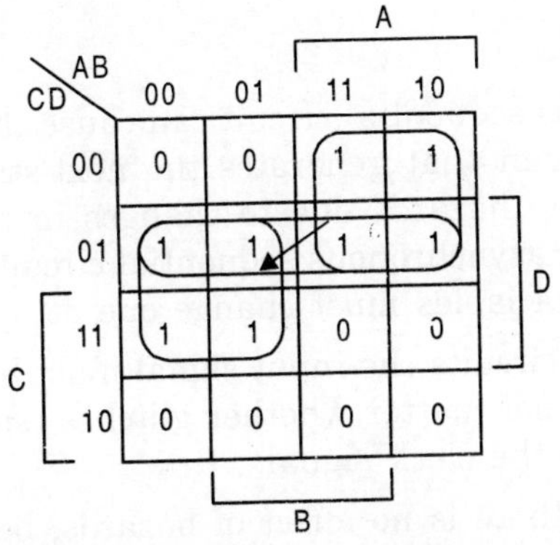

Fig. 4.60 K-Map of the hazards free circuit

Therefore, $F = A\overline{C} + \overline{A}D$ becomes $F = A\overline{C} + \overline{A}D + \overline{C}D$.

Note that when a circuit is implemented in sum of products with AND-OR gates or with NAND gates, the removal of static-1 hazard guarantees that no static-0 hazards or dynamic hazards will occur.

Eliminating a static-0 hazard

Let the example circuit is $F = (\overline{A} + \overline{C})(A + D)$. The K-Map of the circuit is given in Fig. 4.61.

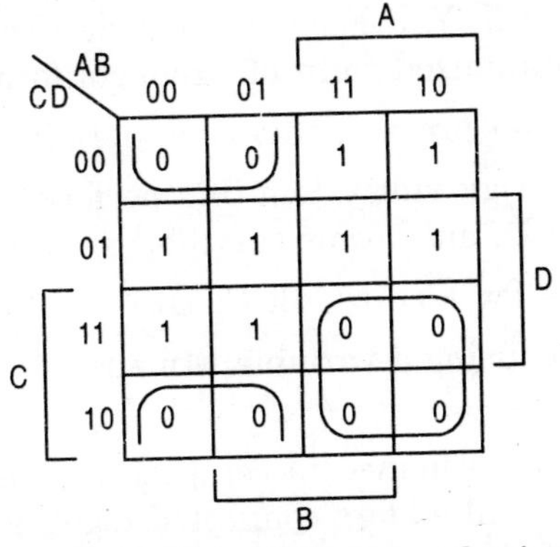

Fig. 4.61 K-Map of the example circuit

By inspecting Karnaugh Map of the required function, we see that occurrence of static-0 hazard from ABCD = 11 10 to 0110. It can be removed by adding the term $(\overline{C}+D)$.

4.3.4 Essential Hazard

Similar to static and dynamic hazards in combinational circuits, essential hazards occur in sequential circuits. Essential hazards is a type of hazard that exists only in asynchronous sequential circuits with two or more feedbacks. Essential hazard occurs normally in toggling type circuits. It is an error generally caused by an excessive delay to a feedback variable in response to an input change, leading to a transition to an improper state. For example, an excessive delay through an inverter circuit in comparison to the delay associated with the feedback path may cause essential hazard. Such hazards cannot be eliminated by adding redundant gates as in static hazards. To avoid essential hazard, each feedback loop must be designed with extra care to ensure that the delay in the feedback path is long enough compared to the delay of other signals that originate from the input terminals.

4.3.5 Significance of Hazards

A glitch in an asynchronous sequential circuit can cause the circuit to enter an incorrect stable state. Therefore, the circuit that generates the next-state variables must be hazard free. It is sufficient to eliminate hazards due to changes in the value of a single variable because the basic premise in an asynchronous sequential circuit is that the values of both the primary inputs and the state variables must change one at a time.

In synchronous sequential circuits the input signal must be stable within the setup and hold times of flip-flops. It does not matter whether glitches (spikes) occur outside the setup and hold times with respect to the clock signal.

In combinational circuits, there is no effect of hazards, because the output of a circuit depends solely on the values of the inputs.

4.4 EXERCISE

1. Develop a minimized Boolean implementation of a "ones count" circuit that works as follows. The subsystem has four binary inputs A, B, C, D and generates a 3-bit output, XYZ, XYZ is 000 if none of the inputs are 1, 001 if one input is 1,010 if two are one, 011 if three inputs are 1, and 100 if all four inputs are 1.

(*a*) Draw the truth tables for XYZ (A, B, C, D).

(*b*) Minimize the functions X, Y, Z, using 4-variable K-maps. Write down the Boolean expressions for the minimized Sum of Products form of each function.

(*c*) Repeat the minimization process, this time deriving Product of Sums form.

2. Consider a combinational logic subsystem that performs a two-bit addition function. It has two 2-bit inputs A B and C D, and forms the 3-bit sum X Y Z.

(*a*) Draw the truth tables for XYZ (A, B, C, D).

(*b*) Minimize the functions using 4-variable K-maps to derive minimized Sum of Products forms.

(*c*) What is the relative performance to compute the resulting sum bits of the 2-bit adder compared to two full adders connected together? (Hint: which has the worst delay in terms of gates to pass through between the inputs and the final outputs, and how many gates is this?).

3. Show how to implement the full adder Sum (A, B, C in) and Carry (A, B, C in) in terms of:

(*a*) Two 8 : 1 multiplexers;

(*b*) Two 4 : 1 multiplexers;

(*c*) If you are limited to 2:1 multiplexers (and inverters) only, how would you use them to implement the full adder and how many 2:1 multiplexers would you need?

4. Design a combinational logic subsystem with three inputs, 13, 12, 11, and two outputs, 01, 01, that behaves as follows. The outputs indicate the highest index of the inputs that is driven high. For example, if 13 is 0, 12 is 1, 11 is 1, then 01, 00 would be (10 (i.e. 12 is the highest input set to 1).

(*a*) Specify the function by filling out a complete truth table.

(*b*) Develop the minimized gate-level implementation using the K-map method.

(*c*) Develop an implementation using two 4 : 1 multiplexers.

(*d*) Compare your implementation for (*b*) and (*c*). Which is better and under what criterion?

5. Design a simple combinational subsystem to the following specification. The system has the ability to pass its inputs directly to its outputs when a control input, S, is not asserted. It interchanges its inputs when the control inputs S is asserted. For example, given four inputs A, B, C, D and four outputs W, X, Y, Z when S = 0, WXYZ = ACD and when S = 1, WXYZ = BCDA. Show how to implement this functionality using building blocks that are restricted to be 2 :1 multiplexers and 2 : 1 demultiplexers.

6. Your task is to design a combinational logic subsystem to decode a hexadecimal digit in the range of 0 through 9. A through F to drive a seven segment display. The hexadecimal numerals are as follows :

0 1 2 3 4 5 6 7

8 9 A b C d E F

Design a minimized implementation in PLA form. That is, look for common terms among the seven output functions.

7. Determine number of days in a month (to control watch display) used in controlling the display of a wrist-watch LCD screen.

Inputs : month, leap year flag.

Outputs : number of days.

8. Consider the following functions, which are five different functions over the inputs A, B, C, D.

(1) F (A, B, C,) = Σm (1, 2, 6, 7)

(2) F (A, B,C,D) = Σm (0, 1, 3, 9, 11, 12, 14, 15)

(3) F' (A, B, C, D) = Σm (2, 4, 5, 6, 7, 8, 10, 13)

(4) F (A, B, C, D) = (ABC + A'B') (C+D)

(5) F (A, B, C, D) = (A + B + C) (A + B + C' + D) (A + B' + C + D') (A' + B')

(*a*) Implement these in a single PLA structure with four inputs, five outputs, and an unlimited number of product terms, how many unique product terms are there in this PLA implementation.

(b) If you are trying to maximize the number of shared product terms across the five functions, rather than minimizing the literal count for each function independently, how many unique terms do you obtain? Draw the new K-maps with your selection of implicants that minimizes the number of unique terms across all five functions.

9. Consider the following Boolean function in Product of Sums form :

F (A, B, C, D) = (A + B' + D) (A' + B' +D) (B' + C' + D') (A' + C + D) (A' + C' + D)

Show how to implement this function with an 8 : 1 multiplexer, with A, B, C on the control inputs and D, its complement, and the constants 0 and 1 available as data inputs.

10. Design a two-bit comparator with the following inputs and outputs:

Inputs : Numbers N1 and N2 to be compared.

N1 = AB

N2 = CD.

Outputs : LT, GT, EQ

LT = 1 when AB < CD

GT = 1 when AB > CD

EQ = 1 when AB = CD

11. Design a 2X2 bit multiplier :

Inputs : Numbers N1 and N2 to be multiplied

N1 = A1 A0

N2 = B1 B0

Outputs ; products : P8, P4, P2, P0

P0 = Product with weighting $2^0 = 1$

P2 = Product with weighting $2^1 = 2$

P4 = Product with weighting $2^2 = 4$

P8 = Product with weighting $2^3 = 8$.

12. Analyse the behaviour of the Circuit below when input A changes from one logic state to another.

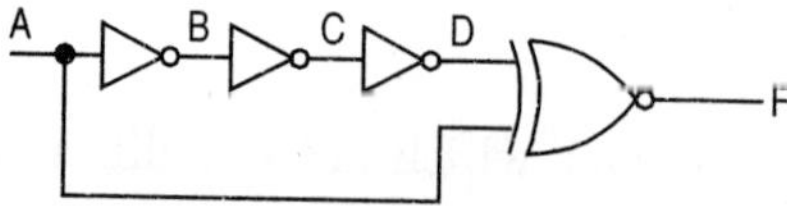

13. Analyse the circuit below for static hazard.

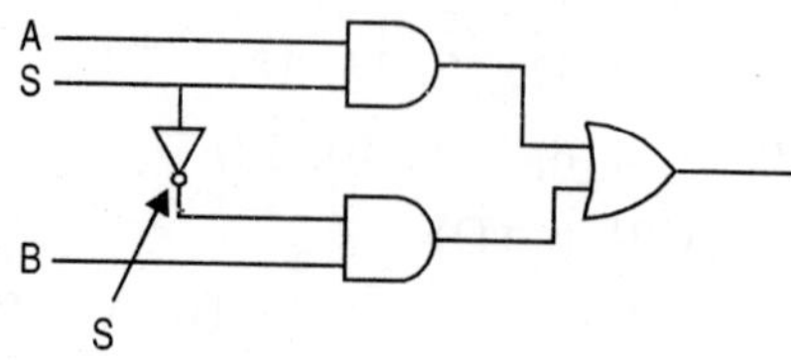

14. Analyse the pulse shaping circuit below:

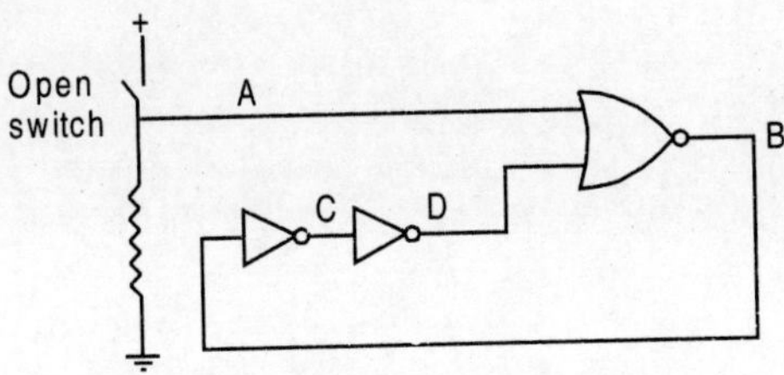

15. Which of the components below can be used to build an inverter?

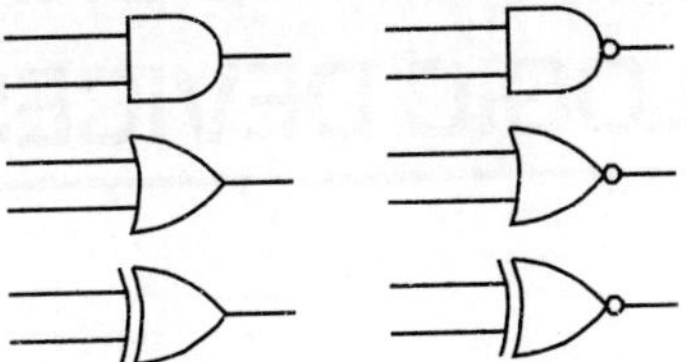

16. Consider the equation :

Z = A'B'C'D + A'B' CD' + A'BC' D' + A' BCD + ABC'D + ABCD' + AB'C'D' + AB'CD.

Implement this using 2-1 multiplexers.

17. Use an 8-input multiplexer to generate the function

$$Y = \overline{A}B + A\overline{D} + C\overline{D} + \overline{B}\overline{C}$$

18. Implement the following function with a multiplexer.

$$y = \Sigma m(0,1,5,7,10,14,15)$$

19. Design a 32:1 multiplexer using two 16:1 multiplexers.

20. Implement a 64 output demultiplexer tree using 1 × 4 DEMUX.

21. Realize the following functions of four variables using

(*i*) 8 : 1 multiplexers (*ii*) 16 : 1 multiplexers.

22. Design a BCD-to-Gray code converter using

(*i*) 8:1 multiplexers (*ii*) dual 4 : 1 multiplexers and some gates.

23. Design a Gray-to-BCD code converter using

(*i*) two dual 4 : 1 multiplexers and some gates.

(*ii*) one 1 : 16 demultiplexer and NAND gates.

24. Design a 40:1 multiplexer using 8 : 1 multiplexers.

25. Implement the following combinational logic circuit using a 4 to 16 line decoder.

$$Y_1 = \Sigma m\ (2, 3, 9)$$
$$Y_2 = \Sigma m\ (10, 12, 13)$$
$$Y_3 = \Sigma m\ (2, 4, 8)$$
$$Y_4 = \Sigma m\ (1, 2, 4, 7, 10, 12)$$

5

MEMORY AND PROGRAMMABLE LOGIC DEVICES

5.0 INTRODUCTION TO PLD

Digital circuit construction with small-scale integrated (SSI) and medium-scale integrated (MSI) logic has long been a basis of introductory digital logic design (refer chap. 3). In recent times, designs using complex programmable logic such as programmable array logic (PLA) chips and field programmable gate arrays (FPGAs) have begun replacing these digital circuits.

This chapter deals with devices that can be programmed to realize specified logical functions. Since evolution of programmable logic devices (PLDs) started with programmable ROM, it introduces ROMs and show how they can be used as a universal logic device and how simple programmable logic devices can be derived from ROMs. It also gives an overview of Complex Programmable Logic Devices (CPLDs) and Field Programmable Gate Arrays (FPGAs).

5.1 READ ONLY MEMORY (ROM)

A Read Only Memory (ROM) as shown in Fig. 5.1 is a matrix of data that is accessed one row at a time. It consists of k input lines and n output lines. Each bit combination of output lines is called a word while each bit combination of input variables is called an address. The number of bits per word is equal to the number of output lines n. A ROM is described by the number of words 2^k, the number of bits per word n.

The A inputs are address lines used to select one row (called a word) of the matrix for access. If A = i, then row i is selected and its data appears on the output terminals D. In this case we say that the contents of row i are read from the memory.

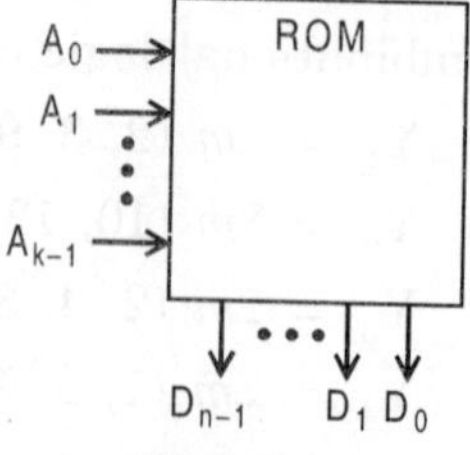

Fig. 5.1

If there are k address lines, then there are 2^k words in the ROM .The number of bits per word is called the word size. The data values of the words are called the contents of the memory and are said to be stored in the memory. The term read only refers to the property that once data is stored in a ROM, either it cannot be changed, or it is not changed very often.

ROM can be viewed as a combinational circuit with AND gates connected as a decoder and number of OR gates equal to the number of outputs. Internally a ROM contains a decoder and a storage array as shown in the Fig. 5.2.

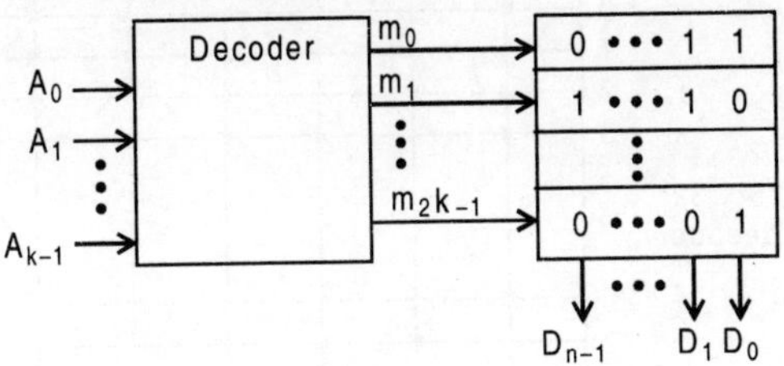

Fig. 5.2

When the address is i, the ith output of the decoder m_i is activated selecting row i of the data array. Functionally the data array can be viewed as a programmable OR array. Each column acts as a stack of OR gates as shown in the Fig. 5.3.

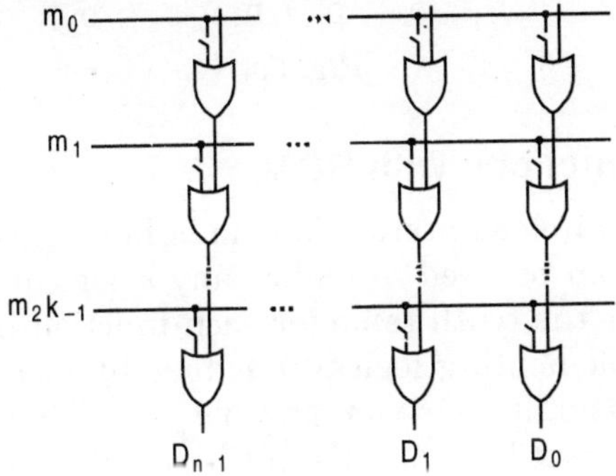

Fig. 5.3

Depending on the stored value (0/1) switch is open or closed. If a 0 is stored in a row, the switch is open and if a 1 is stored, the switch is closed. The type of ROM is determined by the way the switches are set or reset (i.e., programmed).

(I) *Mask programmed ROMs:* In the mask programmed ROMs, switch is realized at the time the ROM is manufactured. Either a connection is made by putting in a wire, or the connection is left open by not putting in a wire.

(II) *Field programmable ROMs (PROMs):* In the field programmable ROMs, switch is realized by a fuse. When the ROM is manufactured all switches are closed since all the fuses are intact. To open a switch the fuse is blown by sending a larger than usual current through it. Once a fuse is blown, it cannot be reinstalled.

(III) *Erasable ROMs (EROMs):* In the erasable ROMs switch is realized by a special kind of fuse that can be restored to its usual closed state, usually by the insertion of extra energy (e.g., shining ultraviolet light on the fuse). All fuses are reset when this is done.

(IV) *Electrically Programmable ROMs (EPROMs):* In the electrically programmable ROMs fuses are reset by the application of larger than usual currents. Sometimes subsections of the ROM can be reset without resetting all fuses.

Consider a 32×8 ROM as shown in Fig. 5.4. The ROM consists of 32 words of bit size 8. That is, there are 32 distinct words stored which may be available through eight output lines. The five inputs are decoded into 32 lines and each output of the decoder represents one of the minterms of a function of five variables. Each one of the 32 addresses selects only one output from the decoder. The 32 outputs of the decoder are connected through links to each OR gate.

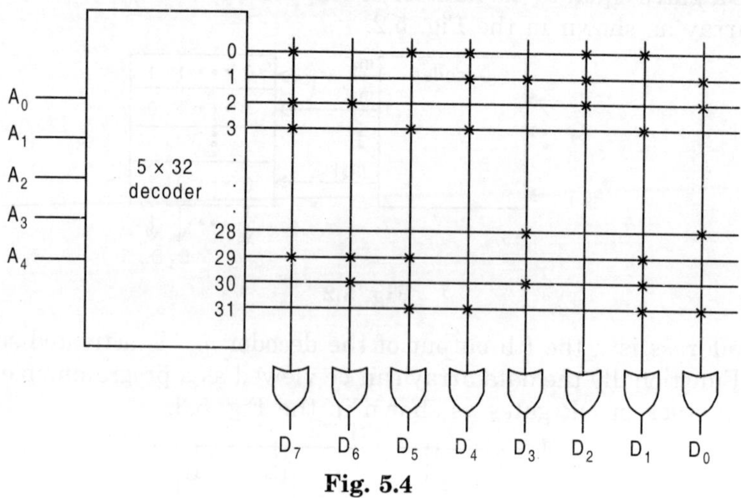

Fig. 5.4

5.1.1 Realizing Logical Functions with ROM

The ROM is a two-level implementation in sum of minterms form. ROMs with k address lines and n data terminals can be used to realize any n logical functions of k variables. For this one have to simply store the truth table for each function in a column of the ROM data array. The advantage of implementing logical functions by means of some form of programmable ROM, we have the possibility of reconfigurable logic. That is, the same hardware being reprogrammed to realize different logic at different times. But the disadvantage of using large ROMs to realize logical functions is that, the ROMs are much slower than gate realizations of the functions. The following example explains the procedure for realizing logical functions.

Example 1. Design a combinational circuit using a ROM that accepts a 2-bit number and generates an output binary number equal to the square of the input number.

Step 1: Derive the truth table for the combinational circuit. For the given example the truth table is

Table 5.1

Inputs		*Outputs*				*Equivalent decimal*
A_1	A_0	B_3	B_2	B_1	B_0	
0	0	0	0	0	0	0
0	1	0	0	0	1	1
1	0	0	1	0	0	4
1	1	1	0	0	1	9

Step 2: If possible, reduce the truth table for the ROM by using certain properties in the truth table of the combinational circuit. For the given example, two inputs and four outputs are needed to accommodate all possible numbers.

Since output B_0 is always equal to input A_0, therefore there is no need to generate B_0 with a ROM. Also B_1 is known as it is always 0. Therefore we need to generate only two outputs with the ROM; the other two are easily obtained.

Step 3: Find the minimum size of ROM from step 2.

The minimum size ROM needed must have two inputs and two outputs. Two inputs specify four word, so the ROM size must be 4×2. The two inputs specify four words of two bits each. The other two outputs of the combinational circuit are equal to 0 (B_1) and A_0(B_0).

Table 5.2 ROM truth table

A_1	A_0	B_3	B_2
0	0	0	0
0	1	0	0
1	0	0	1
1	1	1	0

The ROM truth table (Table 5.2) specifies complete information for ROM programming and the required connections are shown in Fig. 5.5.

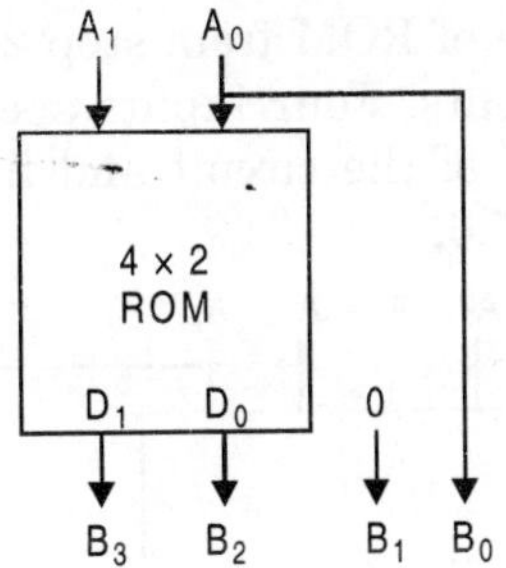

Fig. 5.5 Block diagram

Example 2. *Show that a 16 × 6 ROM can be used to implement a circuit that generates the binary square of an input 4-bit number with $B_0 = A_0$ and $B_1 = 0$ as shown in* Fig. 5.6(a)*. Draw the block diagram of the circuit and specify the first four and last four entries of the ROM truth table.*

Solution. Step 1. Draw the truth table for the combinational circuit. For the given example the truth table is

Table 5.3

Inputs				*Outputs*								*Equivalent Decimal*
A_3	A_2	A_1	A_0	B_7	B_6	B_5	B_4	B_3	B_2	B_1	B_0	
0	0	0	0	0	0	0	0	0	0	0	0	0
0	0	0	1	0	0	0	0	0	0	0	1	1
0	0	1	0	0	0	0	0	0	1	0	0	4
0	0	1	1	0	0	0	0	1	0	0	1	9
0	1	0	0	0	0	0	1	0	0	0	0	16

0	1	0	1	0	0	0	1	1	0	0	1	25
0	1	1	0	0	0	1	0	0	1	0	0	36
0	1	1	1	0	0	1	1	0	0	0	1	49
1	0	0	0	0	1	0	0	0	0	0	0	64
1	0	0	1	0	1	0	1	0	0	0	1	81
1	0	1	0	0	1	1	0	0	1	0	0	100
1	0	1	1	0	1	1	1	1	0	0	1	121
1	1	0	0	1	0	0	1	0	0	0	0	144
1	1	0	1	1	0	1	0	1	0	0	1	169
1	1	1	0	1	1	0	0	0	1	0	0	196
1	1	1	1	1	1	1	0	0	0	0	1	225

Step 2. If possible, reduce the truth table for the ROM by using certain properties in the truth table of the combinational circuit. For the given example, four inputs and eight outputs are needed to accommodate all possible numbers.

Since, $B_0 = A_0$ and $B_1 = 0$ as shown from truth table in step 1. We need only six outputs to generate with the ROM; the other two are easily obtained.

Step 3. Find the minimum size of ROM from step 2. The minimum size of ROM needed must have four inputs and six outputs. Four inputs specify sixteen word so the ROM size must be 16 × 6. The block diagram of the circuit and first four and last four entries of the ROM truth table is shown in Fig. 5.6.

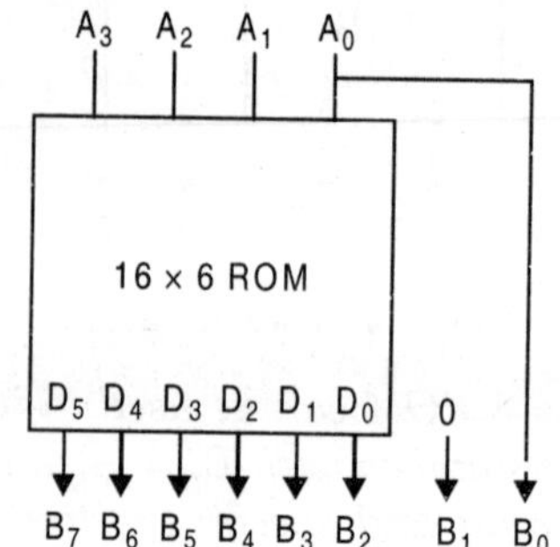

Fig. 5.6 (*a*) Block diagram

ROM truth table

	A_3	A_2	A_1	A_0	D_5	D_4	D_3	D_2	D_1	D_0
0	0	0	0	0	0	0	0	0	0	0
1	0	0	0	1	0	0	0	0	0	0
2	0	0	1	0	0	0	0	0	0	1
3	0	0	1	1	0	0	0	0	1	0
	⋮						⋮			
12	1	1	0	0	1	0	0	1	0	0
13	1	1	0	1	1	0	1	0	1	0
14	1	1	1	0	1	1	0	0	0	1
15	1	1	1	1	1	1	1	0	0	0

Fig. 5.6 (*b*) ROM truth table

Example 3. For the function

$$F_1 = \Sigma(0, 2, 5, 6)$$
$$F_2 = \Sigma(0, 2, 4, 6, 7)$$
$$F_3 = \Sigma(0, 2, 4, 7)$$
$$F_0 = \Sigma(1, 2, 3, 5, 7)$$

Design a three bit addressable ROM.

Solution. From truth table, the minimum size of ROM needed having three inputs and four outputs. Three inputs specify $2^3 = 8$ word so the ROM size must be 8 × 4.

Table 5.4

Minterms	*Inputs*			*Outputs*			
	A_2	A_1	A_0	F_3	F_2	F_1	F_0
0	0	0	0	1	1	1	0
1	0	0	1	0	0	0	1
2	0	1	0	1	1	1	1
3	0	1	1	0	0	0	1
4	1	0	0	1	1	0	0
5	1	0	1	0	0	1	1
6	1	1	0	0	1	1	0
7	1	1	1	1	1	0	1

The implementation of functions are given in Fig. 5.7.

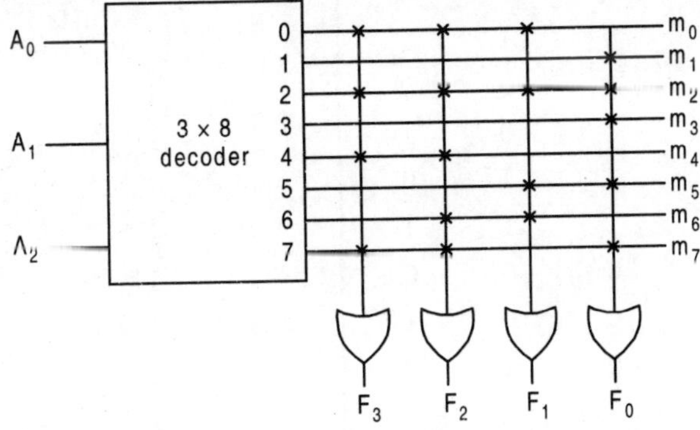

Fig. 5.7

Example 4. A ROM is to be used to implement the Boolean functions given below:

$$F_1\ (A, B, C, D) = ABCD + \overline{A}\,\overline{B}\,\overline{C}\,\overline{D}$$

$$F_2\ (A, B, C, D) = (A + B)\ (\overline{A} + \overline{B} + C)$$

$$F_3\ (A, B, C, D) = \Sigma 13, 15 + \Sigma 3, 5$$

(*a*) What is the minimum size of the ROM required?

(*b*) Determine the data in each location of the ROM.

Solution. *(a)*

$$F_1\ (A, B, C, D) = ABCD + \bar{A}\,\bar{B}\,\bar{C}\,\bar{D}$$

$$F_2\ (A, B, C, D) = (A + B)\,(\bar{A} + \bar{B} + \bar{C})$$

$$= A\bar{A} + A\bar{B} + AC + \bar{A}B + B\bar{B} + BC$$

$$= \bar{A}B + A\bar{B} + AC + BC$$

$$F_3\ (A, B, C, D) = \Sigma(13, 15) + \Sigma(3, 5)$$

To simplify F_3 (using K-Map).

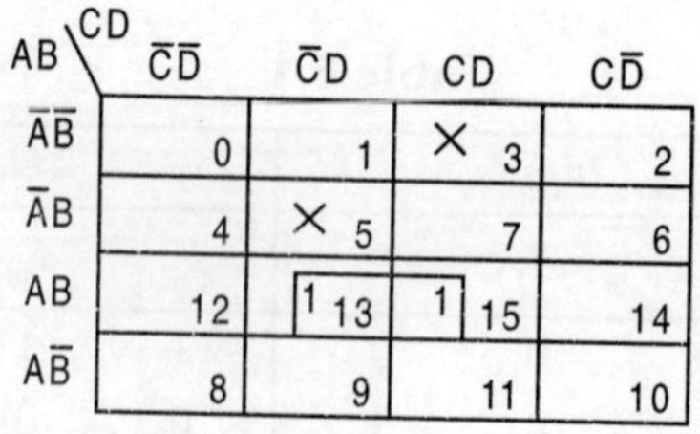

AB \ CD	$\bar{C}\bar{D}$	$\bar{C}D$	CD	$C\bar{D}$
$\bar{A}\bar{B}$	0	1	× 3	2
$\bar{A}B$	4	× 5	7	6
AB	12	1 13	1 15	14
$A\bar{B}$	8	9	11	10

$$F_3\ (A, B, C, D) = ABD$$

Table 5.5 Truth table of ROM

Inputs				*Outputs*		
A	*B*	*C*	*D*	F_3	F_2	F_1
0	0	0	0			1
0	0	0	1			-
0	0	1	0			-
0	0	1	1			-
0	1	0	0		1	-
0	1	0	1		1	-
0	1	1	0		1	-
0	1	1	1		1	-
1	0	0	0		1	-
1	0	0	1		1	-
1	0	1	0		1	-
1	0	1	1		1	-
1	1	0	0			-
1	1	0	1	1		-
1	1	1	0		1	-
1	1	1	1	1	1	1

The size of the ROM required is 4 inputs and 3 outputs. So, the ROM size is $2^4 = 16 \times 3$. The four inputs specify 16 words of 3-bit each.

(b) The data in each location of the ROM is given by F_1, F_2, F_3 as in truth table.

5.2 PROGRAMMABLE LOGIC ARRAYS

The first type of user-programmable chip that could implement logic circuits was the Programmable Read-Only Memory (PROM), in which address lines can be used as logic circuit inputs and data lines as outputs. Figure 5.8 shows the basic configuration of PROM.

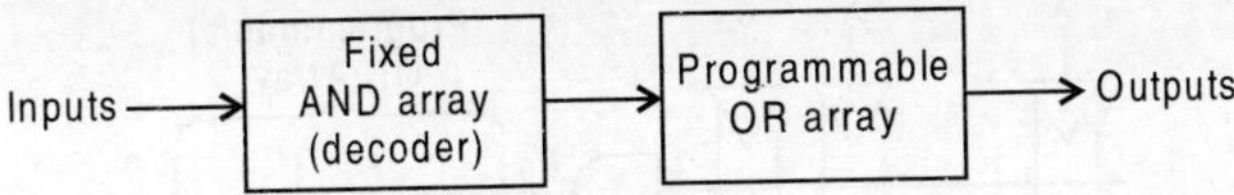

Fig. 5.8 Basic configuration of programmable read-only memory (PROM)

Logic functions, however, rarely require more than a few product terms, and a PROM contains a full decoder for its address inputs. PROMs are thus an inefficient architecture for realizing logic circuits, and so are rarely used in practice for that purpose. The first device developed later specifically for implementing logic circuits was the Field-Programmable Logic Array (FPLA), or simply PLA for short. A PLA consists of two levels of logic gates: a programmable "wired" AND-plane followed by a programmable "wired" OR-plane.

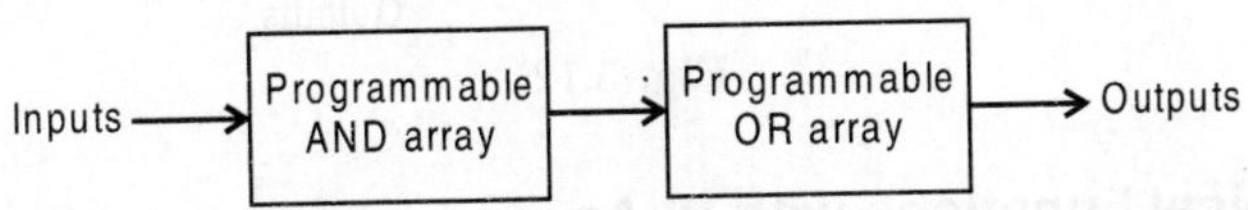

Fig. 5.9 Basic configuration of programmable logic array (PLA)

A PLA is structured so that any of its inputs (or their complements) can be AND'ed together in the AND-plane; each AND-plane output can thus correspond to any product term of the inputs. Similarly, each OR plane output can be configured to produce the logical sum of any of the AND-plane outputs. With this structure, PLAs are well-suited for implementing logic functions in sum-of-products form. They are also quite versatile, since both the AND terms and OR terms can have many inputs (this feature is often referred to as *wide* AND and OR gates).

Programmable Logic Arrays (PLAs) have the same programmable OR array as a ROM, but also have a programmable AND array instead of the decoder as shown in Fig. 5.10. The programmable AND array can be used to produce arbitrary product terms, not just minterms. While the decoder produces all minterms of k variables.

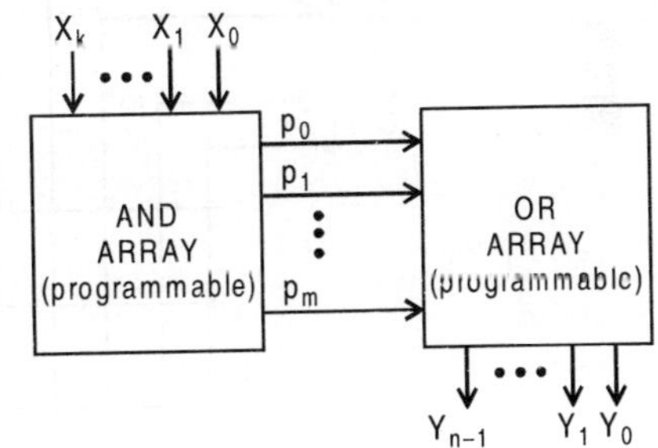

Fig. 5.10

Since, the number of possible product terms m in PLA is much less than the number of possible minterms 2^k, so some functions may not be realizable in PLA.

The structure of a row of the programmable AND array is shown in Fig. 5.11 (*a*).

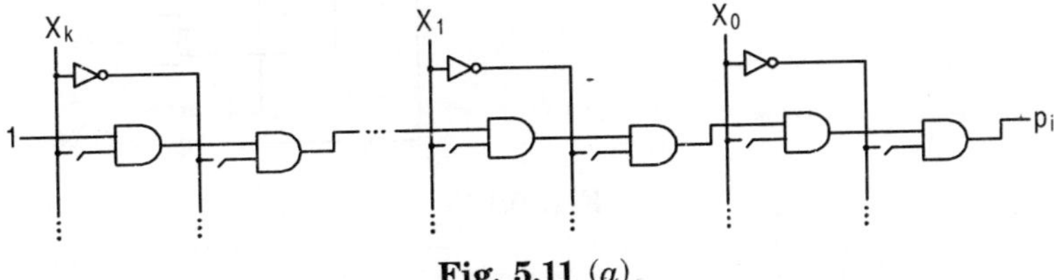

Fig. 5.11 (*a*)

and of a column of the programmable OR array is shown in Fig. 5.11 (*b*).

Fig. 5.11 (*b*)

While the notation of a PLA is shown in Fig. 5.12.

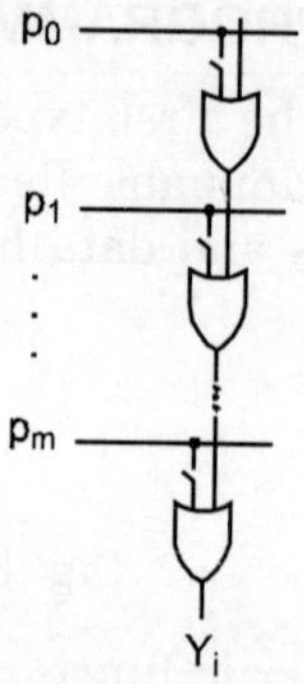

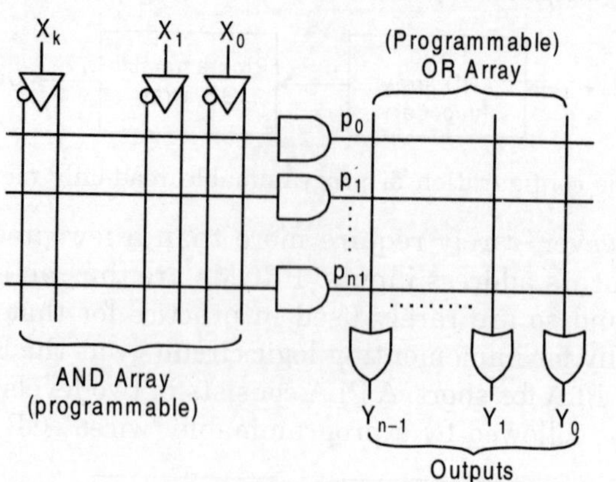

Fig. 5.12

5.2.1 Realizing Logical Functions with PLAs

During implementation (or programming) a dot (or cross) is placed at an intersection if the variable is to be included in the product term or to sum term. For example a PLA with 3 inputs, 4 product terms and 2 outputs is shown in Fig. 5.13.

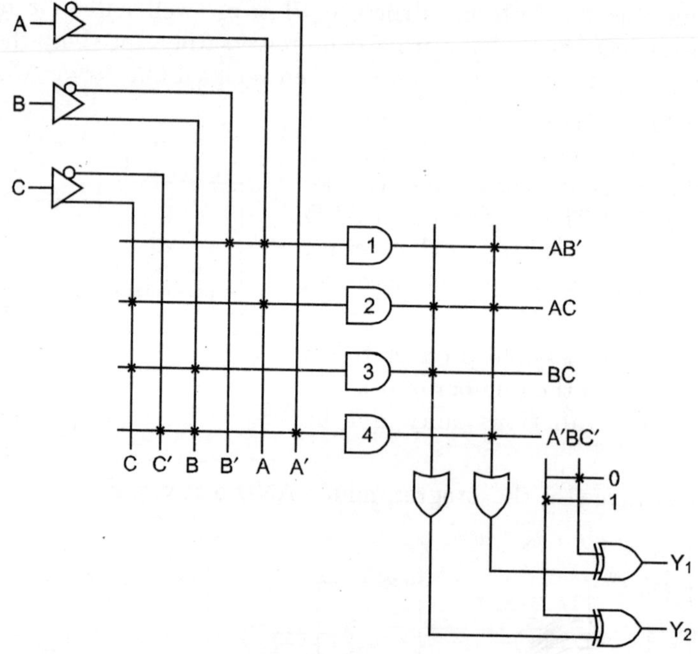

Fig. 5.13

From the exclusive OR operation, we know when one of the input is low the output is same as the other input and when one of the input is high the output is the complement of the other input (Fig. 5.14).

Control input C
Other input X
$Y = CX' + C'X$

Therefore, if $C = 0, \quad Y = X$

$C = 1, \quad Y = X'$

Fig. 5.14

In PLA, the output is present in true form as well as in complement form. In Fig. 5.13, Y_1 is in true form while output Y_2 is complemented.

When realizing (designing) a digital system with a PLA, there is no need to show the internal connections as in Fig. 5.13. The PLA can be programmed with PLA program table which provides appropriate paths.

PLA Program Table

A typical program table consists of three columns. The first column lists the product terms numerically. The second column specifies the required paths between inputs and AND gates. The third column specifies the paths between the AND gates and the OR gates. Under each output variable we write a T (for true) if the output is in true form and C (for complement) if the output is to be complement. For each product term, the inputs are marked as 1, 0 or – (dash). A variable in product term is marked as 1 if it appears in normal form, 0 if it is in complemented form and '–' shows that a variable is not present.

In PLA implementation of combinational circuits the total number of distinct product terms must be reduced by simplifying each function to a minimum number of terms.

Example 1. *Implement the following functions using 3-input, 3 product terms and 2 output PLA.*

$$F_1 = AB' + AC = \Sigma(4, 5, 7)$$
$$F_2 = AC + BC = \Sigma(3, 5, 7)$$

Solution. Step 1. Derive the truth table for the combinational circuit.

Table 5.6

Inputs			*Outputs*	
A	*B*	*C*	F_2	F_1
0	0	0	0	0
0	0	1	0	0
0	1	0	0	0
0	1	1	1	0
1	0	0	0	1
1	0	1	1	1
1	1	0	0	0
1	1	1	1	1

Step 2. Simplify functions using K-map.

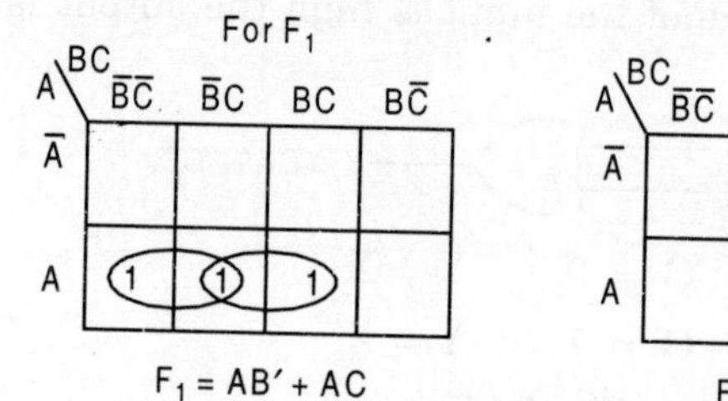

$F_1 = AB' + AC$

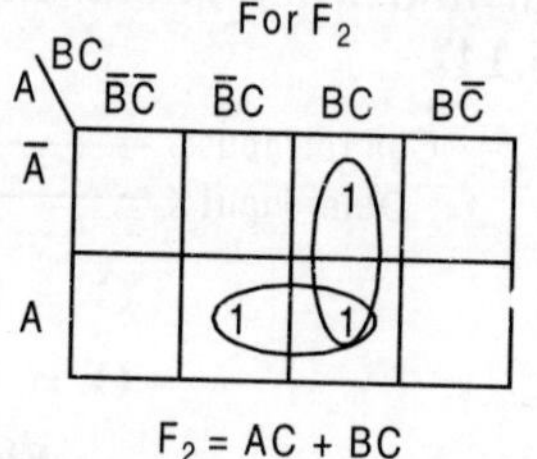

$F_2 = AC + BC$

Step 3. Draw the PLA program table.

Table 5.7

	Product term	*Inputs*			*Outputs*		
		A	*B*	*C*	F_2	F_1	
AB'	1	1	0	-	-	1	
AC	2	1	-	1	1	1	
BC	3	-	1	1	1	-	
					T	T	T/C

Step 4. Draw the logic diagram (Fig. 5.15)

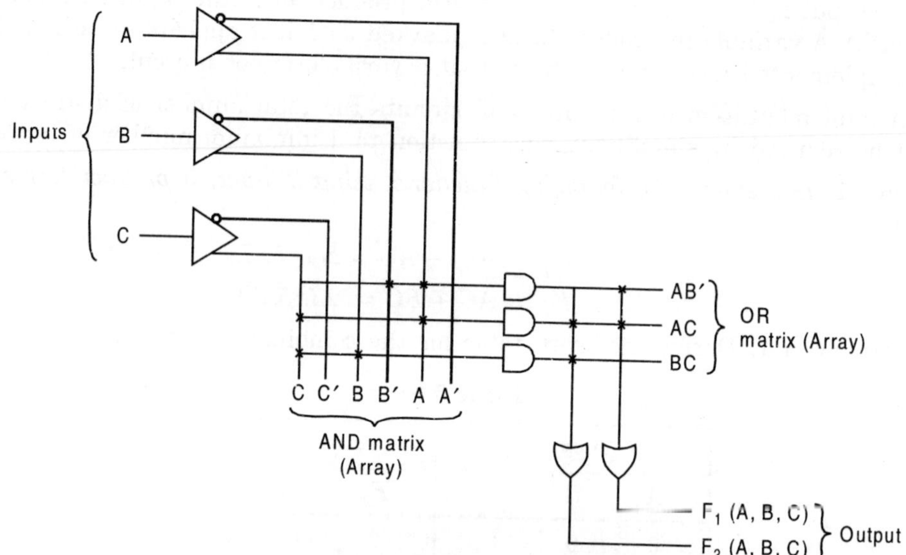

Fig. 5.15

For the implementation with PLA, the PLA program table is necessary and sufficient. Step 4 shows the logic diagram is only to show how PLA is implemented using AND array and OR array. Step 4 is optional.

Example 2. *Implement the following function using PLA.*

$$F_1(A, B, C) = \Sigma(0, 1, 6, 7)$$

$$F_2(A, B, C) = \Sigma(1, 2, 4, 6)$$

$$F_3(A, B, C) = \Sigma(2, 6)$$

Solution. Derive the truth table for the combinational circuit as shown in table 5.8.

Table 5.8

Inputs			*Outputs*		
A	*B*	*C*	F_3	F_2	F_1
0	0	0	0	0	1
0	0	1	0	1	1
0	1	0	1	1	0
0	1	1	0	0	0
1	0	0	0	1	0
1	0	1	0	0	0
1	1	0	1	1	1
1	1	1	0	0	1

Simplify functions using K-map

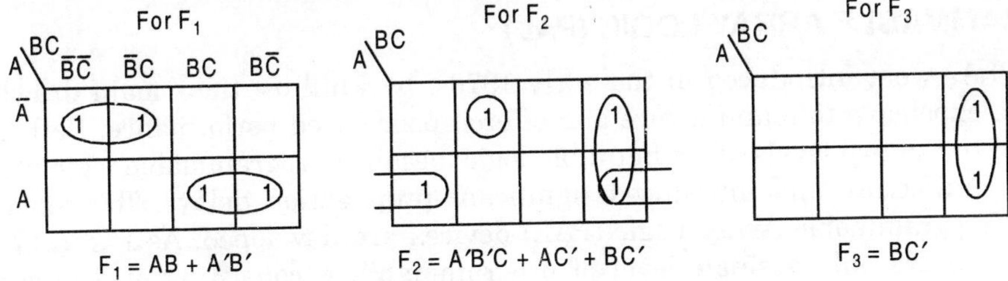

Draw the PLA program table.

Table 5.9

Product term	*Inputs*			*Outputs*		
	A	*B*	*C*	F_3	F_2	F_1
1	0	0	1	-	1	-
2	1	-	0	-	1	-
3	-	1	0	1	1	-
4	0	0	-	-	-	1
5	1	1	-	-	-	1
				T	T	T
						T/C

Example 3. Implement the following with PLA.

P1 = A' • C'

P2 = A' • C

P3 = A • B'

P4 = A • B • C

X = P3 = A • B'

$$Y = P2 + P4 = A' \bullet C + A \bullet B \bullet C$$

$$Z = P1 + P3 = A' \bullet C' + A \bullet B'$$

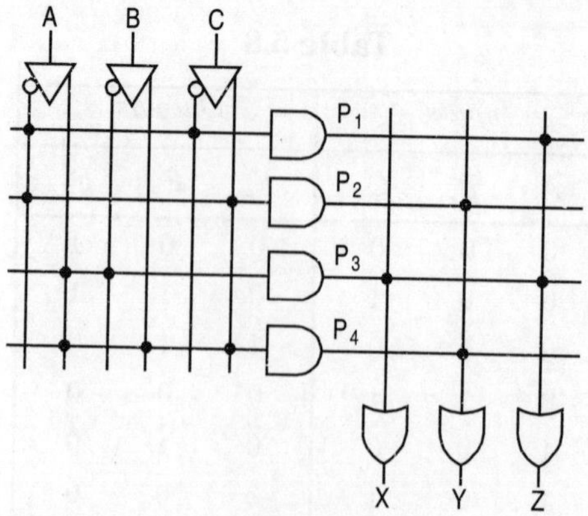

Fig. 5.16

Note: In this implementation, dot is placed at intersection but cross can be used for the same.

5.3 PROGRAMMABLE ARRAY LOGIC (PAL)

When PLAs were introduced in the early 1970s, by Phillips, their main drawbacks are that they are expensive to manufacture and offered poor speed-performance. Both disadvantages are due to the two levels of configurable logic, because programmable logic planes were difficult to manufacture and introduced significant propagation delays. To overcome these weaknesses, Programmable Array Logic (PAL) devices are developed. As Fig. 5.17 (*a*) illustrates, PALs feature only a single level of programmability, consisting of a programmable "wired" AND plane that feeds fixed OR-gates.

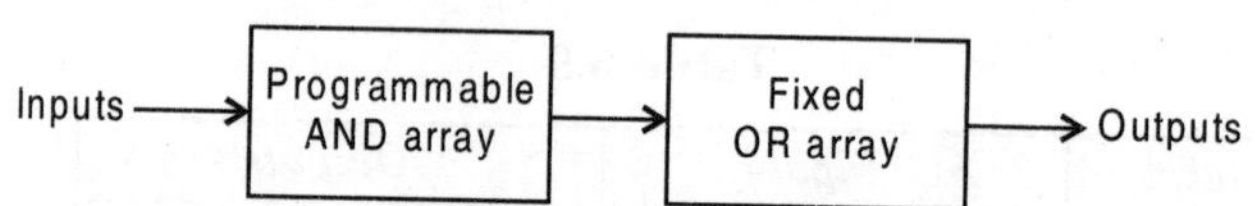

Fig. 5.17 (*a*) Basic configuration of programmable array logic (PAL)

To compensate for lack of generality incurred because the OR outputs plane is fixed, several variants of PALs are produced, with different numbers of inputs and outputs, and various sizes of OR-gates. PALs usually contain flip-flops connected to the OR-gate outputs so that sequential circuits can be realized. PAL devices are important because when introduced they had a profound effect on digital hardware design, and also they are the basis for more sophisticated architectures. Variants of the basic PAL architecture are featured in several other products known by different acronyms. All small PLDs, including PLAs, PALs, and PAL-like devices are grouped into a single category called simple PLDs (SPLDs), whose most important characteristics are low cost and very high pin-to-pin speed-performance.

While the ROM has a fixed AND array and a programmable OR array, the PAL has a programmable AND array and a fixed OR array. The main advantage of the PAL over the PLA and the ROM is that it is faster and easier to fabricate.

Figure 5.17 (*b*) represents a segment of an unprogrammed PAL.

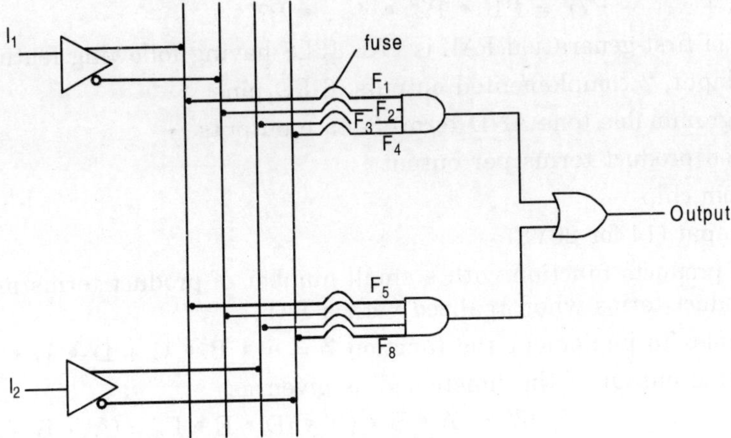

Fig. 5.17 (*b*) PAL segment

The symbol

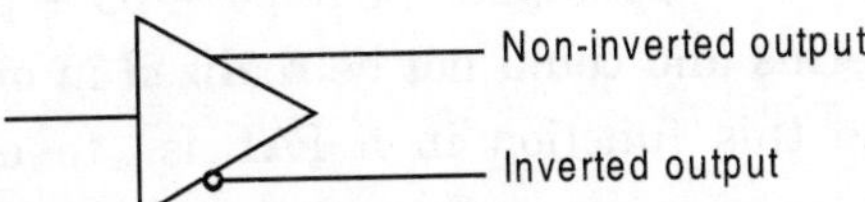

Represents an input buffer which is logically equivalent to

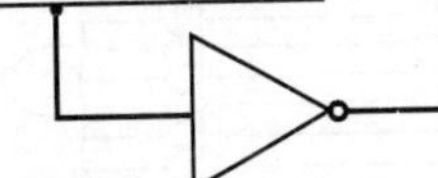

A buffer is used because each PAL input have to drive many AND gate inputs. When the PAL is programmed, the fusible links (F_1, F_2, ... F_8) are selectively blown to leave the desired connection to the AND gate inputs. Connections to the AND gate inputs in a PAL are represented by X's as shown.

Figure 5.17 (*c*) shows the use of PAL segment of Fig. 5.17 (*b*) to realize the function $I_1I_2' + I_1'I_2$. The X's indicate that the I_1 and I_2' lines are connected to the first AND gate, and the I_1', and I_2 lines are connected to the other gate

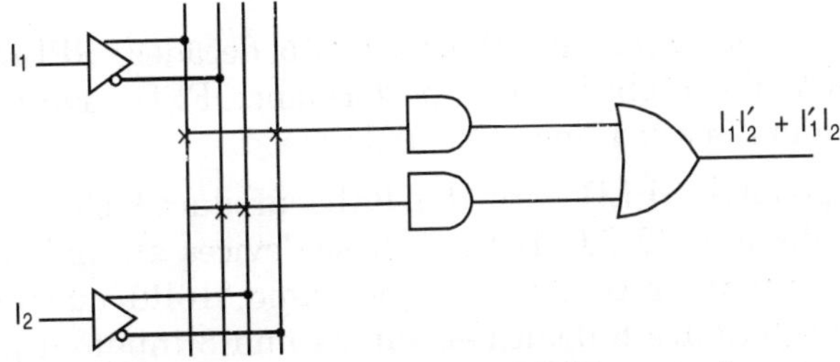

Fig. 5.17 (*c*) Programmed PAL example

In early PALs, only seven product terms could be summed into an OR gate. Therefore, not all functions could be realized with these PLAs. Also, the output was inverted in these early PALs so that what was really realized is

(P1 + P2 + . . . + P7)′ = P1′ • P2′ • . . . • P7′

Example of first-generation PAL is PAL 16L8 having following features.

- 10 input, 2 complemented outputs, 6 I/O pins
- Programmable (one AND term) 3-state outputs
- Seven product terms per output
- 20 pin chip
- 10 input (14 for 20V8)

A sum of products function with a small number of product terms may require a large number of product terms when realized with a PAL.

For example: To implement the function Z = A • B • C + D • E • F

The inverted output of the function Z' is given as

$$
\begin{aligned}
Z' &= (A \bullet B \bullet C)' \bullet (D \bullet E \bullet F)' = (A' + B' + C') \bullet (D' + E' + F') \\
&= A' \bullet D' + A' \bullet E' + A' \bullet F' + B' \bullet D' + B' \bullet E' + B' \bullet F' \\
&\quad + C' \bullet D' + C' \bullet E' + C' \bullet F' \\
&= p1 + p2 + p3 + p4 + p5 + p6 + p7 + p8 + p9
\end{aligned}
$$

This has nine product terms and could not be realized in one pass with the early PALs.

The only way to realize this function in a PAL is to use two passes as shown in Fig. 5.18.

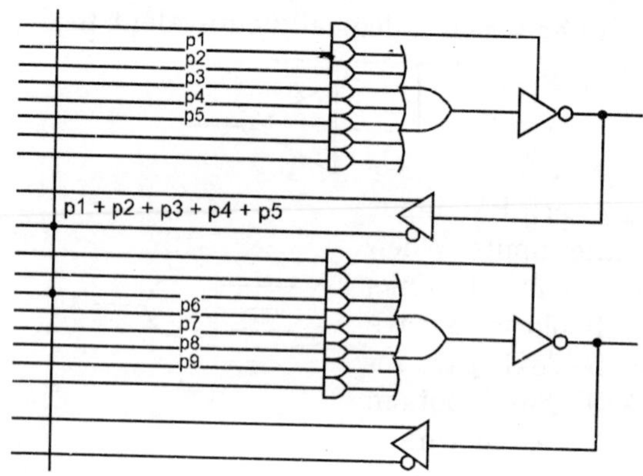

Fig. 5.18

5.3.1 Commercially Available SPLDs

For digital hardware designers for the past two decades, SPLDs are very important devices. SPLDs represent the highest speed-performance FPDs available, and are inexpensive. They are also straightforward and well understood.

Two of the most popular SPLDs are the PALs produced by Advanced Micro Devices (AMD) known as the 16R8 and 22V10. Both of these devices are industry standards and are widely second-sourced by various companies. The name "16R8" means that the PAL has a maximum of 16 inputs (there are 8 dedicated inputs and 8 input/outputs), and a maximum of 8 outputs. The "R" refers to the type of outputs provided by the PAL and means that each output is "registered" by a D flip-flop. Similarly, the "22V10" has a maximum of 22 inputs and 10 outputs. Here, the "V" means each output is "versatile" and can be configured in various ways, some configurations registered and some not.

Another widely used and second sourced SPLD is the Altera Classic EP610. This device is similar in complexity to PALs, but it offers more flexibility in the way that outputs are produced and has larger AND- and OR-planes. In the EP610, outputs can be registered and the flip-flops are configurable as any of D, T, JK, or SR.

In addition to the SPLDs mentioned above many other products are commercially available. All SPLDs share common characteristics, like some sort of logic planes (AND, OR, NOR, or NAND), but each specific product offers unique features that may be particularly suitable for some applications.

5.3.2 Generic Array Logic (GAL)

Generic Array Logic (GAL) is a programmable logic device that can be configured to emulate many earlier PLDs including those with internal flip-flops. GAL 16V8C and 20V8C are examples of Generic Array Logic. The only difference between the two is that the 16V8 is a 20-pin chip and the 20V8 is a 24-pin chip, which uses the extra pins for inputs. The characteristics of these devices are:

- 10 input (14 for 20V8)
- Programmable (one AND term) 3-state outputs
- Seven or eight product terms per output
- Programmable output polarity
    - Realize either true or complemented output signal
    - Realize either POS or SOP directly

When using GAL as a combinational device, all outputs can be programmed to one of the following three configurations except that the two end outputs have some minor limitations as illustrated by Fig. 5.19.

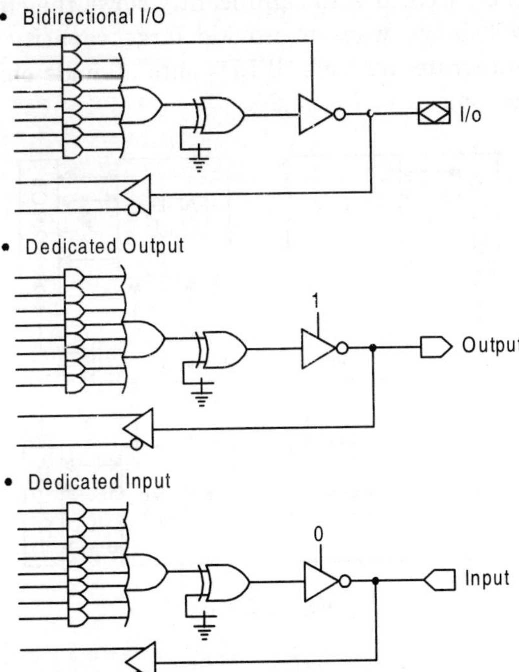

Fig. 5.19

For example, GAL 22V10C is a 24-pin chip having 12 input terminals and 10 input/output terminals. Among outputs, two of the outputs can have up to 8 product terms, two have 10, two have 12, two have 14 and two have 16, except the output buffer control.

The combinational configurations for GAL 22V10C is

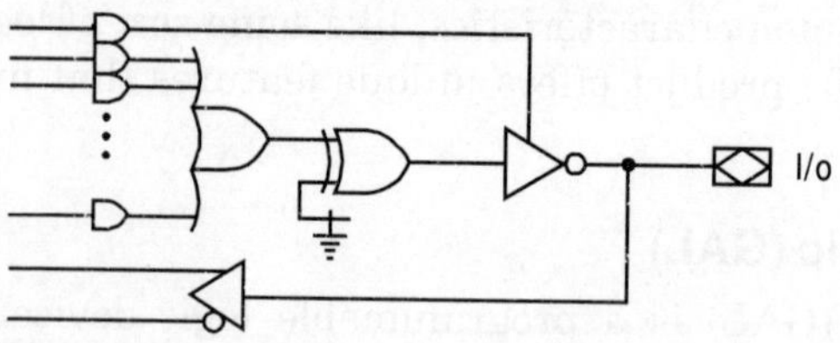

5.3.3 Applications of PLDs

PLDs are often used for address decoding, where they have several clear advantages over the 7400-series TTL parts that they replaced. First, of course, is that one chip requires less board area, power, and wiring than several do. Another advantage is that the design inside the chip is flexible, so a change in the logic doesn't require any rewiring of the board. Rather, the decoding logic can be altered by simply replacing that one PLD with another part that has been programmed with the new design.

5.4 COMPLEX PROGRAMMABLE LOGIC DEVICES (CPLD)

As technology has advanced, it has become possible to produce devices with higher capacity than SPLDs (PALs). The difficulty with increasing capacity of a strict SPLD architecture is that the structure of the programmable logic-planes grows too quickly in size as the number of inputs is increased. It also significantly slows the chip down due to long rows of AND gates. The only feasible way to provide large capacity devices based on SPLD architectures is then to integrate multiple SPLDs onto a single chip, and are referred to as complex PLDs (CPLDs) as shown in Fig. 5.20.

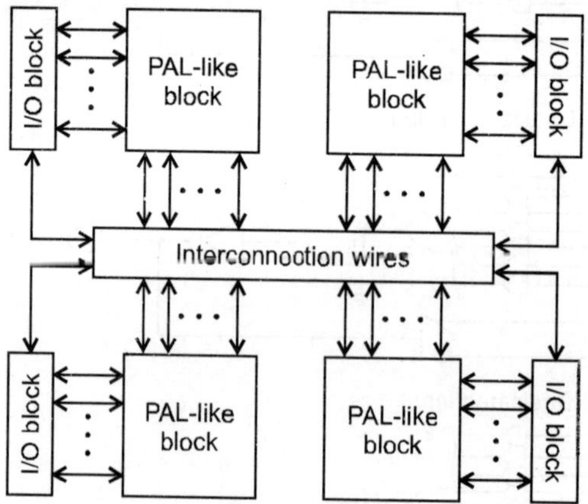

Fig. 5.20 *(a)*

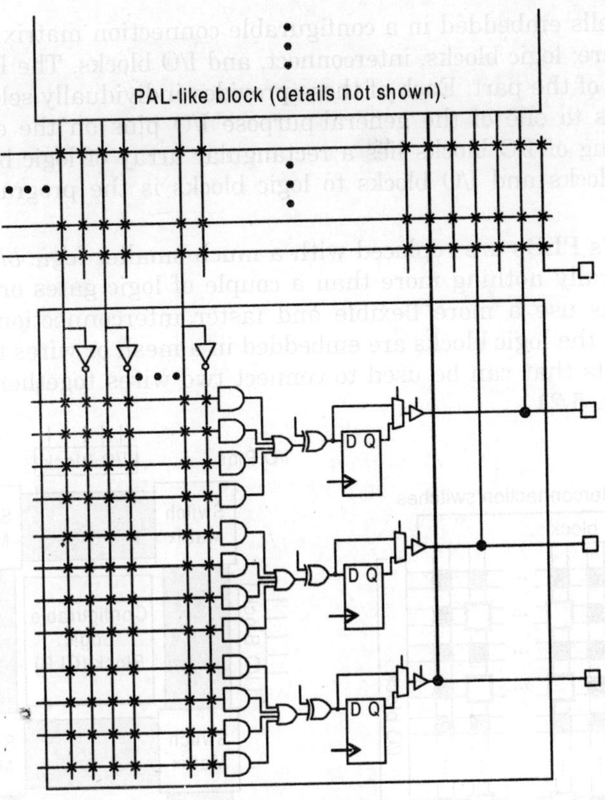

Fig. 5.20 (*b*)

5.4.1 Applications of CPLDs

Because CPLDs offer high speeds and a range of capacities, they are useful for a very wide range of applications, from implementing random glue logic to prototyping small gate arrays. One of the most common uses in industry at this time, and a strong reason for the large growth of the CPLD market, is the conversion of designs that consist of multiple SPLDs into a smaller number of CPLDs.

CPLDs can realize reasonably complex designs, such as graphics controller, LAN controllers, UARTs, cache control, and many others. As a general rule-of-thumb, circuits that can exploit wide AND/OR gates, and do not need a very large number of flip-flops are good candidates for implementation in CPLDs. A significant advantage of CPLDs is that they provide simple design changes through re-programming (all commercial CPLD products are re-programmable). With programmable CPLDs it is even possible to re-configure hardware (an example might be to change a protocol for a communication circuit) without power-down.

Designs often partition naturally into the SPLD-like blocks in a CPLD. The result is more predictable speed-performance than would be the case if a design were split into many small pieces and then those pieces were mapped into different areas of the chip. Predictability of circuit implementation is one of the strongest advantages of CPLD architectures.

5.5 FIELD PROGRAMMABLE GATE ARRAYS (FPGA)

Field Programmable Gate Arrays (FPGAs) are flexible, programmable devices with a broad range of capabilities. Their basic structure consists of an array of universal, programmable logic cells embedded in a configurable connection matrix. There are three key parts of FPGA structure: logic blocks, interconnect, and I/O blocks. The I/O blocks form a ring around the outer edge of the part. Each of these provides individually selectable input, output, or bi-directional access to one of the general-purpose I/O pins on the exterior of the FPGA package. Inside the ring of I/O blocks lies a rectangular array of logic blocks and connecting logic blocks to logic blocks and I/O blocks to logic blocks is the programmable interconnect wiring.

In FPGAs, CPLD's PLDs are replaced with a much smaller *logic block*. The logic blocks in an FPGA are generally nothing more than a couple of logic gates or a look-up table and a flip-flop. The FPGAs use a more flexible and faster interconnection structure than the CPLDs. In the FPGAs, the logic blocks are embedded in a mesh or wires that have programmable interconnect points that can be used to connect two wires together or a wire to a logic block as shown in Fig. 5.21.

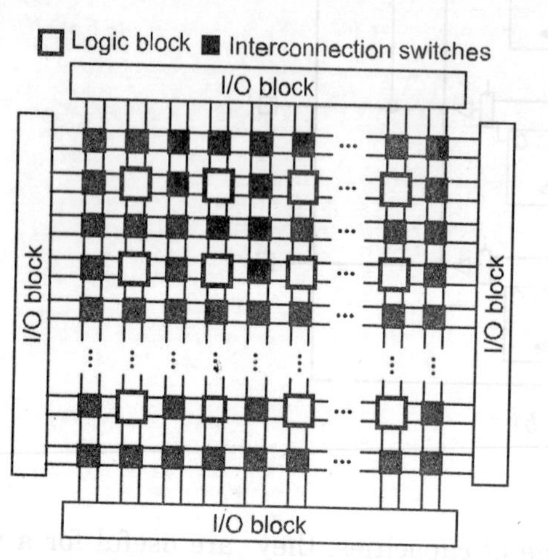

Fig. 5.21

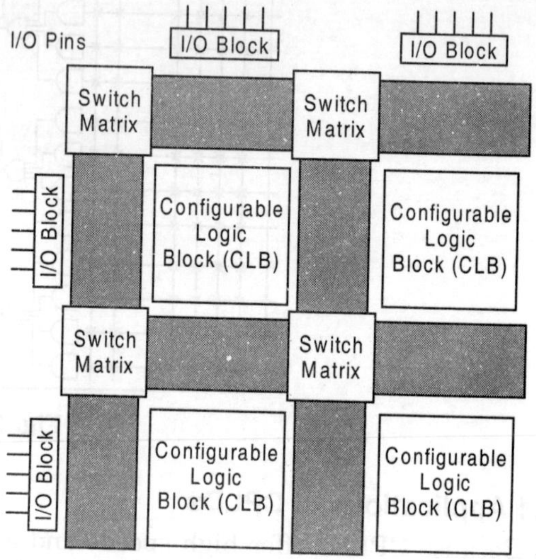

Fig. 5.22

There are several architectures for FPGAs available but the two popular architectures are that, used by Xilinx and Altera. The Xilinx chips utilize an "island-type" architecture, where logic functions are broken up into small islands of 4–6 term arbitrary functions, and connections between these islands are computed. Figure 5.22 illustrates the basic structure of the Xilinx FPGA. Altera's architecture ties the chip inputs and outputs more closely to the logic blocks, as shown in Fig. 5.23. This architecture places the logic blocks around one central, highly connected routing array.

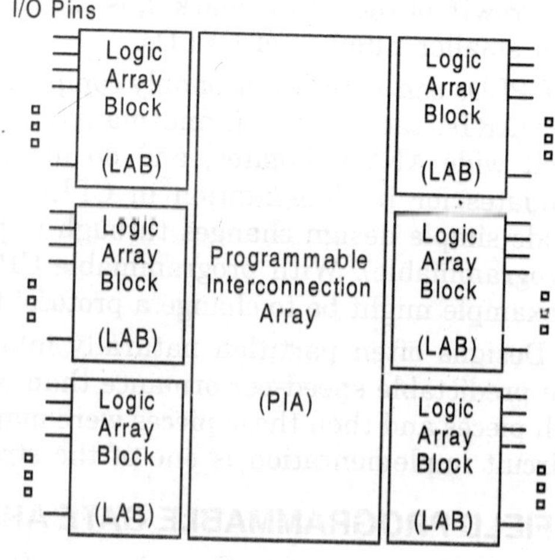

Fig. 5.23

The circuits used to implement combinational logic in logic blocks are called lookup tables (LUT). For example the LUT in the Xilinx XC4000 uses three ROMs to realize the LUTs and generate the following classes of logical functions:

- Any two different functions of 4 variables each plus any other function of 3 variables.
- Any function of 5 variables. How?
- Any function of 4 variables, plus some (but not all) functions of 6 variables.
- Some (but not all) functions of up to 9 variables.

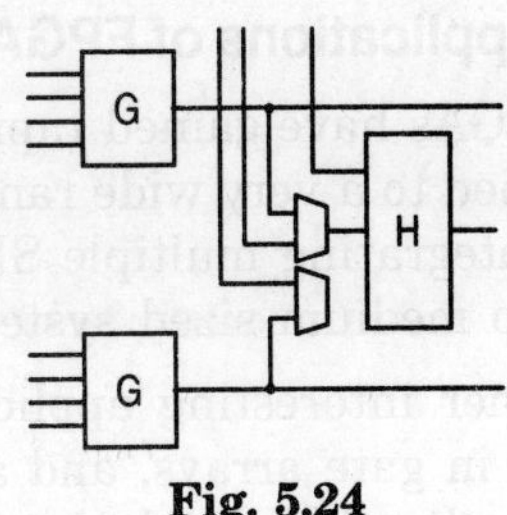

Fig. 5.24

The Fig. 5.24 shows the LUT's structure in Xilinx XC4000. The boxes G and H are ROMs with 16 1-bit words and H is a ROM with 8 1-bit words. While the structure of LUT's in Altera FPGAs are as shown in Fig. 5.25.

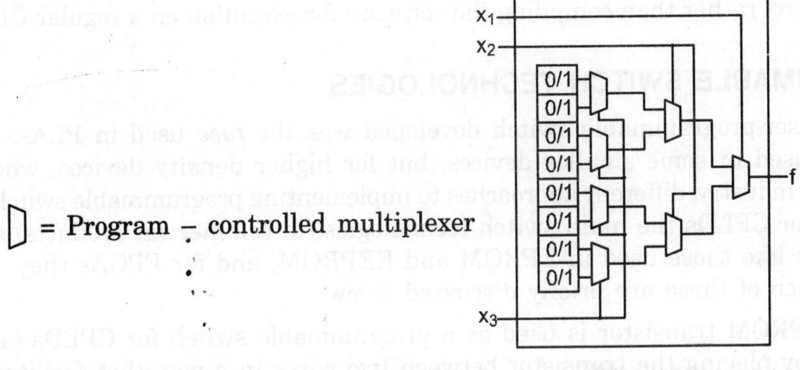

Fig. 5.25

- Example of programmed FPGA

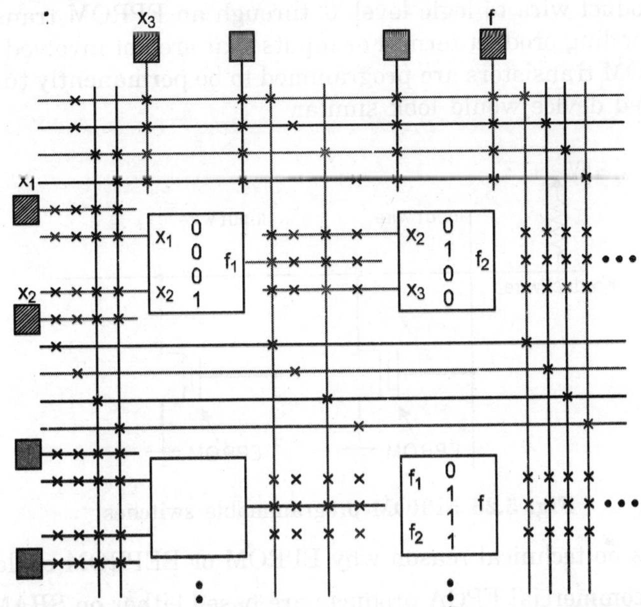

5.5.1 Applications of FPGAs

FPGAs have gained rapid acceptance and growth over the past decade because they can be applied to a very wide range of applications. A list of typical applications includes: random logic, integrating multiple SPLDs, device controllers, communication encoding and filtering, small to medium sized systems with SRAM blocks, and many more.

Other interesting applications of FPGAs are prototyping of designs later to be implemented in gate arrays, and also emulation of entire large hardware systems. The former of these applications might be possible using only a single large FPGA (which corresponds to a small Gate Array in terms of capacity), and the latter would involve many FPGAs connected by some sort of interconnect.

Another promising area for FPGA application, which is only beginning to be developed, is the usage of FPGAs as custom computing machines. This involves using the programmable parts to "execute" software, rather than compiling the software for execution on a regular CPU.

5.6 USER-PROGRAMMABLE SWITCH TECHNOLOGIES

The first type of user-programmable switch developed was the *fuse* used in PLAs. Although fuses are still used in some smaller devices, but for higher density devices, where CMOS dominates the IC industry, different approaches to implementing programmable switches have been developed. For CPLDs the main switch technologies (in commercial products) are floating gate transistors like those used in EPROM and EEPROM, and for FPGAs they are SRAM and antifuse. Each of these are briefly discussed below.

An EEPROM or EPROM transistor is used as a programmable switch for CPLDs (and also for many SPLDs) by placing the transistor between two wires in a way that facilitates implementation of wired-AND functions. This is illustrated in Fig. 5.18, which shows EPROM transistors as they might be connected in an AND-plane of a CPLD. An input to the AND-plane can drive a product wire to logic level '0' through an EPROM transistor, if that input is part of the corresponding product term. For inputs that are not involved for a product term, the appropriate EPROM transistors are programmed to be permanently turned off. A diagram for an EEPROM based device would look similar.

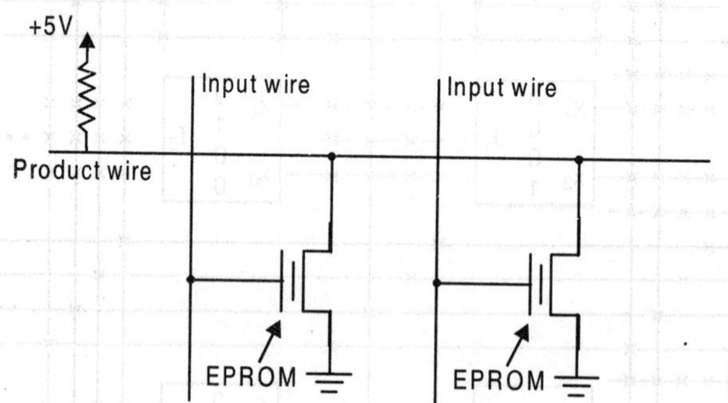

Fig. 5.26 EPROM programmable switches

Although there is no technical reason why EPROM or EEPROM could not be applied to FPGAs, current commercial FPGA products are based either on SRAM or antifuse technologies, as discussed below.

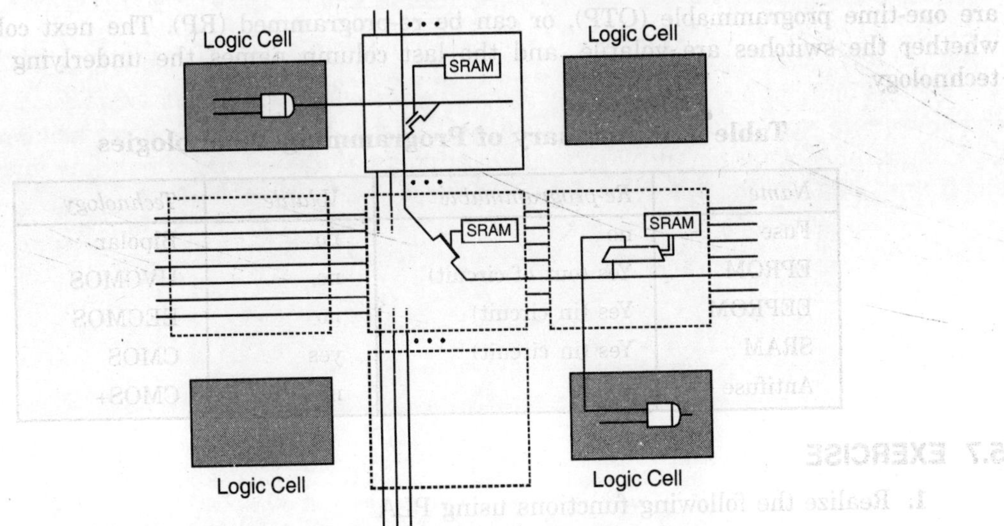

Fig. 5.27 SRAM-controlled programmable switches

An example of usage of SRAM-controlled switches is illustrated in Fig. 5.27, showing two applications of SRAM cells: for controlling the gate nodes of pass-transistor switches and to control the select lines of multiplexers that drive logic block inputs.

The figures gives an example of the connection of one logic block (represented by the AND-gate in the upper left corner) to another through two pass-transistor switches, and then a multiplexer, all controlled by SRAM cells. Whether an FPGA uses pass-transistors or multiplexers or both depends on the particular product.

The other type of programmable switch used in FPGAs is the antifuse. Antifuses are originally open-circuits and take on low resistance only when programmed. Antifuses are suitable for FPGAs because they can be built using modified CMOS technology. As an example, Actel's antifuse structure, known as PLICE, is depicted in Fig. 5.28. The Fig. 5.28 shows that an antifuse is positioned between two interconnect wires and physically consists of three sandwiched layers: the top and bottom layers are conductors, and the middle layer is an insulator. When unprogrammed, the insulator isolates the top and bottom layers, but when programmed the insulator changes to become a low-resistance link. PLICE uses Poly-Si and n+ diffusion as conductors and ONO as an insulator, but other antifuses rely on metal for conductors, with amorphous silicon as the middle layer.

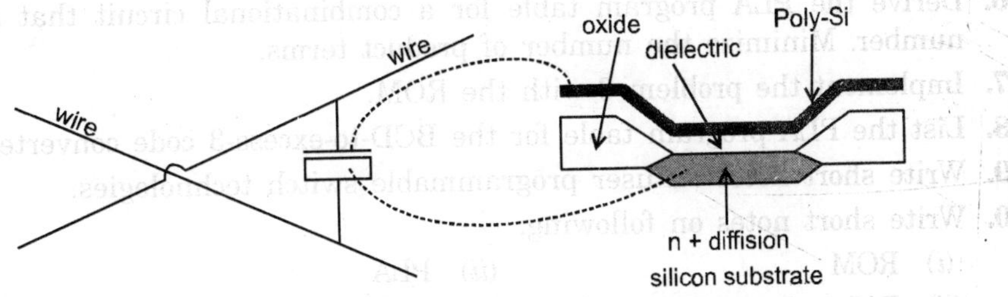

Fig. 5.28 Actel antifuse structure

Table lists the most important characteristics of the programming technologies discussed in this section. The left-most column of the table indicates whether the programmable switches are one-time programmable (OTP), or can be re-programmed (RP). The next column lists whether the switches are volatile, and the last column names the underlying transistor technology.

Table 5.10 Summary of Programming Technologies

Name	*Re-programmable*	*Volatile*	*Technology*
Fuse	no	no	Bipolar
EPROM	Yes (out of circuit)	no	UVCMOS
EEPROM	Yes (in circuit)	no	EECMOS
SRAM	Yes (in circuit)	yes	CMOS
Antifuse	no	no	CMOS+

5.7 EXERCISE

1. Realize the following functions using PLA

$$f_1 \text{ (A, B, C)} = \Sigma \text{ (0, 2, 4, 5)}$$
$$f_2 \text{ (A, B, C)} = \Sigma \text{ (1, 5, 6, 7)}$$

2. Realize the following functions using PLA

$$f_1 = \Sigma \text{ (1, 2, 3, 5)}$$
$$f_2 = \Sigma \text{ (2, 5, 6, 7)}$$
$$f_3 = \Sigma \text{ (0, 4, 6)}$$

3. What is a PLA? Describe its uses.
4. What is ROM? Describe using block diagram. What size ROM would it take to implement a binary multiplier that multiplies two 4 bit-numbers.
5. Implement the combinational circuit specified by the truth table given

Inputs		*Outputs*	
A_1	A_0	F_1	F_2
0	0	0	1
0	1	1	0
1	0	1	1
1	1	1	0

6. Derive the PLA program table for a combinational circuit that squares a 3-bit number. Minimize the number of product terms.
7. Implement the problem 6 with the ROM.
8. List the PLA program table for the BCD-to-excess-3 code converter.
9. Write short notes on user programmable switch technologies.
10. Write short notes on following:

(*i*) ROM (*ii*) PLA
(*iii*) PAL (*iv*) GAL
(*v*) CPLD (*vi*) FPGA

6

SYNCHRONOUS (CLOCKED) SEQUENTIAL CIRCUITS

6.0 INTRODUCTION

In the earlier chapters, we studied the digital circuits whose output at any instant of time are entirely dependent on the input present at that time. Such circuits are called as combinational circuits on the other hand **sequential circuits** are those in which the output at any instant of time is determined by the applied input and past history of these inputs (i.e. present state). Alternately, sequential circuits are those in which output at any given time is not only dependent on the input, present at that time but also on previous outputs. Naturally, such circuits must record the previous outputs. This gives rise to memory. Often, there are requirements of digital circuits whose output remain unchanged, once set, even if the inputs are removed. Such devices are referred as "memory elements", each of which can hold 1-bit of information. These binary bits can be retained in the memory indefinitely (as long as power is delivered) or untill new information is feeded to the circuit.

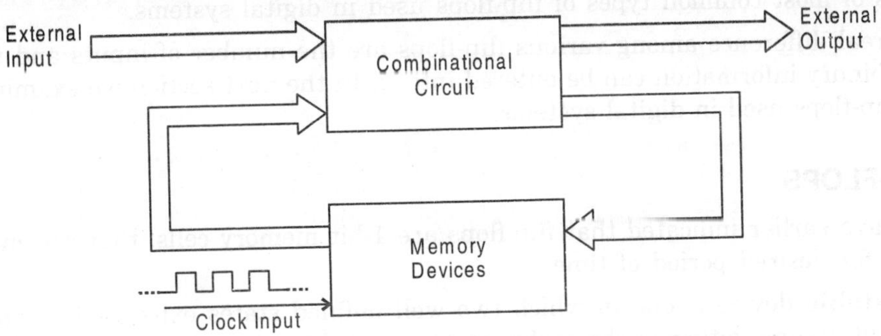

Fig. 6.1 Block diagram of a sequential circuit

A block diagram of a sequential circuit is shown in Fig. 6.1. A **Sequential circuit** can be regarded as a collection of memory elements and combinational circuit, as shown in Fig. 6.1. A feedback path is formed by using memory elements, input to which is the output of combinational circuit. The binary information stored in memory element at any given time is defined as the **state** of sequential circuit at that time. Present contents of memory elements is referred as the **present state.** The combinational circuit receive the signals from external input and from the memory output and determines the external output. They also determine the condition and binary values to change the state of memory. The new contents of the memory elements are referred as **next state** and depend upon the external input and present state. Hence, a sequential circuit can be completely specified by a time sequence of inputs, outputs and the internal states. In general, clock is used to control the operation. The clock frequency determines the speed of operation of a sequential circuit.

There exist two main category of sequential circuits, namely synchronous and asynchronous sequential circuits.

A sequential circuit whose behaviour depends upon the sequence in which the inputs are applied, are called **Asynchronous Sequential Circuits**. In these circuits, outputs are affected whenever a change in inputs are detected. Memory elements used in asynchronous circuits mostly, are time delay devices. The memory capability of time delay devices are due to the propagation delay of the devices. Propagation delay produced by the logic gates are sufficient for this purpose. Hence "An Synchronous sequential circuit can be regarded as a combinational circuit with feedback". However feedback among logic gates make the asynchronous sequential circuits, often susceptible to instability. As a result they may become unstable. This makes the design of asynchronous circuits very tedious and difficult.

A **Synchronous Sequential Circuit** may be defined as a sequential circuit, whose state can be affected only at the discrete instants of time. The synchronization is achieved by using a timing device, termed as **System Clock Generator**, which generates a periodic train of clock pulses. The clock pulses are feeded to entire system in such a way that internal states (i.e. memory contents) are affected only when the clock pulses hit the circuit. A synchronous sequential circuit that uses clock at the input of memory elements are referred as **Clocked Sequential circuit.**

The clocked sequential circuits use a memory element known as **Flip-Flop**. A flip-flop is an electronic circuit used to store 1-bit of information, and thus forms a 1-bit memory cell. These circuits have two outputs, one giving the value of binary bit stored in it and the other gives the complemented value. In this chapter it is our prime concern to discuss the characteristics of most common types of flip-flops used in digital systems.

The real difference among various flip-flops are the number of inputs and the manner in which binary information can be entered into it. In the next section we examine the most general flip-flops used in digital systems.

6.1 FLIP-FLOPS

We have earlier indicated that flip-flops are 1-bit memory cells, that can maintain the stored bit for desired period of time.

A **Bistable** device is one in which two well defined states exist, and at any time the device could assume either of the stable states. A **stable state** is a state, once reached by a device does not changes untill and unless something is done to change it. A toggle switch has two stable states, and can be regarded as a bistable device. When it is closed, it remains closed (A stable state) untill some one opens it. When it is open, it remains open (2nd stable state) untill some one closes it i.e. make it to return to its first stable state. So it is evident that the switch may be viewed as 1-bit memory cell, since it maintains its state (either open or close). Infact any bistable device may be referred as 1-bit memory cell.

A **Flip-Flop** may also be defined as a bistable electronics device whose two stable states are 0V and + 5V corresponding to Logic 0 and Logic 1 respectively. The two stable states and flip-flop as a memory element is illustrated in Fig. 6.2. Figure 6.2 (*a*) shows that the flip-flop is in 'State 0' as output is 0V. This can be regarded as storing Logic 0. Similarly flip-flop is said to be in 'State 1', see Fig. 6.2 (*b*), when the output is 5 V. This can be regarded as storing logic 1. Since at any given time flip-flop is in either of two states the flip-flop may also be regarded as Bistable Multivibrator. Since the state once reached is maintained untill it is deliberately changed, the flip-flop is viewed as memory element.

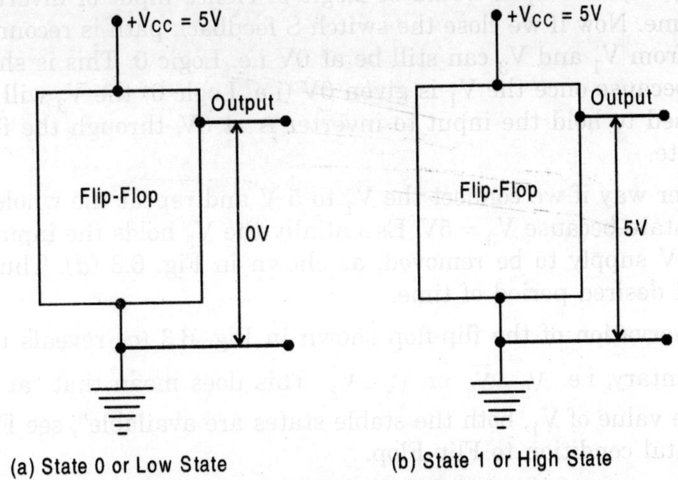

Fig. 6.2 Flip-flop as bistable device

The basic memory circuit or flip-flop can be easily obtained by connecting two inverters (Not gates) in series and then connecting the output of second inverter to the input of first inverter through a feedback path, as shown in Fig. 6.3(a).

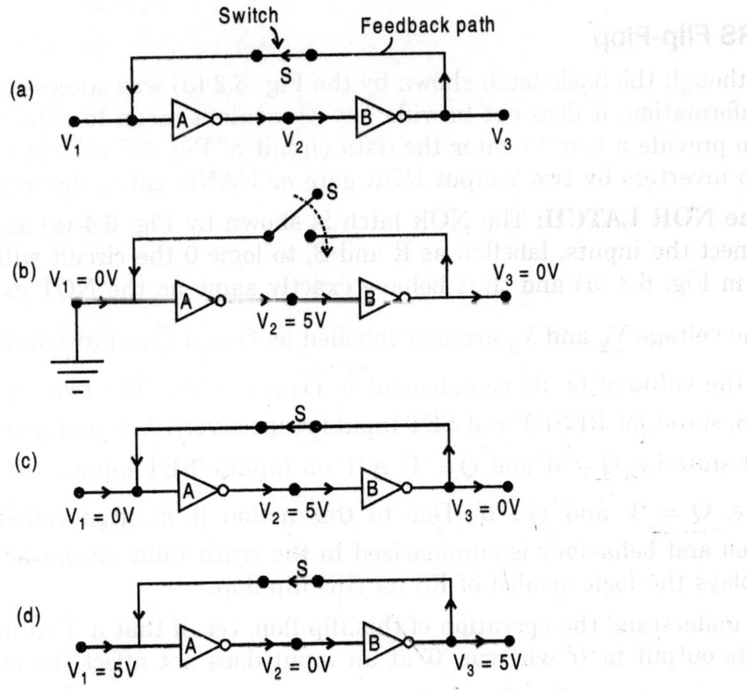

Fig. 6.3 Basic flip-flop or latch 'logic 0' = 0V, and 'logic 1' = 5 V

It is evident from Fig. 6.3 that V_1 and V_3 will always be same, due to very nature of inverters.

Let us define Logic 0 = 0V and Logic 1 = 5 V. Now open the switch 'S' to remove the feedback and connect V_1 to ground, as shown in Fig. 6.3 (b). Thus input to inverter A is Logic 0 and its output would be Logic 1 which is given to the input of inverter B. Since input to inverter B is Logic 1, its output would be Logic 0. Hence input of inverter A and output of inverter B are same. Now if we close the switch S feedback path is reconnected, then ground can be removed from V_1 and V_3 can still be at 0V i.e. Logic 0. This is shown in Fig. 6.3 (c). This is possible because once the V_1 is given 0V (i.e. Logic 0) the V_3 will also be at 0V and then it can be used to hold the input to inverter A at 0V, through the feedback path. This is first stable state.

In the simpler way if we connect the V_1 to 5 V and repeat the whole process, we reach to second stable state because V_3 = 5V. Essentially the V_3 holds the input to inverter. A (i.e. V_1), allowing + 5V supply to be removed, as shown in Fig. 6.3 (*d*). Thus V_3 = 5 V can be maintained untill desired period of time.

A simple observation of the flip-flop shown in Fig. 6.3 (*a*) reveals that V_2 and V_3 are always complementary, i.e. $V_2 = \overline{V_3}$ or $V_3 = \overline{V_2}$. This does mean that "at any point of time, irrespective of the value of V_1, both the stable states are available", see Figs. 6.3 (*c*) and (*d*). This is fundamental condition to Flip-Flop.

Since the information present at the input (i.e. at V_1) is locked or latched in the circuit, it is also referred as **Latch**.

When the output is in low state (i.e. V_3 = 0 V), it is frequently referred as **Reset State**. Where as when the output is in high state (i.e. V_3 = 5 V), it is conveniently called as **Set State**. Figures 6.3 (*c*) and (*d*) shows the reset and set states, respectively.

6.1.1 RS Flip-Flop

Although the basic latch shown by the Fig. 6.3 (*a*) was successful to memorize (or store) 1-bit information, it does not provide any convenient mean to enter the required binary bit. Thus to provide a way to enter the data circuit of Fig. 6.3 (*a*) can be modified by replacing the two inverters by two 2-input NOR gate or NAND gates, discussed in following articles.

The NOR LATCH: The NOR latch is shown by Fig. 6.4 (*a*) and 6.4 (*b*). Notice that if we connect the inputs, labelled as R and S, to logic 0 the circuit will be same as the circuit shown in Fig. 6.3 (*a*) and thus behave exactly same as the NOT gate latch of Fig. 6.3 (*a*).

The voltage V_2 and V_3 are now labelled as Q and $\overline{Q}$ and are declared as output. Regardless of the value of Q, its complement is $\overline{Q}$ (as $V_3 = \overline{V_2}$). The two inputs to this flip-flop are R and S, stand for RESET and SET inputs respectively. A '1' on input R switches the flip-flop in reset state i.e. Q = '0' and $\overline{Q}$ = '1'. A '1' on inputs (SET input) will bring the latch into set state i.e. Q = '1' and $\overline{Q}$ = '0'. Due to this action it is often called **set-reset latch**. The operation and behaviour is summarized in the truth table shown by Fig. 6.4 (*c*). Figure 6.4 (*d*) displays the logic symbol of RS (or SR) flip-flop.

To understand the operation of this flip-flop, recall that a '1' at any input of a NOR gate forces its output to '0' whereas '0' at an input does not affect the output of NOR gate.

When inputs are S = R = 0, first row of truth tables it does not affect the output. As a result the Latch maintains its state. For example if before application of the inputs S = R = 0, the output was Q = 1, then it remains 1 after S = R = 0 are applied. Thus, when both the inputs are low the flip-flop maintain its last state. Thats why the truth table has entry Q in first row.

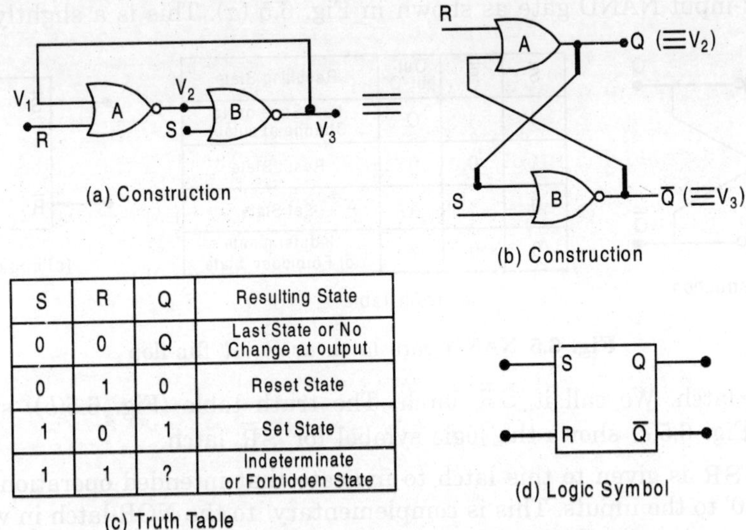

S	R	Q	Resulting State
0	0	Q	Last State or No Change at output
0	1	0	Reset State
1	0	1	Set State
1	1	?	Indeterminate or Forbidden State

(c) Truth Table

Fig. 6.4 Basic NOR gate latch or RS (or SR) flip-flop

Now, if S = 0 and R = 1, output of gate-A goes low i.e. Q = 0. The Q is connected to input of gate-B along with S input. Thus with Q = 0 both the inputs to NOR gate B are LOW. As a result $\overline{Q} = 1$. This Q and $\overline{Q}$ are complementary. Since Q = 0 the flip-flop is said to be in "reset state". This is indicated by the second row of the truth table.

Now if S = 1 and R = 0 output of gate-B is LOW, making both the inputs of gate-A LOW, consequently Q = 1. This is "set state". As Q = 1 and $\overline{Q} = 0$, the two outputs are complementary. This is shown in third row of truth table.

When S = 1 and R = 1, output of both the gates are forced to Logic 0. This conflicts with the definition that both Q and $\overline{Q}$ must be complementary. Hence this condition must not be applied to SR flip-flop. But if due to some reasons S = 1 and R = 1 is applied, then it is not possible to predict the output and flip-flop state is said to be indcterminate. This is shown by the last row of truth table.

It is worth to devote some time to investigate why S = R = 1 results indeterminate state while we said earlier that output of both the gates go LOW for this input. This is true due to the logic function of NOR gate that if any of the input is HIGH output is LOW. In the circuit of Fig. 6.4 (*b*) both Q and $\overline{Q}$ are LOW as long as S and R are HIGH. The problem occurs when inputs S and R go to LOW from High. The two gates can not have exactly same propagation delay. Now the gate having smaller delay will change its state to HIGH earlier than the other gate. And since this output (i.e. Logic 1) is feeded to the second gate, the output of second gate is forced to stay at Logic 0. Thus depending upon the propagation delays of two gates, the flip-flop attains either of the stable states (i.e. either Q = 1 or Q = 0). Therefore it is not possible to predict the state of flip-flop after the inputs S = R = 1 are applied. Thats why the fourth row of truth table contains a question mark (?). For the above reasons the input condition S = R = 1 is forbidden.

The NAND Gate Flip-Flop

The NOR gate latch shown by Fig. 6.4 (*b*) may also be modified by replacing each inverter by a 2-input NAND gate as shown in Fig. 6.5 (*a*). This is a slightly different latch

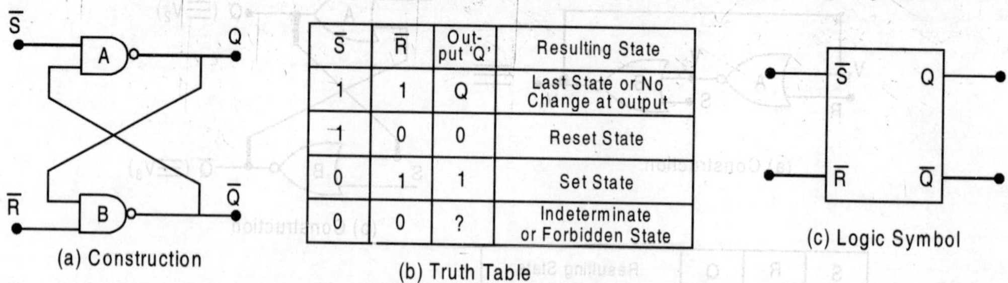

$\overline{S}$	$\overline{R}$	Output 'Q'	Resulting State
1	1	Q	Last State or No Change at output
1	0	0	Reset State
0	1	1	Set State
0	0	?	Indeterminate or Forbidden State

Fig. 6.5 NAND gate latch or $\overline{S}$, $\overline{R}$ flip-flop

from the NOR latch. We call it $\overline{S}\,\overline{R}$ latch. The truth table (Fig. 6.5(*b*)) summarizes the operation and Fig. 6.5(*c*) shows the logic symbol for $\overline{S}\overline{R}$ latch.

The name $\overline{S}\overline{R}$ is given to this latch to indicate that intended operation is achieved on asserting logic '0' to the inputs. This is complementary to the NOR latch in which operation is performed when input is logic '1'.

The explanation of operation of $\overline{S}\overline{R}$ flip-flop lies in the statement that, if any input of NAND gate goes LOW the output is HIGH, whereas a '1' at any NAND input does not affect the output. Moreover, the output will be LOW only and only when all the inputs to NAND gate are HIGH.

When both $\overline{S}$ and $\overline{R}$ are HIGH i.e. $\overline{S} = \overline{R} = 1$, then the NAND output are not affected. Thus last state is maintained. When $\overline{S} = 1$ and $\overline{R} = 0$ then output of gate-B goes HIGH making both the inputs to NAND-A as HIGH. Consequently Q = 0 which is reset state. In the similar way $\overline{S} = 0$ and $\overline{R} = 1$ bring the circuit to set state i.e. Q = 1. When both the inputs are LOW i.e. $\overline{S} = \overline{R} = 0$ both Q and $\overline{Q}$ are forced to stay HIGH which inturns lead to indeterminate state for the similar reasons given for NOR latch.

The $\overline{S}\overline{R}$ flip-flop can be modified further by using two additional NAND gates.

These two gates, labelled as C and D, are connected at $\overline{S}$ and $\overline{R}$ Fig. 6.5 (*a*) inputs to act as NOT gate, as shown in Fig. 6.6 (*a*). This converts $\overline{S}\overline{R}$ latch into a latch that behaves exactly same as the NOR gate flip-flop (i.e. NOR latch), shown in Fig. 6.4 (*b*). Hence this latch also is referred as SR flip-flop. The truth table and logic symbol will be same as that of NOR latch. The truth table may also be obtained by inverting all the input bits in $\overline{S}\overline{R}$ truth table shown in Fig. 6.5 (b) as input to $\overline{S}\overline{R}$ latch is complemented. To understand the operation of this latch, consider the Fig. 6.6 (*a*).

When both S and R inputs are LOW outputs of NAND gates C and D are HIGH. This is applied to the inputs of $\overline{S}\overline{R}$ latch which maintains the last state in response to $\overline{S} = \overline{R} = 1$ (see Fig. 6.5 (*b*)). In the same way, application of S = 0 and R = 1 results $\overline{S} = 1$ and $\overline{R} = 0$, and consequently the latch attains "Reset State". Similarly, the other input combination in truth table can be verified.

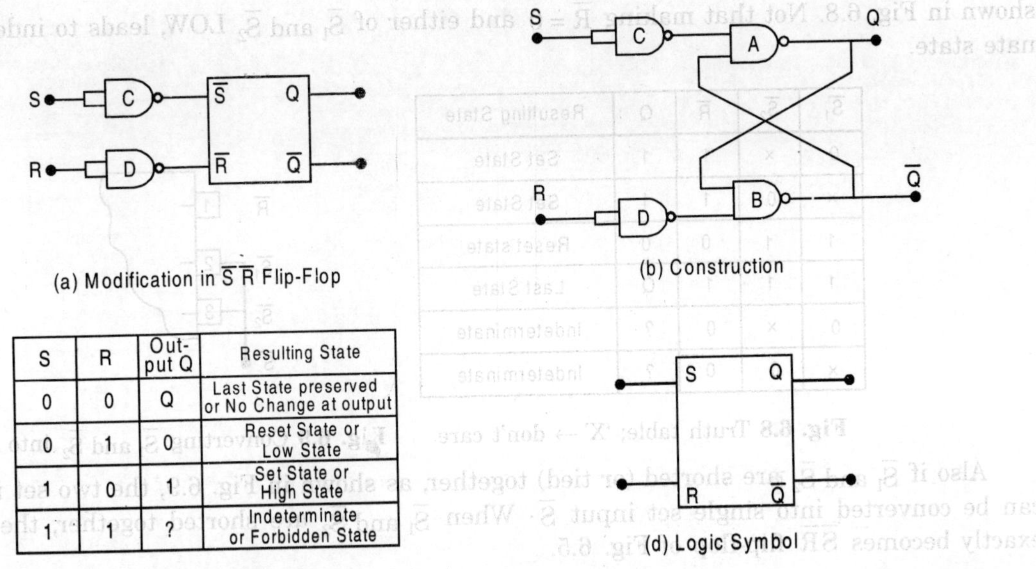

S	R	Output Q	Resulting State
0	0	Q	Last State preserved or No Change at output
0	1	0	Reset State or Low State
1	0	1	Set State or High State
1	1	?	Indeterminate or Forbidden State

(c) Truth Table

Fig. 6.6 NAND gate latch or SR flip-flop

At this point we advice readers to verify the truth table of SR latch through the NAND gate construction of this latch, shown in Fig. 6.6 (*b*).

For the beginners it is worth to go back to the beginning of this article (i.e. start of 6.1.1). Infact the SR (or RS) flip-flop gives the basic building block to study and analyse various flip-flops, and their application in the design of clocked sequential circuits.

Example. *Draw the internal block diagram alongwith pinout for IC 74LS279, a quad set reset latch. Explain its operation in brief with the help of truth table.*

Sol. **Figure 6.7 shows the required block diagram and pinout.**

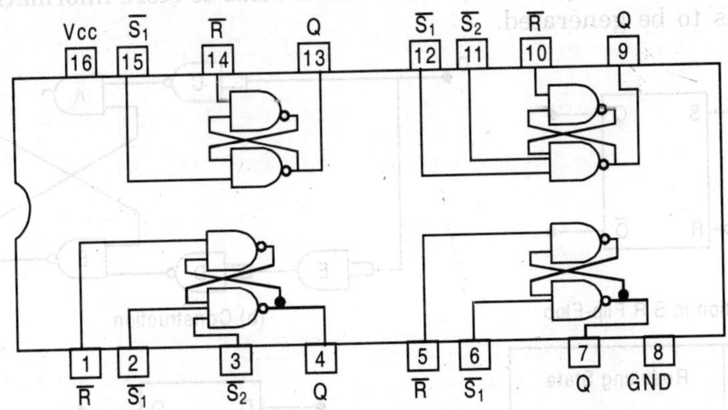

Fig. 6.7 IC 74LS279, a quad set reset latch

From the figure it is evident that the IC contains 4 $\overline{S}\,\overline{R}$ latch shown earlier in Fig. 6.5. Two flip-flops have two inputs, named $\overline{S}_1$, $\overline{R}$ are exact reproduction of $\overline{S}\,\overline{R}$ latch shown in Fig. 6.5. Remaining two flip-flops have three inputs labelled $\overline{S}_1$, $\overline{S}_2$, $\overline{R}$, in which instead of single $\overline{S}$ input we get two set inputs $\overline{S}_1$ and $\overline{S}_2$. Since the latches are constructed by NAND gates a LOW either on $\overline{S}_1$ or on $\overline{S}_2$ will set the latch. Truth table summarizing its operation is shown in Fig. 6.8. Not that making $\overline{R} = 0$ and either of $\overline{S_1}$ and $\overline{S_2}$ LOW, leads to indeterminate state.

$\overline{S_1}$	$\overline{S_2}$	$\overline{R}$	Q	Resulting State
0	×	1	1	Set State
×	0	1	1	Set State
1	1	0	0	Reset state
1	1	1	Q	Last State
0	×	0	?	Indeterminate
×	0	0	?	Indeterminate

Fig. 6.8 Truth table; 'X' → don't care

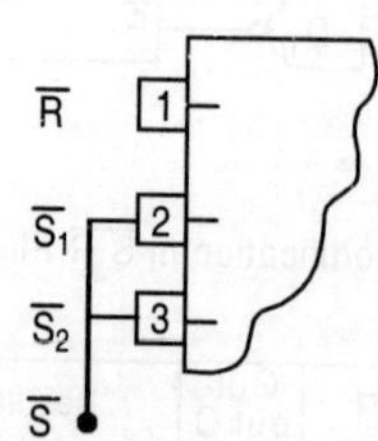

Fig. 6.9 Converting $\overline{S_1}$ and $\overline{S_2}$ into $\overline{S}$

Also if $\overline{S_1}$ and $\overline{S_2}$ are shorted (or tied) together, as shown in Fig. 6.9, the two set inputs can be converted into single set input $\overline{S}$. When $\overline{S_1}$ and $\overline{S_2}$ are shorted together, the latch exactly becomes $\overline{SR}$ flip-flop of Fig. 6.5.

6.1.2 D Flip-Flop

The SR latch, we discussed earlier, has two inputs S and R. At any time to store a bit, we must activate both the inputs simultaneously. This may be troubling in some applications. Use of only one data line is convenient in such applications.

Moreover the forbidden input combination S = R = 1 may occur unintentionally, thus leading the flip-flop to indeterminate state.

In order to deal such issues, SR flip-flop is further modified as shown in Fig. 6.10. The resultant latch is referred as D flip-flop or D latch. The latch has only one input labelled D (called as Data input). An external NAND gate (connected as inverter) is used to ensure that S and R inputs are always complement to each other. Thus to store information in this latch, only one signal has to be generated.

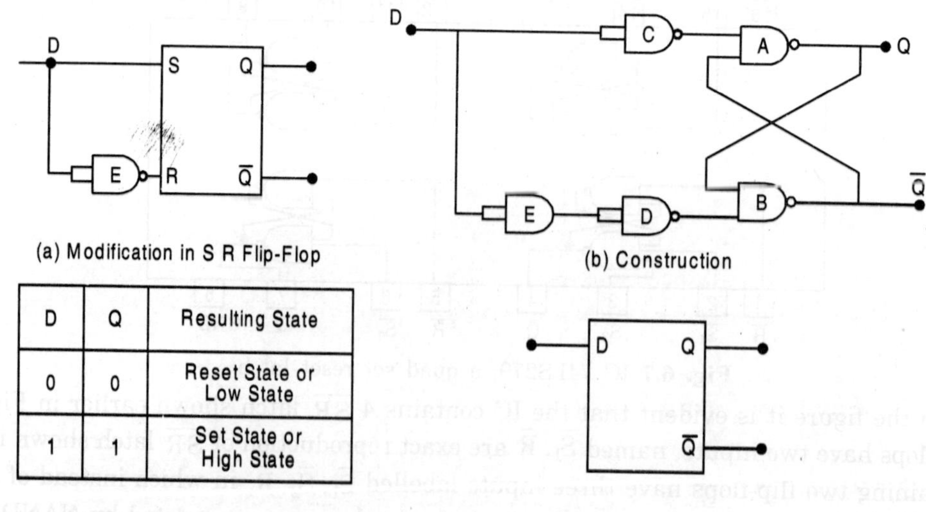

D	Q	Resulting State
0	0	Reset State or Low State
1	1	Set State or High State

(c) Truth Table

Fig. 6.10 D flip-flop or D latch

Operation of this flip-flop is straight forward. At any instant of time the output Q is same as D (i.e. Q = D). Since output is exactly same as the input, the latch may be viewed as a delay unit. The flip-flop always takes some time to produce output, after the input is applied. This is called propagation delay. Thus it is said that the information present at point D (i.e. at input) will take a time equal to the propagation delay to reach to Q. Hence the information is delayed. For this reason it is often called as **Delay (D) Flip-Flop**. To understand the operation of this latch, considr Fig. 6.10 (*a*).

As shown in figure, the D input goes directly to S and its complement is applied to R input. When data input is LOW i.e. D = 0, we get S = 0 and R = 1.50 flip-flop reaches to RESET State where Q = 0. When D = 1 the S input receives 1 and R = 0. Thus the flip-flop goes to SET state, where Q = 1. This operation is summarized in truth table, shown in Fig. 6.10 (c). It is interesting to note that the next state of D flip-flop is independent of present state. It means that if input D = 1 the next state will be SET state, whether presently it is in SET or RESET state.

Furthermore, by Fig. 6.10 (*a*) it is clear that the external inverter ensures that the forbidden condition S = R = 1 will never arrive. The D flip-flops are popularly used as the delay devices and/or latches. In general, simply saying latch means a D flip-flop.

Example. *A logic circuit having a single input labelled X, and two outputs Y_1 and Y_2 is shown in fig. 6.11. Investigate the circuit and find out does this circuit represents a latch? If yes, then name the latch and draw the truth table for this circuit.*

Sol. To investigate the circuit, we find out values of outputs Y_1 and Y_2 for each and every possibility of input X.

A careful inspection of circuit reveals that the portion of the circuit which consist of two NAND gates A and B

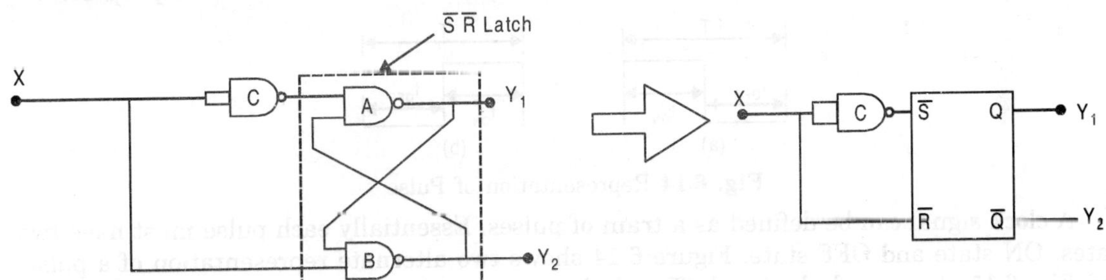

Fig. 6.11 Logic circuit for example 6.2 **Fig. 6.12** Simplified circuit for Fig. 6.11

represent $\overline{S}\overline{R}$ flip-flop. This portion of the circuit is surrounded by dotted lines in Fig. 6.11. The circuit is redrawn in Fig. 6.12 for simplicity. This simplification shows that input to $\overline{S}$ is $\overline{X}$ and input to $\overline{R}$ is X or $\overline{S} = \overline{X}$ and $\overline{R} = \overline{X}$. The outputs Y_1 and Y_2 are nothing but the Q and $\overline{Q}$ outputs $\overline{S}\overline{R}$ latch i.e. $Y_1 = Q$ and $Y = \overline{Q}$. Thus, the outputs Y_1 and Y_2 are always complementary.

Hence, when the input X is LOW i.e. X = 0, it results in $\overline{S} = 1$ and $\overline{R} = 0$. The latch is forced to reset state, in which case Q = 0 and $\overline{Q} = 1$, consequently $Y_1 = 0$ and $Y_2 = 1$. Thus, for X = 0 we get $Y_1 = 0$ and $Y_2 = 1$. In the similar way when X = 1, $\overline{S} = 0$ and $\overline{R} = 1$ making $Y_1 = 1$ and $Y_2 = 0$ which is set state. We now summarize these results in a truth table shown in Fig. 6.13. From the truth table it is clear that $Y_1 = X$ and $Y_2 = \overline{X}$. Thus the given circuit represents a D Latch which gives $Q = D$ and $\overline{Q} = \overline{D}$. In the Figs. 6.11 and 6.12 the input D is renamed as X and the outputs Q and $\overline{Q}$ are named as Y_1 and Y_2 respectively.

X	$\overline{S}$	$\overline{R}$	Q	$\overline{Q}$	$Y_1 = Q$	$Y_2 = \overline{Q}$
0	1	0	0	1	0	1
1	0	1	1	0	1	0

(a)

X	Y_1	Y_2
0	0	1
1	1	0

(b)

Fig. 6.13 Truth table for Fig. 6.12

In Fig. 6.12, if the output of gate C is connected to $\overline{R}$ and input X is directly connected to $\overline{S}$ then $Y_1(= Q) = \overline{X}$ and $Y_2(= \overline{Q}) = X$. Therefore, the circuit may be referred as inverted D latch.

6.1.3 Clocked Flip-Flops

All the flip-flops discussed earlier are said to be **transparent**, because any chance in input is immediately accepted and the output changes accordingly. Since they consist of logic gates along with feedback they are also regarded as **asynchronous flip-flops.**

However, often there are requirements to change the state of flip-flop in synchronism with a train of pulses, called as **Clock**. In fact we need a control signal through which a flip-flop can be instructed to respond to input or not. Use of clock can serve this purpose.

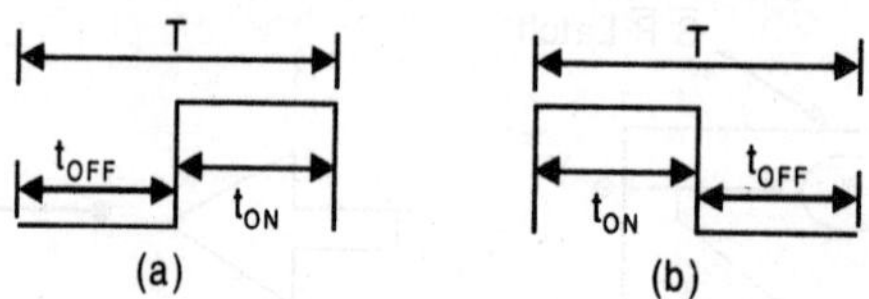

Fig. 6.14 Representation of Pulse

A clock signal can be defined as a train of pulses. Essentially each pulse must have two states, ON state and OFF state. Figure 6.14 shows two alternate representation of a pulse, and Fig. 6.15 shows a clock signal. The clock pulses are characterized by the **duty cycle**, which is representative of ON time in the total time period of pulse, and is given as:

$$\text{Duty Cycle} = D = \frac{t_{ON}}{t_{ON} + t_{OFF}} \text{ or } D = \frac{t_{ON}}{T}$$

In the digital systems we need a clock with duty cycle $D \le 50\%$. The OFF time of a pulse is also referred as **bit-time**. This is the time in which flip-flop remains unaffected in either of two stable states. The state of latch during this time is due to the input signals present during ON time of pulse.

State Q_{n+1} is due to the inputs present during ON time of $(n + 1)$th pulse i.e at $t = n\text{T}$. In the analysis and discussion we adopt the designation Q_n to represent **"present state"** which is the state before the ON time of $(n + 1)$th pulse or state just before the time $t = n\text{T}$ in Fig. 6.15, and Q_{n+1} as **"next state"** i.e. the state just after the ON time of

$(n + 1)$th clock pulse. Thus Q_n represents the state (or output) in bit time n and Q_{n+1} represents the output Q in bit time $n + 1$, as shown in Fig. 6.15.

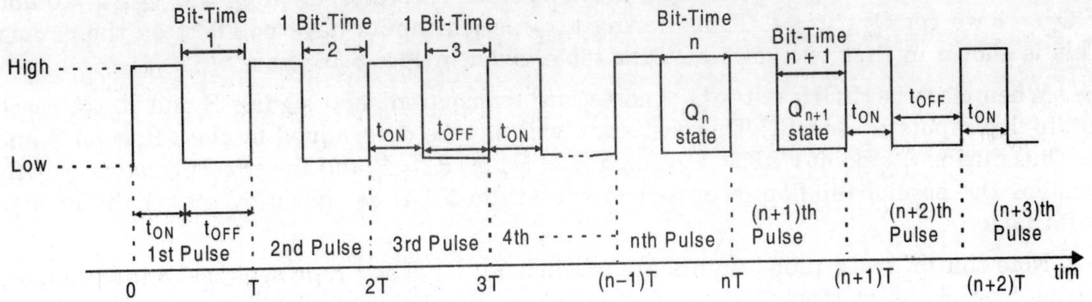

Fig. 6.15 Clock signal-pulse train of shape shown in Fig. 6.14 (b)

As we said earlier, the clock pulse can be used as a control signal. It allows the flip-flop to respond to inputs during t_{ON} period of clock, it is called **enabling** the flip-flop. Where as the flip-flop is instructed, not to respond to inputs during t_{OFF} period of clock, i.e. flip-flop maintains its output irrespective of changes in input. This is called **disabling** the flip-flop.

In this way it is possible to **strobe** or **clock** the flip-flop in order to store the information at any time said alternately clocking allow us to selectively enable or disable the flip-flop which is a necessary requirement in large digital systems.

Clocked SR Flip-Flop: A simple way to get a clocked SR flip-flop is to AND the input signals with clock and then apply them to S and R inputs of flip-flop as shown in Fig. 6.16 (a). For the simplicity SET and RESET inputs of unclocked SR latch are labelled S_1 and R_1 respectively whereas external inputs are labelled as S and R.

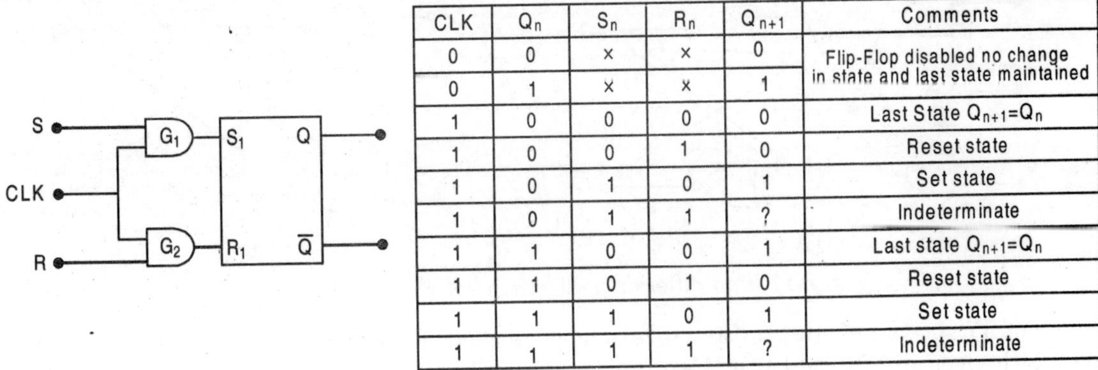

CLK	Q_n	S_n	R_n	Q_{n+1}	Comments
0	0	×	×	0	Flip-Flop disabled no change in state and last state maintained
0	1	×	×	1	
1	0	0	0	0	Last State $Q_{n+1}=Q_n$
1	0	0	1	0	Reset state
1	0	1	0	1	Set state
1	0	1	1	?	Indeterminate
1	1	0	0	1	Last state $Q_{n+1}=Q_n$
1	1	0	1	0	Reset state
1	1	1	0	1	Set state
1	1	1	1	?	Indeterminate

(a) Construction of clocked SR flip-flop.

(b) Truth Table

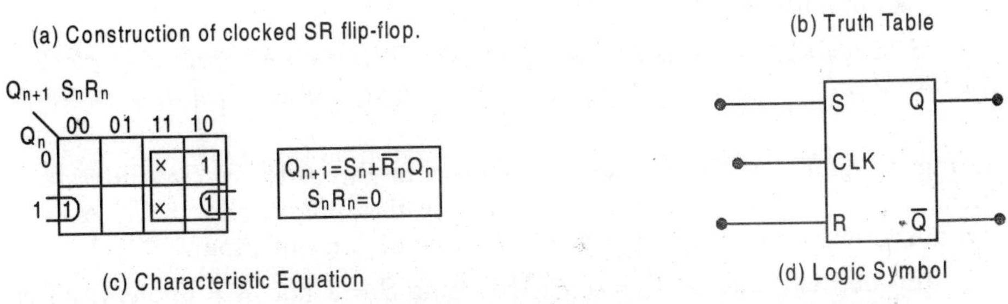

(c) Characteristic Equation

(d) Logic Symbol

$$Q_{n+1} = S_n + \overline{R}_n Q_n, \quad S_n R_n = 0$$

Fig. 6.16 Clocked RS (or SR) flip-flop

When clock (abbreviated as CLK) is LOW, outputs of gates G_1 and G_2 are forced to 0, which is applied to flip-flop inputs. Thus when CLK = 0, S_1 = 0 and R_1 = 0. Since both the inputs are LOW, flip-flop remain in its previous state. Alternatively if $Q_n = 0$, $Q_{n+1} = 0$ and if $Q_n = 1$ we get $Q_{n+1} = 1$. Thus, during t_{OFF} period inputs have no effect on the circuit. This is shown in first two rows of truth table given in Fig. 6.16 (b).

When clock is HIGH, gates G_1 and G_2 are transparent and signals S and R can reach to flip-flop inputs S_1 and R_1. The next state will now be determined by the values of S and R. Thus during t_{ON} point CLK = 1 causing S_1 = S and R_1 = R and the circuit behaves exactly same as the normal flip-flop discussed in subsection 6.1.1, as shown by rest of the rows of truth table.

Note that in truth table, inputs are labelled S_n, R_n. They represent the value of inputs during bit-time '*n*' at the t_{ON} time of $(n + 1)$th pulse or at $t = n$T in Fig. 6.15. The output also is Q_{n+1} not simply Q. Because presence of clock pulses force us to consider two different instants of time : the time before application of pulse, Q_n **(i.e. present state)** and the time after the application of pulse, Q_{n+1} **(i.e. next state).**

Characteristic equation for this flip-flop is obtained from K-map shown in Fig. 6.16 (c), which is an algebric expression for the binary information of the truth table. This expression gives the value of next state as a function of present state and present inputs. The indeterminate conditions are marked "X"-don't care in the map because, depending upon the propagation delay of logic gates, state can be either 1 or 0. Inclusion of relation $S_n.R_n = 0$ as a part of characteristics equation is due to the requirement that both S_n and R_n must not be made 1 simultaneously.

Finally, the logic symbol of clocked SR flip-flop is shown in Fig. 6.16 (d), which now has three inputs named S, R and CLK.

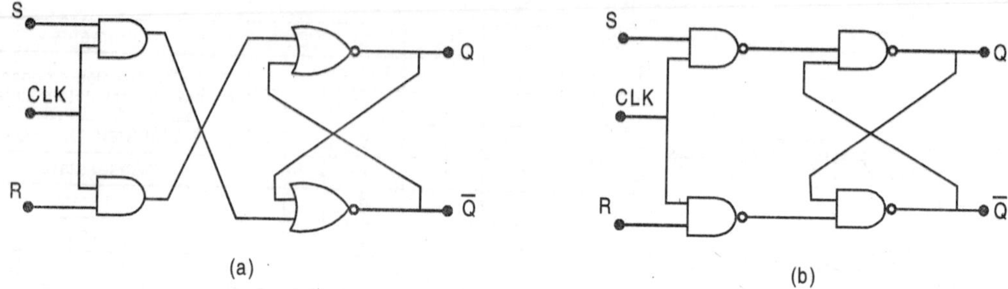

Fig. 6.17 Two different realizations of clocked SR flip-flop

Figure 6.17 shows two alternate realizations of clocked SR flip-flop. Both the realizations are popularly used in MSI and LSI (Medium and Large Scale Integrated) circuits. In many texts the signal CLOCK is also labelled ENABLE or EN.

Example 1. *Fig. 6.18 (a) shows the input waveforms S, R and CLK, applied to clocked SR flip-flop. Obtain the output waveform Q and explain it in brief. Assume flip-flop is resent initially.*

Sol. The resulting waveform at the output Q is shown in Fig. 6.18 (b). To understand the output waveform, recall that the inputs affect the flip-flop only when the clock = 1 otherwise flip-flop maintains its previous output irrespective of present input.

Initially at t = 0 flip-flop is reset i.e. Q = 0. At this time S = 1 and R = 0, but flip-flop remains unaffected since CLK = 0.

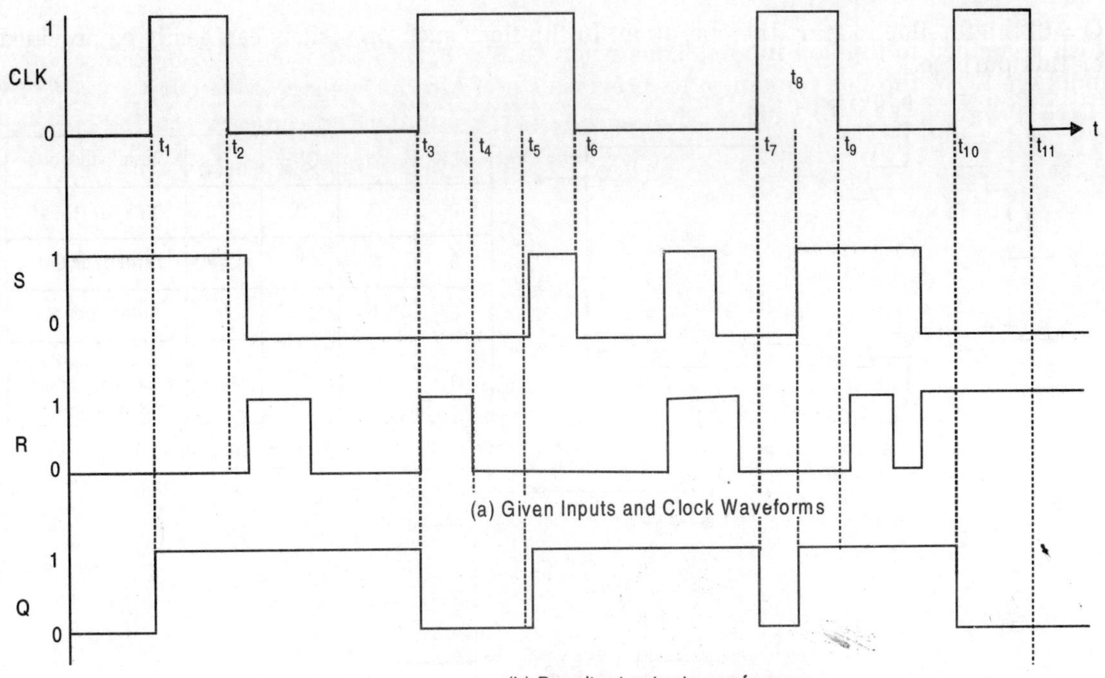

Fig. 6.18 Waveforms for example 6.3

At t_1 clock goes HIGH and output also goes HIGH i.e. Q = 1 since S = 1 and R = 0 at this time. At time t_2 CLK = 0 and flip-flop remain SET untill t_3, irrespective of changes in S and R.

At time $t = t_3$ CLK = 1 and because S = 0 and R = 1 at this instant, we get Q = 0. At t_4 both inputs are LOW while clock is still HIGH. Since S = R = 0 at this instant flip-flop remains reset untill just after time t_5.

Just after t_5 S goes HIGH making Q = 1. At time t_6 clock switches to LOW state. Just before this HIGH to LOW transition of clock S = 1 and R = 0 and Q = 1. Thus flip-flop remains set untill t_7. Changes in R and S does not affect flip-flop during t_6 to t_7 as CLK = 0.

At $t = t_7$ clock goes HIGH, at which time R = 1 and S = 0. So flip-flop enters in reset state. Q = 0 is retained untill t_8 where R switches to 0 and S switches to 1. Therefore Q = 1 at this time.

Clock goes LOW at t_9 and since Q = 1 just before clock goes LOW, flip-flop remains set untill t_{10} where clock goes HIGH again. The inputs S and R changes during time between t_9 to t_{10}, but cannot affect the output since CLK = 0.

At $t = t_{10}$ clock goes HIGH and since R = 1 and S = 0 at t_{10}, the flip-flop attains reset state. At the time $t = t_{11}$ clock goes LOW and still R = 1 and S = 0 was maintained at the input, the flip-flop remains in LOW state beyond t_{11}.

Clocked SR Flip-flop with Clear and Preset

When the power is first applied to the flip-flop, it come up in random state i.e. state of circuit is uncertain. It may be in SET state or in RESET state. This is highly undesired in majority of application. There are requirements that the flip-flop must be in a particular state before the actual operation begins. In practice, it may be required to preset (Q = 1) or clear

(Q = 0) the flip-flop to start the operation. In flip-flops such provisions can easily be provided for this purpose.

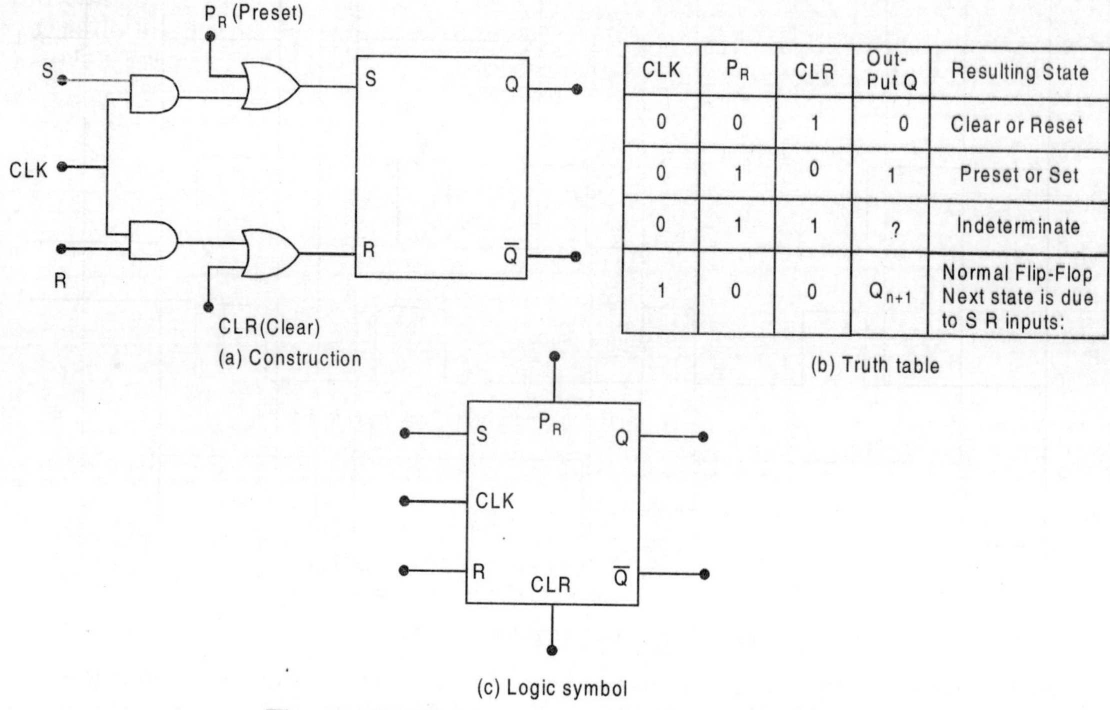

CLK	P_R	CLR	Out-Put Q	Resulting State
0	0	1	0	Clear or Reset
0	1	0	1	Preset or Set
0	1	1	?	Indeterminate
1	0	0	Q_{n+1}	Normal Flip-Flop Next state is due to S R inputs:

(b) Truth table

(c) Logic symbol

Fig. 6.19 SR Flip-Flop with 'CLEAR' and 'PRESET'

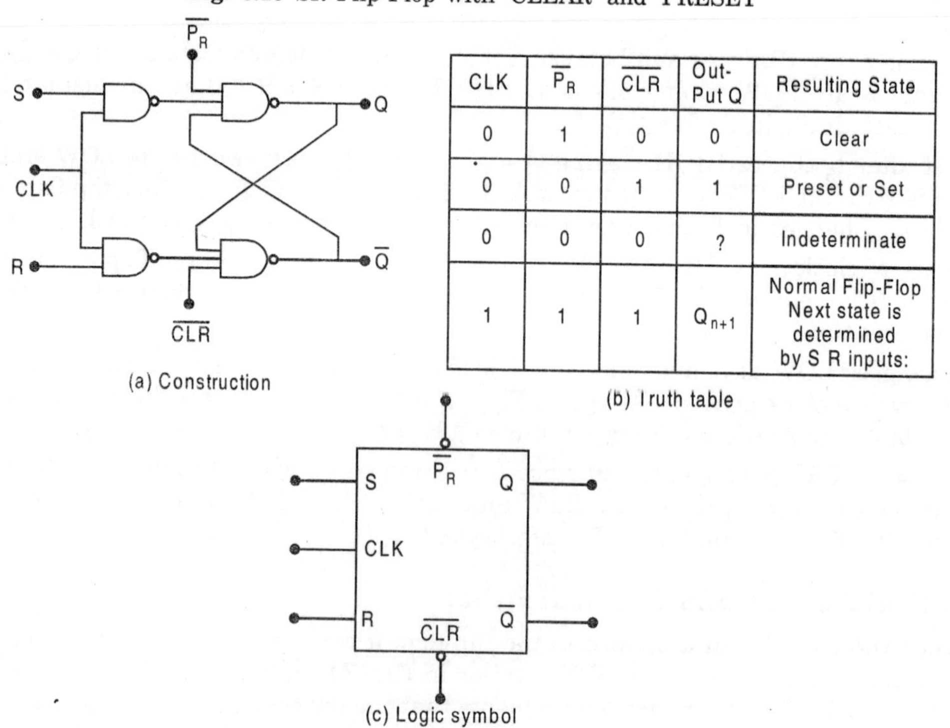

CLK	$\overline{P_R}$	$\overline{CLR}$	Out-Put Q	Resulting State
0	1	0	0	Clear
0	0	1	1	Preset or Set
0	0	0	?	Indeterminate
1	1	1	Q_{n+1}	Normal Flip-Flop Next state is determined by S R inputs:

(b) Truth table

(c) Logic symbol

Fig. 6.20 Alternate realization of 'CLEAR' and 'PRESET' with SR flip-flop

Two different realizations to accommodate a preset (abbreviated as P_R) and a clear (abbreviated as CLR) inputs are shown in Figs. 6.19 and 6.20. The P_R and CLR are direct inputs and are called **asynchronous inputs**; as they don't need the clock to operate. Both of the above circuits require that P_R and CLR inputs should be applied only in absence of clock otherwise unexpected behaviour can be observed. More over both P_R and CLR should not asserted at same time as it leads to indeterminate state. Logic values to be applied to preset and clear are accommodated in the truth tables. Before starting normal clocked operation the two direct inputs must be connected to a fix value as shown by the last entries of corresponding truth tables. The logic symbols for the two circuits are also shown in the figure.

In Fig. 6.19, due to the construction of circuit the flip-flop can be preset or clear on the application of Logic 1 at P_R or CLR respectively. In contrast, realization shown in Fig. 6.20 demands a Logic 0 to be applied at the particular asynchronous input, in order to perform the intended operation. For normal operation these inputs must be connected to Logic 1, as indicated by the last row of truth table shown in Fig. 6.20 (b). Similarly in Fig. 6.19 the P_R and CLR must be connected to Logic 0 to obtain normal flip-flop operation.

The circuits proposed in Figs. 6.19 and 6.20 requires that the P_R and CLR should only be applied when clock is LOW. This imposes a restriction to use these two signals. An improvement may be considered to remove this restriction.

Figure 6.21 shows a clocked SR flip-flop with preset and clear ($\overline{P_R}$ and $\overline{CLR}$) inputs that can override the clock. $\overline{P_R}$ and $\overline{CLR}$ can be safely applied at any instant of time whether clock is present or not.

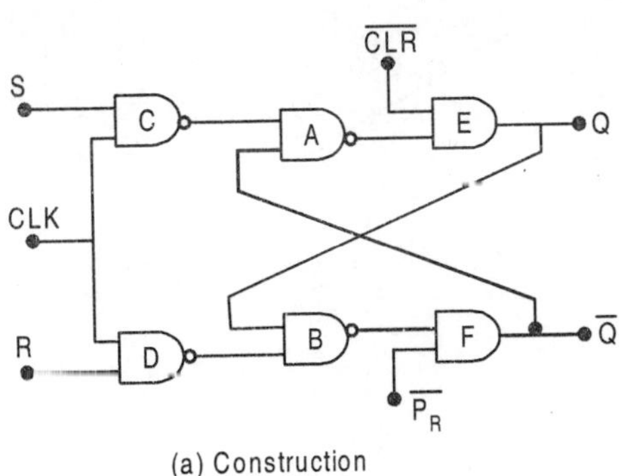

(a) Construction

CLK	$\overline{P_R}$	$\overline{CLR}$	Out-Put Q	Resulting State
×	1	0	0	Clear
×	0	1	1	Preset
×	0	0	?	Indeterminate
1	1	1	Q_{n+1}	Normal Flip-Flop Next state is determined by S and R inputs.

(b) Truth table

Fig. 6.21 SR flip-flop with clock override 'clear' and 'preset'

As shown in figure two AND gates, E and F, are used to accommodate preset ($\overline{P_R}$) and Clear ($\overline{CLR}$) inputs. The truth table summarizes the effect of these asynchronous inputs in the circuit. Output Q is provided through gate E whereas $\overline{Q}$ is output of gate F. Both $\overline{P_R}$ and $\overline{CLR}$ are active LOW signals i.e. asserting Logic '0' at these inputs perform the intended operation.

According to first row of truth table when $\overline{CLR} = 0$ and $\overline{P_R} = 1$ is applied then irrespective of the value of S, R and CLK, output of gate E is 0. Thus, Q = 0 and it is feeded to the input of NAND gate B. So output of NAND-B is 1 which is applied to the input of gate F. Since, we applied $\overline{P_R} = 1$, both the inputs to AND gate F are HIGH. Consequently, the output of gate

F goes HIGH, which is available at $\overline{Q}$ output of flip-flop. Hence, when $\overline{PR} = 1$ and $\overline{CLR} = 0$ is applied, we get Q = 0, and $\overline{Q} = 1$. This is frequently called as **clearing** the flip-flop.

Similarly when $\overline{PR} = 0$ and $\overline{CLR} = 1$ is applied, output of gate F, which is $\overline{Q}$, goes LOW i.e. $\overline{Q} = 0$. This forces the gate-A to give HIGH output. Since $\overline{CLR} = 1$ is already present, output of gate E, which is Q, goes HIGH. Thus, when $\overline{PR} = 0$ and $\overline{CLR} = 1$ is applied we get Q = 1 and $\overline{Q} = 0$, which is required state. This is referred as presetting.

Since both $\overline{PR}$ and $\overline{CLR}$ are active LOW inputs, they must be connected to Logic 1, in order to obtain normal flip-flop operation as shown by fourth row of truth table. Also making both $\overline{PR}$ and $\overline{CLR}$ LOW is forbidden as it results in indeterminate state, for the similar reasons explained earlier.

Clocked D Flip-Flop

In subsection 6.1.2 we obtained a D flip-flop by using an external inverter present at the input of SR latch as shown in Fig. 6.10 (*a*). In the similar way a clocked D flip-flop is obtained by using an external inverter at the input of clocked SR flip-flop. The clocked D flip-flop is shown below in Fig. 6.22. Note that unclocked RS latch of Fig. 6.10 (*a*) is replaced by a clocked RS flip-flop shown in Fig. 6.16 (*d*).

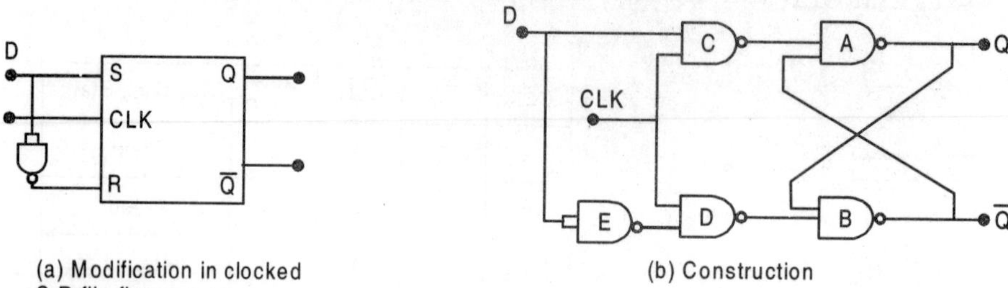

(a) Modification in clocked S R flip-flop

(b) Construction

CLK	Q_n	D_n	Q_{n+1}	Comments
0	0	×	0	Flip-Flop Disabled
0	1	×	1	last state Maintained
1	0	0	0	Rest State
1	0	1	1	Set State
1	1	0	0	Reset State
1	1	1	1	Set State

(c) Truth table

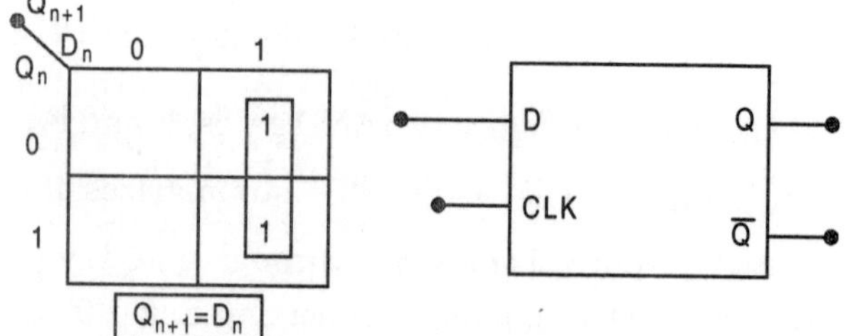

(d) Characteristic equation

(e) Logic symbol

Fig. 6.22 Clocked D flip-flop equation

The characteristic equation is derived from K-map shown in Fig. 6.22 (*d*), which specifies that irrespective of previous state next state will be same as the data input. In truth table D_n represents the data input at the t_{ON} time of *n*th pulse.

The operation of clocked D flip-flop is same as explained in subsection 6.1.2, when CLK = 1. When CLK = 0, the D input has no effect on the flip-flop and the present state is maintained. This is evident by the first two rows of truth table. The Logic symbol of clocked D flip-flop is shown in Fig. 6.22 (*e*).

Similar to clocked SR flip-flops, the clocked D flip-flop may also be accommodated with the asynchronous inputs "preset" and "clear". One particular realization is shown in Fig. 6.23. An alternative arrangement to obtain a D flip-flop, directly from SR flip-flop with clear and preset is shown in Fig. 6.24.

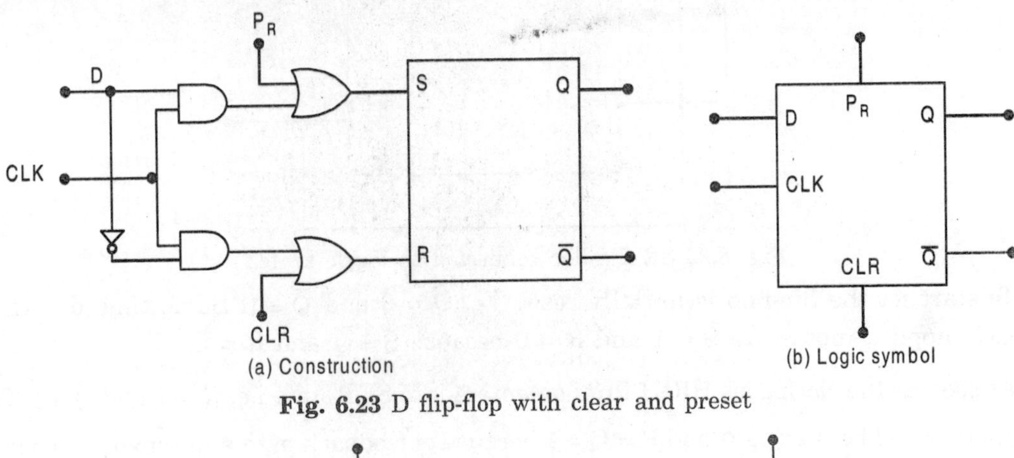

Fig. 6.23 D flip-flop with clear and preset

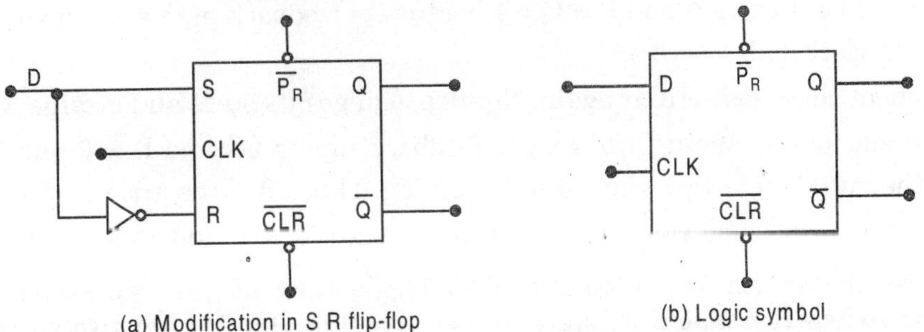

Fig. 6.24 Alternate realization of clear and preset in D flip-flop

On comparing Fig. 6.23 (*a*) with Fig. 6.19 (*a*), we find that both the circuits are same except that, in two external inputs are connected together through an inverter placed between them. Thus both the circuits behave similarly and explanation of Fig. 6.19 (*a*) is equally valid in this case. Similarly, the arrangement shown in Fig. 6.24 (*a*) is built upon the clock SR flip-flop shown in Fig. 6.20 (*c*).

Example 2. *In a certain digital application it is required to connect an SR flip-flop as toggle switch, which changes its state every time when clock pulse hits the system. Show the arrangement and explain in brief how it works as a toggle switch.*

Sol. SR flip-flop can be connected as a toggle switch as shown in Fig. 6.25. On the arrival of CLOCK pulse this arrangement forces the flip-flop either to go to SET state if currently it is RESET or to RESET state if currently it is SET.

As shown, a feedback path connects the output Q to R input while another feedback path connects the $\overline{Q}$ to input S. Recall that the state of flip-flop can be changed only when the CLK = 1 and the last state reached, is maintained while CLK = 0.

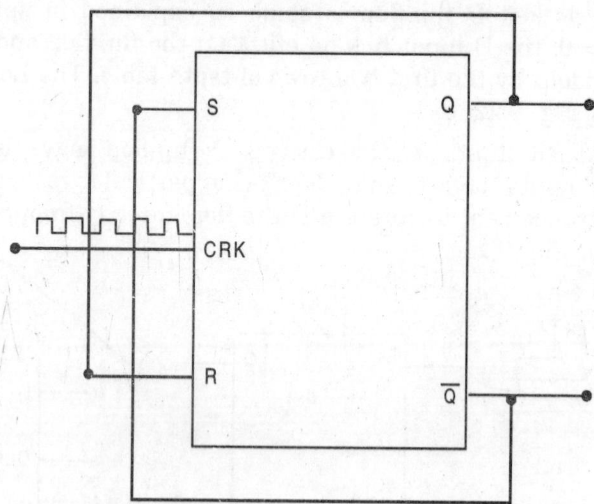

Fig. 6.25 SR flip-flop connected as toggle switch

To start let the flip-flop is initially reset, i.e. Q = 0 and $\overline{Q} = 1$. Same time due to the feedback, applied inputs are S = 1 and R = 0 because $S = \overline{Q}$ and R = Q.

As soon as the clock goes HIGH flip-flop enters into SET state i.e. Q = 1 and $\overline{Q} = 0$. Thus the inputs would be $S = \overline{Q} = 0$ and R = Q = 1 because of feedback path and remain unchanged untill the next clock pulse arrives.

The moment clock goes HIGH again, flip-flop changes its state and attains RESET state where Q = 0 and $\overline{Q} = 1$. Again through the feedback inputs become R = 0 and S = 1. Note that this is the initial state we assumed in beginning. Thus after the arrival of two successive clock pulses, the switch has returned to its initial state Q = 0 and $\overline{Q} = 1$.

Therefore, the switch shown in Fig. 6.25 toggle between the two states with clock. Moreover the switch reproduces its state (either Q = 0 or Q = 1) after two successive clock pulses has arrived. This does mean that it really does not matter whether the initially assumed state is Q = 0 or Q = 1.

In practice, the feedback paths in Fig. 6.25 may lead to uncertainty of the state. In the above paragraphs we assumed that the inputs available at S and R do not change during the t_{ON} period of clock. Thus the change in state can occur only once in a single clock pulse, which is not true. If the propagation delay (t_P) of the flip-flop is smaller than the t_{ON} time, multiple transitions (or state changes more than once) can be observed in a single clock pulse. Hence at the end of clock pulse the state of flip-flop is uncertain. This situation is referred as **race-around condition.**

Race around has resulted because the flip-flop remains transparent as long as CLK = 1 (or for entire t_{ON} time). At this point we refrain ourselves from discussing it further. We address the complete discussion on it, only after examining and identifying the similar situations in various flip-flops.

Example 3. *Realize a toggle switch, that changes its state on the arrival of clock, by using a D flip-flop. Explain its operation briefly.*

Sol. Figure 6.26 shows a D flip-flop configured to work as toggle switch when clock is applied.

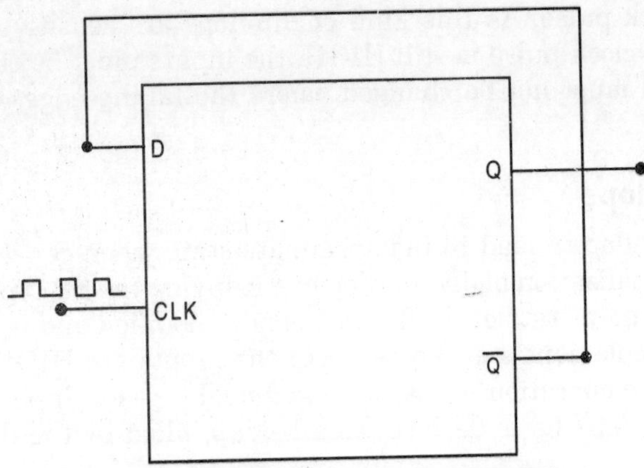

Fig. 6.26 D flip-flop as toggle switch

As shown the output $\overline{Q}$ is connected to D via a feedback path, so that D = $\overline{Q}$ always. To understand the operation recall that "in a D flip-flop output at any instant of time, provided CLK = 1, is same as input." So Q = D always, if CLK = 1. Furthermore output $\overline{Q} = \overline{D}$, so through the feedback path complement of data D is now feeded to input. Thus if initially the flip-flop was reset (Q = 0) the $\overline{Q} = 1$ is applied to input through the feedback path. Consequently D = 1 will be retained untill next clock pulse arrive. As soon as CLK = 1, D = 1 affects the circuit. This results in change in state of flip-flop giving Q = D = 1 and $\overline{Q} = 0$ at the output. But at the same time $\overline{Q}$ is feeded to D input due to which input changes to D = $\overline{Q}$ = 0. On the arrival of next clock pulse the circuit toggles again and change its state in similar way.

It is evident again from Fig. 6.26 that even this switch also, suffers from the "race around" problem, explained in example 6.4. The problem is a consequence of the feedback path present between $\overline{Q}$ and D. The ouput of this switch races for the same reasons as was given for SR flip-flop in example 6.4.

6.1.4 Triggering of Flip-Flops

By a momentarily change in the input signal the state of a flip-flop is switched (0-1-0). This momentarily change in the input signal is called a trigger. There are two types by which flip-flops can be triggered.

Edge trigger

Level (pulse) trigger

An edge-triggered flip-flop responds only during a clock pulse transition i.e. clock pulses switches from one level (reference voltage) to another. In this type of flip-flops, output transitions occur at a specific level of the clock pulse. When the pulse input level exceeds this reference level, the inputs are locked out and flip-flop is therefore unresponsive to further changes in inputs until the clock pulse returns to 0 and another clock pulse occurs. An edge-triggered flip-flop changes states either at the positive edge (rising edge) or at the negative edge (falling edge) of the clock pulse on the control input.

When the triggering occurs on the positive going edge of the clock, it is called positive edge triggering and when the triggering occurs at the trailing edge of the clock this is called as negative edge triggering.

The term pulse-triggered or level triggered means that data are entered into flip-flop on the rising edge of the clock pulse, but the output does not reflect the input state until the falling edge of the clock pulse. As this kind of flip-flops are sensitive to any change of the input levels during the clock pulse is still HIGH, the inputs must be set up prior to the clock pulse's rising edge and must not be changed before the falling edge. Otherwise, ambiguous results will happen.

6.1.5 JK and T Flip-Flops

Problem of SR flip-flop to lead to indeterminate state when S = R = 1, is eliminated in JK flip-flops. A simple realization of JK flip-flop by modifying the SR type is shown in Fig. 6.27 (a). In JK the indeterminate state of SR flip-flop is now modified and is defined as TOGGLED STATE (i.e. its own complement state) when both the inputs are HIGH. Table shown in Fig. 6.27 (b), summarizes the operation of JK flip-flop for all types of input possibilities. Characteristic equation in Fig. 6.27 (c) is derived from K-Map, filled by the data provided by truth table.

Truth table shows the input applied from external world (J and K). It also shows the flip-flop inputs S and R and the whose values are due to the modification and in accordance with the values of J and K. Input signal J is for SET and K is for RESET. Both J and K are ANDED with $\overline{Q}$ and Q respectively to generate appropriate signal for S and R. Since Q and $\overline{Q}$ are always complementary, only one of the AND gates (Fig. 6.27 (a)) is enable at a time. So either only J or only K can reach to one of S and R inputs, thus any one of inputs will receive the data to be stored. While the other AND gate is disabled i.e. its output is 0, thus second input of SR flip-flop will always be 0. Thus, indeterminate state never occurs even when both J and K inputs are HIGH.

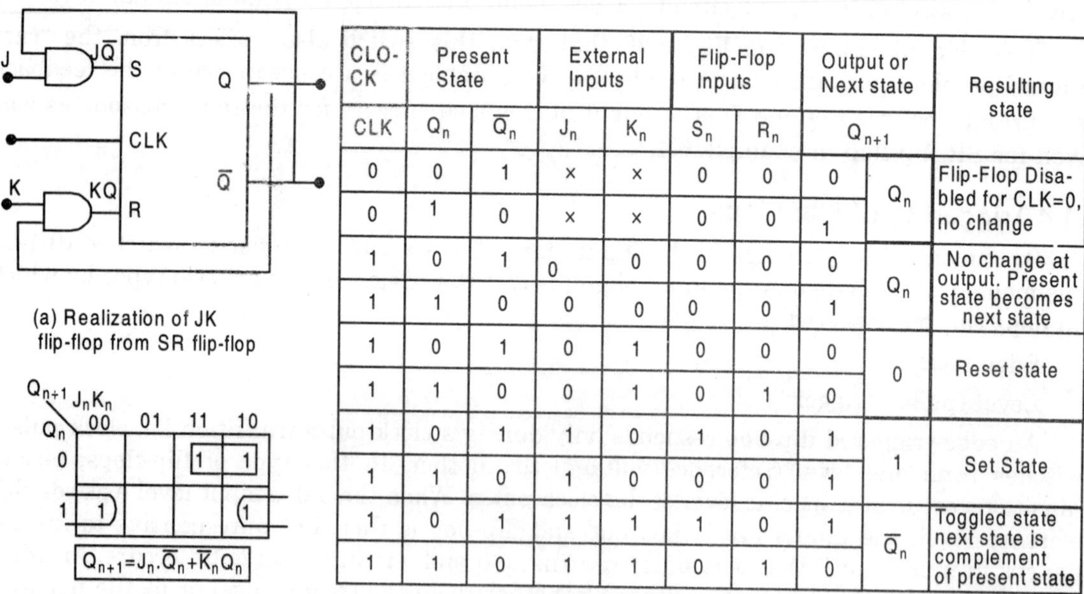

(a) Realization of JK flip-flop from SR flip-flop

CLOCK	Present State		External Inputs		Flip-Flop Inputs		Output or Next state	Resulting state	
CLK	Q_n	$\overline{Q}_n$	J_n	K_n	S_n	R_n	Q_{n+1}		
0	0	1	×	×	0	0	0	Q_n	Flip-Flop Disabled for CLK=0, no change
0	1	0	×	×	0	0	1		
1	0	1	0	0	0	0	0	Q_n	No change at output. Present state becomes next state
1	1	0	0	0	0	0	1		
1	0	1	0	1	0	0	0	0	Reset state
1	1	0	0	1	0	1	0		
1	0	1	1	0	1	0	1	1	Set State
1	1	0	1	0	0	0	1		
1	0	1	1	1	1	0	1	$\overline{Q}_n$	Toggled state next state is complement of present state
1	1	0	1	1	0	1	0		

(c) Characteristic equation

(b) Detailed truth table

Fig. 6.27 JK flip-flop through SR flip-flop

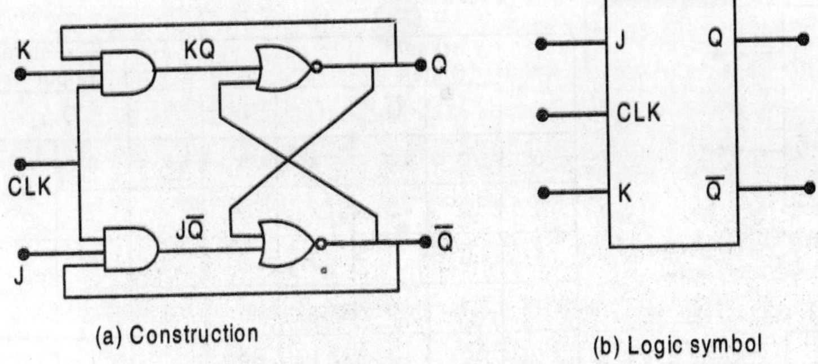

(a) Construction (b) Logic symbol

Fig. 6.28 Clocked JK flip-flop

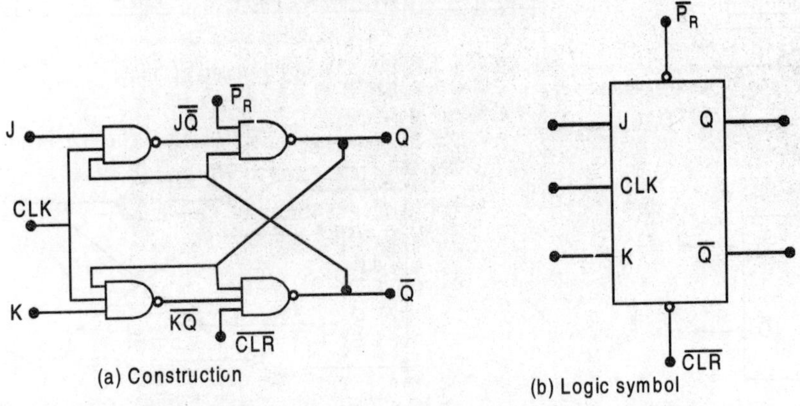

(a) Construction (b) Logic symbol

Fig. 6.29 JK flip-flop with active LOW preset and clear

As an example assume that initially Q_n = 0, $\overline{Q_n} = 1$ and inputs applied are J = 1 and K = 0. So we get S_n = J. $\overline{Q_n} = 1$ and R_n = K Q_n = 0. Application of S = 1 and R = 0, when clock arrives, switches the state of flip-flop HIGH. Now if we assume Q_n = 1. $\overline{Q_n} = 0$ with same J and K, inputs result in $S_n = J_n.\overline{Q_n} = 0$ and $R_n = K_nQ_n$ = 0 applied to SR Flip-flop. When both the S and R inputs are LOW flip-flop does not under go any change of state. So $Q_{n+1} = Q_n$ = 1. These two inputs are shown in 7th and 8th row of truth table. In the similar way entries in other rows of truth table can be verified.

Note that when both inputs are HIGH (i.e. J = 1 and K = 1) then, in Fig. 6.27 (*a*), we find that now S = J = Q and R = K = Q. Thus it is evident that the two external AND gates become full transparent and circuit may be viewed as if $\overline{Q}$ is connected to S input and Q connected to R input. Thus at J = K = 1 flip-flop behaves as an SR toggle switch shown in Fig. 6.25.

Note: Asynchronous inputs $\overline{P}_R$ and $\overline{CLR}$ must not be activate when clock is present, otherwise unexpected behaviour may be observed. Hence all the discussions presented in example 6.4 for toggle switch is equally valid for JK flip-flop when both J and K are HIGH.

Furthermore, when J = K = 1 then due to the two feedback paths the output may start racing around the inputs causing multiple transitions. Thus "Race Around" condition may occur in JK flip-flop when both inputs are HIGH, for the similar reasons given in example 6.4.

Two gate level constructions for the JK flip-flop is shown in Fig. 6.28 (*a*) and 6.29 (*a*) along with their logic symbols shown in Fig. (b) in each figures.

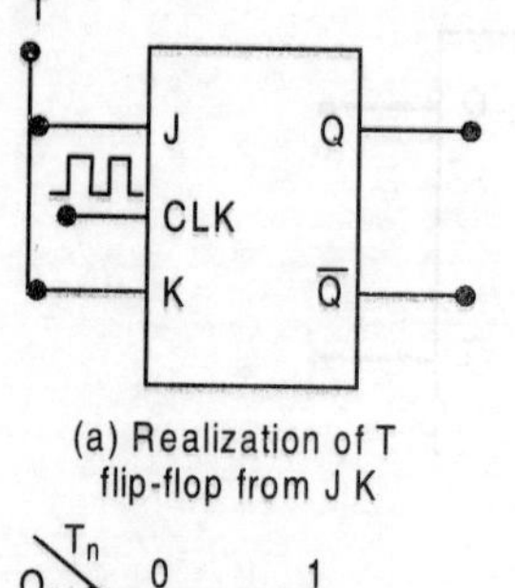

(a) Realization of T flip-flop from J K

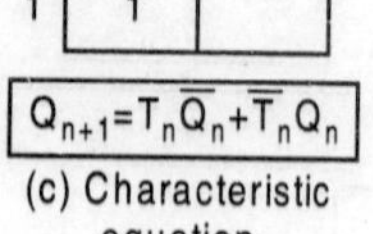

$Q_{n+1} = T_n\overline{Q}_n + \overline{T}_n Q_n$

(c) Characteristic equation

CLK	Present States Q_n	Present States $\overline{Q}_n$	Applied Inputs T_n	Applied Inputs J_n	Applied Inputs K_n	Next state Output Q_{n+1}		Resulting state
0	0	1	×	×	×	0	Q_n	Flip-Flop Disabled for CLK=0, no change next state=present
0	1	0	×	×	×	1		
1	0	1	0	0	0	0	Q_n	Maintain Output state no change in output next state=present state
1	1	0	0	0	0	1		
1	0	1	1	1	1	1	$\overline{Q}_n$	Toggled state or complemented state next state =present state
1	1	0	1	1	1	0		

(b) Detailed truth table

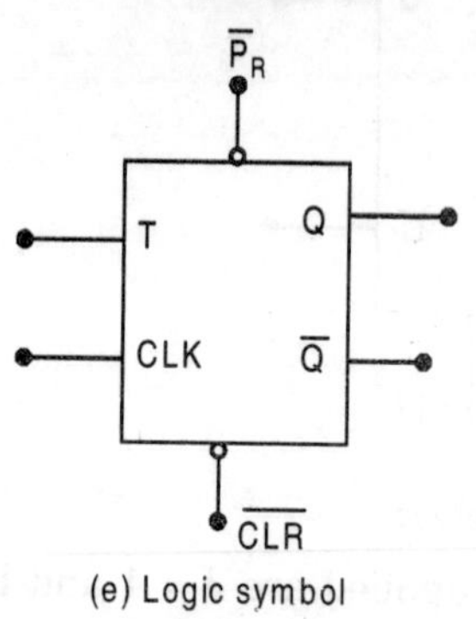

(e) Logic symbol

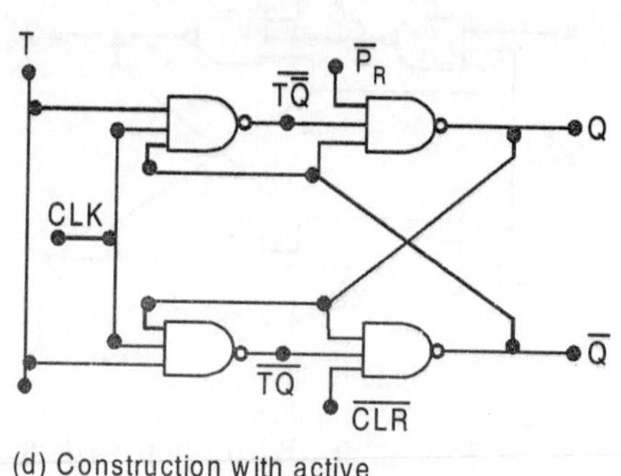

(d) Construction with active low clear & preset

Fig. 6.30 Clocked T (Toggles) flip-flop with active LOW asynchronous inputs

The JK flip-flops are very popular as indeterminate state (as present in SR type) does not exist. Furthermore, due to the toggling capability, when both inputs HIGH, on each arrival of pulse, it forms the basic element for counters. For this purpose JK flip-flop is further modified to provide a T flip-flop as shown in Fig. 6.30.

T flip-flop is a single input version of JK flip-flop, in which the inputs J and K are connected together, as shown, and is provided as a single input labelled as T. The operation is straight forward and easy, and summarized in truth-table given in Fig. 6.30 (b), while characteristic equation is derived in Fig. 6.30 (c).

When the clock is absent, the flip-flop is disable as usual and previously latched output is maintained at output. When the clock is present and T = 0, even though flip-flop is enabled the output does not switch its state. It happens so because for T = 0 we get J = K = 0 and thus next state is same as present state. Thus if either CLK = 0 or T = 0, state does not change next state is always same as present state.

When T = 1 during CLK = 1, it causes J = K = 1 and as earlier discussed it will toggle the output state. Thus when input T is HIGH, flip-flop toggles its output on the arrival of clock, and for this reason input T is called as the **Toggle Input**. Essentially the T flip-flop also, suffer from race around condition, (when input is HIGH) and thus causing multiple transition at output due to same reasons given in example 6.4.

If the race around problem is some how eliminated then T flip-flop can be used as frequency divider circuit. To obtain such operation input T is permanently made HIGH and the frequency to be divided is applied CLK inputs. At the outputs Q and $\overline{Q}$ we receive a square wave whose time period is now doubled due to which frequency reduces to half of the input frequency. Note that Q and $\overline{Q}$ generate square waves complementary to each other.

6.1.6 Race Around Condition and Solution

Whenever the width of the trigger pulse is greater than the propagation time of the flip-flop, then flip-flop continues to toggle 1-0-1-0 until the pulse turns 0. When the pulse turns 0, unpredictable output may result i.e. we don't know in what state the output is whether 0 or 1. This is called race around condition.

In level-triggered flip-flop circuits, the circuit is always active when the clock signal is high, and consequently unpredictable output may result. For example, during this active clock period, the output of a T-FF may toggle continuously. The output at the end of the active period is therefore unpredictable. To overcome this problem, *edge-triggered* circuits can be used whose output is determined by the edge, instead of the level, of the clock signal, for example, the rising (or trailing) edge.

Another way to resolve the problem is the Master-Slave circuit shown in Fig. 6.31.

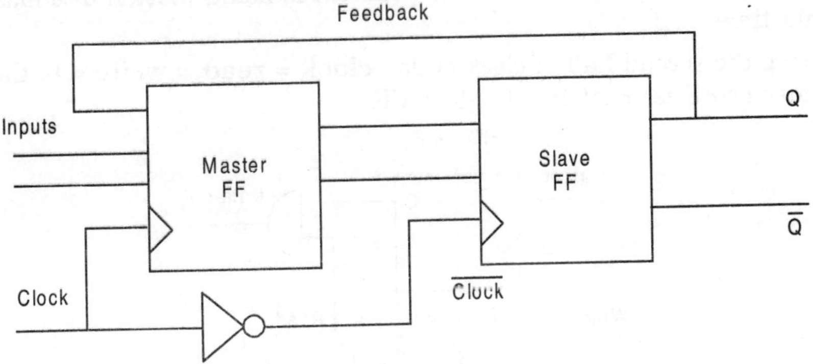

Fig. 6.31 Master slave circuit

The operation of a Master-Slave FF has two phases as shown in Fig. 6.32.

- During the high period of the clock, the master FF is active, taking both inputs and feedback from the slave FF. The slave FF is de-activated during this period by the negation of the clock so that the new output of the master FF won't effect it.
- During the low period of the clock, the master FF is deactivated while the slave FF is active. The output of the master FF can now trigger the slave FF to properly set its output. However, this output will not effect the master FF through the feedback as it is not active.

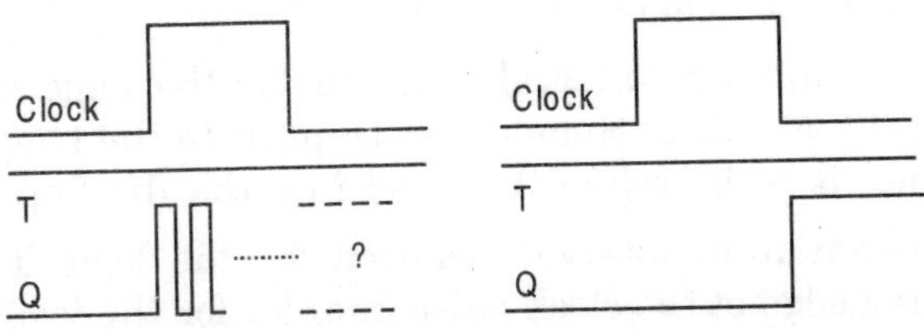

Fig. 6.32 Master slave operation

It is seen that the trailing edge of the clock signal will trigger the change of the output of the Master-Slave FF. The logic diagram for a basic master-slave S-R flip-flop is shown in Fig. 6.33.

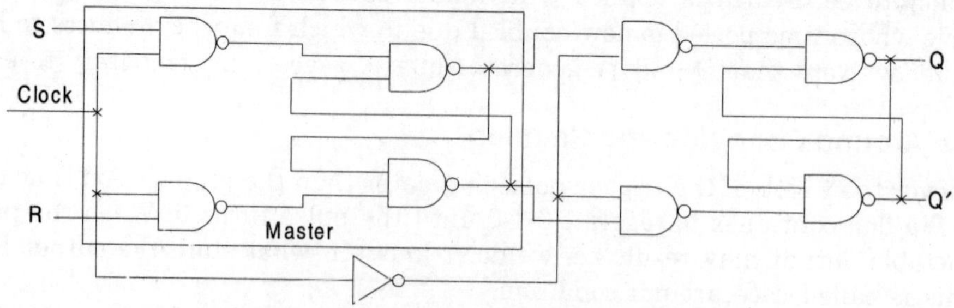

Fig. 6.33 S-R master slave flip-flop

Flip-flops are generally used for storing binary information. One bit of information can be written into a flip-flop, and later read out from it. If a master-slave FF is used, both read and write operations can take place during the same clock cycle under the control of two control signals **read** and **write** as shown in Fig. 6.34.

- During the first half of clock cycle : **clock = read = write = 1,** the old content in slave-FF is read out, while the new content is being written into master-FF at the same time.
- During the second half of clock cycle : **clock = read = write = 0,** the new content in master-FF is written into slave-FF.

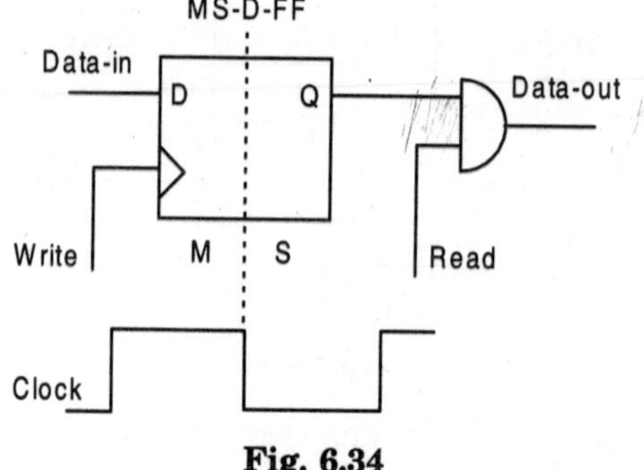

Fig. 6.34

6.1.7 Operating Characteristics of Flip-flops

The operation characteristics specify the performance, operating requirements, and operating limitations of the circuits. The operation characteristics mentioned here apply to all flip-flops regardless of the particular form of the circuit.

Propagation Delay Time—is the interval of time required after an input signal has been applied for the resulting output change to occur.

Set-up Time—is the minimum interval required for the logic levels to be maintained constantly on the inputs (J and K, or S and R, or D) prior to the triggering edge of the clock pulse in order for the levels to be reliably clocked into the flip-flop.

Hold Time—is the minimum interval required for the logic levels to remain on the inputs after the triggering edge of the clock pulse in order for the levels to be reliably clocked into the flip-flop.

Maximum Clock Frequency—is the highest rate that a flip-flop can be reliably triggered.

Power Dissipation—is the total power consumption of the device.

Pulse Widths—are the minimum pulse widths specified by the manufacturer for the Clock, SET and CLEAR inputs.

6.1.8 Flip-Flop Applications

Frequency Division

When a pulse waveform is applied to the clock input of a J-K flip-flop that is connected to toggle, the Q output is a square wave with half the frequency of the clock input. If more flip-flops are connected together as shown in the figure below, further division of the clock frequency can be achieved as shown in Fig. 6.35.

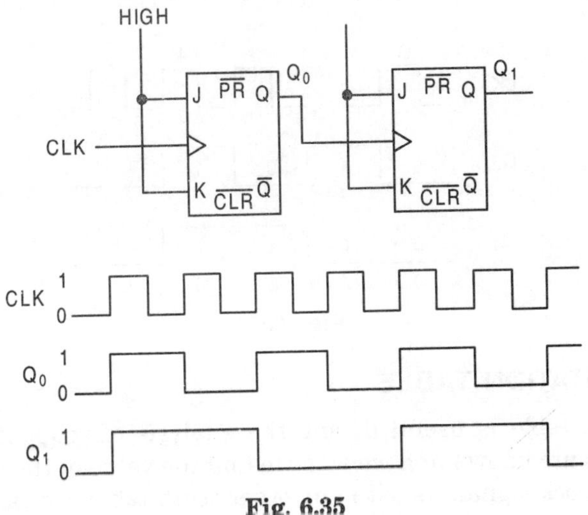

Fig. 6.35

The Q output of the second flip-flop is one-fourth the frequency of the original clock input. This is because the frequency of the clock is divided by 2 by the first flip-flop, then divided by 2 again by the second flip-flop. If more flip-flops are connected this way, the frequency division would be 2 to the power *n*, where *n* is the number of flip-flops.

Parallel Data Storage

In digital systems, data are normally stored in groups of bits that represent numbers, codes, or other information. So, it is common to take several bits of data on parallel lines and store them simultaneously in a group of flip-flops. This operation is illustrated in the Fig. 6.36.

Each of the three parallel data lines is connected to the D input of a flip-flop. Since, all the clock inputs are connected to the same clock, the data on the D inputs are stored simultaneously by the flip-flops on the positive edge of the clock.

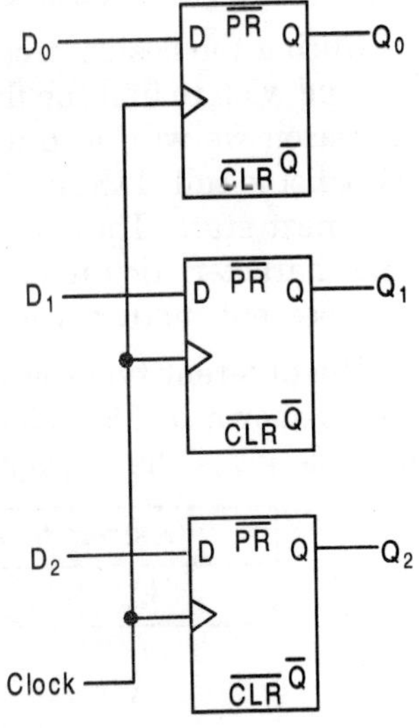

Fig. 6.36

Another very important application of flip-flops is in digital counters, which are covered in detail in chapter 7. A counter that counts from 0 to 2 is illustrated in the timing diagram given in Fig. 6.37. The two-bit binary sequence repeats every four clock pulses. When it counts to 3, it recycles back to 0 to begin the sequence again.

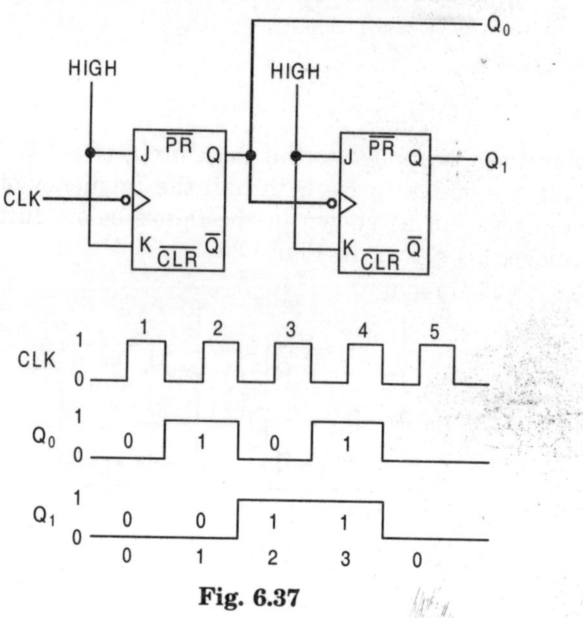

Fig. 6.37

6.2 FLIP-FLOP EXCITATION TABLE

The characteristic table is useful during the analysis of sequential circuits when the value of flip-flop inputs are known and we want to find the value of the flip-flop output Q after the rising edge of the clock signal. As with any other truth table, we can use the map method to derive the characteristic equation for each flip-flop.

During the design process we usually know the transition from present state to the next state and wish to find the flip-flop input conditions that will cause the required transition. For this reason we will need a table that lists the required inputs for a given change of state. Such a list is called the *excitation table.* There are four possible transitions from present state to the next state. The required input conditions are derived from the information available in the characteristic table. The symbol X in the table represents a "don't care" condition, that is, it does not matter whether the input is 1 or 0.

The different types of flip flops (RS, JK, D, T) can also be described by their excitation, table as shown in Fig. 6.38. The left side shows the desired transition from Q_n to Q_{n+1}, the right side gives the triggering signals of various types of FFs needed for the transitions.

Desired transition		*Triggering signal needed*					
Q_n	Q_{n+1}	*S*	*R*	*J*	*K*	*D*	*T*
0	0	0	x	0	x	0	0
0	1	1	0	1	x	1	1
1	0	0	1	x	1	0	1
1	1	x	0	x	0	1	0

Fig. 6.38 Excitation table

6.3 FLIP-FLOP CONVERSIONS

This section shows how to convert a given type A FF to a desired type B FF using some conversion logic.

The key here is to use the excitation table of Fig. 6.38 which shows the necessary triggering signal (S, R, J, K, D and T) for a desired flip flop state transition $Q_n \rightarrow Q_{n+1}$ is reproduced here.

Q_n	Q_{n+1}	S	R	J	K	D	T
0	0	0	x	0	x	0	0
0	1	1	0	1	x	1	1
1	0	0	1	x	1	0	1
1	1	x	0	x	0	1	0

Example 1. *Convert a D-FF to a T-FF:*

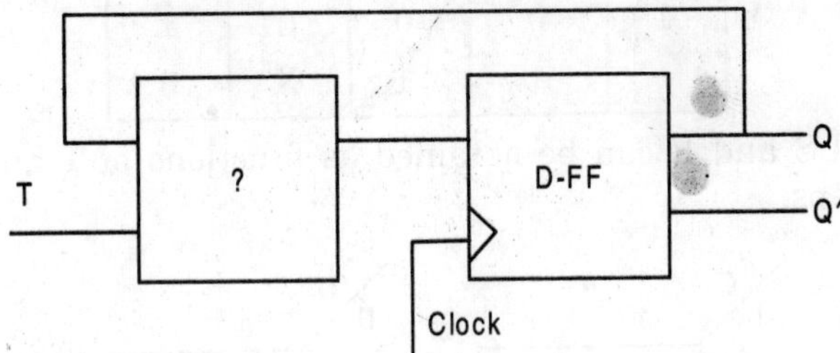

We need to design the circuit to generate the triggering signal D as a function of T and Q : D = f (T, Q)

Consider the excitation table:

Q_n	Q_{n+1}	T	D
0	0	0	0
0	1	1	1
1	0	1	0
1	1	0	1

Treating D as a function of T and current FF state Q Q_n we have

$$D = T'Q + TQ = T \oplus Q$$

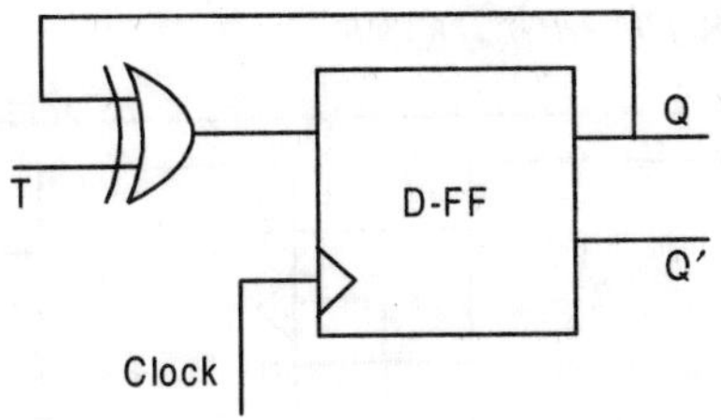

Example 2. *Convert a RS-FF to a D-FF:*

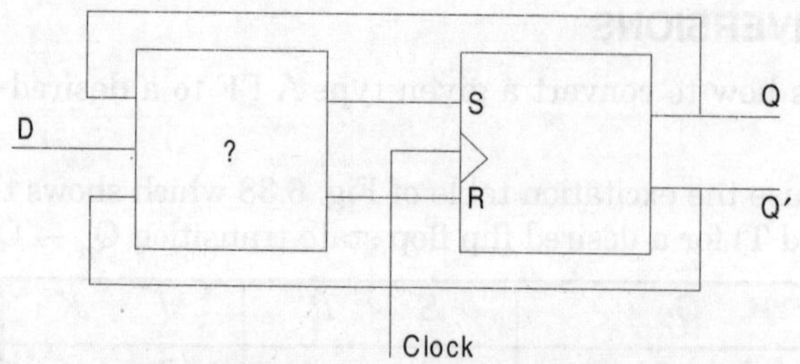

We need to design the circuit to generate the triggering signals S and R as functions of D and Q. Consider the excitation table:

Q_n	Q_{n+1}	D	S	R
0	0	0	0	X
0	1	1	1	0
1	0	0	0	1
1	1	1	X	0

The desired signal S and R can be obtained as functions of T and current FF state Q from the Karnaugh maps:

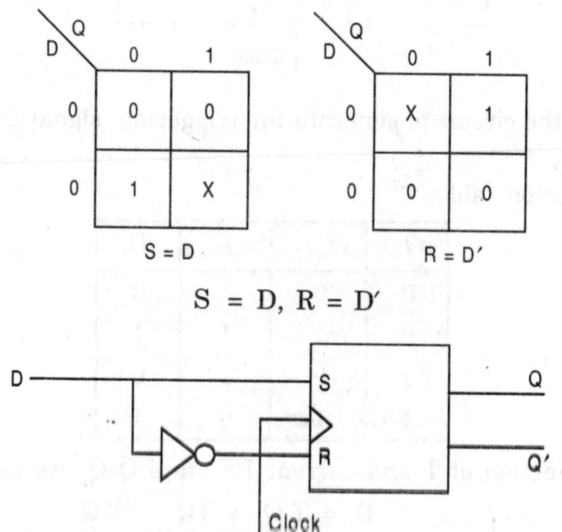

$$S = D, R = D'$$

Example 3. *Convert a RS-FF to a JK-FF.*

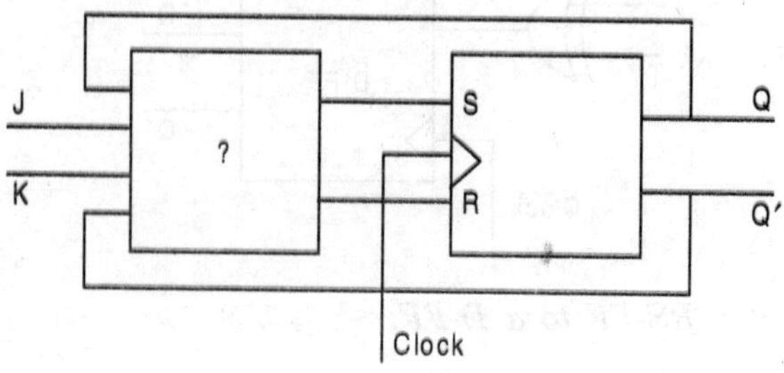

We need to design the circuit to generate the triggering signals S and R as functions of J, K and Q. Consider the excitation table.

Q_n	Q_{n+1}	J	K	S	R
0	0	0	x	0	x
0	1	1	x	1	0
1	0	x	1	0	1
1	1	x	0	x	0

The desired signals S and R as function J, K and current FF state Q can be obtained from the Karnaugh maps:

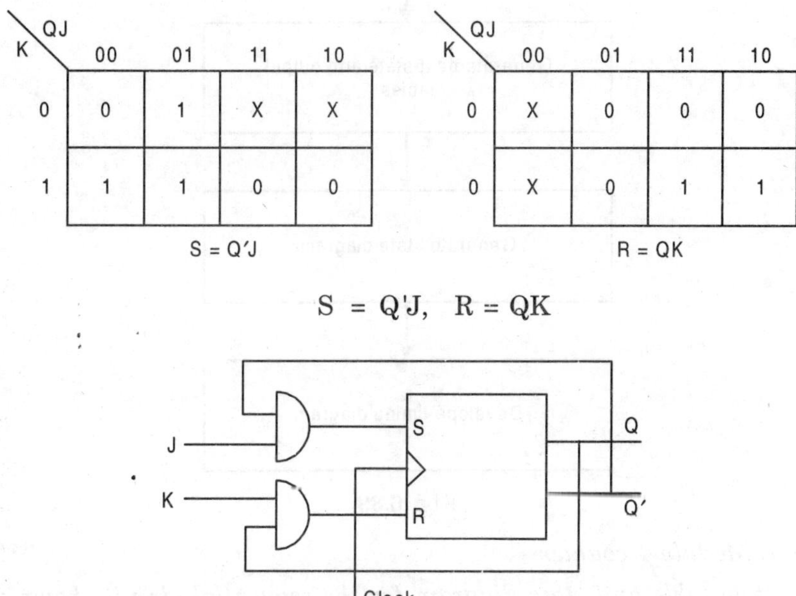

$$S = Q'J, \quad R = QK$$

6.4 ANALYSIS OF CLOCKED SEQUENTIAL CIRCUITS

The behaviour of a sequential circuit is determined from the inputs, the outputs and the states of its flip-flops. Both the output and the next state are a function of the inputs and the present state.

The suggested analysis procedure of a sequential circuit is depicted through Fig. 6.39.

We start with the logic schematic from which we can derive excitation equations for each flip-flop input. Then, to obtain next-state equations, we insert the excitation equations into the characteristic equations. The output equations can be derived from the schematic, and once we have our output and next-state equations, we can generate the next-state and output tables as well as state diagrams. When we reach this stage, we use either the table or the state diagram to develop a timing diagram which can be verified through simulation.

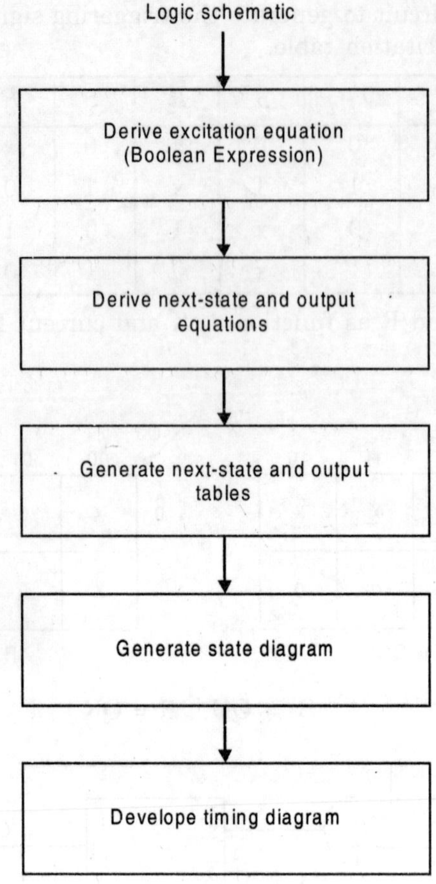

Fig. 6.39

Example 1. *Modulo-4 counter*

Derive the state table and state diagram for the sequential circuit shown in Fig. 6.40.

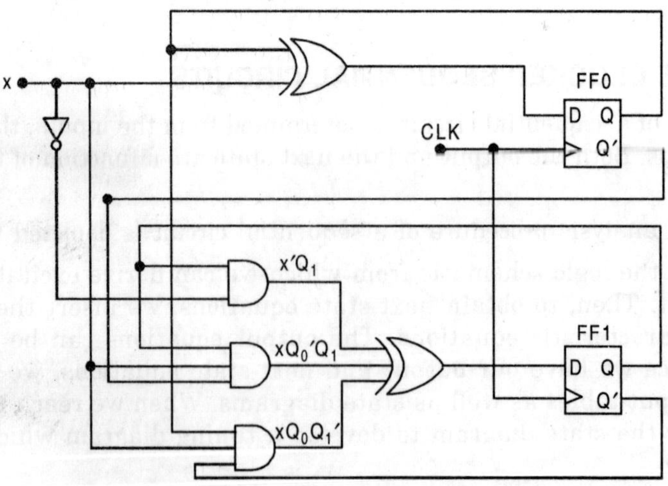

Fig. 6.40 Logic schematic of a sequential circuit

Solution.

Step 1 : First we derive the Boolean expressions for the inputs of each flip-flops in the schematic, in terms of external input X and the flip-flop outputs Q_1 and Q_0. Since there are two D flip-flops in this example, we derive two expressions for D_1 and D_0 :

$$D_0 = x \oplus Q_0 = x'Q_0 + xQ_0'$$

$$D_1 = x'Q_1 + xQ_1'Q_0 + xQ_1Q_0'$$

These Boolean expressions are called excitation equations since they represent the inputs to the flip-flops of the sequential circuit in the next clock cycle.

Step 2 : Derive the next-state equations by converting these excitation equations into flip-flop characteristic equations. In the case of D flip-flops, Q(next) = D. Therefore the next state equal the excitation equations.

$$Q_0(\text{next}) = D_0 = x'Q_0 + xQ_0'$$

$$Q_1(\text{next}) = D_1 = x'Q_1 + xQ_1'\,Q_0'Q_0 + xQ_1Q_0'$$

Step 3 : Now convert these next-state equations into tabular form called the next-state table (Fig. 6.41).

Present State	*Next State*	
Q_1Q_0	$x = 0$	$x = 1$
0 0	0 0	0 1
0 1	0 1	1 0
1 0	1 0	1 1
1 1	1 1	0 0

Fig. 6.41 Next-state table

Each row is corresponding to a state of the sequential circuit and each column represents one set of input values. Since we have two flip-flops, the number of possible states is four, *i.e.*, Q_1Q_0 can be equal to 00, 01, 10, or 11. These are present states as shown in the table.

For the next state part of the table, each entry defines the value of the sequential circuit in the next clock cycle after the rising edge of the CLK. Since this value depends on the present state and the value of the input signals, the next state table will contain one column for each assignment of binary values to the input signals. In this example, since there is only one input signal, x, the next-state table shown has only two columns, corresponding to $x = 0$ and $x = 1$.

Note that each entry in the next-state table indicates the values of the flip-flops in the next state if their value in the present state is in the row header and the input values in the column header.

Each of these next-state values has been computed from the next-state equations in STEP 2.

Step 4 : The state diagram is generated directly from the next-state table, shown in Fig. 6.42.

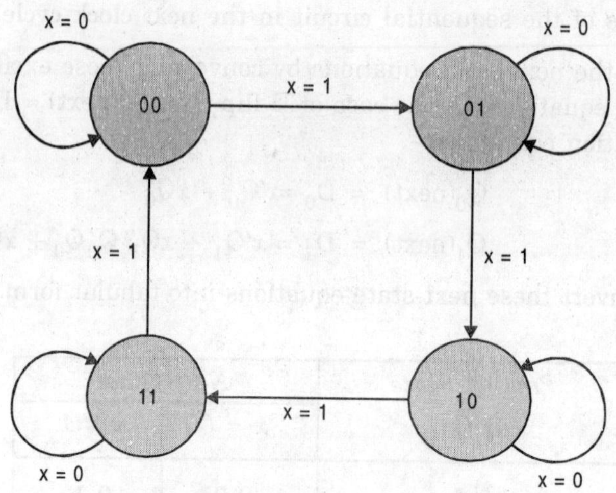

Fig. 6.42 State diagram

Each arc is labelled with the values of the input signals that cause the transition from the present state (the source of the arc) to the next state (the destination of the arc).

In general, the number of states in a next-state table or a state diagram will equal $2m$ where m is the number of flip-flops. Similarly, the number of arcs will equal $2^m \times 2^k$, where k is the number of binary input signals. Therefore, in the state diagram, there must be four states and eight transitions. Following these transition arcs, we can see that as long as $x = 1$, the sequential circuit goes through the states in the following sequence : 0, 1, 2, 3, 0, 1, 2, ... On the other hand, when $x = 0$, the circuit stays in its present state until x changes to 1, at which the counting continues.

Since, this sequence is characteristic of modulo-4 counting, we can conclude that the sequential circuit in Fig. 6.40 is a modulo-4 counter with one control signal, x, which enables counting when $x = 1$ and disables it when $x = 0$.

Example 2. *Derive the next state, the output table and the state diagram for the sequential circuit shown in Fig. 6.43.*

Solution. The input combinational logic in Fig. 6.43 is the same as in Fig. 6.40, so the excitation and the next-state equations will be same as in previous example.

Excitation equations :

$$D_0 = x \oplus Q_0 = x'Q_0 + xQ_0'$$

$$D_0 = x'Q_1 + xQ_1'Q_0 + xQ_1Q_0'$$

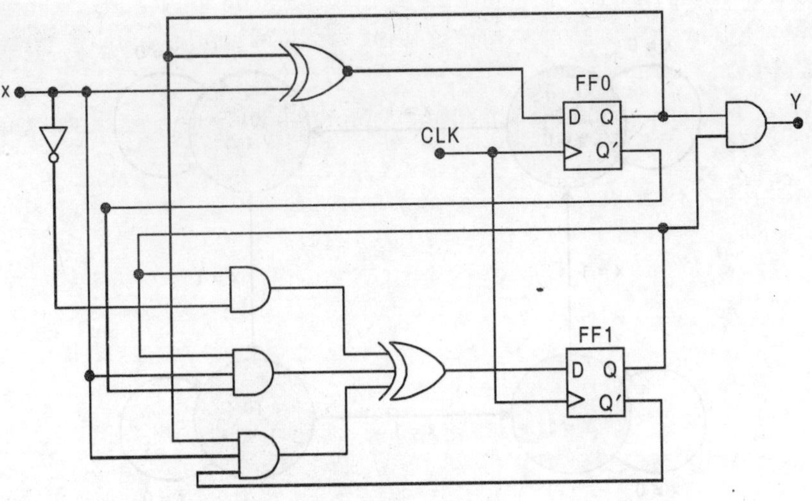

Fig. 6.43 Logic schematic of a sequential circuit

Next-state equations :

$$Q_0(\text{next}) = D_0 = x'Q_0 + xQ_0'$$

$$Q_1(\text{next}) = D_0 = x'Q_1 + xQ_1'Q_0 + xQ_1Q_0'$$

In addition, however, we have computed the output equation.

Output equation : $Y = Q_1 Q_0$

As this equation shows, the output Y will equal to 1 when the counter is in state $Q_1Q_0 = 11$, and it will stay 1 as long as the counter stays in that state.

Next-state and output table (Fig. 6.44):

Present State	*Next State*		*Output*
Q_1Q_0	$x = 0$	$x = 1$	Y
00	00	01	0
01	01	10	0
10	10	11	0
11	11	00	1

Fig. 6.44

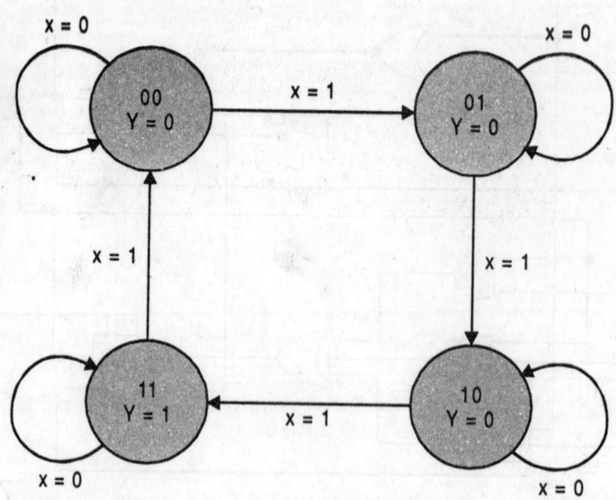

Fig. 6.45 State diagram of sequential circuit in Fig. 6.43

6.5 DESIGN OF CLOCKED SEQUENTIAL CIRCUITS

The design of a synchronous sequential circuit starts from a set of specifications and culminates in a logic diagram or a list of Boolean functions from which a logic diagram can be obtained. In contrast to a combinational logic, which is fully specified by a truth table, a sequential circuit requires a state table for its specification. The first step in the design of sequential circuits is to obtain a state table or an equivalence representation, such as a state diagram.

The recommended steps for the design of sequential circuits are depicted through Fig. 6.46.

A synchronous sequential circuit is made up of flip-flops and combinational gates. The design of the circuit consists of choosing the flip-flops and then finding the combinational structure which, together with the flip-flops, produces a circuit that fulfils the required specifications. The number of flip-flops is determined from the number of states needed in the circuit.

State Table

The state table representation of a sequential circuit consists of three sections labelled *present state, next state* and *output*. The present state designates the state of flip-flops before the occurrence of a clock pulse. The next state shows the states of flip-flops after the clock pulse, and the output section lists the value of the output variables during the present state.

State Diagram

In addition to graphical symbols, tables or equations, flip-flops can also be represented graphically by a state diagram. In this diagram, a state is represented by a circle, and the transition between states is indicated by directed lines (or arcs) connecting the circles.

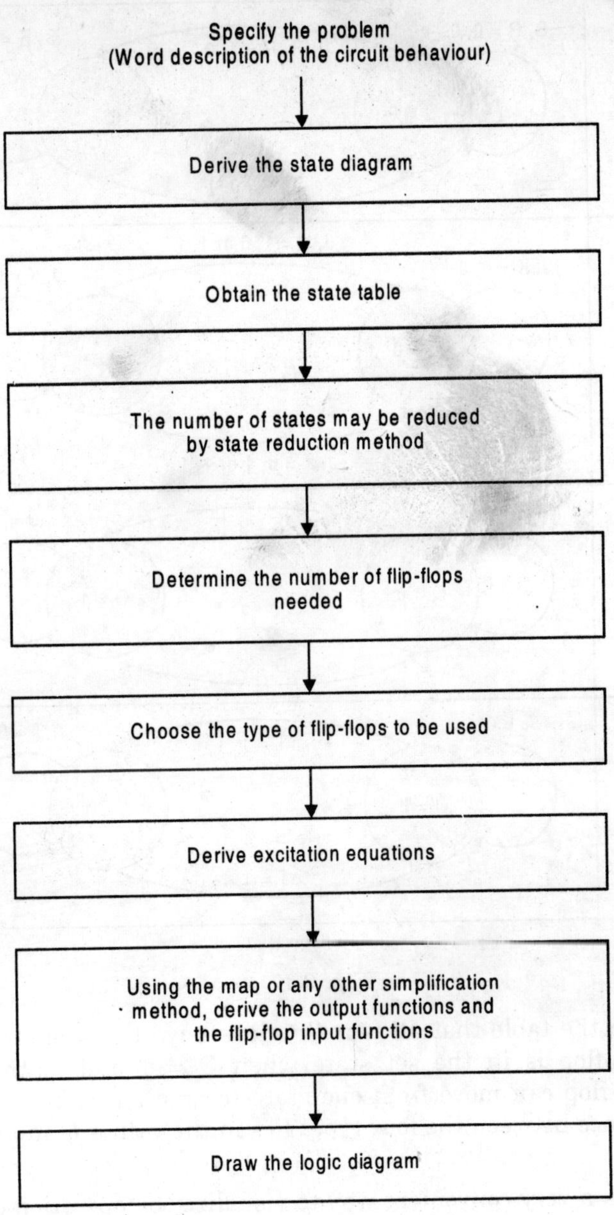

Fig. 6.46

State Diagrams of Various Flip-Flops

Table of Fig. 6.47 shows the state diagrams of the four types of flip-flops.

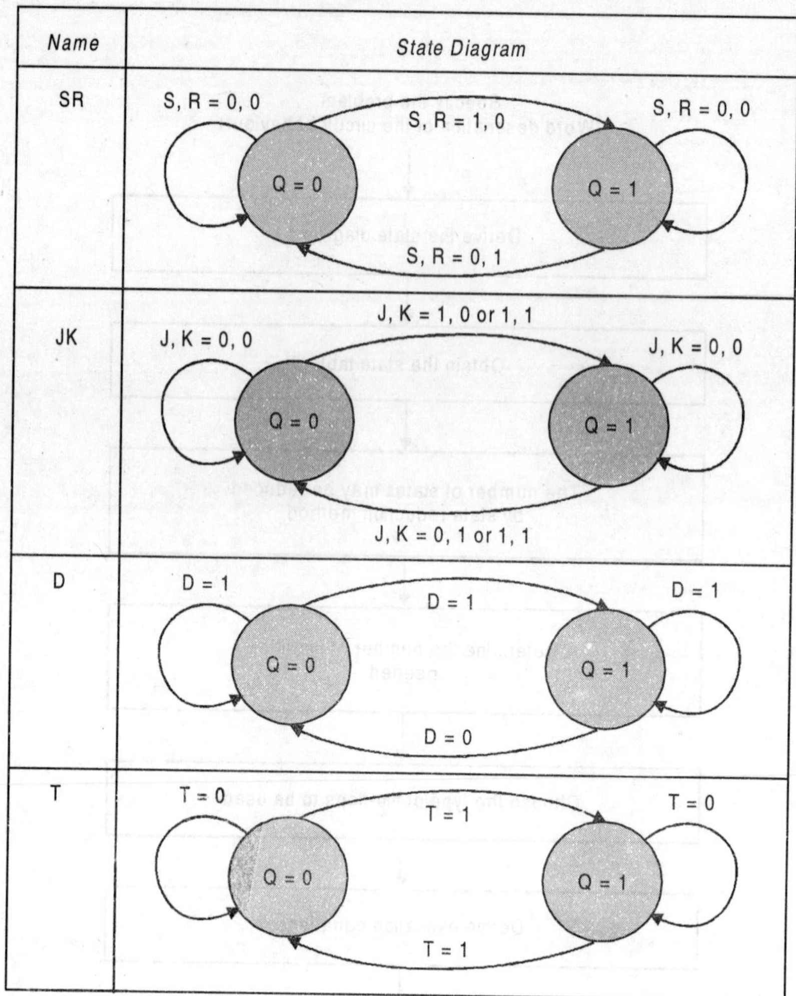

Fig. 6.47

One can see from the table that all four flip-flops have the same number of states and transitions. Each flip-flop is in the set state when Q = 1 and in the reset state when Q = 0. Also, each flip-flop can move from one state to another, or it can re-enter the same state. The only difference between the four types lies in the values of input signals that cause these transitions.

A state diagram is a very convenient way to visualize the operation of a flip-flop or even of large sequential components.

State Reduction

Any design process must consider the problem of minimizing the cost of the final circuit. The two most obvious cost reductions are reductions in the number of flip-flops and the number of gates.

The number of states in a sequential circuit is closely related to the complexity of the resulting circuit. It is therefore desirable to know when two or more states are equivalent in all aspects. The process of eliminating the equivalent or redundant states from a state table/diagram is known as *state reduction.*

Example. Let us consider the state table of a sequential circuit shown in Fig. 6.48.

State table

Present State	*Next State*		*Output*	
	$x = 0$	$x = 1$	$x = 0$	$x = 1$
A	B	C	1	0
B	F	D	0	0
C	D	E	1	1
D	F	E	0	1
E	A	D	0	0
F	B	C	1	0

Fig. 6.48

It can be seen from the table that the present state A and F both have the same next states, B (when $x = 0$) and C (when $x = 1$). They also produce the same output 1 (when $x = 0$) and 0 (when $x = 1$). Therefore states A and F are equivalent. Thus one of the states, A or F can be removed from the state table. For example, if we remove row F from the table and replace all F's by A's in the columns, the state table is modified as shown in Fig. 6.49.

State F removed

Present State	*Next State*		*Output*	
	x = 0	*x = 1*	*x = 0*	*x = 1*
A	B	C	1	0
B	A	D	0	0
C	D	E	1	1
D	A	E	0	1
E	A	D	0	0

Fig. 6.49

It is apparent that states B and E are equivalent. Removing E and replacing E's by B's results in the reduce table shown in Fig. 6.50.

Reduced state table

Present State	*Next State*		*Output*	
	x = 0	*x = 1*	*x = 0*	*x = 1*
A	B	C	1	0
B	A	D	0	0
C	D	B	1	1
D	A	B	0	1

Fig. 6.50

The removal of equivalent states has reduced the number of states in the circuit from six to four. Two states are considered to be *equivalent* if and only if for every input sequence the circuit produces the same output sequence irrespective of which one of the two states is the starting state.

6.6 DESIGN EXAMPLES

Serial Binary Adder

The serial binary adder is capable of adding two binary numbers (say A and B) serially, *i.e.,* bit by bit. In the addition process, the output at a time instant t_i, depends not only in A_i and B_i bits but also on carry (C_{i-1}) generated in the previous addition of A_{i-1} and B_{i-1} bits. Therefore, the C_{i-1} must be memorised and fed back to the input at time t_i, *i.e.,* after a time delay of a unit. Hence, the adder must be able to preserve or memories the carry generated at any instant of time (say t_{i-1}) upto time t_i.

From the above word description, the block diagram of the serial binary adder can be drawn as shown in Fig. 6.51.

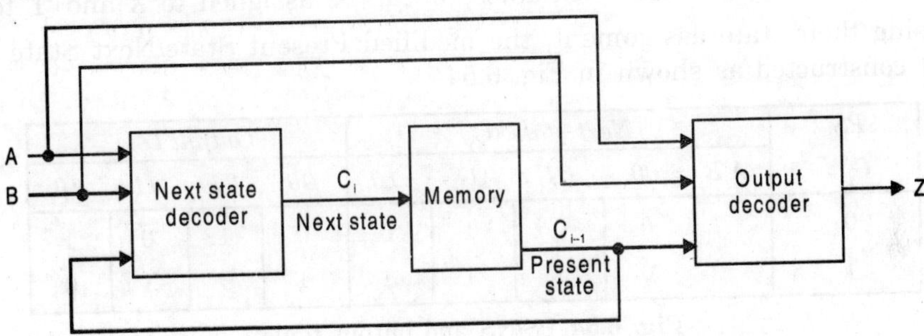

Fig. 6.51 Block diagram of serial binary adder

In the case of serial adder, the carry generated at any instant of time will either be '0' or 1. Let us designate the carry as the state of the adder. Let X be the state of the adder at time t_i if a carry is generated at time t_{i-1} and Y be the state if '1' is generated as carry at time t_{i-1}. The state of the adder at the time (t_i) when the present inputs (A_i and B_i) are applied is referred to as Next State (NS) because this is the state to which the adder goes due to the new carry. The output of the memory element is called Present State (PS) of the adder. The output z at time t_i depends on inputs at that time (A_i and B_i) and the state of the adder at that time.

Primitive State Diagram

Now, a primitive state diagram of serial binary adder can be constructed as shown in Fig. 6.52.

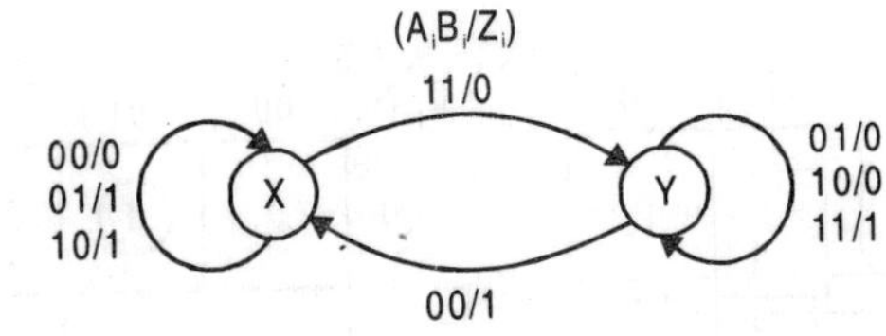

Fig. 6.52 State diagram of serial binary adder

Primitive State Table

From the primitive state diagram, the primitive state table can be constructed as shown in Fig. 6.53.

Present State (PS)	Next state, Output (NS, Z)			
	A_iB_i = 00	01	11	10
X	X, 0	X, 1	Y, 0	X, 1
Y	X, 1	Y, 0	Y, 1	Y, 0

Fig. 6.53 State table of serial binary adder

There are only two states X and Y which are not redundant since states X and Y are not equivalent states.

State Assignment

Since there are only two state variables, '0' can be assigned to X and '1' to Y.

Now, using their state assignment, the modified Present State/Next State and output table can be constructed as shown in Fig. 6.54.

PS C_{i-1}	Next state (C_i)				Output (Z)			
	A_iB_i = 00	01	11	10	00	01	11	10
0	0	0	1	0	0	1	0	1
1	0	1	1	1	1	0	1	0

Fig. 6.54 PS/NS and output table

Next step is to select the memory element, *i.e.*, flip-flop, to be used. Suppose D flip-flop is used as a memory element in serial adder, then the excitation table can be obtained from PS/NS table using excitation table of D flip-flop as shown in Fig. 6.55.

C_{i-1}	D			
	A_iB_i = 00	01	11	10
0	0	0	1	0
1	0	1	1	1

Fig. 6.55 Excitation table of serial binary adder

The excitation maps for flip-flop input 'D' and output map can be drawn as shown in Fig. 6.56.

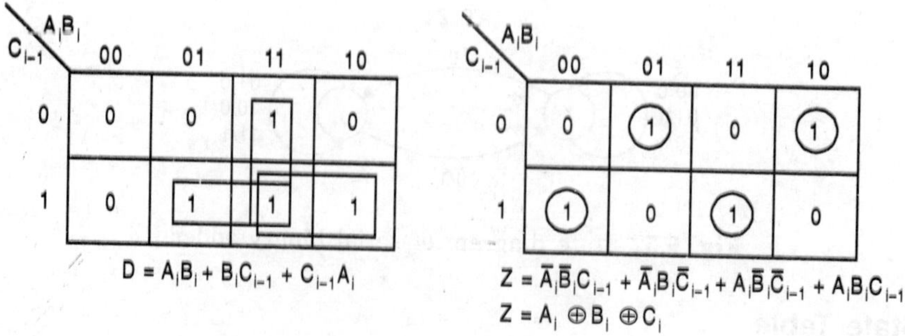

Fig. 6.56 K-map for excitation and output functions

Using simplified excitation and output function, the circuit diagram for serial adder can be implemented as shown in Fig. 6.57.

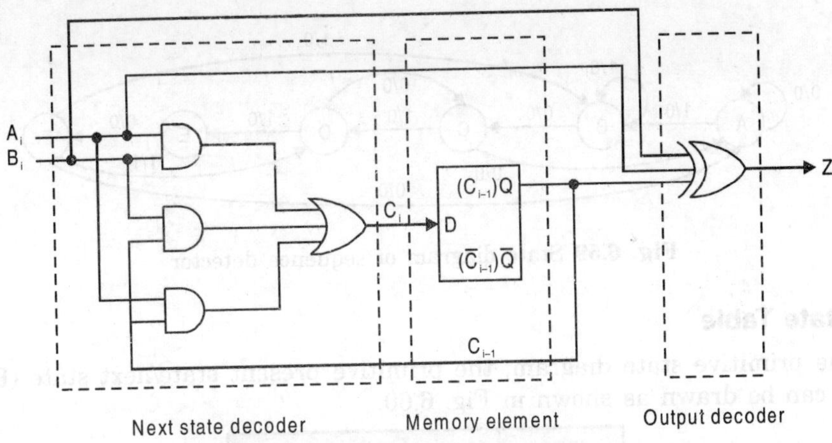

Fig. 6.57 Circuit diagram of serial binary adder

Sequence Detector

The desired sequence can be detected using a logic circuit called sequence detector. Once we know the sequence which is to be detected, we can design the circuit using state diagram for it. For a sequence detector that produces and output 1 whenever the sequence 101101 is detected the designing involves following steps.

From the word description of the problem, one can understand that the sequence detector is a single input circuit that will accept a stream of bits and generate an output '1' whenever the sequence 101101 is detected. Otherwise, output '0' is generated. For example, for the input 0101101101, the output 0000001001 will be generated, and for the input 10110101101 the output is 00000100001.

From the above word description, the sequence detector is a block with one input (X) and output (Z) as shown in Fig. 6.58.

Fig. 6.58 Block diagram of sequence detector

Primitive State Diagram

Considering the input sequence 101101 to be detected, let:

A be the initial arbitrary state;

B be the state when the last received one symbol in the input sequence is 1;

C be the state when the last received two symbols in the input sequence are 10;

D be the state when the last received three symbols in the input sequence are 101;

E be the state when the last received four symbols in the input sequence are 1011; and F be the state when the last received five symbols in the input sequence are 10110.

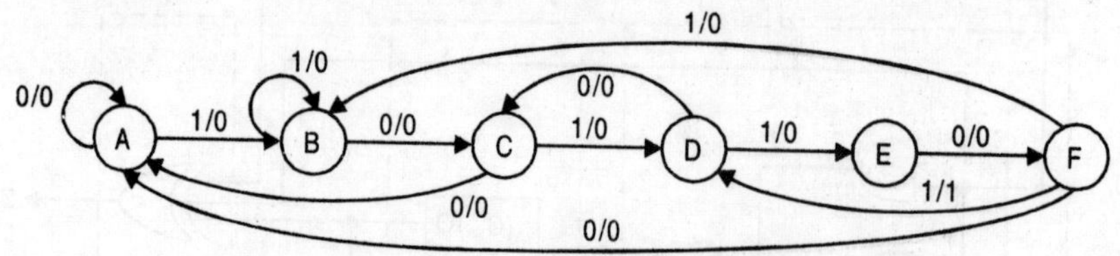

Fig. 6.59 State diagram of sequence detector

Primitive State Table

From the primitive state diagram, the primitive present state/next state (PS/NS) and output table can be drawn as shown in Fig. 6.60.

PS	*NS, Z*	
	X = 0	*X = 1*
A	A, 0	B, 0
B	C, 0	B, 0
C	A, 0	D, 0
D	C, 0	E, 0
E	F, 0	B, 0
F	A, 0	D, 1

Fig. 6.60 Primitive PS/NS and output table

From the primitive state table, one can understand that no state is redundant since no two states are equivalents states. Therefore, the state table cannot be reduced further.

State Assignment

In this step, the following state assignments can be made aribitrarily to the states A to F. Since there are six states, at least three state variables are required. There are three possible binary state assignments such as binary, gray code and one-hot state assignments. Here Binary assignment is considered.

A – 000;

B – 001;

C – 010;

D – 011;

E – 100;

F – 101;

Now, using the above state assignment the Present state/Next state and output table can be modified as shown in Fig. 6.61.

PS	NS		Z	
$Y_3\ Y_2\ Y_1$	$Y_3\ Y_2\ Y_1$			
	$X = 0$	$X = 1$	$X = 0$	$X = 1$
0 0 0	000	001	0	0
0 0 1	010	001	0	0
0 1 0	000	011	0	0
0 1 1	010	100	0	0
1 0 0	101	001	0	0
1 0 1	000	011	0	1

Fig. 6.61 PS/NS and output table

Next step is to select the memory element, *i.e.*, flip-flop to be used. If delay flip-flops are used as memory elements then using the excitation table of delay flip-flop, the excitation table can be drawn as shown in Fig. 6.62.

PS	$D_3D_2D_1$	
$y_3\ y_2\ y_1$	$x = 0$	$x = 1$
0 0 0	000	001
0 0 1	010	001
0 1 0	000	011
0 1 1	010	100
1 0 0	101	001
1 0 1	000	011

Fig. 6.62 Excitation table of sequence detector

The excitation maps for flip-flop inputs D_3, D_2, D_1 and output map can be drawn as shown in Fig. 6.63 and the excitation functions and output function can be simplified.

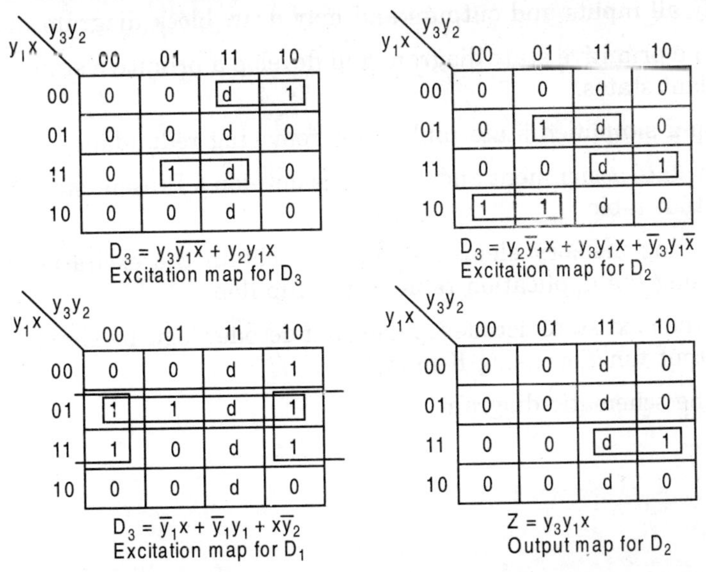

Fig. 6.63 K-map simplification for excitation and output functions

Using the simplified expression for excitation and output functions, the circuit diagram of sequence detector can be drawn as shown in Fig. 6.64.

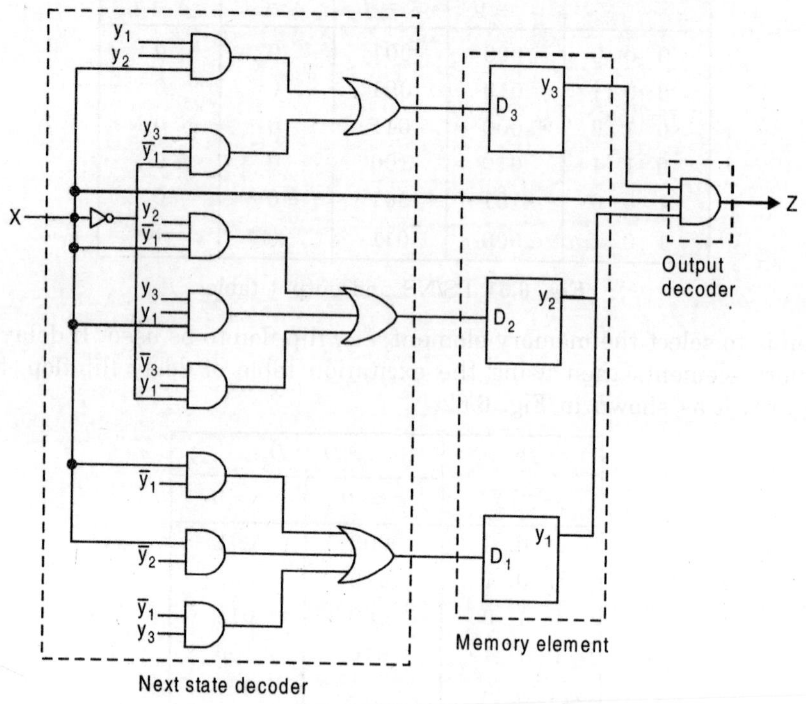

Fig. 6.64 Circuit diagram of sequence detector

The design examples of serial binary adder and sequence detector follow the following steps:

- Obtain the design specifications.
- Identify all inputs and outputs and may draw block diagram model.
- Draw the primitive state diagram and develop a primitive state table and check for redundant states.
- Develop a simplified state table after removing redundancy.
- Make a state assignment and using this develop a Present state/Next state (PS/NS) and output table.
- By selecting memory elements (flip-flop) obtain the excitation table form PS/NS table using the application table of the flip-flop.
- Derive the next state decoder and output decoder logic by simplifying the excitation and output function using K-maps.
- Draw the schematic diagram.

6.7 SOLVED EXAMPLES

Example 1. *For the state diagram shown in Fig. 6.65. Write state table & reduced state table.*

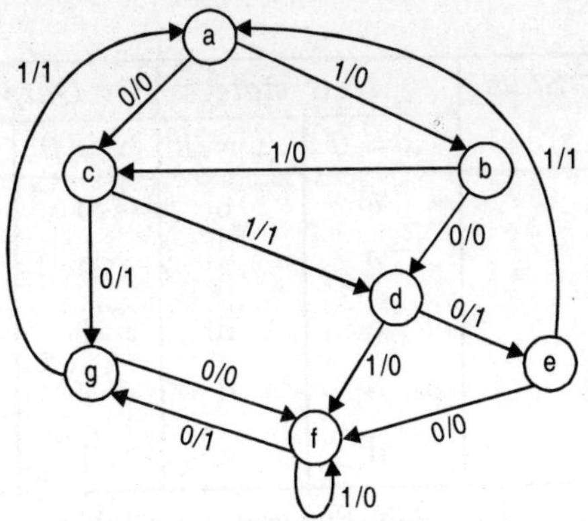

Fig. 6.65

Solution. From the state diagram, a state table is prepared as shown in Fig. 6.66.

Present State	*Next state*		*Output*	
	$x = 0$	$x = 1$	$x = 0$	$x = 1$
a	c	b	0	0
b	d	c	0	0
c	g	d	1	1
d	e	f	1	0
e	f	a	0	1
f	g	f	1	0
g	f	a	0	1

Fig. 6.66

It has been shown from the table of Fig. 6.66 that the present state *e* and *g* both have the same next states *f* (when $x = 0$) and *a* (when $x = 1$). They also produce the same output 0 (when $x = 0$) and 1 (when $x = 1$). Therefore, states *e* and *g* are equivalent. Thus one of the states, e or g can be removed from the state table. For example, if we remove row *g* from the table and replace all *g*'s by *e*'s in the columns, the state table is modified as shown in Fig. 6.67.

State g removed

Present State	*Next state*		*Output*	
	$x = 0$	$x = 1$	$x = 0$	$x = 1$
a	c	b	0	0
b	d	c	0	0
c	e	d	1	1
d	e	f	1	0
e	f	a	0	1
f	e	f	1	0

Fig. 6.67

Similarly state *d* and *f* are also equivalent, therefore one of them say *f*, can be eliminated. After replacing all *f*'s by *d*'s in the columns, the reduced state table is given in Fig. 6.68.

Present State	*Next state*		*Output*	
	x = 0	*x = 1*	*x = 0*	*x = 1*
a	c	b	0	0
b	d	c	0	0
c	e	d	1	1
d	e	d	1	0
e	d	a	0	1

Fig. 6.68 Reduced state table

Example 2. *A sequential circuit has two flip-flops say A and B, two inputs say X and Y, and an output say Z. The flip-flop input functions and the circuit output function are as follows :*

$$JA = xB + y'B'$$

$$JB = xA'$$

$$Z = xy\,A + x'y'B$$

$$KA = xy'B'$$

$$KB = xy' + A$$

Obtain state table, state diagram and state equations.

Solution.

State table for the problem is shown in Fig. 6.69.

Present state		*Next state*								*Output z*			
		xy = 00		*xy = 01*		*xy = 10*		*xy = 11*		*xy = 00*	*xy = 01*	*xy = 10*	*xy = 11*
A	B	A	B	A	B	A	B	A	B				
0	0	1	0	0	0	1	1	0	1	0	0	0	0
0	1	0	1	0	1	1	0	1	1	1	0	0	0
1	0	1	0	1	0	0	0	1	0	0	0	0	1
1	1	1	0	1	0	1	0	1	0	1	0	0	1

Fig. 6.69 State table

With the help of state table we can draw the state diagram as shown in Fig. 6.70.

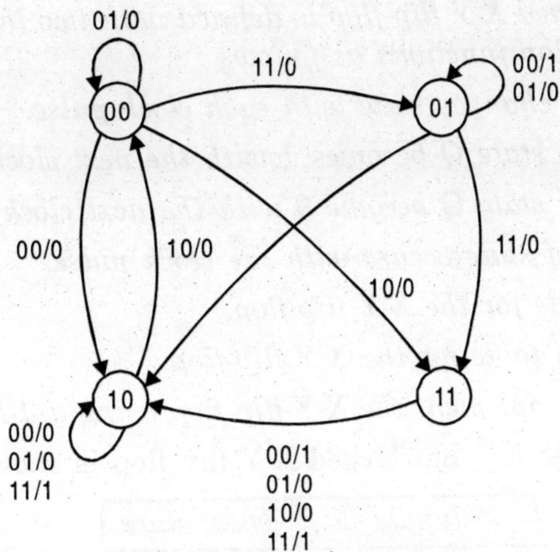

Fig. 6.70 State diagram

The state equation will be

$$A(t+1) = xB + Y' + YA + x'A$$

$$B(t+1) = xA'B' + x'AB + YA'B$$

Example 3. *A clocked sequential circuit has three states, A, B and C and one input X. As long as the input X is 0, the circuit alternates between the states A and B. If the input X becomes 1 (either in state A or in state B), the circuit goes to state C and remains in the state C as long as X continues to be 1. The circuit returns to state A if the input becomes 0 once again and from then one repeats its behaviour. Assume that the state assignments are A = 00, B = 01 and C = 10.*

(*a*) *Draw the state diagram.*

(*b*) *Give the state table for the circuit.*

Solution. (*a*) First draw circles for 3 states A, B, C and write state assignments.

The directed line indicate the transition and input on the directed line are those causes the change on the line. Figure 6.71 shows the state diagram.

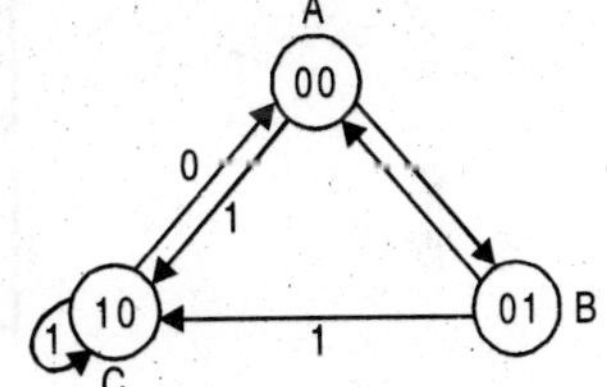

Fig. 6.71

(*b*) From the state diagram, a state table is drawn as shown in Fig. 6.72.

Present State	*Next state*		*Output*	
	$x = 0$	$x = 1$	$x = 0$	$x = 1$
A	B	C	0	1
B	A	C	0	1
C	A	C	0	1

Fig. 6.72 State table

Given the state assignments A = 00, B = 01, C = 10.

Example 4. *A new clocked X-Y flip-flop is defined with two inputs X and Y in addition to the clock input. The flip-flop functions as follows :*

If XY = 00, the flip-flop changes state with each clock pulse.

If XY = 01, the flip flop state Q becomes 1 with the next clock pulse.

If XY = 10, the flip flop state Q become 0 with the next clock pulse.

If XY = 11, no change of state occurs with the clock pulse.

(a) Write the truth table for the X-Y flip-flop.

(b) Write the excitation table for the X-Y flip-flop.

(c) Draw a circuit 40 implement the X-Y flip-flop using a J-K flip-flop.

Solution. (*a*) Truth table for the clocked X-Y flip flop is shown in Fig. 6.73.

Inputs		*Next state*
X	*Y*	Q_{n+1}
0	0	Q_n
0	1	1
1	0	0
1	1	Q_n

Fig. 6.73

(*b*) The Excitation table for the X-Y flip flop is shown in Fig. 6.74.

Q_n	Q_{n+1}	*X*	*Y*
0	0	1	x
0	1	0	x
1	0	x	0
1	1	x	1

X = don't care

Fig. 6.74

(*c*) On comparing Excitation table of X-Y flip-flop with JK flip-flop

$$X = \bar{J} : Y = \bar{K}$$

Therefore, the X-Y flip-flop can be implemented using J-K flip-flop as shown in Fig. 6.75.

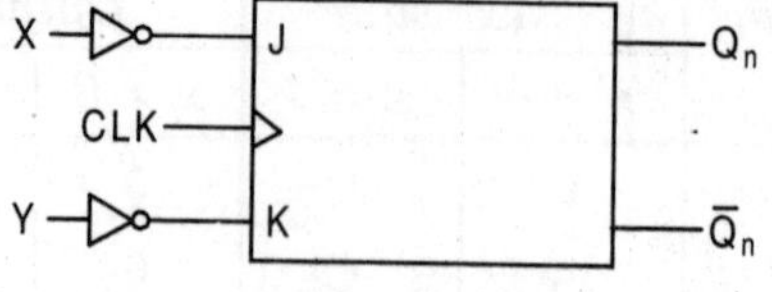

Fig. 6.75

Example 5. *For the digital circuit shown in the Fig. 6.76. Explain what happens at the nodes N_1, N_2, F and $\bar{F}$, when* *(i)* C_K =1 and 'A' changes from '0' to '1'.

(ii) A = 1 and 'C_K' changes from '1' to '0'.

(iii) C_K = 0 and 'A' changes from '1' to '0'.

(iv) Initially, C_K = 0 and 'A' changes from 0 to 1, and then C_K goes to 1.

(v) What is the circuit performing?

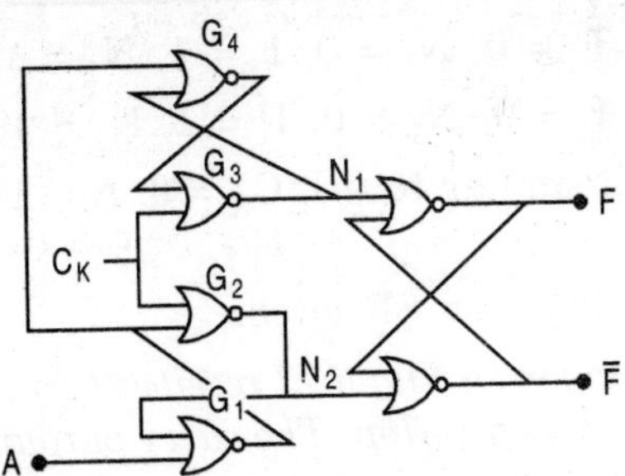

Fig. 6.76

Solution.

(i)

$N_{1(n)}$	$N_{2(n)}$	C_K	A	G_1	G_2	G_3	G_4	$N_{2(n+1)}$	$N_{1(n+1)}$
0	0	1	1	0	0	0	1	0	0
1	0	1	1	0	0	0	0	0	0
0	1	1	1	0	0	0	1	0	0
1	1	1	1	0	0	0	0	0	0

Initially if N_1, N_2 are 0, it remains at 0. If at 1, they go to 0.

As N_1, N_2 are at 0, initially, if F is 1, F is 0 or initially if F = 0, F is 1.

That is F, $\bar{F}$ do not change their initial states.

(ii)

$N_{1(n)}$	$N_{2(n)}$	C_K	A	G_1	G_2	$N_{2(n+1)}$	G_3	G_4	$N_{1(n+1)}$
0	0	0	1	0	1	1	0	1	0
1	0	0	1	0	1	1	1	0	1
0	1	0	1	0	1	1	0	1	0
1	1	0	1	0	1	1	1	0	1

N_2 continues to be 1 whatever be its initial state. N_1 remains at 0 if initially 0, and 1 if initially 1.

If F = 0, $\bar{F}$ = 0

If N_1 = 0, F will become 1 and $\bar{F}$ = 0

If N_1 = 1, F will be 0 and $\bar{F}$ also 0, which is prohibited state.

(*iii*)

$N_{1(n)}$	$N_{2(n)}$	C_K	A	G_1	G_2	$N_{2(n+1)}$	G_3	G_4	$N_{1(n+1)}$
0	0	0	0	1	0	0	1	0	1
1	0	0	0	1	0	0	1	0	1
0	1	0	0	0	1	1	0	1	0
1	1	0	0	0	1	1	1	0	1

$N_2 = 0,\ F = 1,\ \bar{F} = 0,\ N_1 = 0,\ F = 1,\ N_1 = 1,\ F = 0,\ \bar{F} = 1$

$N_2 = 1,\ F = 1,\ F = 0,\ N_1 = 0,\ F = 1,\ N_1 = 1,\ F = 0,\ \bar{F} = 1$

(*iv*) The change of state is similar to (*ii*), $C_K = 0$, $A = 1$ initially and finally as at (*i*), $C_K = 1$, $A = 1$.

(*v*) The circuit is functioning as a SR latch.

Example 6. *The full adder given in Fig. 6.77 receives two external inputs x & y; the third input Z comes from the output of a D flip-flop. The carry output is transferred to the flip-flop, every clock pulse. The external S output gives the sum of x, y and z. Obtain the state table and state diagram of the sequential circuit.*

Solution.

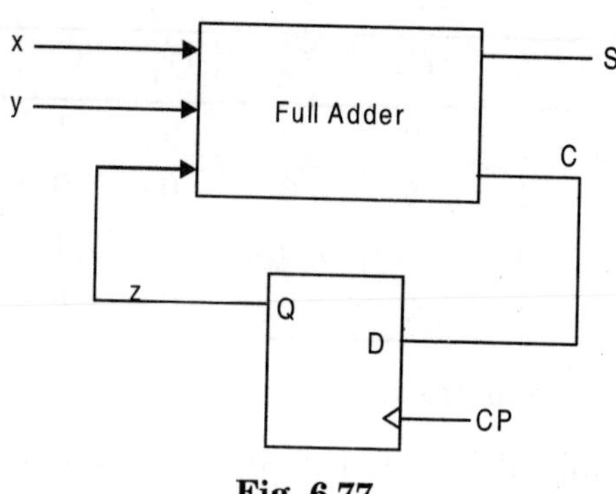

Fig. 6.77

The state table for the given circuit is shown in Fig. 6.78.

Present state	*Inputs*		*Next state*	*Output*
z	*x*	*y*	*z*	*S*
0	0	0	0	0
0	0	1	0	1
0	1	0	0	1
0	1	1	1	0
1	0	0	0	1
1	0	1	1	0
1	1	0	1	0
1	1	1	1	1

Fig. 6.78 State table

The state diagram corresponding to state-table of Fig. 6.79.

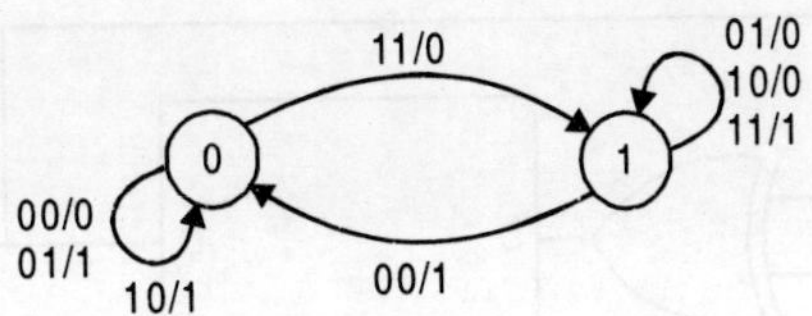

Fig. 6.79 State diagram

6.8 EXERCISE

1. An R-S latch can be based on cross-coupled NOR gates. It is also possible to contruct an R'-S' latch using cross-coupled NAND gates.
   (*a*) Draw the R'-S' latch, labelling the R' and S' inputs and the Q and Q' outputs.
   (*b*) Show the timing behaviour across the four configurations of R' and S'. Indicate on your timing diagram the behaviour in entering and leaving the forbidden state when R' = S' = 0.
   (c) Draw the state diagram that shows the complete input/output and state transition behaviour of the R'-S' latch.
   (*d*) What is the characteristic equation of the R'-S' latch?
   (*e*) Draw a simple schematic for a gated R-S latch with an extra enable input, using NAND gates only.
2. Consider a D-type storage element implemented in five different ways :
   (*a*) D-latch (i.e., D wired to the S-input and D' wired to the R-input of an R-S latch);
   (*b*) Clock Enabled D-latch;
   (c) Master-Slave Clock Enabled D-Flip-flop;
   (*d*) Positive Edge-triggered Flip-flop;
   (*e*) Negative Edge-triggered Flip-flop,

   Complete the following timing charts indicating the behaviour of these alternative storage elements. Ignore set-up and hold time limitations (assume all constraints are meant):

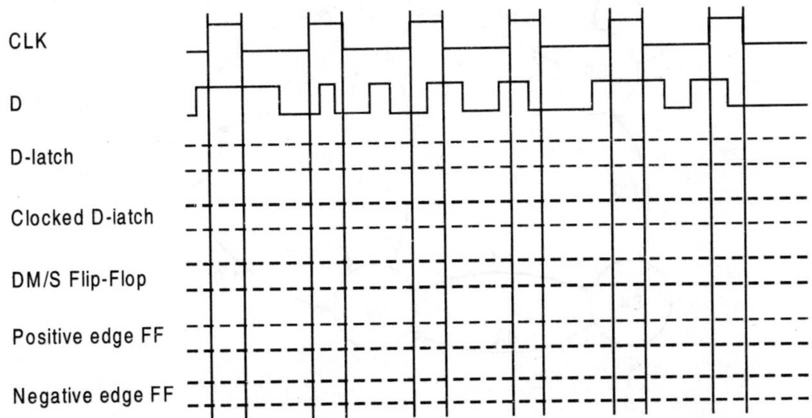

3. Complete the timing diagram for this circuit.

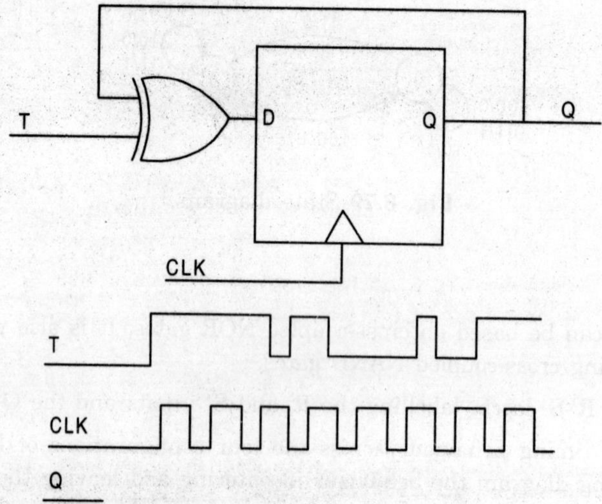

4. Design a circuit that implements the state diagram.

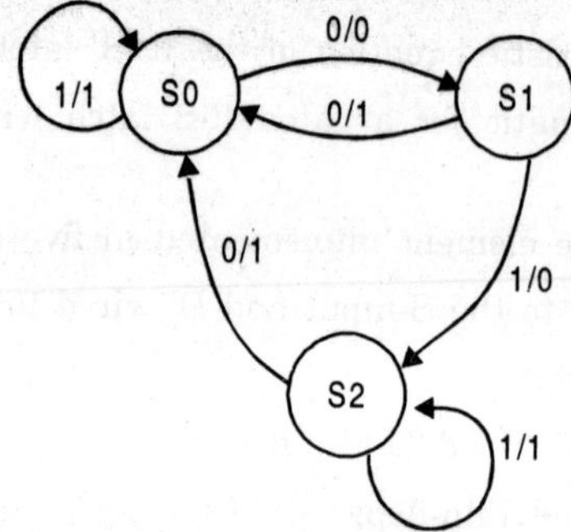

5. Design a circuit that implements the state diagram.

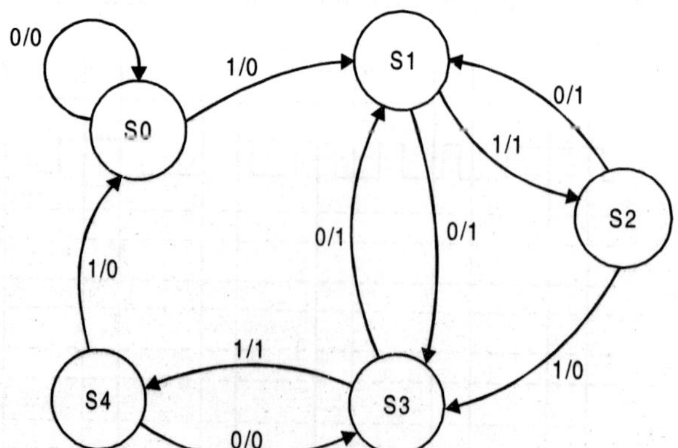

6. A sequential network has one input X and two outputs S and V. X represent a four bit binary number N, which is input least significant bit first. S represents a four bit binary number equal to N + 2, which is output least significant bit first. At the time the fourth input is sampled, V = 1, in N + 2 is too large to be represented by four bits; otherwise V = 0.

   Derive a Mealy state graph and table with a minimum number of states.

7. A sequential network has one input X and two outputs S and V. X represent a four bit binary number N, which is input least significant bit first. S represents a four bit binary number equal to N – 2, which is output least significant bit first. At the time the fourth input is sampled, V = 1, in N – 2 is too small to be represented by four bits; otherwise V = 0.

   Derive a Mealy state graph and table with a minimum number of states.

8. Design a synchronous circuit using negative edge-trigered D flip-flops that provides an output signal Z which has one-fifth the frequency of the clock signal. Draw a timing diagram to indicate the exact relationship between the clock signal and the output signal Z. To ensure illegal state recovery, force all unused or illegal states to go to 0.

9. Consider the design of a sequence detector finite state machine that will assert a 1 when the current input equals the just previously seen input.

   (*a*) Draw as simple state diagrams for a MEALY MACHINE and a MOORE MACHINE implementation as you can (minimization is not necessary). The MEALY MACHINE should have fewer states. Briefly explain why.

   (*b*) If the Mealy Machine is implemented as a SYNCHRONOUS MEALY MACHINE, draw the timing diagram for the example input/output sequence described above.

   (*c*) If the timing behaviours are different for the MOORE, MEALY, and SYNCHRONOUS MEALY machines, explain the reason why.

10. A sequential circuit is specified by the following flip-flop input functions. Draw the logic diagram of the circuit.

    JA = Bx' KA = Bx

    JB = x KB = A⊕x

11. Design the circuit and draw the logic diagram of the sequential circuit specified by the following state diagram. Use an RS flip-flop.

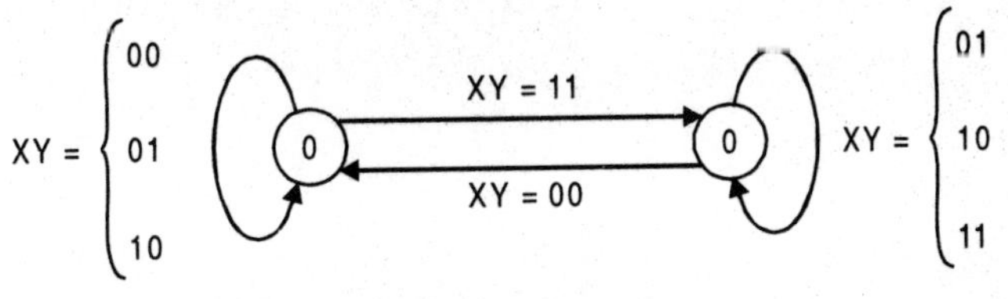

12. Complete the truth table for the latch constructed from 2 NOR gates.

S	R	Q	Q'	
1	0			
0	0			(after S = 1, R = 0)
0	1			
0	0			(after S = 0, R = 1)
1	1			

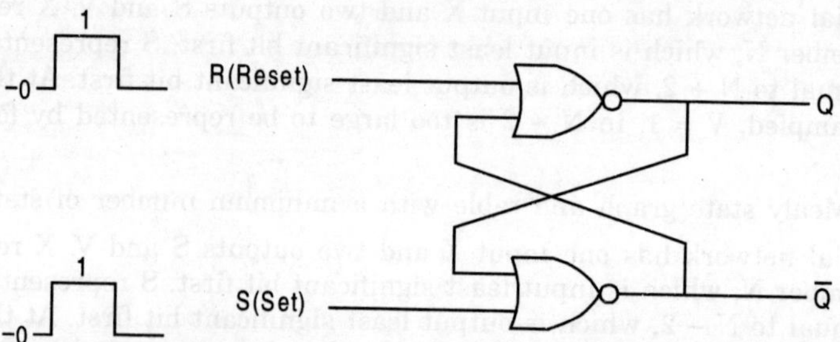

13. Construct a logic diagram of a clocked D flip-flop using AND and NOR gates.

14. Explain the master-slave flip-flop constructed from two R-S flip-flop.

15. Draw the logic diagram of a master-slave D flip-flop using NAND gates.

7

SHIFT REGISTERS AND COUNTERS

7.0 INTRODUCTION

Registers are the group of flip-flops (single bit storage element). The simplest type of register is a data register, which is used for the temporary storage of data. In its simplest form, it consists of a set of N D flip-flops, all sharing a common clock. All of the digits in the N bit data word are connected to the data register by an N line "data bus". Figure 7.0 shows a four bit data register, implemented with four D flip-flops.

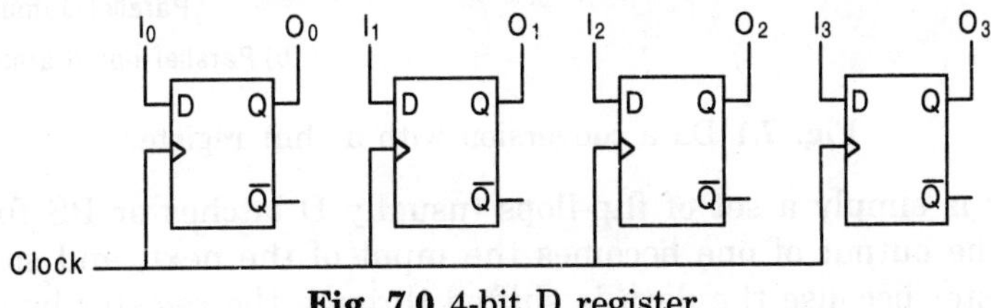

Fig. 7.0 4-bit D register

The data register is said to be a synchronous device, because all the flip-flops change state at the same time.

7.1 SHIFT REGISTERS

A common form of register used in many types of logic circuits is a shift register. Registers like counters, are sequential circuits and as they employ flip-flops they possess memory; but memory is not the only requirement of a shift register. The function of storage of binary data can be very well performed by a simple register. Shift registers are required to do much more than that. They are required to store binary data momentarily until it is utilized for instance, by a computer, microprocessor, etc. Sometimes data is required to be presented to a device in a manner which may be different from the way in which it is fed to a shift register. For instance, shift register can present data to a device in a serial or parallel form, irrespective of the manner in which it is fed to a shift register. Data can also be manipulated within the shift register, so that it is presented to a device in the required form. These devices can also shift left or right and it is this capability which gives them the name of shift register. Figure 7.1 show the many ways in which data can be fed into a shift register and presented by it to a device.

Shift registers have found considerable application in arithmatic operations. Since, moving a binary number one bit to the left is equivalent to multiplying the number by 2 and moving the number one bit position to the right amounts to dividing the number by 2. Thus, multiplications and divisions can be accomplished by shifting data bits. Shift registers find considerable application in generating a sequence of control pulses.

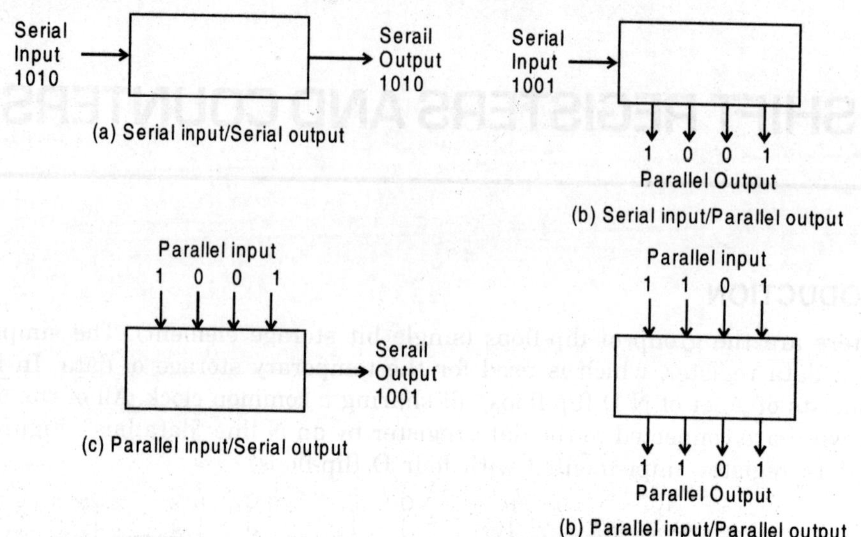

Fig. 7.1 Data conversion with a shift register

Shift register is simply a set of flip-flops (usually D latches or RS flip-flops) connected together so that the output of one becomes the input of the next, and so on in series. It is called a shift register because the data is shifted through the register by one bit position on each clock pulse. Figure 7.2 shows a four bit shift register, implemented with D flip-flops.

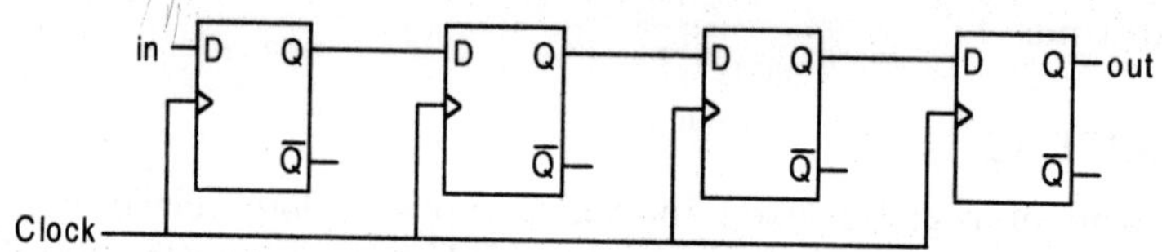

Fig. 7.2 4-bit serial-in serial-out shift register

- On the leading edge of the first clock pulse, the signal on the DATA input is latched in the first flip-flop. On the leading edge of the next clock pulse, the contents of the first flip-flop is stored in the second flip-flop, and the signal which is present at the DATA input is stored in the first flip-flop, etc. Because the data is entered one bit at a time, this is called a serial-in shift register. Since there is only one output, and data leaves the shift register one bit at a time, then it is also a serial out shift register. (Shift registers are named by their method of input and output; either serial or parallel.) Parallel input can be provided through the use of the preset and clear inputs to the flip-flop. The parallel loading of the flip-flop can be synchronous (i.e., occurs with the clock pulse) or asynchronous (independent of the clock pulse) depending on the design of the shift register. Parallel output can be obtained from the outputs of each flip-flop as shown in Fig. 7.3.

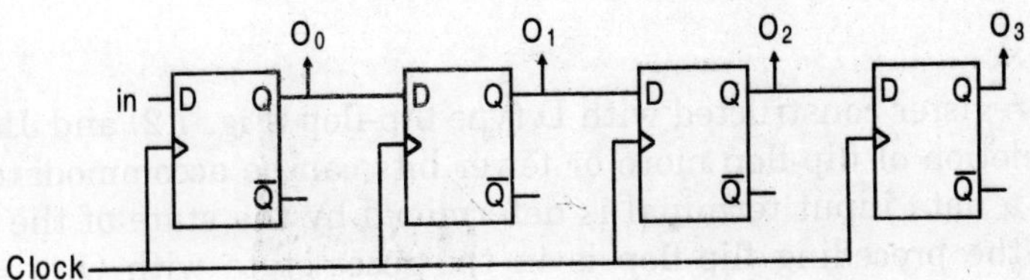

Fig. 7.3 4-bit serial-in parallel-out shift register

Communication between a computer and a peripheral device is usually done serially, while computation in the computer itself is usually performed with parallel logic circuitry. A shift register can be used to convert information from serial form to parallel form, and vice versa. Many different kinds of shift registers are available, depending upon the degree of sophistication required.

Here we deal with the basic characteristics of shift registers and their applications. Normally shift registers are obtained through D-Flip-Flops. However, if required other flip-flops may also be used. D-Flip-Flops are used because of simplicity that data presented at input is available at the output. Throughout the chapter it is our strategy to discuss all the shift registers using D-flip-flops only. If one need to use some other Flip-Flop, say JK Flip-Flop, then we recommend following procedure–

1. Design the shift register using D-flip-flops only.
2. Take JK Flip-Flop and convert it into D Flip-Flop.
3. Replace each of the D-Flip-Flop of step1 by the flip-flop obtained in step 1 after conversion.

To elaborate this let us consider the shift register shown in Fig. 7.2.

Step 1: It is readily obtained in Fig. 7.2.

Step 2: Convert JK into D-Flip-Flop. It is shown below in Fig. 7.4

Step 3: Replace each D-Flip-Flop of Fig. 7.2 by the one shown in Fig. 7.4.

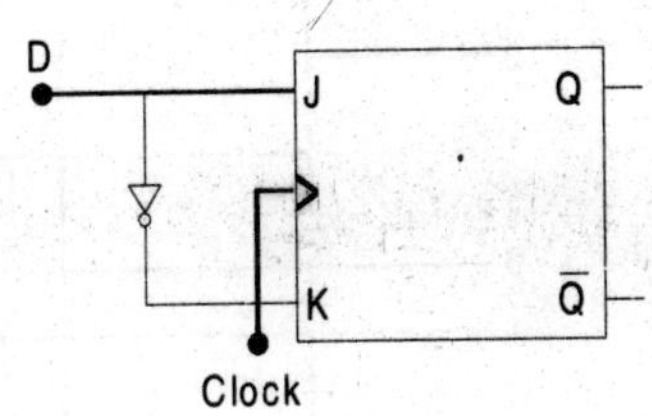

Fig. 7.4 JK Flip-flop converted into D-Flip-Flop

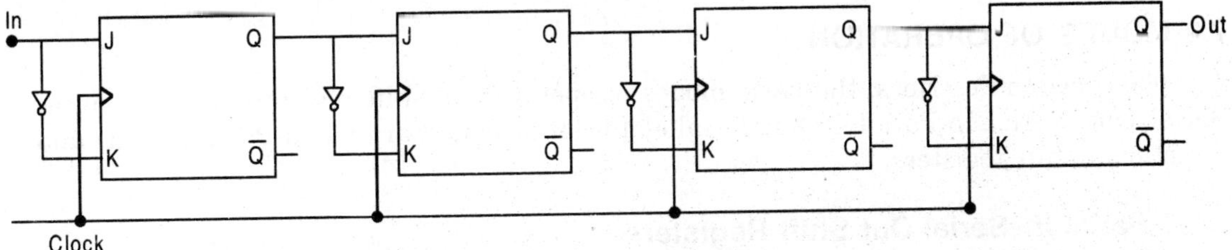

Fig. 7.5 *(a)* 4-bit serial in-serial out shift register using JK Flip-Flop

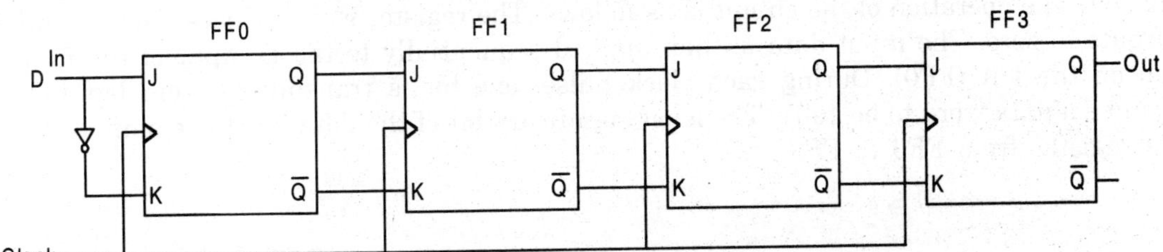

Fig. 7.5 *(b)* 4-bit serial in–serial out shift register using JK Flip-flop

OPERATION

A 4-bit shift register constructed with D type flip-flop (Fig. 7.2) and JK flip-flop (Fig. 7.5). By addition or deletion of flip-flop more or fewer bits can be accommodated. Except for FF0, the logic level at a data input terminal is determined by the state of the preceding flip-flop. Thus, D_n is 0 if the preceding flip-flop is in the reset state with $Q_{n-1} = 0$, and $D_n = 1$ if $Q_{n-1} = 1$. The input at FF0 is determined by an external source.

From the characteristic of D-flip-flop we know that immediately after the triggering transition of the clock, the output Q of flip-flop goes to the state present at its input D just before this clock transition. Therefore, at each clock transition, pattern of bits, 1s and 0s, is shifted one flip-flop to the right. The bit of the last flip-flop (FF3 in Fig. 7.6) is lost, while the first flip-flop (FF0) goes to the state determined by its input D_0. This operation is shown in Fig. 7.6. We have assumed that the flip-flop triggers on the positive-going transition of the clock waveform, and initially we have D0 = 0, FF0 = 1 and FF2 = FF3 = FF4 = 0.

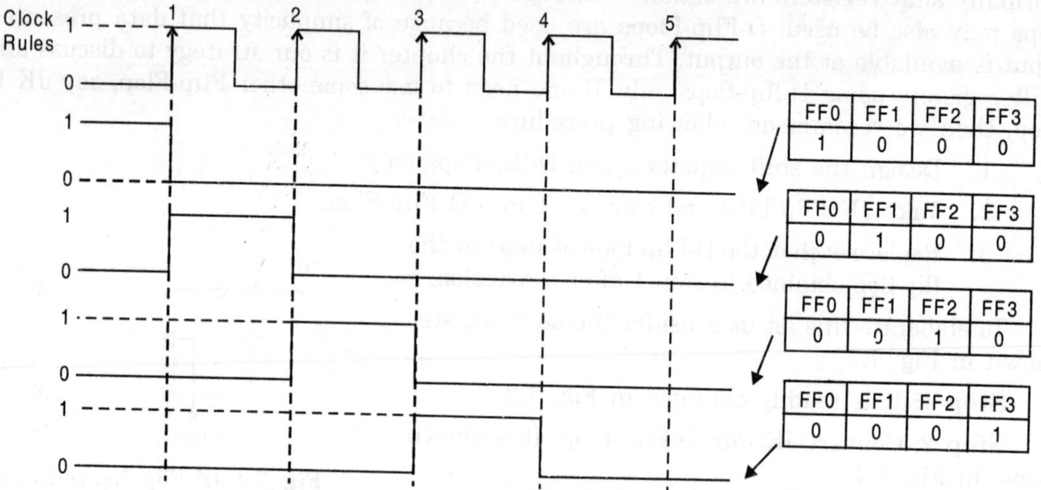

Fig. 7.6 A 4-bit shift register operation

7.2 MODES OF OPERATION

This section describes, the basic modes of operation of shift registers such as Serial In-Serial Out, Serial In-Parallel Out, Parallel In-Serial Out, Parallel In-Parallel Out, and bi-directional shift registers.

7.2.1 Serial In–Serial Out Shift Registers

A basic four-bit shift register can be constructed using four D-flip-flops, as shown in Fig. 7.7. The operation of the circuit is as follows. The register is first cleared, forcing all four outputs to zero. The input data is then applied sequentially to the D input of the first flip-flop on the left (FF0). During each clock pulse, one bit is transmitted from left to right. Assume a data word to be 1001. The least significant bit of the data has to be shifted through the register from FF0 to FF3.

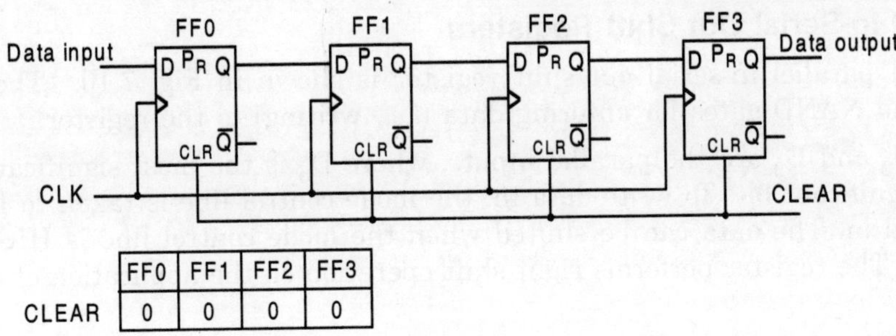

Fig. 7.7

In order to get the data out of the register, they must be shifted out serially. This can be done destructively or non-destructively. For destructive readout, the original data is lost and at the end of the read cycle, all flip-flops are reset to zero.

	FF0	FF1	FF2	FF3	
0000	1	0	0	1	0000

To avoid the loss of data, an arrangement for a non-destructive reading can be done by adding two AND gates, an OR gate and an inverter to the system. The construction of this circuit is shown in Fig. 7.8.

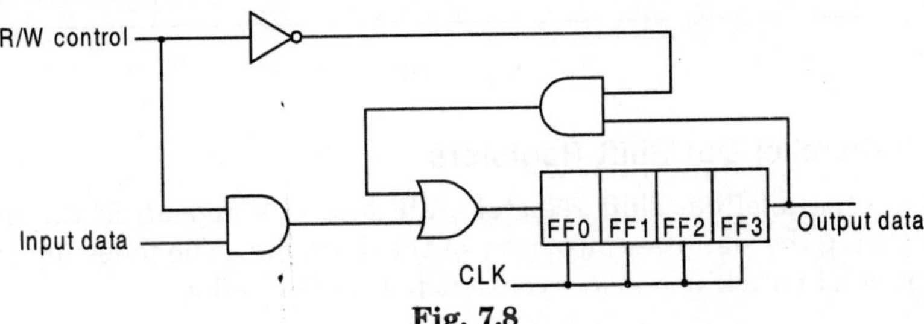

Fig. 7.8

The data is loaded to the register when the control line is HIGH (*i.e.* WRITE). The data can be shifted out of the register when the control line is LOW (*i.e.* READ).

7.2.2 Serial In-Parallel Out Shift Registers

For this kind of register, data bits are entered serially in the same manner as discussed in the last section. The difference is the way in which the data bits are taken out of the register. Once the data are stored, each bit appears on its respective output line, and all bits are available simultaneously. A construction of a four-bit serial in-parallel out register is shown in Fig. 7.9.

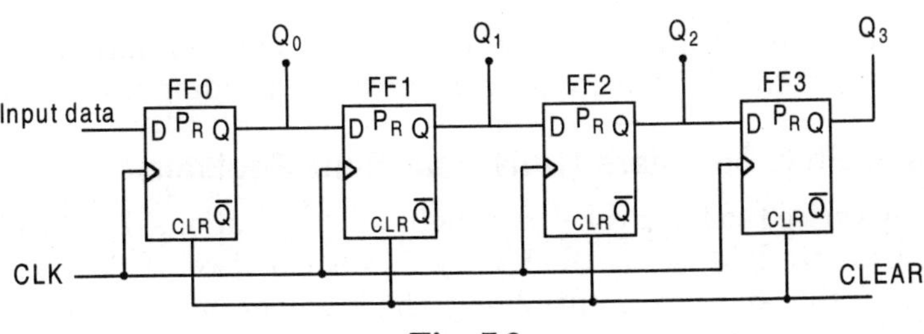

Fig. 7.9

7.2.3 Parallel In-Serial Out Shift Registers

A four-bit parallel in-serial out shift register is shown in Fig. 7.10. The circuit uses D-flip-flops and NAND gates for entering data (*i.e.*, writing) to the register.

D_0, D_1, D_2 and D_3 are the parallel inputs, where D_0 is the most significant bit and D_3 is the least significant bit. To write data in, the mode control line is taken to LOW and the data is clocked in. The data can be shifted when the mode control line is HIGH as SHIFT is active high. The register performs right shift operation on the application of a clock pulse.

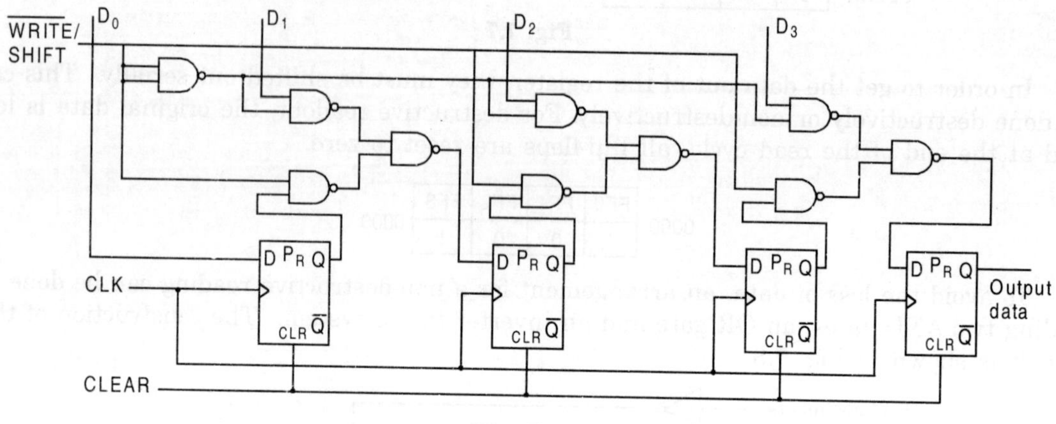

Fig. 7.10

7.2.4 Parallel In-Parallel Out Shift Registers

For parallel in-parallel out shift registers, all data bits appear on the parallel outputs immediately following the simultaneous entry of the data bits. The following circuit is a four-bit parallel in-parallel out shift register constructed by D-flip-flops.

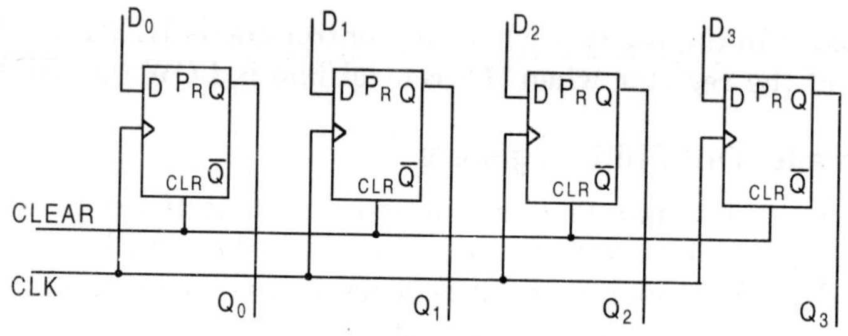

Fig. 7.11

The D's are the parallel inputs and the Q's are the parallel outputs. Once the register is clocked, all the data at the D inputs appear at the corresponding Q outputs simultaneously.

7.2.5 Bidirectional Shift Registers (Universal Shift Register)

The registers discussed so far involve only right shift operations. Each right shift operation has the effect of successively dividing the binary number by two. If the operation is reversed (left shift), this has the effect of multiplying the number by two. With suitable gating arrangement a serial shift register can perform both operations.

A bi-directional, or reversible shift register is one in which the data can be shift either left or right. A four-bit bi-directional shift register using D-flip-flops is shown in Fig. 7.12.

Here a set of NAND gates are configured as OR gates to select data inputs from the right or left adjacent bistables, as selected by the $\overline{\text{LEFT}}$/RIGHT control line.

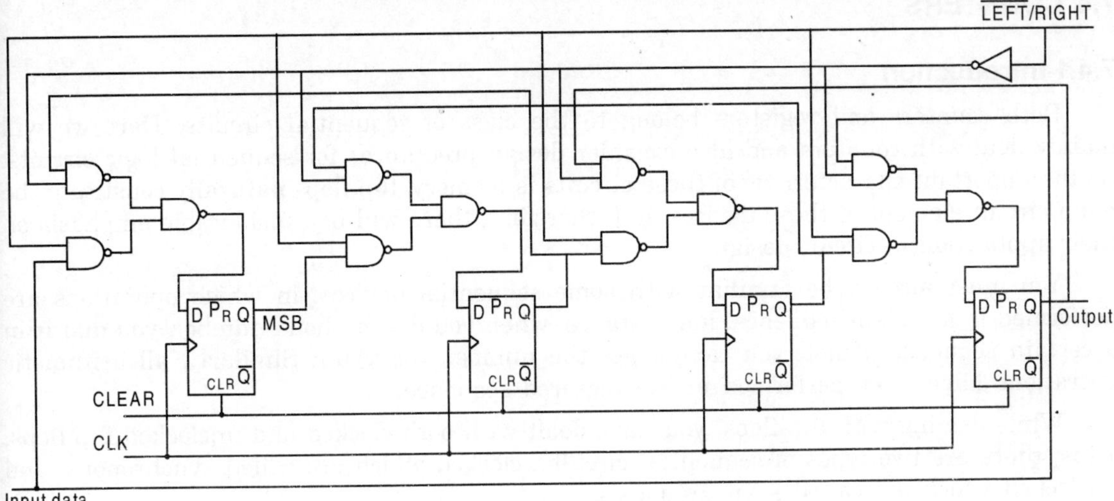

Fig. 7.12

7.3 APPLICATIONS OF SHIFT REGISTERS

Shift registers can be found in many applications. Here is a list of a few.

7.3.1 To Produce Time Delay

The serial in-serial out shift register can be used as a time delay device. The amount of delay can be controlled by:

- the number of stages in the register (N)
- the clock frequency (f)

The time delay ΔT is given by

$$\Delta T = N^* f$$

7.3.2 To Simplify Combinational Logic

The ring counter technique can be effectively utilized to implement synchronous sequential circuits. A major problem in the realization of sequential circuits is the assignment of binary codes to the internal states of the circuit in order to reduce the complexity of circuits required. By assigning one flip-flop to one internal state, it is possible to simplify the combinational logic required to realize the complete sequential circuit. When the circuit is in a particular state, the flip-flop corresponding to that state is set to HIGH and all other flip-flops remain LOW.

7.3.3 To Convert Serial Data to Parallel Data

A computer or microprocessor-based system commonly requires incoming data to be in parallel format. But frequently, these systems must communicate with external devices that send or receive serial data. So, serial-to-parallel conversion is required. As shown in the previous sections, a serial in-parallel out register can achieve this.

7.4 COUNTERS

7.4.1 Introduction

Both counters and registers belong to the class of sequential circuits. Here we will mainly deal with counters and also consider design procedures for sequential logic circuits. As the important characteristic of these circuits is memory, flip-flops naturally constitute the main circuit element of these devices and, therefore, there will be considerable emphasis on their application in circuit design.

You must already be familiar with some sequential devices, in which operations are performed in a certain sequence. For instance, when you dial a phone number, you dial it in a certain sequence, if not, you cannot get the number you want. Similarly, all arithmetic operations have to be performed in the required sequence.

While dealing with flip-flops, you have dealt with both clocked and unclocked flip-flops. Thus, there are two types of sequential circuits, clocked which are called synchronous, and unclocked which are called asynchronous.

In asynchronous devices, a change occurs only after the completion of the previous event. A digital telephone is an example of an asynchronous device.

If you are dialing a number, say 6354, you will first punch 6 followed by 3, 5 and 4. The important point to note is that, each successive event occurs after the previous event has been completed.

Sequential logic circuits find application in a variety of binary counters and storage devices and they are made up of flip-flops. A binary counter can count the number of pulses applied at its input. On the application of clock pulses, the flip-flops incorporated in the counter undergo a change of state in such a manner that the binary number stored in the flip-flops of the counter represents the number of clock pulses applied at the input. By looking at the counter output, you can determine the number of clock pulses applied at the counter input.

Digital circuits use several types of counters which can count in the pure binary form and in the standard BCD code as well as in some special codes. Counters can count up as well as count down. In this section we will be looking at some of the counters in common use in digital devices.

Another area of concern to us will be the design of sequential circuits. We will be considering both synchronous and asynchronous sequential circuits.

7.4.2 Binary Ripple Up-Counter

We will now consider a 3-bit binary up-counter, which belongs to the class asynchronous counter circuits and is commonly known as a ripple counter. Figure 7.13 shows a 3-bit counter, which has been implemented with three T-type (toggle) flip-flops. The number of states of which this counter is capable is 2^3 or 8. This counter is also referred to as a modulo 8 (or divide by 8) counter. Since a flip-flop has two states, a counter having n flip-flops will have 2^n states.

When clock pulses are applied to a ripple counter, the counter progresses from state to state and the final output of the flip-flop in the counter indicates the pulse count. The circuit recylces back to the starting state and starts counting all over again.

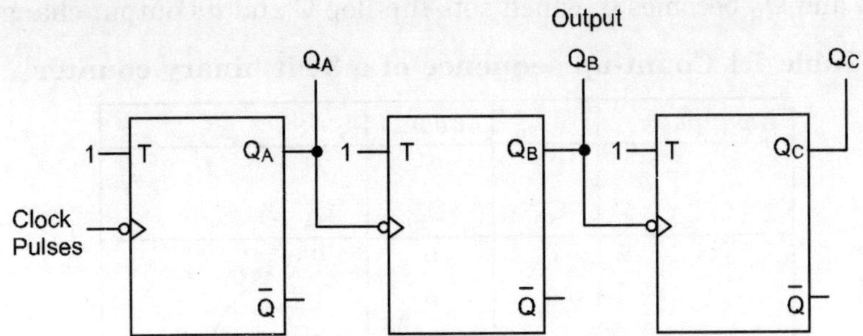

Fig. 7.13 3-Bit binary up-counter

There are two types of ripple counters, (*a*) asynchronous counters and (*b*) synchronous counters. In asynchronous counters all flip-flops are not clocked at the same time, while in synchronous counters all flip-flops are clocked simultaneously.

You will notice from the diagram that the normal output, Q, of each flip-flop is connected to the clock input of the next flip-flop. The T inputs of all the flip-flops, which are T-type, are held high to enable the flip-flops to toggle (change their logic state) at every transition of the input pulse from 1 to 0. The circuit is so arranged that flip-flop B receives its clock pulse from the Q_A output of flip-flop A and, as a consequence, the output of flip-flop B will change its logic state when output Q_A of flip-flop A changes from binary 1 to 0. This applies to all the other flip-flops in the circuit. It is thus an asynchronous counter, as all the flip-flops do not change their logic state at the same time.

Let us assume that all the flip-flops have been reset, so that the output of the counter at the start of the count is 0 0 0 as shown in the first row of Table 7.1. Also refer to Fig. 7.14 which shows the output changes for all the flip-flops at every transition of the input pulse from 1 $\rightarrow$ 0.

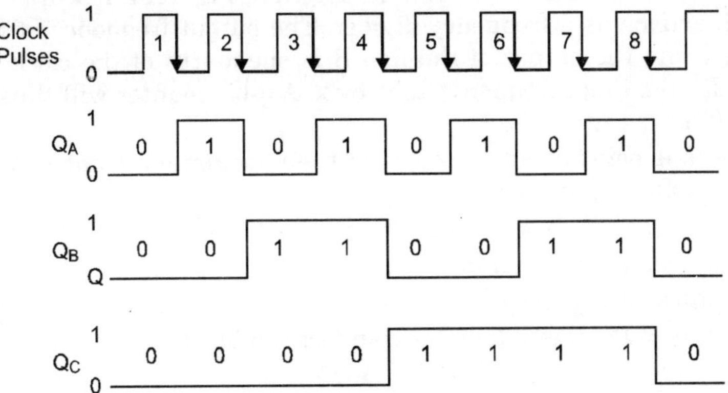

Fig. 7.14 Waveform for 3-bit binary ripple up-counter

When the trailing edge of the first pulse arrives, flip-flop A sets and Q_A becomes 1, which does not affect the output of flip-flop B. The counter output now is as shown in row 2 of the table. As a result of the second clock pulse, flip-flop A resets and its output Q_A changes from 1 to 0, which sets flip-flop B and the counter output now is as shown in row 3 of the table.

When the third clock pulse arrives, flip-flop A sets and its output Q_A becomes 1, which does not change the state of the B or the C flip-flop. The counter output is now as shown in row 3 of the table. When the fourth pulse occurs, flip-flop A resets and Q_B becomes 0 which in turn resets flip-flop B and Q_B becomes 0, which sets flip-flop C and its output changes to 1.

Table 7.1 Count-up sequence of a 3-bit binary counter

Input pulse		*Count*		
	2^2 Q_C	2^1 Q_B	2^0 Q_A	
0	0	0	0	
1	0	0	1	
2	0	1	0	
3	0	1	1	RECYCLE
4	1	0	0	
5	1	0	1	
6	1	1	0	
7	1	1	1	

When the 5th clock pulse arrives, flip-flop A sets and Q_A becomes 1; but the other flip-flops remain unchanged. The number stored in the counter is shown in the 6th row of the table. The 6th pulse resets flip-flop A and at the same time flip-flop B and C are set. The 7th pulse sets all the flip-flops and the counter output is now shown in the last row of the table.

The next clock pulse will reset all the flip-flops, as the counter has reached its maximum count capability. The counter has in all 8 states. In other words, it registers a count of 1 for every 8 clock input pulses. It means that it divides the number of input pulses by 8. It is thus a divide by 8 counter.

Count Capability of Ripple Counters

If you refer to Table 7.1 and the waveform diagram, Fig. 7.14, it will be apparent to you that the counter functions as a frequency divider. The output frequency of flip-flop A is half the input frequency and the output of flip-flop B is one-fourth of the clock input frequency. Each flip-flop divides the input frequency to it by 2. A 3-bit counter will thus divide the clock input frequency by 8.

Another important point about counters is their maximum count capability. It can be calculated from the following equation

$$N = 2^n - 1$$

where N is the maximum count number and

n is the number of flip-flops.

For example, if $n = 12$, the maximum count capability is

$$N = 2^{12} - 1 = 4095$$

If you have to calculate the number of flip-flops required to have a certain count capability, use the following equation :

$$n = 3.32 \log_{10} N$$

For example, if the required count capability is 5000

$$n = 3.32 \log_{10} 5000 = 12.28$$

which means that 13 flip-flops will be required.

Counting Speed of Ripple Counters

The primary limitation of ripple counters is their speed. This is due to the fact that each successive flip-flop is driven by the output of the previous flip-flop. Therefore, each flip-flop in the counter contributes to the total propagation delay. Hence, it takes an appreciable time for an impulse to ripple through all the flip-flops and change the state of the last flip-flop in the chain. This delay may cause malfunction, if all the flip-flops change state at the same time. In the counter we have just considered, this happens when the state changes from 011 to 100 and from 111 to 000. If each flip-flop in the counter changes state during the course of a counting operation, and if each flip-flop has a propagation delay of 30 nanoseconds, a counter having three flip-flops will cause a delay of 90 ns. The maximum counting speed for such a flip-flop will be less than

$$\frac{1}{90} \times 10^9 \text{ or } 11.11 \text{ MHz.}$$

If the input pulses occur at a rate faster than 90 ns, the counter output will not be a true representation of the number of input pulses at the counter. For reliable operation of the counter, the upper limit of the clock pulses of the counter can be calculated from

$$f = \frac{1}{nt} \times 10^9$$

where, n is the number of flip-flops and

t is the propagation delay of each flip-flop.

7.4.3 4-Bit Binary Ripple Up-Counter

A 4-bit binary ripple up-counter can be built with four T-type flip-flops. The diagram will follow the same pattern as for a 3-bit up-counter. The count-up sequence for this counter is given in Table 7.2 and a waveform diagram is given in Fig. 7.15. After the counter has counted up to

Table 7.2 Count-up sequence of a 4-bit binary up-counter

Input pulse	*Count*			
	2^3 Q_D	2^2 Q_C	2^1 Q_B	2^0 Q_A
0	0	0	0	0
1	0	0	0	1
2	0	0	1	0
3	0	0	1	1
4	0	1	0	0
5	0	1	0	1
6	0	1	1	0
7	0	1	1	1
8	1	0	0	0
9	1	0	0	1
10	1	0	1	0
11	1	0	1	1
12	1	1	0	0
13	1	1	0	1
14	1	1	1	0
15	1	1	1	1
16 or 0	0	0	0	0

RECYCLE

1111, it recycles to 0000 like the 3-bit counter. You must have observed that each flip-flop divides the input frequency by, 2 and the counter divides the frequency of the clock input pulses by 16.

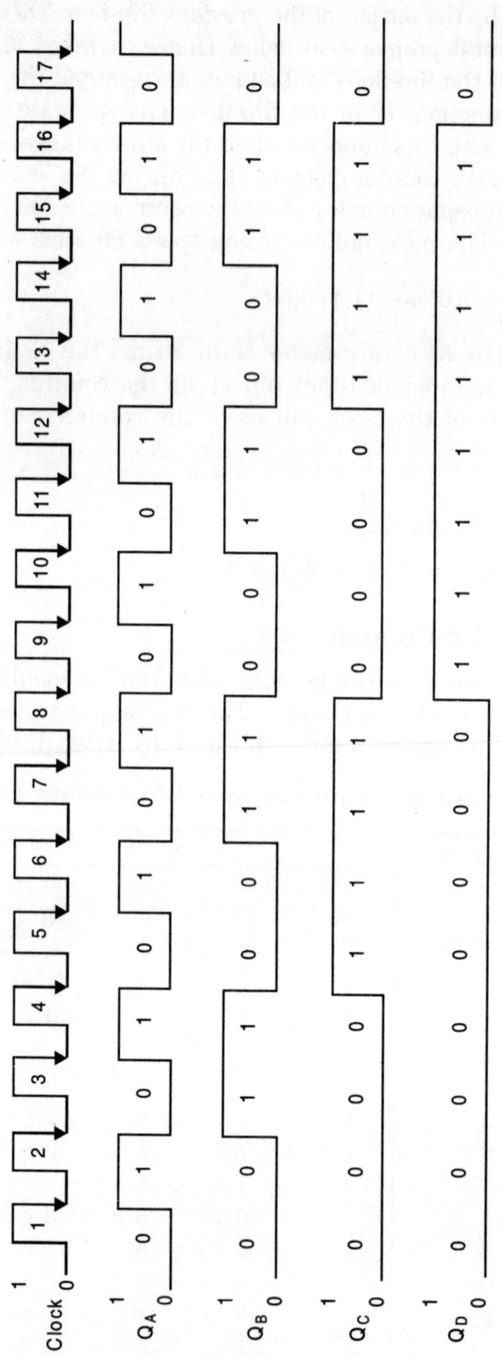

Fig. 7.15 Waveform for 4-bit binary up-counter

7.4.4 3-Bit Binary Ripple Down Counter

The binary ripple up-counter we have just considered increases the count by one, each time a pulse occurs at its input. The binary ripple down counter which we are going to consider in this section decreases the count by one, each time a pulse occurs at the input. A circuit for a 3-bit down counter is given in Fig. 7.16. If you compare this counter with the up-counter in Fig. 7.13 the only difference you will notice is that, in the down counter in Fig. 7.16 the complement output $\overline{Q}$, instead of the normal output, is connected to the clock input of the next flip-flop. The counter output which is relevant even in the down counter is the normal output, Q, of the flip-flops.

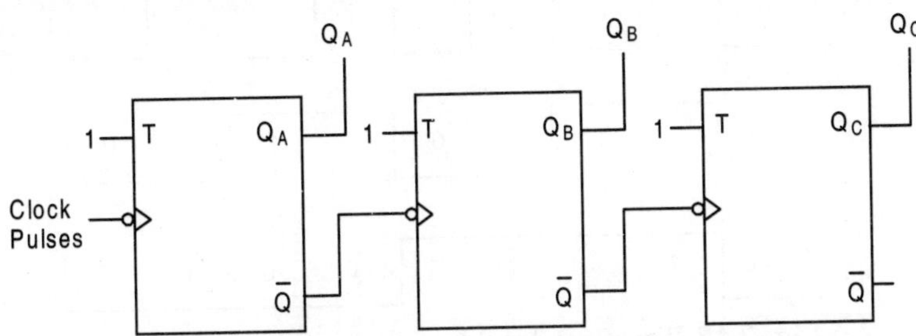

Fig. 7.16 3-bit binary ripple down counter

We can now analyse the circuit and examine its operation. It will help you to follow the operation of the counter, if you refer to Table 7.3 and waveform of the counter given in Fig. 7.17 for each input pulse count. Let us assume that the counter is initially reset, so that the counter output is 0 0 0. When the first input pulse is applied, flip-flop A will set, and its complement output will be 0. This will set flip-flop B, as there will be a 1 → 0 transition at the clock input. The counter output will now be 1 1 1.

Table 7.3 Count-down sequence of a 3-bit binary counter

Clock pulse	*Count*		
	2^2 Q_C	2^1 Q_B	2^0 Q_A
0	0	0	0
1	1	1	1
2	1	1	0
3	1	0	1
4	1	0	0 RECYCLE
5	0	1	1
6	0	1	0
7	0	0	1
8	0	0	0

When the second clock pulse is applied, flip-flop A will reset and its complement output will become 1, which will not affect the other flip-flops. The counter output will now be 1 1 0 as shown in row 3 of the Table 7.3.

When the third clock pulse occurs, flip-flop A will set and its complement output will become 0, which will reset flip-flop B, its output becomes 0, and the complement output will be 1, which will not affect the other flip-flops. The counter will now show an output of 1 0 1, as in the fourth row of the table.

You will notice that every clock pulse decrements the counter by 1. After the eighth clock pulse, the counter output will be 0 0 0 and the counter will recycle thereafter.

The waveform for this 3-bit down counter is given in Fig. 7.17.

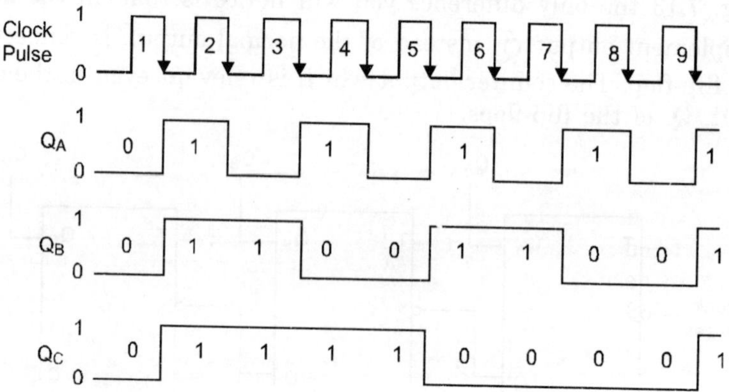

Fig. 7.17 Waveform for 3-bit binary down-counter

7.4.5 Up-Down Counters

The counters which we have considered so far can only count up or down; but they cannot be programmed to count up or down. However, this facility can be easily incorporated by some modification in the circuitry. You might recall that in an up-counter the normal output of a flip-flop is connected to the clock input of the following flip-flop, and in a down counter it is the complement output which is connected to the clock input of the following flip-flop. The change from normal to complement connection to the clock input of the following flip-flop can be easily managed. A circuit for this purpose is shown in Fig. 7.18.

The normal and complement outputs of flip-flops are connected to AND gates D and E and the output of the AND gates goes to the clock input of the next flip-flop via OR gates F. When the up-down control is binary 1, gates D and F are enabled and the normal output of each flip-flop is coupled via OR gates F to the clock input of the next flip-flop. Gates E are inhibited, as one input of all these gates goes low because of the Inverter. The counter, therefore, counts up.

When the up-down control is binary 0, gates D are inhibited and gates E are enabled. As a consequence the complement output of each flip-flop is coupled via OR gates F to the clock input of the next flip-flop. The counter, therefore, counts down.

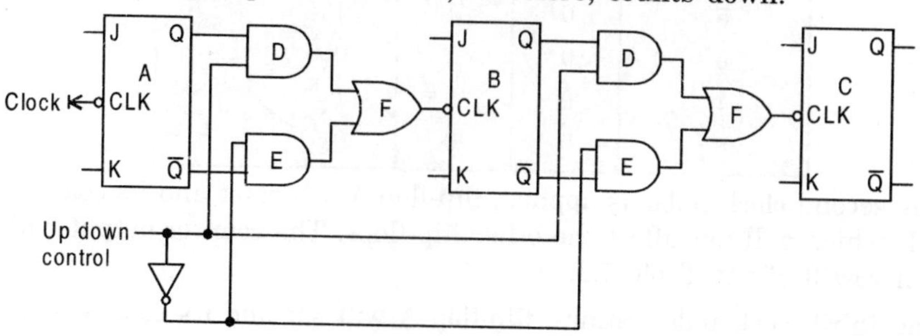

Fig. 7.18 Up-down counter

7.4.6 Reset and Preset Functions

Reset and Preset functions are usually necessary in most counter applications. When using a counter you would, in most cases, like the counter to begin counting with no prior counts stored in the counter. Resetting is a process by which all flip-flops in a counter are cleared and they are thus in a binary O state. JK flip-flops have a CLEAR or RESET input and you can activate them to reset flip-flops. If there are more than one flip-flop, the reset inputs of all flip-flops are connected to a common input line as shown in Fig. 7.19.

You will notice that the reset inputs of all the flip-flops in the counter are active low, and therefore, to reset the counter you take the reset input line low and then high. The output of the counter will then be 0 0 0 0.

At times you may want the counter to start the count from a predetermined point. If you load the required number into the counter, it can start counting from that point. This can be easily accomplished by using the arrangement shown in diagram. The preset inputs of all the flip-flops are connected to NAND gate outputs. One input of each NAND gate is connected to a common PRESET line and the desired number is fed into the other inputs of the NAND gates. To load a number into the counter, first clear the counter and then feed the required number into the NAND gates as indicated in the diagram. When you take the PRESET line high momentarily, the output of NAND gates 1 and 4 will be 1, so flip-flops A and D will remain reset. The output of gates 2 and 3 will be 0 and so flip-flops B and C will be set. The number stored in the counter will now be 0 1 1 0, which is the number required to be loaded in the counter.

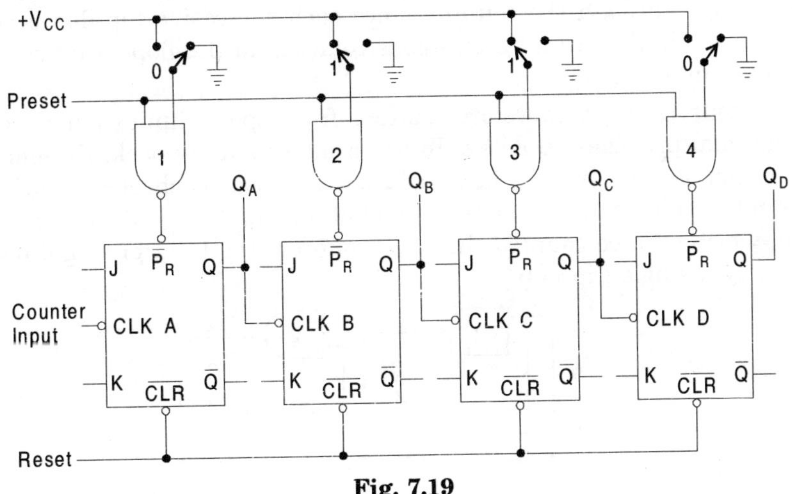

Fig. 7.19

It is also possible to load a number in a counter in a single operation, by using the arrangement shown in Fig. 7.20.

The arrangement for data transfer, which is a single pulse operation makes use of the Preset and Clear inputs of the flip-flops. When the clock pulse is low, the output of both NAND gates 1 and 2 is high, which has no effect on the Preset and Clear inputs of the flip-flop and there is no change in its output. If the D_0 input is high, the output of NAND gate 1 will go low when the clock pulse goes high. This will result in output Q_A going high at the same time. Since one input of NAND gate 2 will be low at this time, the clear input to the flip-flop remains high.

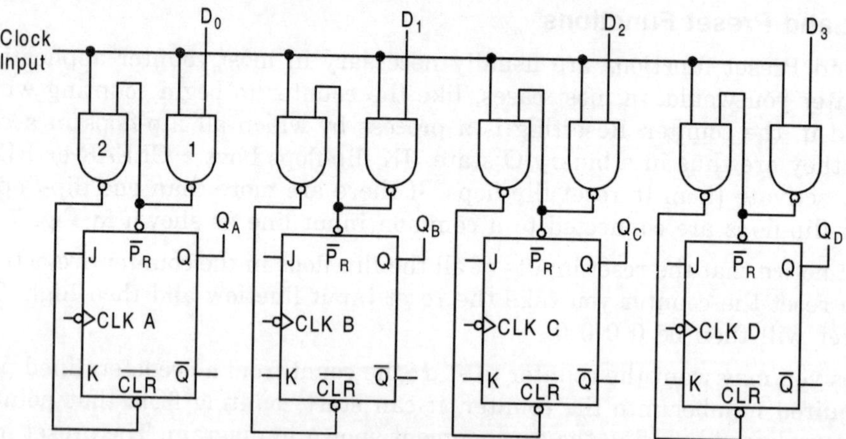

Fig. 7.20 Single pulse data transfer

If the D_0 input is low and the clock pulse goes high, the output of NAND gate 1 will remain high, which will have no effect on the Preset input. The output of NAND gate 2 will go low, which will clear the flip-flop and Q_A will go low.

7.4.7 Universal Synchronous Counter Stage

The up and down counters which we have considered so far are asynchronous counters, also known as ripple counters, for the simple reason that, following the application of a clock pulse, the count ripples through the counter, since each successive flip-flop is driven by the output of the previous flip-flop. In a synchronous counter, all flip-flops are driven simultaneously by the same timing signal.

The asynchronous counter, therefore, suffers from speed limitation as each flip-flop contributes to the total propagation delay. To overcome this draw-back, flip-flops with lower propagation delay can be used; but the ideal solution is to use synchronous counters. In these counters the circuit is so arranged that triggering of all flip-flops is done simultaneously by the input signal, which is to be counted. In these counters the total propagation delay is the delay contributed by a single flip-flop.

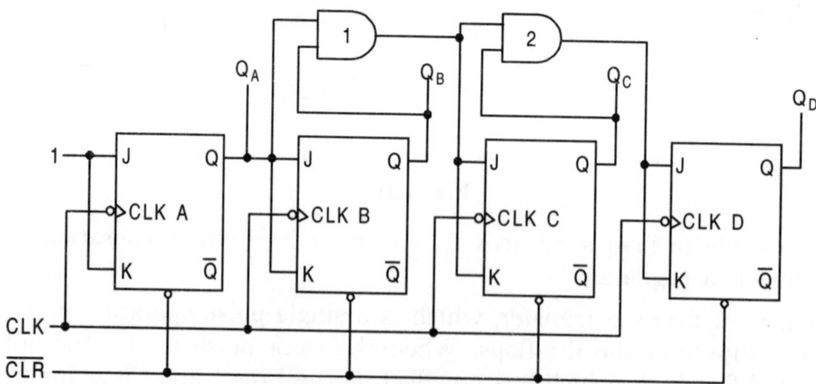

Fig. 7.21 (*a*) Synchronous counter

The design concept used in the synchronous counter shown in Fig. 7.21 (*a*) uses counter stage blocks and this design concept lends itself to building large synchronous counters. Counter modules of the type used in this circuit and also shown separately in Fig. 7.21 (*b*) can be interconnected to build counters of any length.

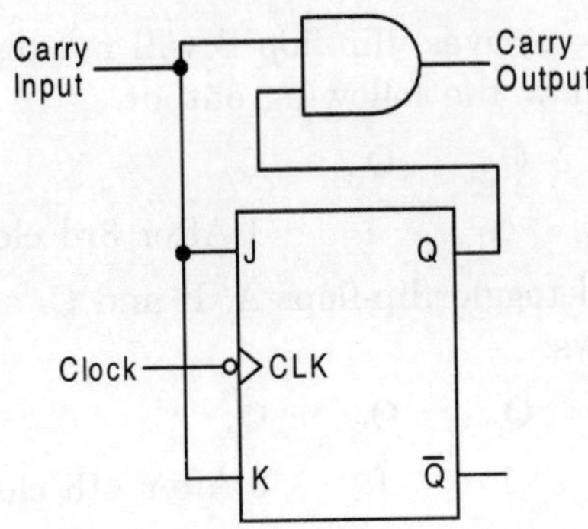

Fig. 7.21 (*b*) Universal counter stage block

Let us consider the synchronous counting circuit shown in Fig. 7.21(*a*). It is a 4-bit counter and the clock inputs of all the flip-flops are connected to a common clock signal, which enables all flip-flops to be triggered simultaneously. The clear inputs are also connected to a common clear input line. The J and K inputs of each flip-flop are connected together, so that they can toggle when the JK input is high. The JK input of flip-flop A is held high. Also notice the two AND gates 1 and 2, and the way they are connected. Gate 1 ensures that the JK input to flip-flop C will be binary 1 when both inputs Q_A and Q_B are binary 1. AND gate 2 ensures that the JK input to flip-flop D will be binary 1 only when outputs Q_A, Q_B and Q_C are binary 1.

We can now look into the output states required for the flip-flops to toggle. This has been summarized below :

1. Flip-flop A toggles on negative clock edge.
2. Flip-flop B toggles when Q_A is 1
3. Flip-flop C toggles when Q_A and Q_B are 1
4. Flip-flop D toggles when Q_A, Q_B are Q_C are 1

This means that a flip-flop will toggle only if all flip-flops preceding it are at binary 1 level.

We can now look into the counting process of this counter. We begin by resetting the counter, which is done by taking CLR temporarily low.

MSB			LSB
Q_D	Q_C	Q_B	Q_A
0	0	0	0

Since, Q_A is low and J and K are high, the first negative clock edge will set flip-flop A. The counter output will now be as follows:

Q_D	Q_C	Q_B	Q_A
0	0	0	1 After 1st clock pulse.

When the second negative clock edge occurs, both A and B flip-flops will toggle and the counter output will change to the following:

Q_D	Q_C	Q_B	Q_A
0	0	1	0 After 2nd clock pulse.

When the third clock pulse arrives, flip-flop B will not toggle as Q_A is 0 but flip-flop A will toggle. The counter will show the following output.

Q_D	Q_C	Q_B	Q_A
0	0	1	1 After 3rd clock pulse.

The fourth clock pulse will toggle flip-flops A, B and C, as both Q_A and Q_B are 1. The counter output is now as follows:

Q_D	Q_C	Q_B	Q_A
0	1	0	0 After 4th clock pulse.

The counter will continue to count in the binary system until the counter output registers 1 1 1 1, when it will be reset by the next clock pulse and the counting cycle will be repeated.

7.4.8 Modulus Counters

The modulus of a counter, as discussed before, is the nùmber of discrete states a counter can take up. A single flip-flop can assume only two states 0 and 1, while a counter having two flip-flops can assume any one of the four possible states. A counter with three flip-flops will have 8 states and so on. In short the number of states is a multiple of 2. With n flip-flops the number of possible states will be 2^n. Thus by building counters which count in the normal binary sequence, we can build counters with modulus of 2, 4, 8, 16 etc. In these counters the count increases or decreases by 1 in pure binary sequence. The problem arises in building counters whose modulus is 3, 5, 7, 9 etc. For instance, if we need, a counter with a modulus of 3, we have to use a counter with a modulus of 4 and so arrange the circuit that it skips one of the states. Similarly, for a counter with a modulus of 5 we require 2^3 or 8 states and arrange the circuit so that it skips 3 states to give us a modulus of $2^n - 3$ or 5 states. Thus for a modulus N counter the number n of flip-flops should be such that n is the smallest number for which $2^n > N$. It, therefore, follows that for a decade (mod-10) counter the number of flip-flops should be 4. For this counter we shall have to skip $2^4 - 10$ or 6 states. Which of these states are to be skipped is a matter of choice, which is largely governed by decisions which will make the circuit as simple as possible.

Many methods have been developed for designing such counters. We will consider the following:

(1) Counter Reset Method

In this method, the counter is reset after the desired count has been reached and the count cycle starts all over again from the reset state.

(2) Logic Gating Method

This method provides the exact count sequence required without any need to reset the counter at some stage.

(3) Counter Coupling Method

This method is used to implement counters of the required modulus. For instance we can interconnect mod-2 and mod-3 counters to implement a modulus 3 × 2 or mod-6 counter.

7.4.9 Asynchronous Counters (Counter Reset Method)

Let us first consider the typical case of a counter which has 3 states as shown in Fig. 7.22.

Mod-3 Counter

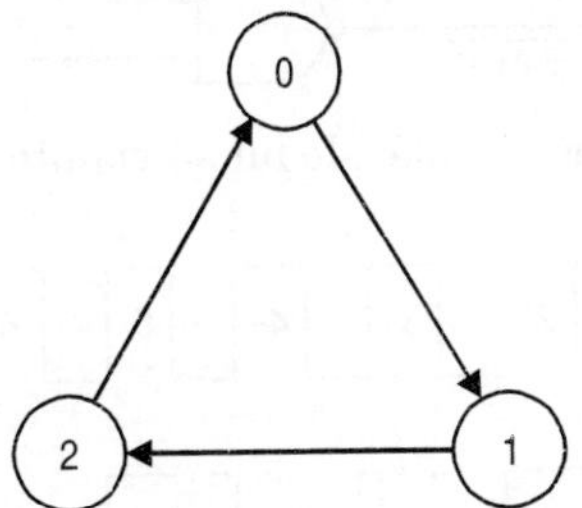

Fig. 7.22 State diagram for a mod-3 counter

It is obvious that a mod-3 counter will require two flip-flops which, when connected as a counter, will provide four states as shown in Table 7.4.

Table 7.4 States for a two flip-flop counter

Q_A *LSB*	Q_B	*Count value (Decimal)*
0	0	0
1	0	1
0	1	2
1	1	3
0	0	0

This counter counts in the binary sequence 0, 1, 2, 3 and then it returns to 0, the starting point. Each count is referred to as a state. If we are building a mod-3 counter, the most convenient solution is to skip state 3 and then return to state 0 from state 2 and then again go through states 0, 1, 2 before returning to state 0. What we need is a combinational logic circuit, which will feed a reset pulse to the counter during state 3, and immediately after state 2, which is the last desired state. This reset pulse is applied to the CLR inputs which resets the counter to 0 after state 2.

A circuit diagram for a mod-3 counter together with the required combinational logic is given in Fig. 7.23.

When both outputs Q_A and Q_B are 1, the output of the NAND gate, which provides the reset pulse, goes low and both the flip-flops are reset. The counter returns to state 0 and it starts counting again in 0, 1, 2, 0 sequence. The waveforms for this counter are given in Fig. 7.24.

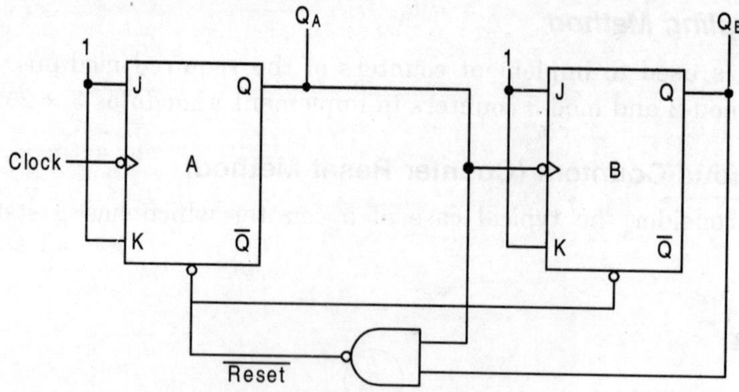

Fig. 7.23 Modulo-3 counter

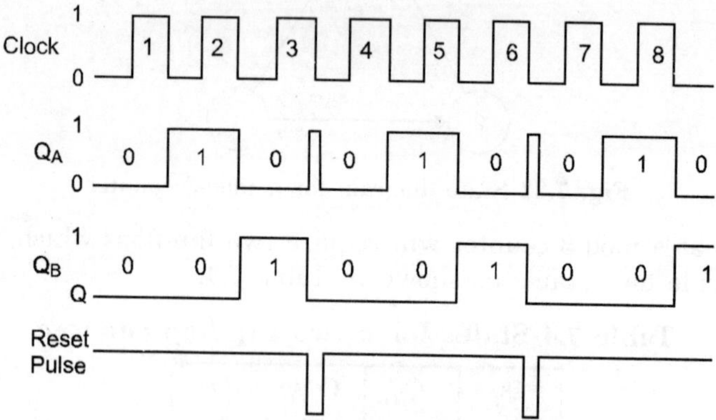

Fig. 7.24 Waveform for mod-3 counter

Mod-5 Counter

The minimum number of flip-flops required to implement this counter is three. With three flip-flops, the number of states will be 8. A modulo-5 counter will have only 5 states. A state diagram for this counter is given in Fig. 7.25. It will progress from state 000 through 100. The truth table for this counter, which will determine the stage at which the reset pulse should be applied, is given in Table 7.5.

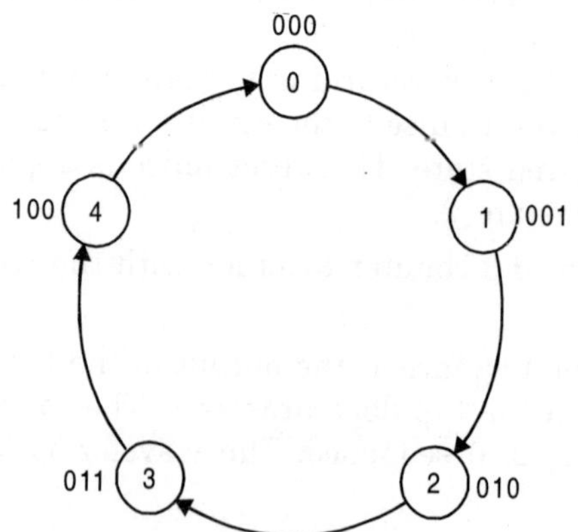

Fig. 7.25 State diagram for mod-5 counter

The truth table shows that state 5 will be the reset state and that states 6 and 7 will be the don't care states. The next step is to plot the states on a map as shown in Fig. 4.39.

Table 7.5 Truth table for Mod-5 Counter

Q_A *LSB*	Q_B	Q_C	*State*
0	0	0	0
1	0	0	1
0	1	0	2
1	1	0	3
0	0	1	4
1	0	1	5
0	1	1	6 X
1	1	1	7 X

X, Don't care states

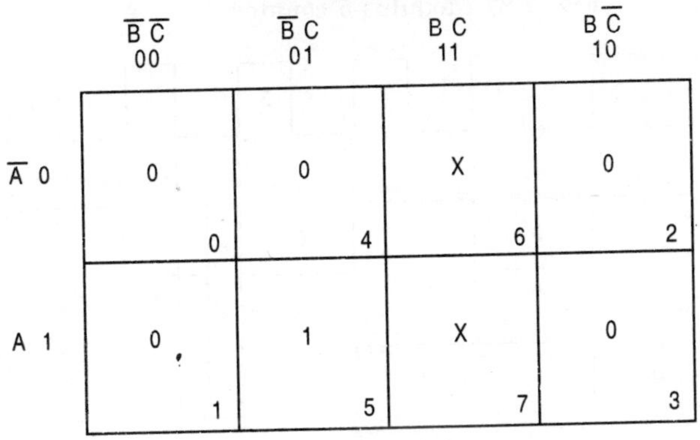

	$\overline{B}\,\overline{C}$ 00	$\overline{B}\,C$ 01	B C 11	$B\,\overline{C}$ 10
$\overline{A}$ 0	0 (0)	0 (4)	X (6)	0 (2)
A 1	0 (1)	1 (5)	X (7)	0 (3)

Fig. 7.26

The map shows that the reset pulse is determined by $R = Q_A.\overline{Q_B}.Q_C$. The logic diagram for this counter is given in Fig. 7.27. The diagram shows that a reset pulse will be applied when both A and C are 1. You may have noticed that the reset pulse shown in Fig. 7.24 for the Mod-3 counter was very narrow and in some cases it may not be suitable to control other logic devices associated with the counter. The Mod-5 counter circuit (Fig. 7.27) incorporates an RS flip-flop, which produces a reset pulse, the width of which is equal to the duration for which the clock pulse is low. The way it works is like this. State 5 is decoded by gate D, its output goes low, the RS flip-flop is set, and output $\overline{Q}$ goes low, which resets all the flip-flops. The leading edge of the next clock pulse resets the RS flip-flop, $\overline{Q}$ goes high which removes the reset pulse. The counter thus remains reset for the duration of the low time of the clock pulse. When the trailing edge of the same clock pulse arrives, a new cycle is started. The waveform for Mod-5 counter is given in Fig. 7.28.

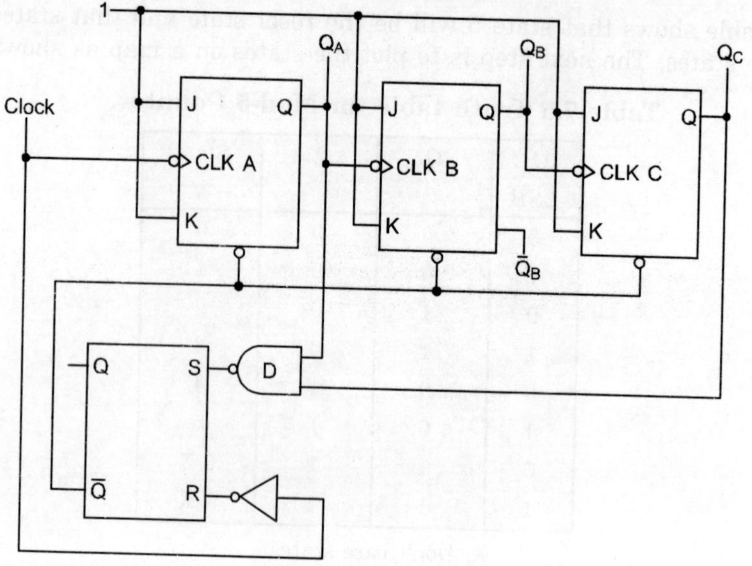

Fig. 7.27 Modulus-5 counter

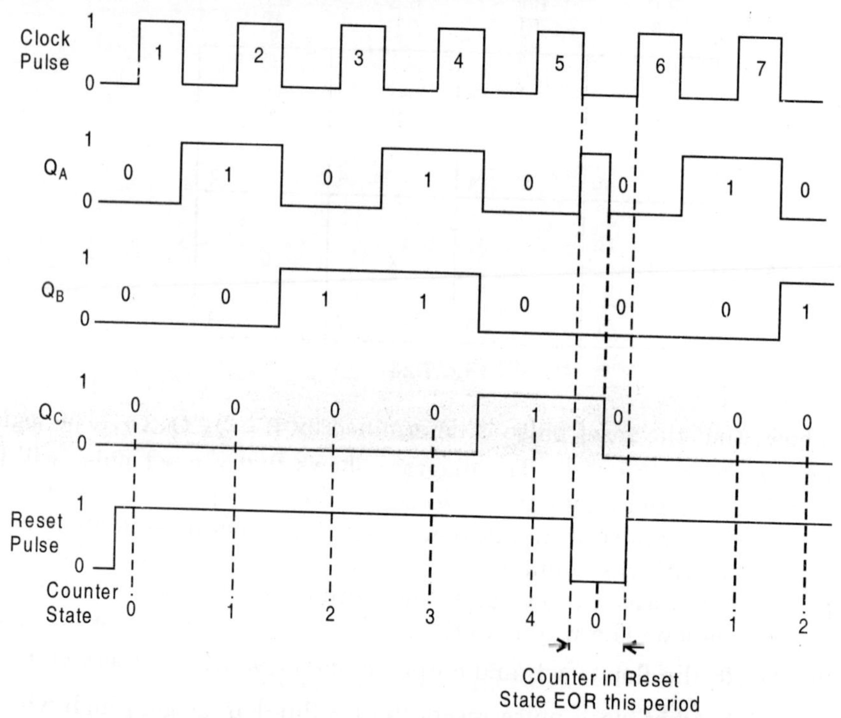

Fig. 7.28 Waveform for modulus-5 asynchronous counter

Mod-10 (Decade) Counter

The decade counter discussed here is also an asynchronous counter and has been implemented using the counter reset method. As the decade counter has ten states, it will require four flip-flops to implement it. A state diagram for this counter is given in Fig. 7.29 and the truth table is given in Table 7.6.

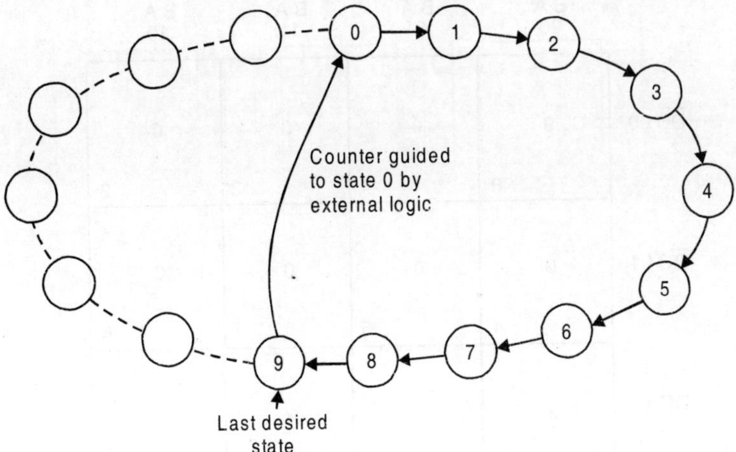

Fig. 7.29 State diagram for decade counter

Table 7.6 Truth table for decade counter

Q_A LSB	Q_B	Q_C	Q_D	*State*
0	0	0	0	0
1	0	0	0	1
0	1	0	0	2
1	1	0	0	3
0	0	1	0	4
1	0	1	0	5
0	1	1	0	6
1	1	1	0	7
0	0	0	1	8
1	0	0	1	9
0	1	0	1	10
1	1	0	1	11 X
0	0	1	1	12 X
1	0	1	1	13 X
0	1	1	1	14 X
1	1	1	1	15 X

The table shows that state 9 will be the last desired state and state 10 will be the reset state. State 11, 12, 13, 14 and 15 will be the don't care states. The next step is to plot the states on a map to determine the reset pulse. This has been done in Fig. 7.30.

The map shows that the reset pulse is determined by the following expression:

$$R = \overline{Q}_A . Q_B . \overline{Q}_C . Q_D$$

	$\overline{B}\,\overline{A}$ 00	$\overline{B}A$ 01	BA 11	$B\overline{A}$ 10
$\overline{D}\,\overline{C}$ 00	0 (0)	0 (1)	0 (3)	0 (2)
$\overline{D}C$ 01	0 (4)	0 (5)	0 (7)	0 (6)
DC 11	3 (12)	3 (13)	X (15)	3 (14)
$D\overline{C}$ 10	0 (8)	0 (9)	X (11)	1 (10)

Fig. 7.30

The decade counter circuit Fig. 7.31 is essentially a binary ripple counter, which can count from 0000 to 1111; but since a decade counter is required to count only from 0000 to 1001, a reset pulse is applied at count 10 when the counter output is $\overline{Q}_A.Q_B.\overline{Q}_C.Q_D$. In order to have control over the reset pulse width, a 4-input NAND gate is used to decode state 10.

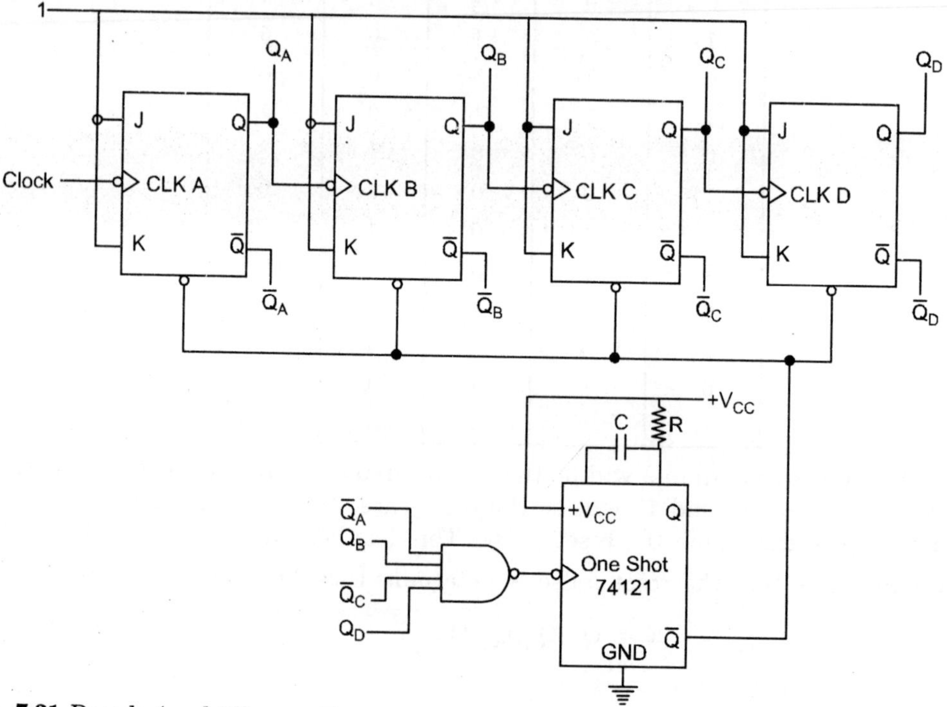

Fig. 7.31 Decade (mod-10) asynchronous counter using count reset and pulse width control

To decode count 10, logic inputs that are all one at the count of 10, are used to feed the NAND gate. At this count the NAND gate output goes low providing a 1 → 0 change which triggers the one-shot unit. The $\overline{Q}$ output of the one shot unit is used, as it is normally high and it goes low during the one-shot timing period, which depends on the RC constants of the circuit. The timing period of the one-shot can be adjusted, so that the slowest counter state resets. Although only A and D flip-flops need to be reset, the reset pulse is applied to all the flip-flop to make doubly sure that all flip-flops are reset.

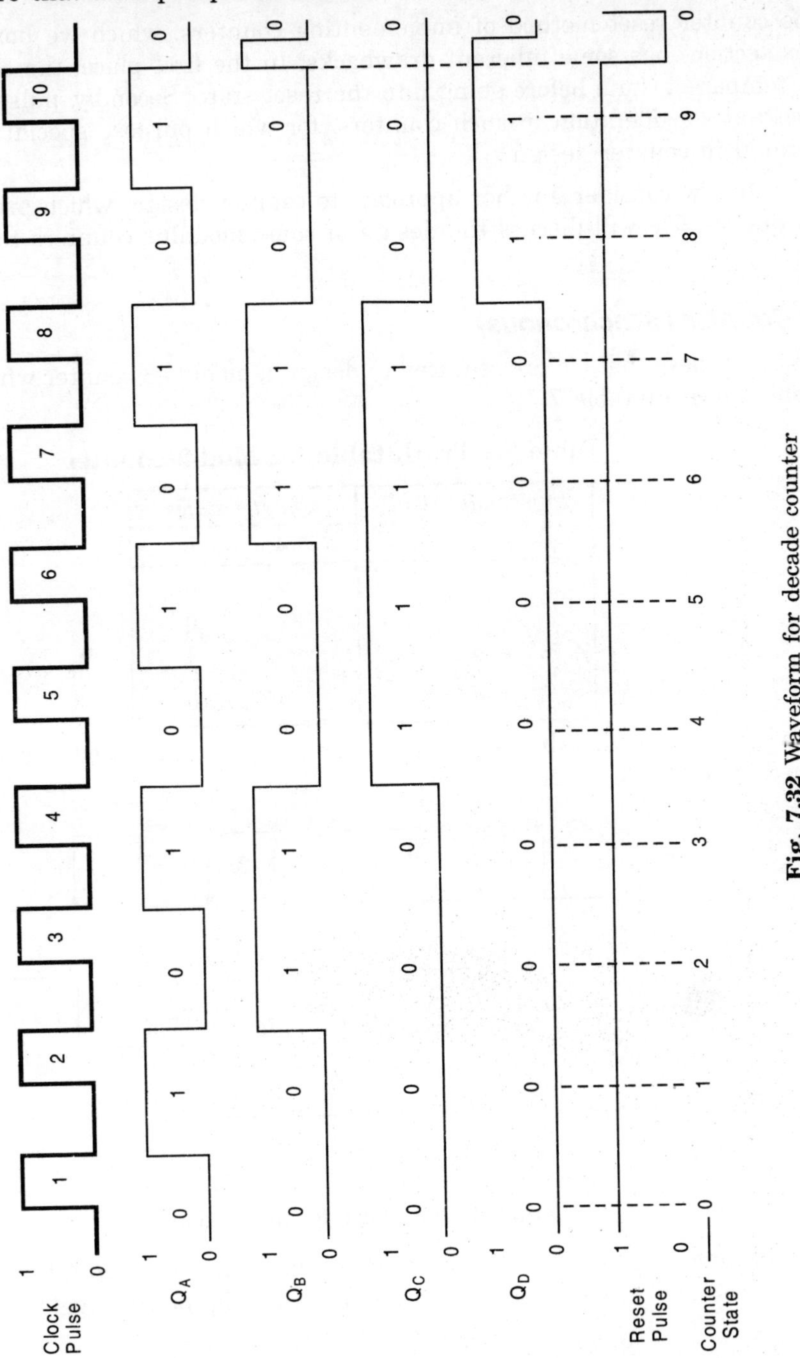

Fig. 7.32 Waveform for decade counter

Since decade (Modulus-10) counters have 10 discrete starts, they can be used to divide the input frequency by 10. You will notice that at the output of the D-flip-flop, there is only one output pulse for every 10 input pulses. These counters can be cascaded to increase count capability.

The waveform for this counter is shown in Fig. 7.32.

7.4.10 Logic Gating Method

The counter reset method of implementing counters, which we have discussed in the previous section, has some inherent drawbacks. In the first place, the counter has to move up to a temporary state before going into the reset state. Secondly, pulse duration timing is an important consideration in such counters, for which purpose special circuits have to be incorporated in counter design.

We will now consider another approach to counter design, which provides for the exact count sequence. We will discuss the design of some modulus counters to illustrate the procedures.

Mod-3 Counter (Synchronous)

Let us suppose that we are required to design a modulo-3 counter which conforms to the truth table given in Table 7.7.

Table 7.7 Truth table for Mod-3 counter

Input pulse count	*Counter states*	
	A	*B*
0	0	0
1	1	0
2	0	1
3	0	0

Based on this truth table, the output waveform for this Mod-3 counter should be as shown in Fig. 7.33.

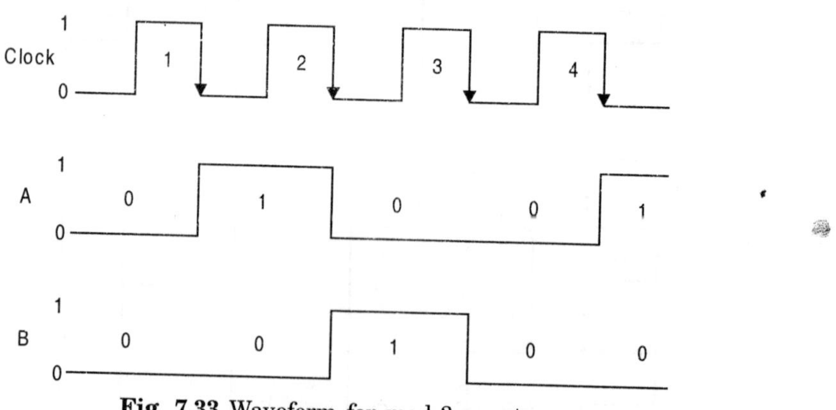

Fig. 7.33 Waveform for mod-3 counter

You will notice from the waveform of the counter, that flip-flop A toggles on the trailing edge of the first and second pulses. Also observe that flip-flop B toggles only on the second and third clock pulses. We have to bear this in mind, in figuring out logic levels for the J and K inputs of the flip-flops.

Suppose that initially both the flip-flops are reset. Since flip-flop A has to toggle when the trailing edges of the first and the second clock pulses arrive, its J and K inputs should be at logic 1 level during this period. This is achieved by connecting the K input to logic 1 level and the J input to the complement output of flip-flop B, as during this period the $\overline{B}$ output of flip-flop B is at a high logic level. In this situation, the first clock pulse produces a logic 1 output and the second clock pulse produces a logic 0 output.

The J input of flip-flop B is connected to the normal output of flip-flop A. Therefore, when the first clock pulse arrives, the J input of flip-flop B is low. Its output will remain low as you will notice from the truth table and the waveform. The second pulse is required to toggle this flip-flop and its K input is, therefore held high. When the second clock pulse arrives, the flip-flop will toggle as both the J and K inputs are high. The output will go high. At the same time its complement output will be low, which makes the J input of flip-flop A low.

When the third clock pulse arrives, the output of flip-flop A will go low. Since after the second clock pulse the output of flip-flop A was already low, the third clock pulse produces a low output at flip-flop B. Both the A and B flip-flops are now reset and the cycle will be repeated.

A logic diagram for the Mod-3 counter is given in Fig. 7.34.

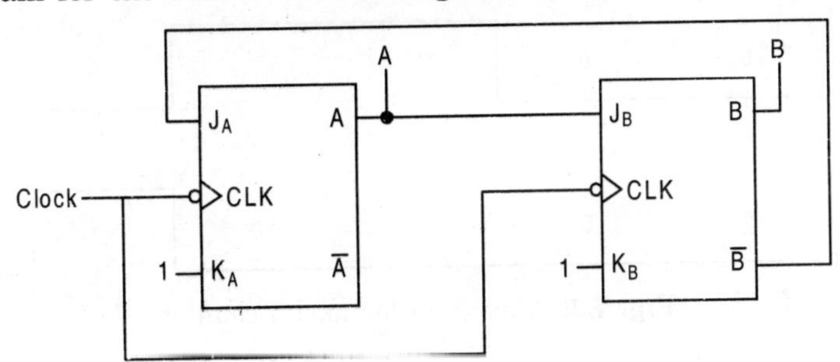

Fig. 7.34 Mod-3 counter (synchronous)

Mod-5 Counter (Asynchronous)

We will use the same procedure to design a mod-5 counter as before. The truth table required for this counter is given in Table 7.8.

Table 7.8 Truth table for Mod-5 counter

Input pulse count	*Counter states*		
	A	*B*	*C*
0	0	0	0
1	1	0	0
2	0	1	0
3	1	1	0
4	0	0	1
5	0	0	0

The waveform for this counter based on this truth table is given in Fig. 7.35. You will notice from the truth table and the waveform that the A flip-flop complements each input pulse, except when the normal output of flip-flop C is logic 1, which is so after the trailing edge of the 4th, clock pulse. It, therefore, follows that the K input of flip-flop A should be a constant logic 1 and the J input should be connected to the complement output of flip-flop will be 0 when C is 1 so that the output of flip-flop A remains low after the trailing edge of the 5th clock pulse.

If you observe the changing pattern of the output of the B flip-flop, you will notice that it toggles at each transition of the A output from 1 → 0. It is, therefore, obvious that the A output should be connected to the clock input of the B flip-flop and the J and K inputs of this flip-flop should be at logic 1 level.

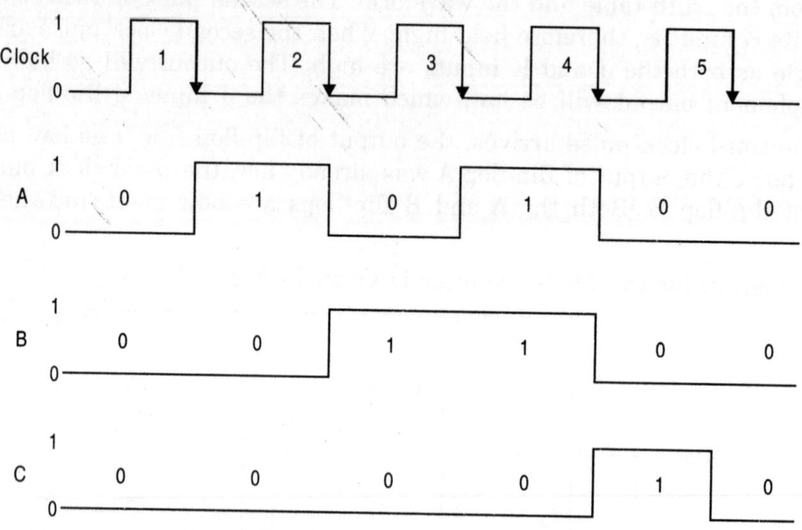

Fig. 7.35 Waveform for mod-5 counter

After the 3rd clock pulse, the outputs of A and B flip-flops are 1. An AND gate is used to make the J input to flip-flop C as 1 when both A and B are 1. The K input to flip-flop C is also held at logic 1 to enable it to toggle. The clock is also connected to the clock input to flip-flop C, which toggles it on the 4th, clock pulse and its output becomes 1. When the 5th, clock pulse arrives, the J input to flip-flop C is 0 and it resets on the trailing edge of this clock pulse. Thereafter the cycles are repeated. The logic diagram for the mod-5 counter is given in Fig. 7.36.

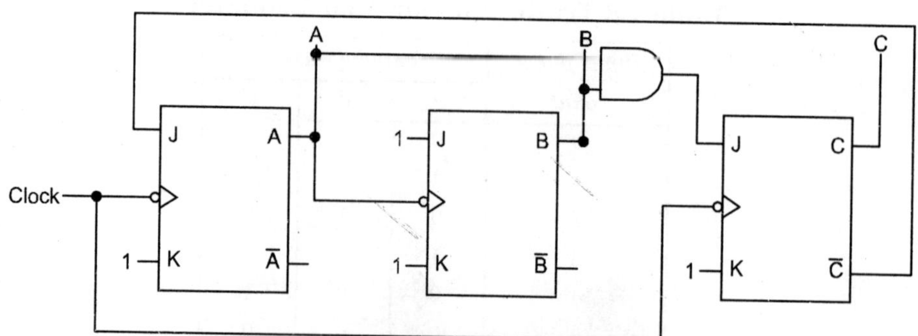

Fig. 7.36 Logic diagram for mod-5 counter

Mod-10 (Decade) Counter (Asynchronous)

The truth table for a Decade counter is given in Table 7.9.

Table 7.9 Truth table for decade counter

Input pulse count	*Counter states*			
	A	*B*	*C*	*D*
0	0	0	0	0
1	1	0	0	0
2	0	1	0	0
3	1	1	0	0
4	0	0	1	0
5	1	0	1	0
6	0	1	1	0
7	1	1	1	0
8	0	0	0	1
9	1	0	0	1
10 (0)	0	0	0	0

The waveform for this counter based on this truth table is given in Fig. 7.37.

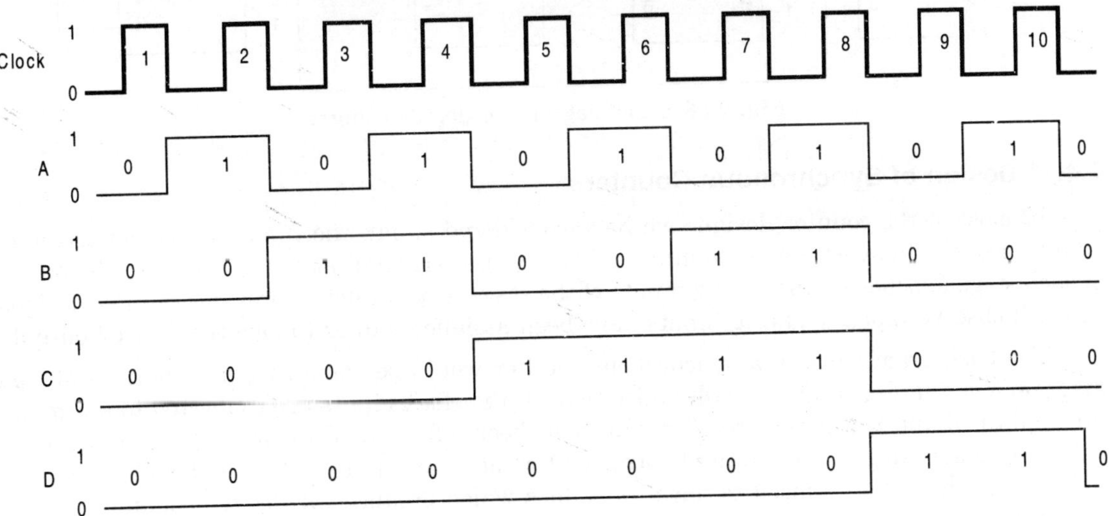

Fig. 7.37 Waveform for decade counter

If you compare truth Table 7.9 for the Decade counter with Table 7.2 which gives the count-up sequence for a 4-bit binary up-counter, you will notice a close similarity between the two, up to input pulse 8. You will also notice a close resemblance between waveforms of Fig. 7.37 and Fig. 7.15 up to a certain point.

The count ripples through the A, B and C flip-flops for the first seven input pulses, as in the standard 4-bit binary up-counter. At this point the counter will show an output of 1 1 1 0 (decimal 7). On the application of the 8th pulse, flip-flops A, B and C must reset and the D output should be 1, that is the counter state should change from 1 1 1 0 to 0 0 0 1. In order that the J input to flip-flop D is 1, so that when K is 1 the D flip-flop output goes from 0 to

1; B and C outputs are applied to the input of an AND gate and its output goes to the J input. In order that the B and C outputs are 0, when D output is 1 for the 8th and the 9th count, the complement output of the D flip-flop which will be 0 when D is 1, is connected to the J input of the B flip-flop.

After the trailing edge of the 8th pulse D becomes 1 and A, B and C become 0, the 9th pulse is required to change the output from 0 0 0 1 to 1 0 0 1. Since no change is required in the D output, the D-flip-flop is triggered by the A output. When the 9th pulse arrives, the A output changes from 0 to 1, but this causes no change in the D output. When the 10th input pulse arrives, it changes the A output from 1 to 0, which changes the D output from 1 to 0. The counter output changes from 1 0 0 1 to 0 0 0 0. During the 9th and the 10th pulses, the B and C outputs will remain unchanged.

A logic diagram for the Decade counter is given in Fig. 7.38.

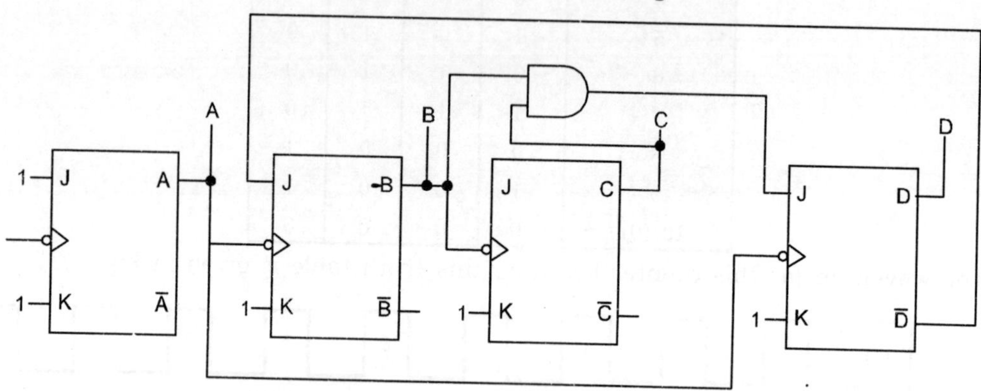

Fig. 7.38 Logic diagram for decade counter

7.4.11 Design of Synchronous Counters

In most of the counter designs we have considered so far, the flip-flops are not triggered simultaneously. In synchronous counters all stages are triggered at the same time. The output of each stage depends on the gating inputs of the stage. If you refer to previous counter designs, you will observe that the gating inputs have been assigned values to give the desired outputs.

The basic framework of a synchronous counter would be somewhat like the partial logic diagram given in Fig. 7.39. You will notice that all the clock inputs are connected to a common line and the J and K inputs of the flip-flops have been left open. They are required to have the values necessary to give the required outputs after each input pulse. The J and K inputs of each flip-flop are therefore required to have the values which produce the desired counter states at each input pulse. The entire purpose of the exercise is to determine the input values for each stage. A typical design procedure can be summed up in the following steps.

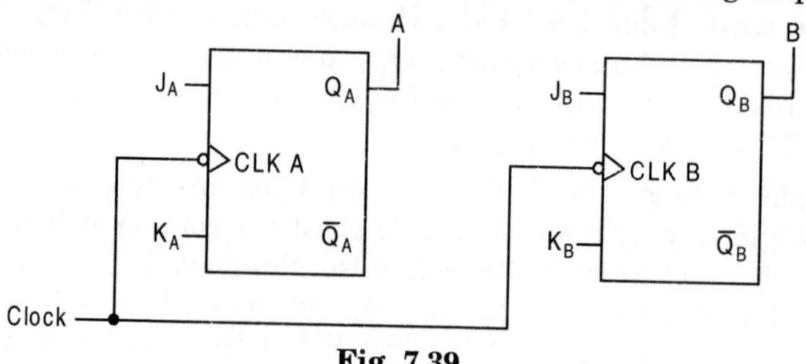

Fig. 7.39

(*a*) Write the desired truth table for the counter.
(*b*) Write the counter transition table which should list the starting state and the subsequent states the counter is required to take up.
(*c*) With the help of the excitation table and using the counter transition table, write down the input values for the J and K inputs to enable each flip-flop to attain the output state as required by the transition table.
(*d*) Prepare Karnaugh maps for the J and K inputs of each stage.
(*e*) Derive Boolean algebra expressions for each of the inputs to the flip-flops.
(*f*) Draw the synchronous counter circuit incorporating the J and K input values obtained from the above steps.

We will take up a specific case to illustrate the above procedure.

Mod-3 Synchronous Counter

We have implemented a Mod-3 synchronous counter as described in Sec. 7.4.10.1. We will implement the same counter by the procedure described here. We will follow the truth table given in Table 7.7. For your convenience the excitation table for JK flip-flops is reproduced here.

Table 7.10 Excitation table for JK flip-flop

Present state	*Next state*	*J*	*K*
0	0	0	X
0	1	1	X
1	0	X	1
1	1	X	0

We now prepare a counter design table listing the two flip-flops and their states and also the four inputs to the two flip-flops as in table 7.11.

Table 7.11 Counter design table

Counter state		*Flip-flop inputs*			
A	*B*	*A*		*B*	
		J_A	K_A	J_B	K_B
0	0	1	X	0	X
1	0	X	1	1	X
0	1	0	X	X	1
0	0				

The table shows that if the counter is in the state A = 0, B = 0 and a clock pulse is applied, the counter is required to step up to A = 1, B = 0. When the counter is in the state A = 1, B = 0 and a clock pulse is applied, the counter has to step up to A = 0, B = 1. Lastly when another clock pulse is applied the counter has to reset.

From the excitation table for JK flip-flops we can determine the states of the J and K inputs, so that the counter steps up as required. For instance for the A flip-flop to step up from 0 to 1, J_A should be 1 and K_A should be X. Similarly, the J and K input values of both the flip-flops for the remaining counter states have been worked out as shown in the table.

The next step is to derive boolean algebra expressions for each of the inputs to the flip-flops. In the above exercise, our effort was to generate flip-flop inputs in a given row, so that when the counter is in the state in that row, the inputs will take on the listed values, so that the next clock pulse will cause the counter to step up to the counter state in the row below.

We now form boolean algebra expressions from this table for the J_A, K_A, J_B and K_B inputs to the flip-flops and simplify these expressions using Karnaugh maps. Expressions for these inputs have been entered in Karanaugh maps in Fig. 7.40 (*a*), (*b*), (*c*) and (*d*). The simplified expressions obtained for the inputs are also indicated under the maps.

The counter circuit when drawn up with the following resultant data will be the same as worked out before in Fig. 7.34.

$$J_A = \overline{B}$$
$$K_A = 1$$
$$J_B = A$$
$$K_B = 1$$

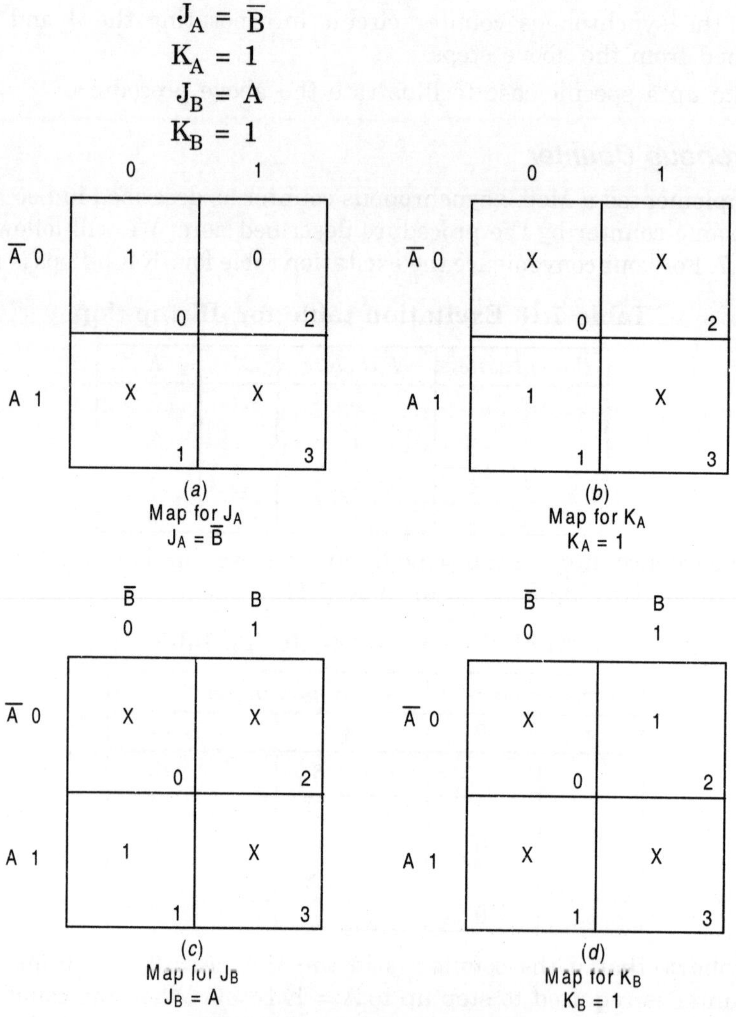

Fig. 7.40 (*a*), (*b*), (*c*) and (*d*)

Mod-5 Counter (Synchronous)

The Mod-5 counter we are going to implement will be a synchronous counter, but it will have the same counter states as given earlier in Table 7.8. The counter design table for this counter lists the three flip-flops and their states as also the six inputs for the three flip-flops. The flip-flop inputs required to step up the counter from the present to the next state have been worked out with the help of the excitation table (Table 7.10). This listing has been shown in Table 7.12.

	$\overline{B}\overline{C}$ 00	$\overline{B}C$ 01	BC 11	$B\overline{C}$ 10
$\overline{A}$ 0	1 0	0 4	X 6	1 2
A 1	X 1	X 5	X 7	X 3

(*a*) Map for J_A
$J_A = \overline{C}$

	$\overline{B}\overline{C}$ 00	$\overline{B}C$ 01	BC 11	$B\overline{C}$ 10
$\overline{A}$ 0	X 0	X 4	X 6	X 2
A 1	1 1	X 5	X 7	1 3

(*b*) Map for K_A
$K_A = 1$

	$\overline{B}\overline{C}$ 00	$\overline{B}C$ 01	BC 11	$B\overline{C}$ 10
$\overline{A}$ 0	0 0	0 4	X 6	X 2
A 1	1 1	X 5	X 7	X 3

(*c*) Map for J_B
$J_B = A$

	$\overline{B}\overline{C}$ 00	$\overline{B}C$ 01	BC 11	$B\overline{C}$ 10
$\overline{A}$ 0	X 0	X 4	X 6	0 2
A 1	X 1	X 5	X 7	1 3

(*d*) Map for K_B
$K_B = A$

	$\overline{B}\overline{C}$ 00	$\overline{B}C$ 01	BC 11	$B\overline{C}$ 10
$\overline{A}$ 0	0 0	X 4	X 6	0 2
A 1	0 1	X 5	X 7	1 3

(*e*) Map for J_C
$J_C = AB$

	$\overline{B}\overline{C}$ 00	$\overline{B}C$ 01	BC 11	$B\overline{C}$ 10
$\overline{A}$ 0	X 0	1 4	X 6	X 2
A 1	X 1	X 5	X 7	X 3

(*f*) Map for K_C
$K_C = 1$

Fig. 7.41

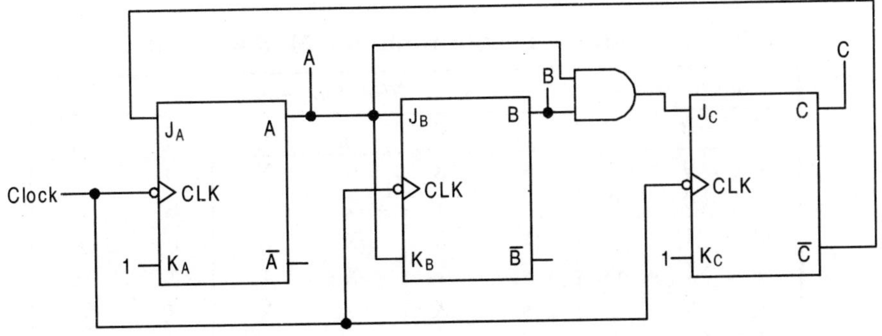

Fig. 7.42 Synchronous mod-5 counter

Table 7.12 Counter design table for Mod-5 counter

Input pulse count	*Counter states*			*Flip-flop inputs*					
	A	*B*	*C*	J_A	K_A	J_B	K_B	J_C	K_C
0	0	0	0	1	X	0	X	0	X
1	1	0	0	X	1	1	X	0	X
2	0	1	0	1	X	X	0	0	X
3	1	1	0	X	1	X	1	1	X
4	0	0	1	0	X	0	X	X	1
5 (0)	0	0	0						

The flip-flop inputs have been determined with the help of the excitation table 7.10. Some examples are as follow:

A flip-flop

The initial state is 0. It changes to 1 after the clock pulse. Therefore J_A should be 1 and K_A may be 0 or 1 (that is X).

B flip-flop

The initial state is 0 and it remains unchanged after the clock pulse. Therefore J_B should be 0 and K_B may be 0 or 1 (that is X).

C flip-flop

The state remains unchanged. Therefore J_C should be 0 to K_B should be X.

The flip-flop input values are entered in Karnaugh maps Fig. 7.41 [(*a*), (*b*), (*c*), (*d*), (*e*) and (*f*)] and a boolean expression is formed for the inputs to the three flip-flops and then each expression is simplified. As all the counter states have not been utilized, Xs (don't) are entered to denote un-utilized states. The simplified expressions for each input have been shown under each map. Finally, these minimal expressions for the flip-flop inputs are used to draw a logic diagram for the counter, which is given in Fig. 7.42.

Mod-6 Counter (Synchronous)

The desired counter states and the JK inputs required for counter flip-flops are given in the counter design table (Table 7.13).

Table 7.13 Counter design table for Mod-6 counter

Input pulse count	*Counter states*			*Flip-flop inputs*					
	A	*B*	*C*	J_A	K_A	J_B	K_B	J_C	K_C
0	0	0	0	1	X	0	X	0	X
1	1	0	0	X	1	1	X	0	X
2	0	1	0	1	X	X	0	0	X
3	1	1	0	X	1	X	1	1	X
4	0	0	1	1	X	0	X	X	0
5	1	0	1	X	1	0	X	X	1
6 (0)	0	0	0						

As before, the JK inputs required for this have been determined with the help of the excitation table, (Table 7.10). These input values have been entered in Karnaugh maps Fig. 7.43 and a boolean expression is formed for the inputs to the three flip-flops and then each expression is simplified. Xs have been entered in those counter states which have not been utilized. The simplified expressions for each input have been shown under each map and finally a logic diagram based on these expressions has been drawn, as given in Fig. 7.44.

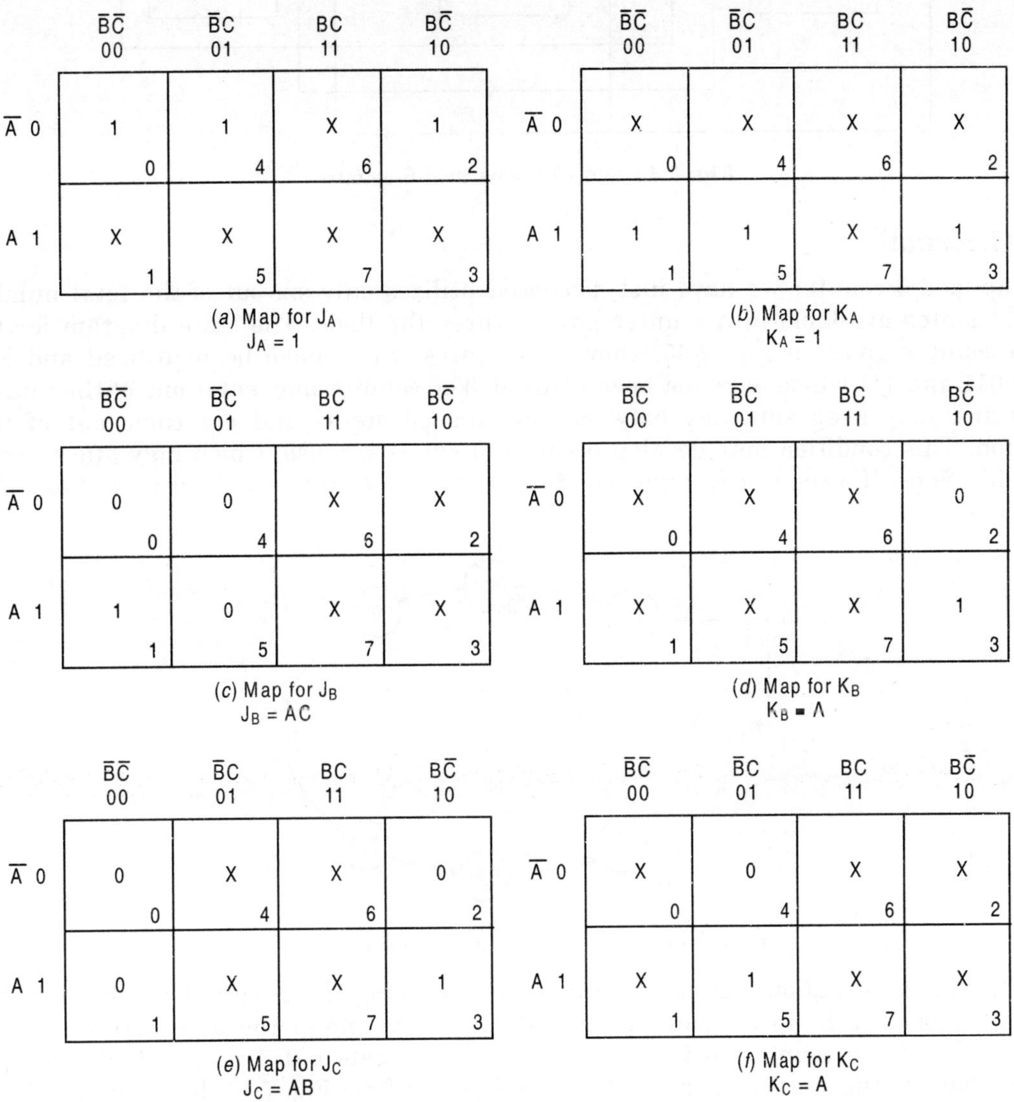

(*a*) Map for J_A
$J_A = 1$

(*b*) Map for K_A
$K_A = 1$

(*c*) Map for J_B
$J_B = AC$

(*d*) Map for K_B
$K_B = A$

(*e*) Map for J_C
$J_C = AB$

(*f*) Map for K_C
$K_C = A$

Fig. 7.43

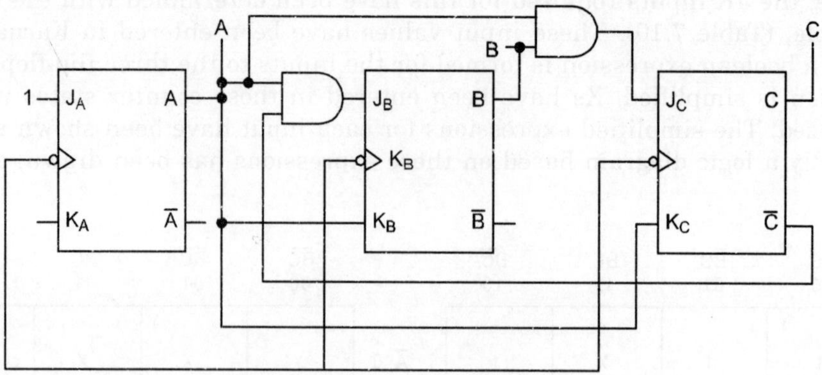

Fig. 7.44 Synchromous mod-6 counter

7.4.12 Lockout

The mod-6 counter we have just discussed utilizes only six out of the total number of eight states available in a counter having three flip-flops. The state diagram for the mod-6 counter given in Fig. 7.45, shows the states which have been utilized and also states 011 and 111 which have not been utilized. The counter may enter one of the unused states and may keep shuttling between the unused states and not come out of this situation. This condition may develop because of external noise, which may affect states of the flip-flops. If a counter has unused states with this characteristic, it is said to suffer from lockout.

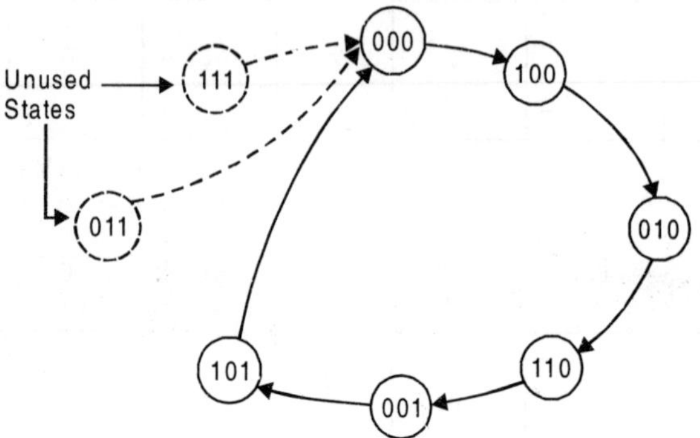

Fig. 7.45 State diagram for mod-6 counter

The lockout situation can be avoided by so arranging the circuit that whenever the counter happens to be in an unused state, it reverts to one of the used states. We will redesign the mod-6 counter so that whenever it is in state 0 1 1 or 1 1 1, the counter swithces back to the starting point 0 0 0. You will notice from Fig. 7.43 that Js and Ks were marked X in squares corresponding to the unused states. We will now assign values for Js and Ks for the unused states, so that the counter reverts to state 0 0 0. This has been done in Table 7.14.

Table 7.14

Counter states			Flip-flop inputs					
A	B	C	J_A	K_A	J_B	K_B	J_C	K_C
0	1	1	0	X	X	1	X	1
1	1	1	X	1	X	1	X	1
0	0	0						

These values of Js and Ks have been entered in K-maps for those counter states where Xs had been entered previously. K-maps for the revised values of Js and Ks are given in Fig. 7.46. Boolean expressions are formed for the inputs to the three flip-flops and the expressions so obtained are simplified. The expressions for each input have been shown under each map and the logic diagram for the improved mod-6 counter is given in Fig. 7.47.

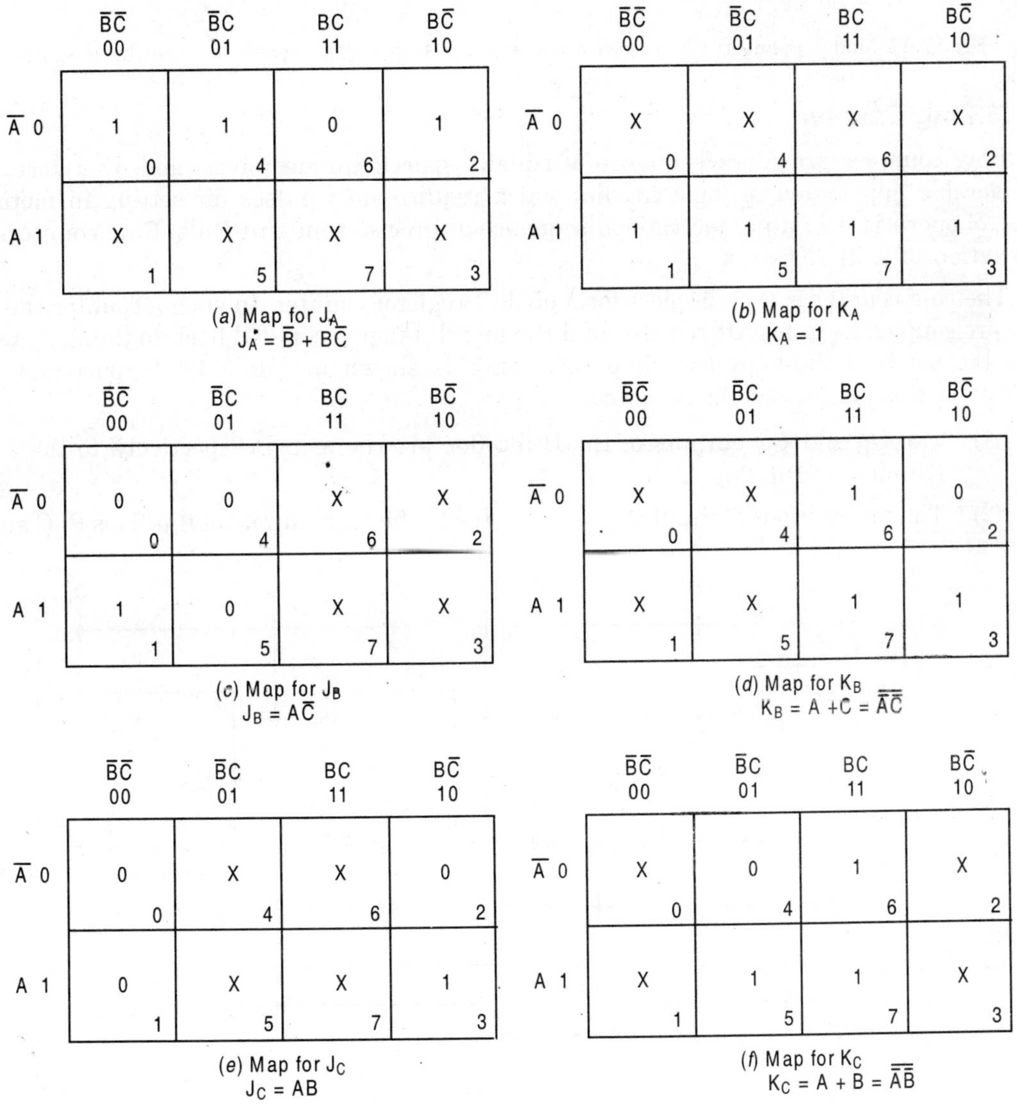

(a) Map for J_A
$J_A = \bar{B} + B\bar{C}$

(b) Map for K_A
$K_A = 1$

(c) Map for J_B
$J_B = A\bar{C}$

(d) Map for K_B
$K_B = A + C = \overline{\bar{A}\bar{C}}$

(e) Map for J_C
$J_C = AB$

(f) Map for K_C
$K_C = A + B = \overline{\bar{A}\bar{B}}$

Fig. 7.46

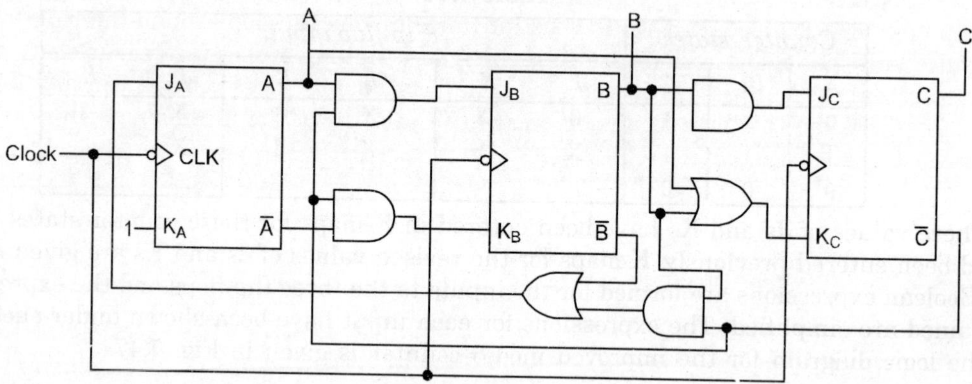

Fig. 7.47 Mod-6 counter which will reset when it happens to reach an unutilized state

7.4.13 Ring Counter

Ring counters provide a sequence of equally spaced timing pulses and, therefore, find considerable application in logic circuits which require such pulses for setting in motion a series of operations in a predetermined sequence at precise time intervals. Ring counters are a variation of shift registers.

The ring counter is the simplest form of shift register counter. In such a counter the flip-flops are coupled as in a shift register and the last flip-flop is coupled back to the first, which gives the array of flip-flops the shape of a ring as shown in Fig. 7.48. In particular two features of this circuit should be noted.

(1) The Q_D and $\overline{Q}_D$ outputs of the D flip-flop are connected respectively, to the J and K inputs of flip-flop A.

(2) The preset input of flip-flop A is connected to the reset inputs of flip-flops B, C and D.

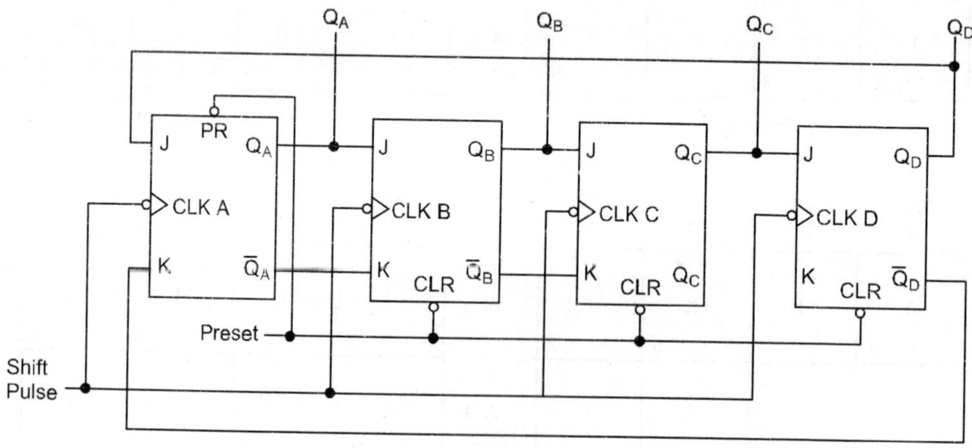

Fig. 7.48 Ring counter

If we place only one of the flip-flops in the set state and the others in the reset state and then apply clock pulses, the logic 1 will advance by one flip-flop around the ring for each clock pulse and the logic 1 will return to the original flip-flop after exactly four clock pulses, as there are only four flip-flops in the ring. The ring counter does not require any decoder, as we can determine the pulse count by noting the position of the flip-flop, which is set. The total cycle length of the ring is equal to the number of flip-flop stages in the counter. The ring counter has the advantage that it is extremely fast and requires no gates for decoding the count. However it is uneconomical in the number of flip-flops. Whereas a mod-8 counter will require four flip-flops, a mod-8 ring counter will require eight flip-flops.

The ring counter is ideally suited for applications where each count has to be recognized to perform some logical operation.

We can now consider how the modified shift register shown in Fig. 7.48 operates. When the preset is taken low momentarily, flip-flop A sets and all other flip-flops are reset. The counter output will now be as follows:

Q_A	Q_B	Q_C	Q_D
1	0	0	0

At the negative clock edge of the 1st pulse, flip-flop A resets Q_A becomes 0, Q_B becomes 1 and Q_C and Q_D remain 0. The counter output is now as follows:

Q_A	Q_B	Q_C	Q_D
1	0	0	0

After the 4th clock pulse, the counter output will be as follows:

Q_A	Q_B	Q_C	Q_D
1	0	0	0

You will notice that this was the position at the beginning of the operation, when the preset input was activated. A single logic 1 has travelled round the counter shifting one flip-flop position at a time and has returned to flip-flop A. The states of the flip-flops have been summarized in Table 7.15.

Table 7.15 Ring counter states

States	*Counter output*			
	Q_A	Q_B	Q_C	Q_D
1	1	0	0	0
2	0	1	0	0
3	0	0	1	0
4	0	0	0	1
5	1	0	0	0

The relevant waveforms are shown in Fig. 7.49.

If preset and clear inputs are not available, it is necessary to provide the required gating, so that the counter starts from the initial state. This can be simply arranged by using a NOR gate as shown in Fig. 7.50.

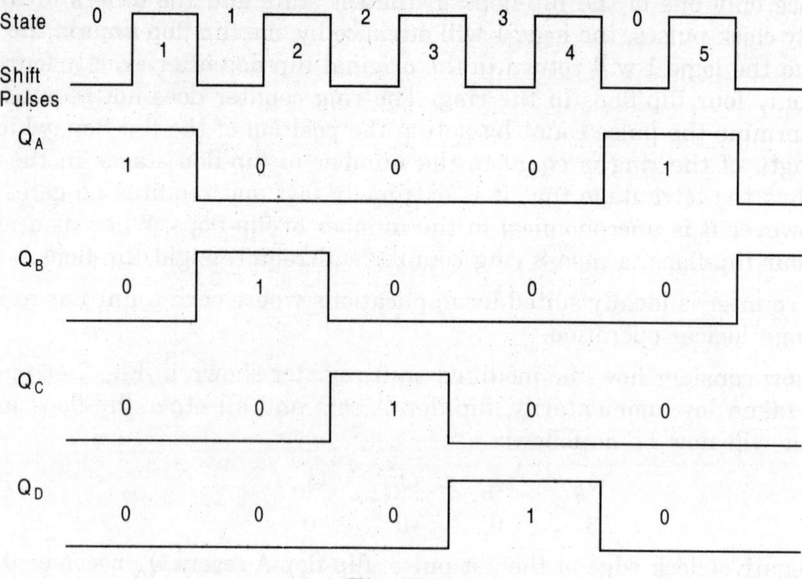

Fig. 7.49

The NOR gate ensures that the input to flip-flop A will be 0 if any of the outputs of A, B, C flip-flops is a logic 1. Now, on the application of clock pulses 0s will be moved right into the counter until all A, B and C flip-flops are reset. When this happens, a logic 1 will be shifted into the counter, and as this 1 is shifted right through the A, B and C flip-flops it will be preceded by three more 0s, which will again be followed by a logic 1 from the NOR gate when flip-flops, A, B and C are all reset.

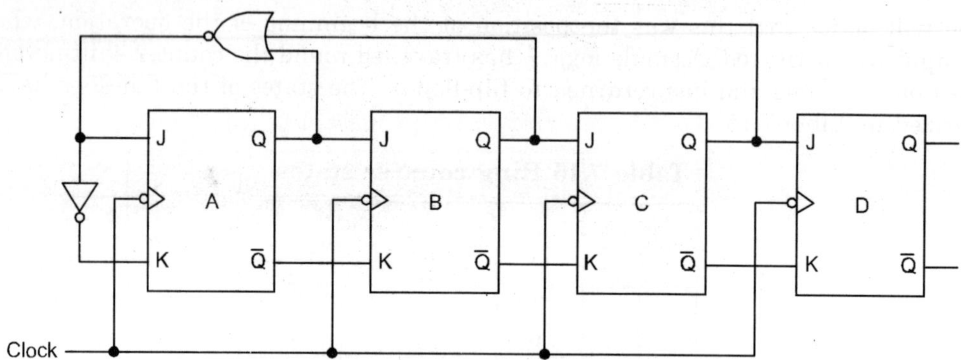

Fig. 7.50 Ring counter with correcting circuit

7.4.14 Johnson Counter

The ring counter can be modified to effect an economy in the number of flip-flops used to implement a ring counter. In modified form it is known as a switchtail ring counter or Johnson counter. The modified ring counter can be implemented with only half the number of flip-flops.

In the ring counter circuit shown in Fig. 7.48, the Q_D and $\overline{Q}_D$ outputs of the D-flip-flop were connected respectively, to the J and K inputs of flip-flop A. In the Johnson counter, the outputs of the last flip-flop are crossed over and then connected to the J and K inputs of the first flip-flop. Figure 7.51 shows a Johnson counter using four JK flip-flops in the shift register configuration, shown Q_D and $\overline{Q}_D$ outputs connected respectively, to the K and J inputs of flip-flop A. Because of this cross-connection, the Johnson counter is sometimes referred to as a twisted ring counter.

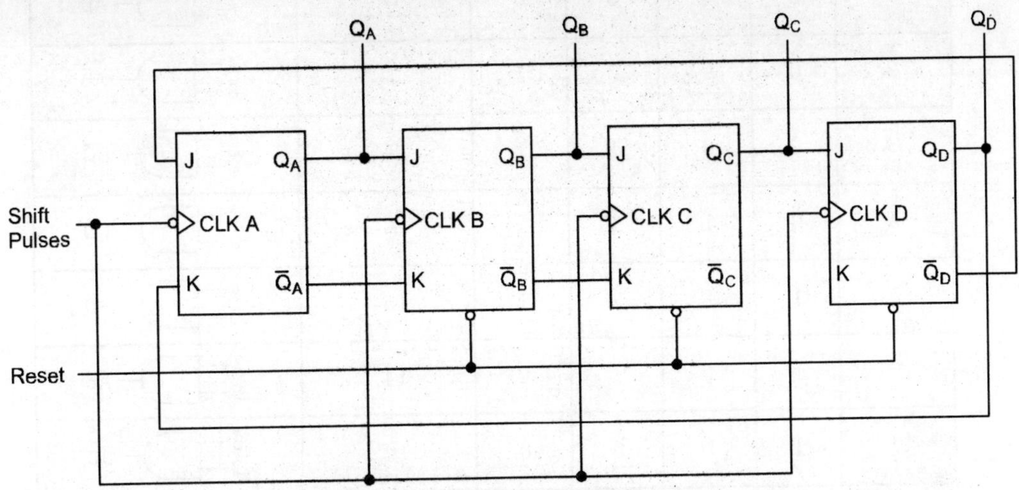

Fig. 7.51 Four-stage Johnson counter

To enable the counter to function according to the desired sequence, it is necessary to reset all the flip-flops. Initially therefore, Q_D is 0 and Q_A is 1, which makes the J input of flip-flop A logic 1. We will now study how shift pulses alter the counter output.

(1) Since the J input of flip-flop A is 1, the 1st shift pulse sets the A flip-flop and the other flip-flops remain reset as the J inputs of these flip-flops are 0 and K inputs are 1.

(2) When the 2nd shift pulse is applied, since Q_D is still 1, flip-flop A remains set and flip-flop B is set, while flip-flop C and D remain reset.

(3) During the 3rd shift pulse, flip-flop C also sets, while flip-flops A and B are already set; but flip-flop D remains reset.

(4) During the 4th pulse, flip-flop D also sets while flip-flops A, B and C are already set.

(5) During the 5th pulse as $\overline{Q}_D$ is 0, flip-flop A resets, while flip-flops B, C and D remain set.

The entire sequence of states, which are 8 in all, is as shown in Table 7.16.

You will notice from Table 7.16 that Johnson counter with four flip-flops has eight valid states. Since four flip-flops have been used, the total number of states is 16, out of which 8 are invalid, which have been listed in Table 7.17.

The valid states require decoding, which is different from normal decoding used for standard pure binary count sequence. You will notice that state 1 is uniquely defined, when the outputs of flip-flops A and D are low. Thus a 2-input AND gate with inputs as shown in the table can decode state 1. State 2 is also fully defined by A high and B low. Similarly, the other outputs can be decoded by the gates with inputs as shown in Table 7.16.

Table 7.16

State	Q_D	Q_C	Q_B	Q_A	*Binary equivalent*	*Output decoding*
1	0	0	0	0	0	$\bar{A}$, $\bar{D}$ → $\bar{A}\bar{D}$
2	0	0	0	1	1	A, $\bar{B}$ → $A\bar{B}$
3	0	0	1	1	3	B, $\bar{C}$ → $B\bar{C}$
4	0	1	1	1	7	C, $\bar{D}$ → $C\bar{D}$
5	1	1	1	1	15	A, D → AD
6	1	1	1	0	14	$\bar{A}$, B → $\bar{A}B$
7	1	1	0	0	12	$\bar{B}$, C → $\bar{B}C$
8	1	0	0	0	8	$\bar{C}$, D → $\bar{C}D$

Table 7.17 Invalid states

Q_D	Q_C	Q_B	Q_A	*Binary equivalent*
0	1	0	0	4
1	0	0	1	9
0	0	1	0	2
0	1	0	1	5
1	0	1	1	11
0	1	1	0	6
1	1	0	1	13
1	0	1	0	10

In order to ensure that the counter counts in the prescribed sequence given in Table 7.16, an initial reset pulse may be applied, which will reset all the flip-flops. If this is not done, there is no surety that the counter will revert to the valid counting sequence. If the counter should find itself in an unused state, it may continue to advance from one disallowed state to another. The solution to the problem lies in applying extra feedback, so that the counter reverts to the correct counting sequence. For this purpose, the self-correcting circuit given in Fig. 7.53 may be used. The input to the AND gate is $Q_A\ \bar{Q}_B\ \bar{Q}_C\ Q_D$ and thus it decodes the word 1 0 0 1, which overrides the input, which is 0 and the counter produces an output of 1 1 0 0, which is a part of the allowed counting sequence. From then onwards the counter functions in the desired sequence.

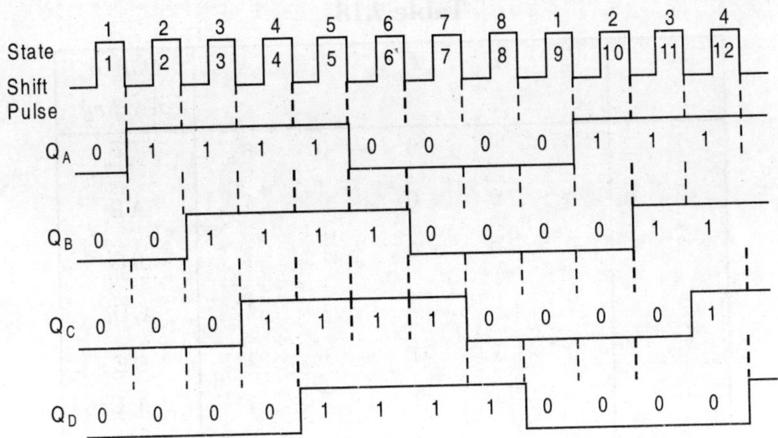

Fig. 7.52 Waveforms for a 4-stage Johnson counter

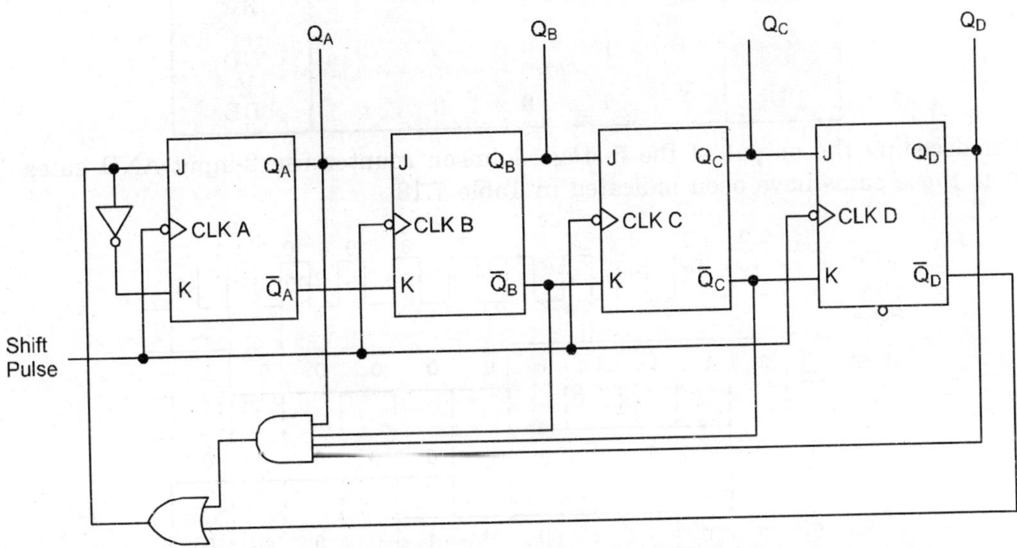

Fig. 7.53 Self-starting and self-correcting Johnson counter

Five-stage Johnson Counter

While discussing the 4-stage Johnson counter, you must have observed that this counter divides the clock frequency by 8. Therefore, a Johnson counter with n flip-flops will divide the clock frequency by $2n$ or, in other words, there will be $2n$ discrete states. If we have five flip-flops connected as a Johnson counter, we will have 10 discrete states. Consequently, we will have a decade counter. However, it should be noted that this counter will have in all 32 states, out of which the desired count sequence will utilize only 10 states and the remaining 22 will have to be disallowed. As in the case of a four flip-flop Johnson counter, some form of feedback will have to be incorporated, to disallow the illegal states. A self-correcting circuit like the one shown in Fig. 7.53 may be used with this counter. Table 7.18 shows the sequence of the ten allowed states for this counter. The waveforms are shown in Fig. 7.54.

Table 7.18

State	*E*	*D*	*C*	*B*	*A*	*Output decoding*
1	0	0	0	0	0	$\bar{A}\bar{E}$
2	0	0	0	0	1	$A\bar{B}$
3	0	0	0	1	1	$B\bar{C}$
4	0	0	1	1	1	$C\bar{D}$
5	0	1	1	1	1	$D\bar{E}$
6	1	1	1	1	1	$A\,E$
7	1	1	1	1	0	$\bar{A}B$
8	1	1	1	0	0	$\bar{B}C$
9	1	1	0	0	0	$\bar{C}D$
10	1	0	0	0	0	$\bar{D}E$

For decoding the output of the 5-stage Johnson counter use 2-input AND gates. The inputs to these gates have been indicated in Table 7.18.

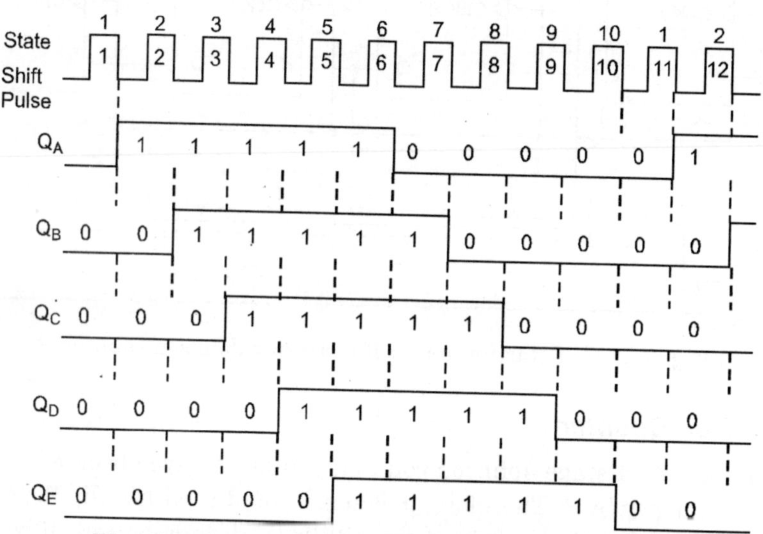

Fig. 7.54 Waveform for a 5-stage Johnson counter

7.4.15 Ring Counter Applications

Ring counters find many applications as

(1) Frequency dividers

(2) Counters

(3) Code generators and

(4) Period and sequence generators

Frequency dividers

If you look at the waveform Fig. 7.49 of the 4-stage ring counter shown in Fig. 7.48, you will notice that the B flip-flop produces one output pulse for two input pulses, that is it divides the frequency of the shift pulse by 2. Similarly, flip-flop C produces one output pulse for every three input pulses, that is it divides the input frequency by 3, and flip-flop D divides the input frequency by 4. If there are n flip-flops they will divide the shift pulse by n. Thus, a shift register connected as a ring counter can be used as a frequency divider.

Counters

A shift register, when connected as a ring counter, can also be used as a counter. For instance, the flip-flop outputs of the ring counter in Fig. 7.48 also give an indication of the number of pulses applied and, therefore counting requires no decoding.

Sequence generators

Sequence generators are circuits which generate a prescribed sequence of bits in synchronism with a clock. By connecting the outputs of flip-flops in a ring counter to the logic circuits whose operations are to be controlled according to a certain sequence, a ring counter can perform a very useful function. Since ring counters are activated by fixed frequency clocks, the timing intervals between the logic circuits to be controlled can be very precise.

This is of particular importance in computers where instructions have to be executed at the right time and in the correct sequence.

Feedback Counters

The ring counters which we have considered so far have a cycle length which is the same as the number of flip-flops in the counter. For instance, the ring counter in Fig. 7.48 has a cycle length of 4. It is possible to design a ring counter which produces a longer cycle length of 2^n–1, where n is the number of flip-flops in the ring counter. The trick lies in decoding the outputs of the shift register and feeding the decoded output back to the input. This technique can be used to develop a wide variety of count sequences and output waveforms. To achieve a cycle length of $2^n - 1$, an exclusive-OR gate may be used as the feedback element, which provides a feedback term from an even number of stages to the first stage. Table 7.19 intended for counters up to 12 stages, shows the stages the outputs of which are to be fed back to the first flip-flop in the chain.

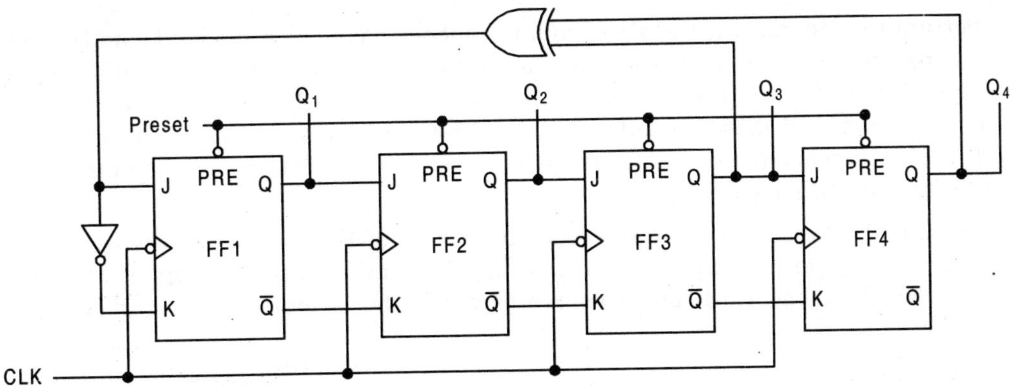

Fig. 7.55 Four-stage feedback counter

This table can be used for designing counters of the type shown in Fig. 7.55, when the feedback element consists of a single XOR gate. The count sequence for this 4-stage counter is given in Table 7.20. When you refer to Table 7.19, you will notice that the feedback term for a 4-stage counter using an XOR gate as the feedback element is $F = (Q_3 \oplus Q_4)$. The truth table for an XOR gate reproduced below will enable you to determine the input to the first stage in the counter.

Input		*Output*
A	*B*	*F*
0	0	0
0	1	1
1	0	1
1	1	0

Table 7.19 Feedback terms for counter design

No. of stage	*Feedback stage*			
2		Q_1	Q_2	
3		Q_2	Q_3	
4		Q_3	Q_4	
5		Q_3	Q_5	
6		Q_5	Q_6	
7		Q_6	Q_7	
8	Q_4	Q_5	Q_6	Q_8
9		Q_5	Q_9	
10		Q_7	Q_{10}	
11		Q_9	Q_{11}	
12	Q_6	Q_8	Q_{11}	Q_{12}

In determining the counter states, all that is necessary is to determine the feedback input to the first flip-flop and, since JK flip-flops have been used, the input to the first flip-flop will be the same as the output of the XOR gate, which depends on the outputs of FF3 and FF4. Table 7.20 has been prepared on this basis.

It is important to note that the 0 state of count sequence has to be excluded by additional gating or by using the preset input. If you refer to the first row of the table, you will observe that both outputs Q_3 to Q_4 are 1 and therefore F = 0. Consequently, the input to the first flip-flop is also 0, which will make its output on the first clock pulse 0. The outputs of FF2 and FF3 will remain unchanged on the first clock pulse. You can determine the outputs in the remaining rows on this basis.

A close look at the table will show you that the output of FF2 resembles the output of FF1, but it is delayed by one clock pulse from that of FF1. Similarly, the outputs of FF3 and FF4 are also delayed by one clock pulse as compared to the outputs of the immediately preceding flip-flops.

Table 7.20 Count sequence for 4-stage feedback counter

Clock input	*Output*			
	Q_1	Q_2	Q_3	Q_4
0	1	1	1	1
1	0	1	1	1
2	0	0	1	1
3	0	0	0	1
4	1	0	0	0
5	0	1	0	0
6	0	0	1	0
7	1	0	0	1
8	1	1	0	0
9	0	1	1	0
10	1	0	1	1
11	0	1	0	1
12	1	0	1	0
13	1	1	0	1
14	1	1	1	0
15	1	1	1	1

This procedure can be used for designing counters which are required to cycle through a large number of states. For instance a counter which uses 8 flip-flops will cycle through $2^8 - 1$ or 255 states. We have used only a single XOR gate as the feedback element, but the feedback logic can be designed differently to sequence through any desired sequence or waveform.

Sequence generators

Here we are concerned with pseudo-random sequence generators. They will be random in the sense that the output generated will not cycle through the normal binary count. The sequence is termed pseudo, as it is not random in the real sense, because it will sequence through all the possible states once every $2^n - 1$ clock cycles. The random sequence generator given in Fig. 7.56 has n stages and it will therefore sequence through $2^n - 1$ values before it repeats the same sequence of values.

Let us consider the sequence 100110011001. The bit sequence in this number has a length of 4 that is 1001, if you read it from the first bit on the left. You can also read the sequence from the 2nd and 3rd bits on the left, when the bit patterns will appear to be 0011 and 0110. No matter how you read it, the bit length does not change, nor does the sequence of bits change. You can describe the pattern of bits as 1001, 0011 or 0110.

We can now consider the structure of a sequence generator given in a simple form in Fig. 7.56 using D-type flip-flops connected as in a shift register. The output of the flip-flops are connected through a feedback decoder to the input of the first flip-flop. The output of the decoder is a function of the flip-flop outputs connected to it and the decoder circuitry. We can state this as follows :

$$F = f(Q_1, Q_2, Q_3, Q_n)$$

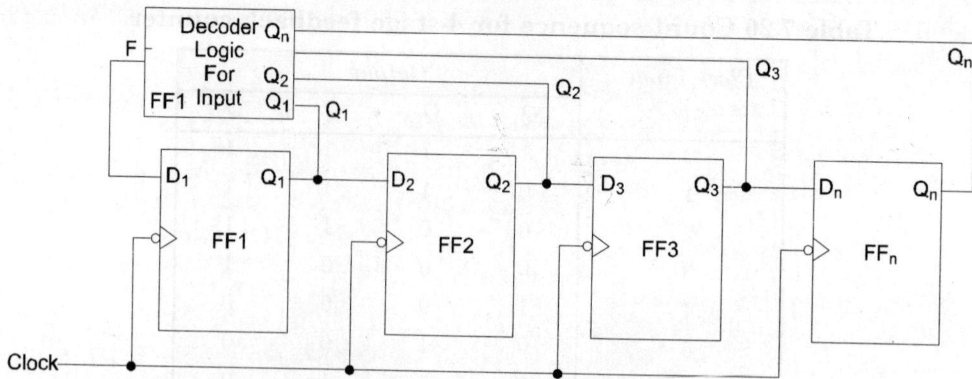

Fig. 7.56 Basic structure of a sequence generator

The desired sequence of bits will appear at the output of each of the flip-flops, but the output of each of the successive flip-flops will show a delay in the appearance of the sequence by one clock interval over the one which precedes it.

The minimum number of flip-flops required to generate a sequence of length S is given by

$$S = 2^n - 1$$

Where, n is the number of flip-flops in the chain.

However, if the minimum number of flip-flops is used, it is not possible to say off hand, that it will be possible to generate a sequence of the required length; but for a given number of flip-flops there is invariably one sequence which has the maximum length.

It is important that in the generation of a sequence no state should be repeated, as that will put a limit on the number of states, because every state determines the development of the future sequence. Besides, the all 0 state has to be excluded, as in this case the input to the first flip-flop in the chain will be 0, which implies that the next state will also be 0, in which case the sequence generator would stop functioning.

We will now consider the steps in the generation of the sequence 1001001 of seven bits. The number of stages that will be required to generate this sequence can be determined as follows:

$$S = 2^n - 1$$

Since, S = 7; n should be 3, that is three flip-flops will be required.

However, there is no guarantee that a 7-bit sequence can be generated in 3 stages. If it is not possible, we can try to implement the sequence by using a 4-stage counter; but in this particular case, as you will see, it will be possible to generate this sequence with three stages. The basic arrangement for generating this sequence is shown in Fig. 7.80, which uses three JK flip-flops. The outputs of FF2 and FF3 constitute the inputs to the logic decoder, which in this case is an XOR gate. The output of the XOR gate, which constitutes the input F to FF1 can be stated as follows:

$$F = (Q_2 \oplus Q_3)$$

You must have noticed that the outputs of FF2 to FF3 are one CLK pulse behind the outputs of flip-flops immediately preceding them. After the first sequence of 7 states has been completed, the sequence is repeated when the 8th (or 1st) CLK pulse arrives. Also observe that no output state has been repeated, which shows that it has been possible to implement the sequence with only 3 flip-flops.

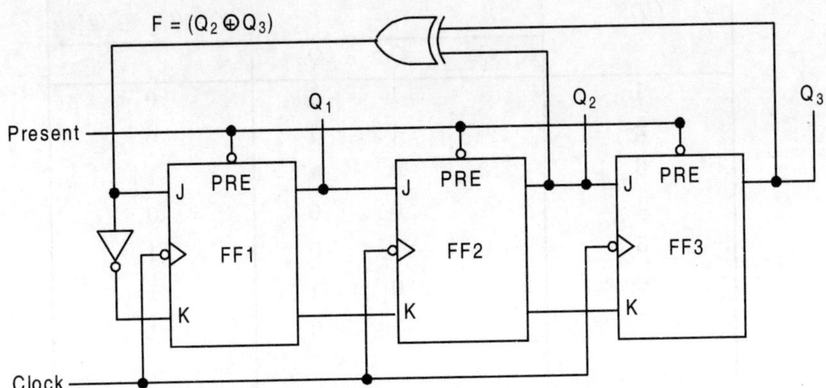

Fig. 7.57 Three-stage sequence generator

When a larger or smaller number of flip-flops are used, the input to the first flip-flop can be worked out on the same basis; but the feedback logic will be different as shown in Table 7.21 for sequence generators using up to 8 stages. For instance for a generator using four flip-flops, F will be as follows:

$$F = (Q_3 \oplus Q_4)$$

Table 7.21 Logic design table for shift register sequences of maximum length ($S = 2^n - 1$)

Clock *n*	*Feedback state*			
2		Q_1	Q_2	
3		Q_2	Q_3	
4		Q_3	Q_4	
5		Q_3	Q_5	
6		Q_5	Q_6	
7		Q_6	Q_7	
8	Q_2	Q_3	Q_4	Q_8

The implementation of sequence 1001011 has been presented in Table 7.22.

The count sequence has been developed as follows: You will notice from the table that at the commencement of the operation, the counter is set as shown against CLK 1. Before CLK 2 is applied at FF1 input, the F input to it should be 0, so that its output changes from 1 to 0. Since Q_2 and Q_3 are both 1, the F input to FF1 will be 0. This condition is, therefore, satisfied. The second clock pulse, therefore, changes Q_1 from 1 to 0 and Q_2 and Q_3 remain on 1 as the inputs to these flip-flops are 1. Since both Q_2 and Q_3 are again 1, the F input to FF1, before the arrival of the 3rd clock pulse will again be 0. Therefore, on the arrival of CLK pulse 3, the output of Q_1 will remain 0, as the input to it is 0. On the same CLK pulse Q_2 will change from 1 to 0 as the input to it is 0 and Q_3 will remain on 1 as the input to Q_3 is still 1. Successive changes in the outputs have been worked out on this basis.

Table 7.22

Clock interval CLK	*Flip-flop outputs*			*Input to FF1* $F = (Q_2 \oplus Q_3)$
	Q_1	Q_2	Q_3	
1	1	1	1	0
2	0	1	1	0
3	0	0	1	1
4	1	0	0	0
5	0	1	0	1
6	1	0	1	1
7	1	1	0	1
.	.	.	.	.
.	.	.	.	.
.	.	.	.	.
.	.	.	.	.
1	1	1	1	1

7.5 EXERCISE

1. Can one store decimal number 12 in an 8-bit shift register.

2. The number stored in a 4-bit binary up-counter is 0101. What will be state of the counter after the following clock pulses?

(*a*) 3rd clock pulse

(*b*) 5th clock pulse

(*c*) 8th clock pulse

(*d*) 12th clock pulse

3. In a 4-bit ripple up-counter how many clock pulses will you apply, starting from state 0 0 0 0, so that the counter outputs are as follows?

(*a*) 0 0 1 0

(*b*) 0 1 1 1

(*c*) 1 0 0 1

(*d*) 1 1 1 0

4. Draw the logic diagram for a binary up-counter using four JK flip-flops and draw the truth table and the output waveforms.

5. Connect four edge-triggered D-type flip-flops to make an asynchronous up-counter.

6. How many JK flip-flops will you require to make the following modulo counters?

(*a*) Mod-4 (*b*) Mod-6

(*c*) Mod-9 (*d*) Mod-11

7. What will be maximum count capability of a counter having 12 JK flip-flops?

8. How many flip-flops will you require to attain a count capability of 8500?
9. An asynchronous counter has four flip-flops and the propagation delay of each flip-flop is 20 ns. Calculate the maximum counting speed of the counter.
10. A synchronous counter has four flip-flops and the propagation delay of each is 20 ns. What is its maximum counting speed?
11. By how much will a ripple down-counter having three flip-flops divide the input frequency?
12. Draw a logic diagram, truth table and output waveforms for a ripple down-counter with four flip-flops.
13. What will be the output states of a four flip-flop binary down-counter, after the following input clock pulses, if the initial state of the counter was 1111?

    (*a*) 4 (*b*) 7 (*c*) 9 (*d*) 14
14. Draw the logic diagram of a presettable down counter with a maximum preset capability of 7.
15. What will be the modulus of IC 74193 in the up-counting mode, if the numbers preset in the counter are as follows?

    (*a*) Decimal 5 (*b*) Decimal 7

    (*c*) Decimal 9 (*d*) Decimal 12
16. What will be the modulus of IC 74193 in the down-counting mode, when the binary numbers preset in the counter are the same as in Problem 15?
17. A 74193 up-counter starts counting up from binary number 1 0 0 0. What will be the state of the counter after the 8th clock pulse?
18. Draw the logic diagram of a Mod-6 counter using the counter reset method. Write its truth table and draw the output waveforms.
19. Show how you will connect two ICs 74193 to build an 8-bit up-down counter.
20. What is the maximum counting capacity of a chain of five BCD counters?
21. A BCD counter is required to have the following states. After how many clock pulses will these states be reached, if the counter was initially reset?

    (*a*) 0 0 1 0

    (b) 0 1 0 0

    (c) 0 1 1 0

    (d) 1 0 0 1
22. Connect two ICs 74193 to make a moduluo-20 divider circuit.
23. Design a mod-10 (Decade) synchronous counter using JK flip-flops.
24. Draw decoding gates for the decade counter in Fig. 4.51.
25. Draw decoding gates for the counter of Fig. 4.49.
26. Redesign the synchronous mod-5 counter circuit discussed in Sec 4.8.12.2 so that whenever the counter reaches the unutilized state 1 0 1, 0 1 1 and 1 1 1 the counter is reset.
27. Design a Mod-7 counter using IC 7490 A.

28. Design a divide-by 120 counter using ICs 7490 A and 7492 A.

29. Design a correcting circuit for a 4-stage ring counter using a NAND gate instead of a NOR gate as used in Fig. 4.80.

30. Determine the maximal length sequence, which can be generated using four JK flip-flops and draw the sequence generated by the first flip-flop in the chain.

31. Draw waveforms to illustrate how a serial binary number 1011 is loaded into a shift register.

32. A binary number is to be divided by 64. By how many positions will you shift the number and in what direction.

33. Describe the working of shift register with PISO/SIPO operation.

34. Design a mod-5 synchronous counter having the states 011, 100, 101, 110, 111 respectively. Obtain a minimal cost design with J-K F/F.

35. Design a shift register counter to generate a sequence length of 8 having self-start feature.

8

MEMORY

8.0 INTRODUCTION

Memories are most essential part of computer and microprocessor systems. They are used to store data and programs for the computer systems. In many general-purpose applications they are also used to store the needed control information for the applications. There are various types of memory devices to accommodate various needs of data storage.

To be used with computer systems mainly we have two types of memories, semiconductor and magnetic memory. The semiconductor devices are fast storage devices and are connected directly to the CPU. For this reason, they are many a times called *main memory* or *primary memory*. Since their storage capacity is relatively small they are generally used for temporary storage. Some of them are also used for permanent storage, as we will see later. The magnetic memories are used for very large data storage for long time, and are called *mass storage devices*. Since they are not connected directly to the CPU they are generally called *secondary memory*.

In the computer systems very first memory elements were magnetic and were referred as core memory. These were very bulky and needed large PCBs. In contrast semiconductor memory is compact and fast. With the latest advances in VLSI, the high-density reliable semiconductor memory devices are available at cheaper costs. Such devices internally use BJTs and MOS devices. Both read/write and read only memories are available with semiconductor technology and influenced the computational world greatly.

The advent of programmable read only memories (PROMs) made programmable logic devices (PLDs) to appear. A read only memory (ROM) consist of an array of AND and OR gates which are fixed. A PROM is one in which these arrays can be programmed but only once. The re-programmable PROMs are also available which greatly enhanced the utilization of these memory devices. The flexibility of programming the AND-OR array provided many complex systems to appear. Programmable logic array (PLA) and complex PLDs (CPLDs) are among them. Consequently, this made possible the appearance of field programmable graphic array (FPGAs) and application specific integrated circuits (ASICs).

PLDs, CPLDs, PROMs, FPGAs, etc. have been presented earlier. In this chapter we will study the semiconductor memory devices extensively.

8.1 MEMORY BASICS

The ability to remember is the most important characteristics of digital circuits. The circuits or systems designed to facilitate storage of digital information are called *memory*.

In its simplest form, a F/F is a basic memory device that can be used to store 1-bit of information. Any type of memory can be supposed to be composed of a no of F/F. The general information that is stored in the memory is called as *Data*. The F/F or circuit arrangement that holds a 1-bit information is conveniently called a *storage cell* or simply memory cell. The total number of bits (or storage cells) that a memory device can store is referred to as its *storage capacity* e.g. a device that can store 1024 bits is said to have a storage capacity of 1024 bits. In memory, data is always stored as a group of bits. In fact a single memory cell is rarely accessed, instead they are accessed in a group of fixed size. The number of bits in a group (of data) is called as the *Word Length* and each group is called as *Word* e.g. in a memory if there are 10 datas each of which are a group of 4-bits, then memory is said to have ability to store 10 words of word length 4-bits. What about storage capacity of this memory device? It is simply 10 words × 4-bits (word length) = 40-bits. So,

$$\text{Storage Capacity (in bits)} = \text{No. of Words} \times \text{Word Length} \quad \text{...(1)}$$

The eq. 1 is the basic relation to determine storage capacity of a memory device. Although word length can be any no of bits (even 1-bit), now a days all have settled down to the basic memory word length as byte.

In a memory device there are several storage cells collected together to form a memory of larger capacity. In order to access them unambiguously each of the storage cell (or group of cell) is associated with a unique name, called *address* of the cell. So each cell may be called an address location or *memory location*, a place where data resides. In an analogy let us consider a residential colony where each house is a memory cell, people living in it are data stored, and house numbers are address of memory cells.

Any memory device may permit two types of operations to be performed on it, the read and the write. A *Read operation*, also called *sensing*, means detecting the status (i.e. information) of memory cell (whether it is 1 or 0) and the *write operation* means setting the status of memory cell either to 0 or to 1 according to the given data.

8.2 MEMORY CHARACTERISTICS

All kind of memory devices (i.e. semiconductor, magnetic, and optical memory), used to build the memory system have some functional characteristics in common. One or more of these characteristics may be used to describe the performance of the memory devices. Following are the important functional characteristics that must be considered for any description of memory.

- ***Storage Capacity.*** It is defined as the total number of bits (or words) that a memory device can store. In general storage capacity is specified in Kilobytes (KB), Megabytes (MB), or Gigabytes (GB). Kilobit is represented as Kb. In binary the kilo means 1024 (not 10^3) and Mega means 1024 K (not 10^9). If storage capacity of device is 1KB it means it has 1024 storage locations and at each location a word of length 1 byte (i.e. 8-bits) can be stored. So the total capacity in bits would be 1024 × 8 = 8192 bits or 8 Kb.
- ***Cost.*** The cost of the memory can be defined as the cost per bit. It is expected that the cost/bit should be as low as possible. If S is the storage capacity, C is the total cost of memory then the cost per bit *Cp/b* can be defined as

Cp/b = [total cost (C)/Storage capacity (S)]

- ***Memory Access Time.*** The performance of a memory device is primarily determined by the rate at which the information can be read or write into the memory. The average time required to read a fixed amount of information from the selected memory location is termed as *Read Access Time* or *Access Time* and is denoted as 't_A'. In precise words, it is average time delay after memory address is issued till the data appears at output. The access time should be kept as low as possible. Similarly the average time required to store a fixed amount of information at the selected memory location is termed as *Write Access Time*.

  Read Access Time > Write Access Time

  The average rate at which memory can be accessed is defined as the *Access Rate* 'b_A' of a memory and is measured in words/sec.

  Access Rate (b_A) = $1/t_A$

- ***Memory Accessing Modes.*** It is defined as the way or sequence in which the information is accessed in the memory. The two types of access modes are
  - Random Access Mode
  - Serial Access Mode

  If the memory cells can be accessed in any order, then the access time is independent of the position of cell. This type of access mode is referred as random access mode, and the memories having such access mode are called *Random Access Memory*. In such memories any storage cell (or location) can be reached in a fixed amount of time and each storage cell can be accessed independently of the positions of other cells. Obviously this type of memories are faster. Semiconductor memories are of this type.

  If the memory cells can be accessed one after another or only in certain predetermined sequences, the access time is dependent of position of cell. Thus the access time will be different for different cell locations. This is called serial access mode and the memories having such access mode are called *Serial Access Memory*. Magnetic memories are of this type.

- ***Memory Cycle Time and Bandwidth.*** The minimum time that must elapse between the initiation of two consecutive memory accesses is defined as the *Memory Cycle Time* t_C. The cycle time t_C can be greater than the access time t_A. This is because after completion of one memory access some memory devices need a small amount of time, t_{DELAY} to make them ready for next access. In general

  $t_C = t_A + t_{DELAY}$ (t_{DELAY} is additional delay due to the physical characteristics of memory)

  Note that t_{DELAY} may not be present in all type of memories.

  Maximum amount of information that can be transferred to or from memory every second is referred to as the maximum *Data Transfer Rate* in words/sec or the *Bandwidth* 'b_C', defined as

  Bandwidth (b_C) = $1/t_C$

- ***Non Destructive Read Out (NDRO).*** It is required that the read process should create a copy (i.e. sense) of the data stored and must not disturb the data. Such a reading is called *Non Destructive Read Out* (NDRO) and is expected from all kind of memory.

In some memories, the reading may destroy the data stored at a location. Such a reading is called *Destructive Read Out* (DRO). In such cases, the read operation must be followed by a write operation to restore the original state of memory. Obviously such memories would be slower ones.

- ***Permanence of Storage.*** It is concerned with how long a stored information can be retained when
  (*a*) *Power is Switched off* If the stored information does not vanish with power off, the memory is referred as *NON VOLATILE* memory. But if the stored information vanishes completely at power off the memory is called as *VOLATILE* memory.
  (*b*) *Power is NOT Switched off.* If the data can be retained indefinitely the memory is called as *STATIC* memory. If it can be retained for some definite small amount of time, the memory is called as *DYNAMIC* memory. In the later case, the data must be rewritten periodically, this is called refreshing.
- ***Alterability.*** Alterability is concerned to the capability of memory devices with which the stored data can be changed or updated when they are in use (i.e. online). Memories, which can be altered online, are referred as *Read-Write Memories*. Memories, which can not be altered online (if at all they can be altered offline) are called erasable read only memory devices. However, some memory devices can not be altered even when they are off line and are called nonerasable read only memory devices.

8.3 MASS STORAGE DEVICES

The devices used to store high volumes of data are referred as mass storage devices. The magnetic memories and Optical memories (CD-ROMs) are such memory devices. The magnetic memories are used for very large data storage for long time, but due to their lower speed they are not connected directly to the CPU. Due to this they are generally called *secondary memory* or *backup storage device*. Optical memories (CD-ROMs) are another type of mass storage device. The optical disks are popular due to large storage capacity, transportability and reliability.

8.3.1 Magnetic Memory

In magnetic memory devices some sort of magnetic coating (usually ferric oxide) runs throughout the base. On this coating direction of magnetic flux or direction of magnetization can be used to store 1 or 0. These devices are called *Serial Access Memory* (SAM) devices, because we can not reach the data directly and to reach a data stored we must go one after another. For example consider an audio tape and if you want to reach song number 3 you must traverse through song numbers 0, 1 and 2.

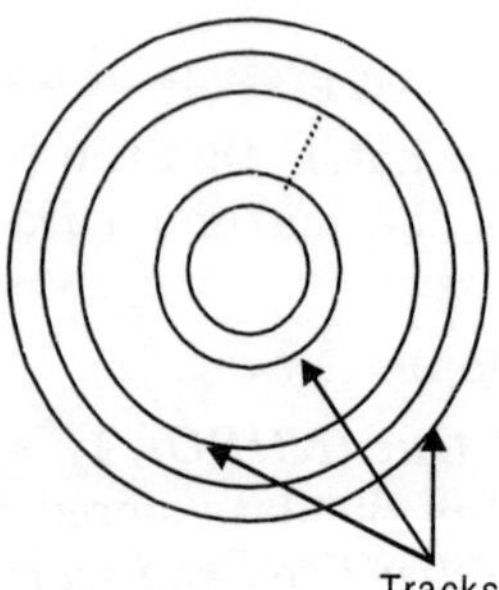

Fig. 8.1 Organization of tracks

Hard disks and floppy disks are common type of magnetic memories used in computers. The storage media of magnetic disks are made up of aluminum or plastic base with a thin coating magnetic on its surface where the datas are actually stored.

In a ***Floppy Disk***, as shown in Fig. 8.1, the whole media is divided into a number of concentric circles, called track. Each track is further divided into a number of sectors. In these sectors bytes of data block are stored. To reach to a stored data first a particular track is selected, and then particular sector is reached on this track, ofcourse starting from the very first sector on this track. To reach to required sector an external rotating mechanism is employed, thus time delay to reach particular sector is called *rotational delay* or *latency time*. The time delay to select a particular track is called *Seek time*. Nowadays the 3.5-inch floppy disk of storage capacity 1.44-MB is popular. It contains a total of 40 tracks where innermost track is numbered as Track '0' and outermost track is numbered as Track '39'. For each of the data access search starts from 1st sector of Track '0'. The main advantage of using floppy disk is portability, but its use is restricted by its small storage capacity.

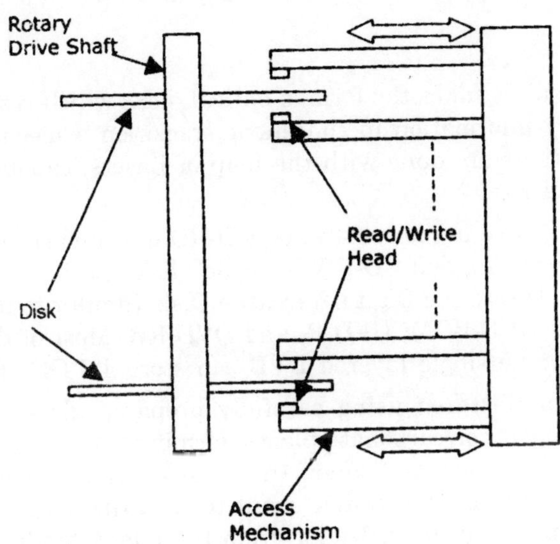

Fig. 8.2 Structure of hard disks

A ***Hard Disk*** can be supposed to have collection of such circular disk surfaces kept in a cylinder on a common shaft, called spindle for rotation. A simple organization of hard disk storage unit is shown in Fig. 8.2. The spindle rotates the disk surfaces at higher speeds, making data access faster than the floppy disks. On each disk surface there are several hundreds of tracks arranged in concentric circle as shown in Fig. 8.1. Each of the tracks is further divided into a number of sectors where data actually resides. Each of the disk surfaces is provided with the separate Read/Write head that may be connected to form a Read/write arm.

During the access operations the disks are rotated continuously by the spindle, at a constant speed. The Read/Write arm the moves in a fixed linear path to select a particular set of tracks and the sectors on these tracks. The disk monitoring system determines which disk surface has to be accessed to find the target data. In comparison with floppy disks the construction of hard disks are much more robust and reliable but they are not portable.

In both the magnetic disks the digital information can be *stored* on disk surface by applying a current pulse to magnetization coil present in Read/Write head. This causes a change in magnetization in the area under the read/write head. The direction of magnetization will be parallel to the direction of applied field. To *read* the magnetically stored information on the disk the Read/Write head has to sense the state of magnetization resulting from the area under the head. Movement of magnetic disk causes change in magnetic field, which induces a voltage in the coil of head. The polarity of induced voltage is monitored to obtain the Magnetization State, which in turns gives the stored binary bit.

The principal disadvantage of magnetic disk memory is relatively slower speed that is attributed to slow rotational mechanism, seek and latency time and the fact that data transfer through such devices are serial. Also presence of mechanically moving parts makes magnetic memory unreliable.

But their advantage lies in two great facts that it offers very large-scale data storage for long time at very low cost per bit. There are other kind of magnetic memory devices also, namely ferrite core memory, magnetic tape memory, magnetic bubble memories etc.

8.3.2 Optical Memory

Optical memories usually takes the form of optical disks which resembles magnetic disks in that they store binary information in concentric tracks on a mechanically rotated disks. The read & write operations are done with the help of Lasers, capable of producing a light spot of the order of 1 micron.

Optical discs are available in many formats: CD-ROM (compact disk-read only memory), CD-R (compact disk-recordable), and CD-RW (compact disk-rewritable). Newer high capacity version of optical disks is DVDs, the Digital Versatile Disk. Similar to the CDs, the DVDs also have different formats as DVD-ROM, DVD-R, and DVD-RW. Most of the CDs can store upto 650 MB where as the simple single layered DVD can store 4.7 GB of data.

The CD-ROM is manufactured using carefully prepared glass master. The master is pressed into injection molded polycarbonate plastic forming the CD-ROM. The resulting CD-ROM contains several small pits and lands (no pits) on the various tracks that can be processed optically. When a computer system want to read the disk, a Laser beam is aimed on the disk from the bottom and the reflection from the pits & lands are interpreted by the photo detector as logic 0s and 1s.

For the purpose of permanent storage (archival storage) the CD-Rs are very popular, and many a times known as WORM (*write once read many*) storage devices. The information can be recorded on CD-R (commonly known as CD burning) by using CD-writers. During the burning process a laser heats gold reflective layer, and a dye layer in the CD-R causing it to have a dull appearance. When read by the CD-R drive the dark burnt areas (like pits) reflect less light. The CD-R reader interprets the shiny areas (lands) and the dull areas (burned) as logic 0s & 1s.

The CD-ROM and DVD-ROM both are manufactured by using the same technology and read data from a spiral track of pits and lands. However the DVD-ROM has a greater storage capacity. It is because the pits and lands on the DVD-ROM are much smaller and more closely packed, thus allowing more information per track. The tracks also are closely spaced allowing more tracks per disk. The DVD-ROMs can store 4.7 GB (single sided layer), 8.4 GB (double sided layer), or 17 GB (double sided double layered), where as the CD-ROMs can store only upto 650 MB.

The CDs are associated with all kind archival storage and have no special standards dictated for the storage of specific type of data on the media. However the DVDs are most commonly associated with the video productions. DVD-video standards are used when disk holds only audio/video (such as movies). The DVD-ROM standards are used when the DVDs are used for data storage as with a computer system.

The *data transfer rate* of a CD-ROM drive is indicated as 1x, 2x, 4x, 16x, 32x etc. A CD-ROM drive with a designation of 1x would have a max data transfer rate of 150 Kbytes/sec. Therefore a 4x drive would have a data transfer speed of 150 Kbytes/sec × 4= 900 Kbytes/sec. These data transfer rates are theoretical and the actual data transfer rates are usually lesser. The DVD-ROMs can provide data transfer rates of 1.38 Mbytes/sec, which is approximately same as the speed of 9x CD-ROM.

The advantages of Optical disks are huge amount of data storage, low cost, easily transferable, a few mechanical parts results in reliability, and high speed. The main disadvantage could be the complexity of Read/Write head.

Although lot more pages can be shed to explain the architecture and characteristics of magnetic & optical memories, but a detailed discussion is beyond the scope of this text. However interested readers can refer to *"Hayes, J.P.: Computer Architecture"*. A glance over magnetic & optical memory was presented to maintain the completeness of the topic. The semiconductor memory is the main consideration of this text so until & unless specified memory must be read as semiconductor memory throughout this text.

8.4 SEMICONDUCTOR MEMORY

Numerous developments in semiconductor technology has emerged into large numbers of LSI and MSI memory devices, called memory *chips*. Besides being faster and compatible with CPU they are economical also. A semiconductor memory is organized as a rectangular array (preferably square array) of storage cells that are integrated on a silicon wafer and are available in DIP packages. Due to this organization any memory cell can be accessed randomly, thus all the semiconductor memories are called *Random Access Memory* (RAM). The basic memory cell may be a flip-flop, or a charge storing element like capacitor that is said to hold logic 1 when charged and logic 0 when discharged. The type of transistors e.g. bipolar, or MOS, or CMOS, used to form a memory cell dictates the storage capacity and speed of a memory chip.

8.4.1 Basic Memory Unit

A simple block diagram of a memory unit can be drawn as shown in Fig. 8.3. It contains data bus, address bus and control signals. A *bus* is a group of parallel conductors whose job is to carry binary information. Each of the conductors carries 1-bit information. The number of bits that a memory data bus carries simultaneously is called *memory bus width*. Usually, but not necessarily, the memory word length and memory bus widths are same.

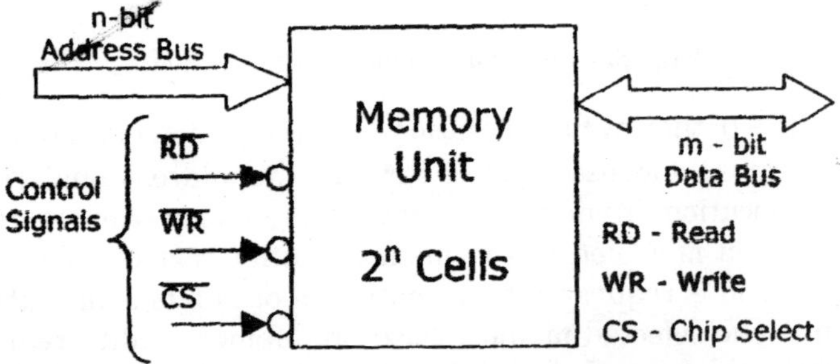

Fig. 8.3 Block diagram of memory unit

The m-bit data bus is used to transfer data to and from the memory. The n-bit address bus carries the address of memory locations. An n-bit address bus can access upto 2^n storage cells i.e. storage capacity is 2^n bits as each cell can store 1-bit. The Read (RD) & Write (WR) signals specify the operation to be performed. As shown the two signals are active low i.e. they are activated when logic '0'. At any time either of the two signals can be activated. In some memory chips read & write controls are available at one signal line. Chip select signal (CS) (sometimes labeled as enable EN also) is used to enable/disable the chip in multiple memory system. It is also a logic low signal. There may be multiple enable signals in a single chip that must be activated simultaneously to enable the chip.

8.4.2 Basic Memory Organization

Basic organization of semiconductor memory is shown in Fig. 8.4 (*a*). It consists of a storage cell array, Read/Write control circuit along with control signals, an address decoder feeded by a register called memory address register, and a buffer register through which data transfer is carried out. In a semiconductor memory, storage cells are organized as a rectangular array. The address bus is connected to internal address decoder through the address register. The RD and WR control line specifies the type of access: If $\overline{RD} = 0$ and $\overline{WR} = 1$ then read operation.

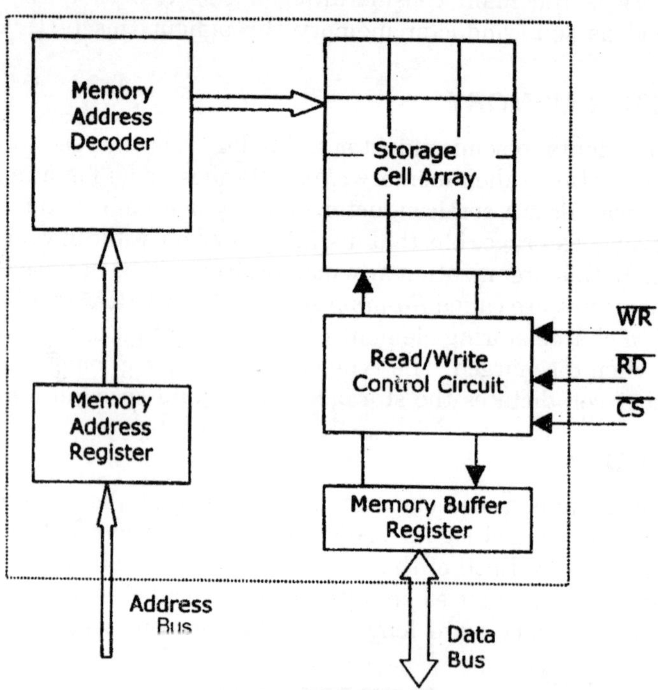

Fig. 8.4 (*a*) Basic memory organization

If $\overline{RD} = 1$ and $\overline{WR} = 0$ then write operation. To access memory, address of the required cell is issued to the memory address register through the address bus. The address decoder selects the particular location in memory array. If read operation is requested then the content of selected location is copied to the memory buffer register which transfer it to the data bus. If write operation is requested then content of data bus is brought to buffer register and then transferred to the selected memory location. Memory buffer register contains bilateral three state buffers, one for each bit of data bus. By combining two three state buffers a

1-bit bi-directional input output line is formed. This is shown in Fig. 8.4 (*b*). The enable or disable signals for the bilateral buffers can be generated through the realization shown in Fig. 8.4 (*c*). These signals are given to all the bilateral buffers of buffer register. Note that if both RD & WR are activated, or CS inactivated, arrangement in Fig. 8.4 (*c*) will generate disable signals for both the buffers i.e. both TSL1 & TSL2 are disabled and data bus enters into high impedance state.

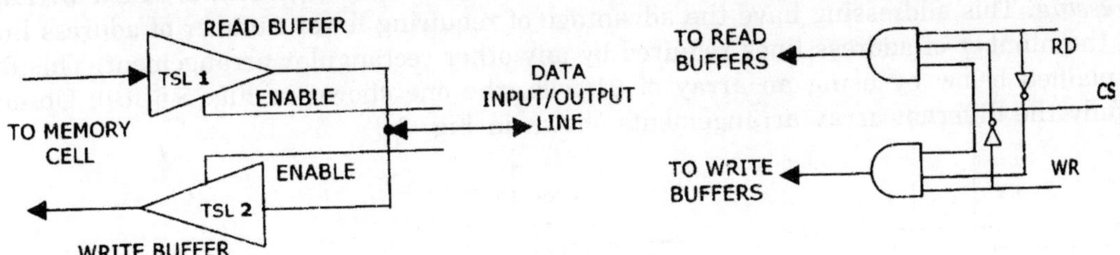

Fig. 8.4 (*b*) Bilateral memory bus buffer **Fig. 8.4** (*c*) Generation of enable/disable signals for buffers

8.4.3 Cell Organization (Memory Addressing)

Memory addressing is concerned with the selection of one particular memory cell from storage cell array. To facilitate selection, the memory cells are organized as rectangular array of m-rows and n-columns as shown in Fig. 8.5(*a*).

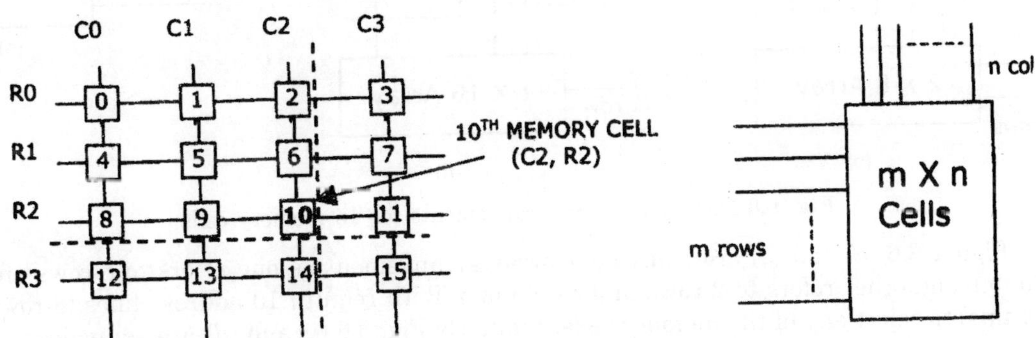

Fig. 8.5 (*a*) An array of 4×4 cells **Fig. 8.5** (*b*) An array of m × n cells

Figure 8.5(a) shows simplified diagram of memory cells array that have 4-rows and 4-columns, thus containing a total of 16 memory cells. The control circuitry associated with memory allows only one row and one column to be activated at a time. In order to select a particular cell we need to specify the appropriate row and column number. Let us consider the selection of 10th cell. It requires column C2 and row R2 to be activated and the cell at the intersection of C2 and R2 is selected which is 10th cell. Dashed lines in Fig. 8.5 (*a*) shows this. So the 10th cell can be designated as the cell C2, R2 (or 2, 2 or 22). This designation is defined as the *address* of the 10th cell. Collectively the lines corresponding to rows and columns are called *address lines*. In a memory array each cell will have a similar type of unique address. Hence any cell can be accessed, just by specifying the row and column number (or address) associated with it. Since any cell can be accessed *randomly*, it is named as *Random Access Memory*. Figure 8.5 (*b*) shows a simple block representation of a rectangular array that we will use in following subsections to understand some important facts. If in Fig. 8.5 (*b*) $m = 4$ and $n = 4$ then it represents the array shown in Fig. 8.5 (*a*).

8.4.3.1 Matrix Addressing

We know that to facilitate selection, the memory cells are organized as rectangular array of m-rows and n-columns. When rows and columns are made equal i.e. $m = n$, then the array becomes a *Square Array* of capacity $n \times n = n^2$. This type of arrangement is called *MATRIX Addressing*. This addressing have the advantage of requiring fewer number of address lines than the number of address lines required by any other rectangular arrangement. This fact is explained below by using an array of 91-cells (the one shown in Fig. 8.5 (*a*)). Observe carefully the different array arrangements shown in Fig. 8.6.

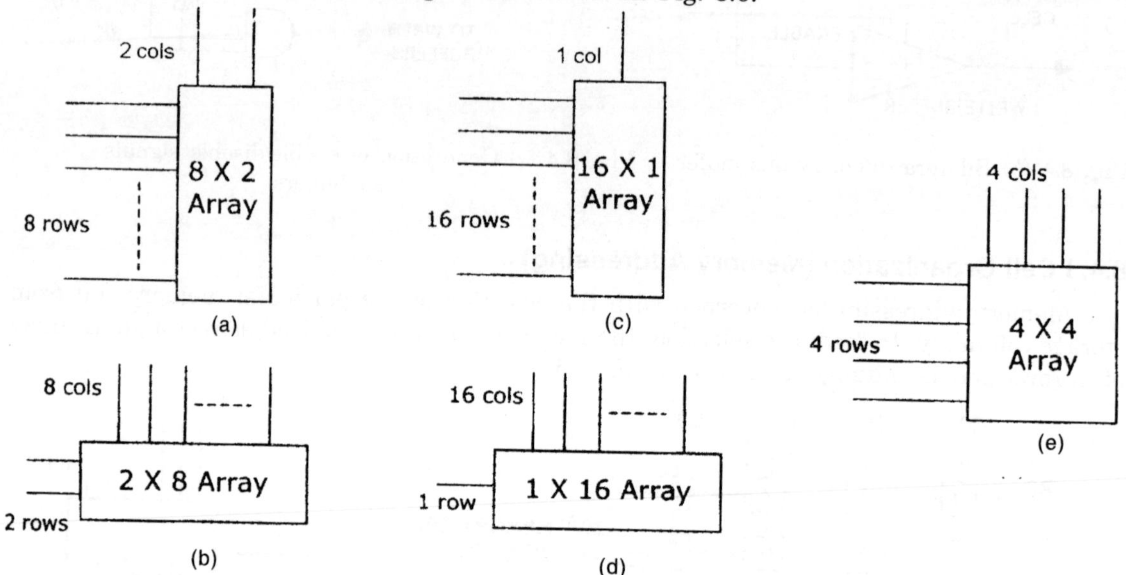

Fig. 8.6 Different array arrangements for 16 memory cells

Figure 8.6 (*a*) and (*b*) are equivalent array arrangement as one refers to 8-rows and 2-columns and other refers to 2-rows and 8-columns. Both require 10 address lines (8-rows+2-columns) to select any of the memory cells. Similarly Fig. 8.6 (*c*) and (*d*) are equivalent array arrangement and require 17 address lines. Infact 111 lines will suffice, as there is only one column/row. Figure 8.6 (*e*) is square array arrangement with equal number of rows and columns and require 8 address lines. So in order to reduce the number of address lines square array arrangement is best.

The arrangement in Fig. 8.6 (*e*) is referred as *Matrix addressing* as defined earlier. In contrast the 16 × 1 array arrangement as in Fig. 8.6 (*c*) is called *Linear Addressing*. It is because there is only one column and to select any of the cells, one needs to specify the row only.

8.4.3.2 The Address Decoding

The address decoding is concerned with the activation of particular row and column to select a memory cell. Here our concern is to see this process exclusively for matrix addressing. Let us reconsider the Fig. 8.6 (*e*) that requires 4-rows and 4-columns. If we use two 2 to 4 line decoders then the address line requirements will be reduced further, as shown in Fig. 8.7 (*a*). The decoders will activate only one of the row & column lines out of 4 lines. As evident from figure now the total number of required address lines are 4 only! The activation of a particular row and column line depends upon the row address and column address. Any column from C0 to C3 can be activated by specifying it as a 2-bit binary number (00 to 9) at column address lines. Similarly, specifying a 2-bit binary number at row address lines can activate any of the four rows. Thus, a total of 4-bit address can access any memory cell unambiguously. Since 4-bit address representation allows $2^4 = 16$ unique addresses, we can say that an n-bit address can be used to define a square array of 2^n memory cells.

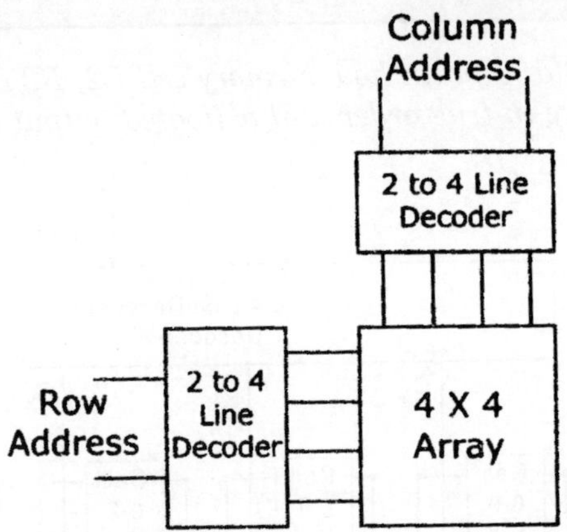

Fig. 8.7 (*a*) Square array of 16 cells with row and column address decoders

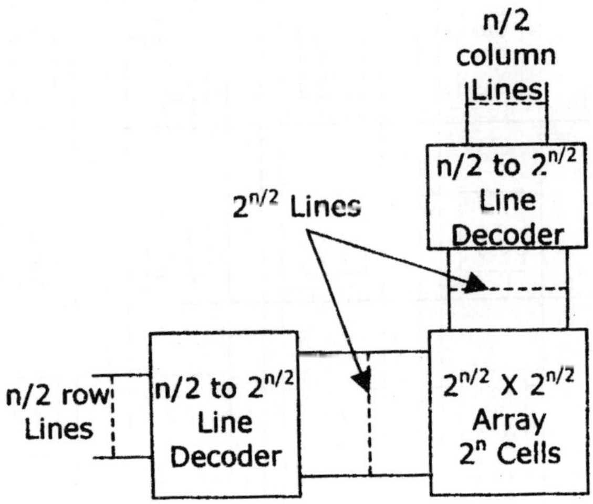

Fig. 8.7 (*b*) Square array of 2^n cells with row and column address decoders

Figure 8.7 (*b*) shows a generalized arrangement of square array with address decoders. The array is organized as n/2 column address lines and *n*/2 row address lines thus making a total of n-bit address lines. It should be noticed that in such arrangements the value of n should an even integer (2, 4, 6, 8 ...). Since at the output of each of the decoder we have $2^{n/2}$ lines, the number of cells (or capacity) in this array would be $2^{n/2} \times 2^{n/2} = 2^n$. This is exactly the reason why commercially available memories have capacity equal to some integer power of 2 i.e. 2^{10}=1024, 2^{12} = 4096 etc. The above discussion concludes that if *n* is the number of address lines then

$$\text{Number of storage cells} = 2^n$$

In general,

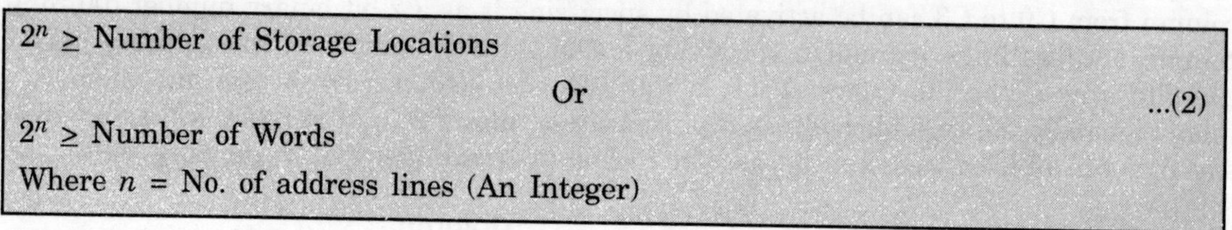

$2^n \geq$ Number of Storage Locations

Or ...(2)

$2^n \geq$ Number of Words

Where n = No. of address lines (An Integer)

Example. *In the Fig. 8.8(a) explain how memory cell C2, R2 can be accessed? Given that the row & column numbers are in true order and activated output line have status logic high.*

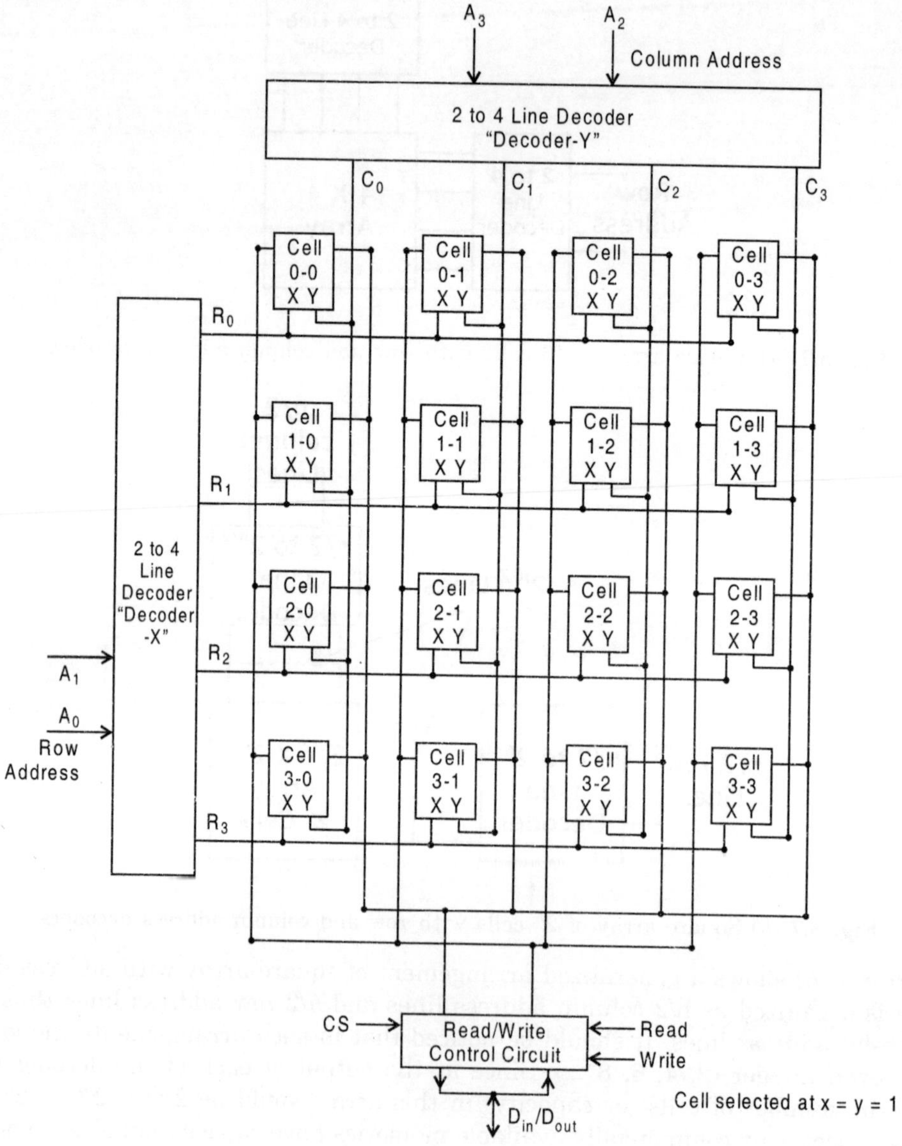

Fig. 8.8 (*a*) 4×4 array organized as 16 × 1 memory

Solution. Figure shows an array of 4×4 having storage capacity of 16-bits. Issuing its address can access the required cell. Since row & column numbers are in true order, the binary equivalent of these numbers will give required memory address.

Memory cell to be accessed C2, R2

Column C2 specified in binary = 10 = A_3 A_2

Row R2 specified in binary = 10 = A_1 A_0

So the complete address A_3 A_2 A_1 A_0 = 1010

Upon receiving the address, the address decoders 'X' and 'Y' will activate row R2 and column C2, respectively. All other rows and columns will be deactivated at this time i.e. R3 = R1 = R0 = 0 and C3 = C1 = C0 = 0. Now referring to Fig. 8.8(*a*), we can see that only the cell 2-2 has got both *x* and *y* input high which is present at the intersection of R2 and C2. Thus, the cell C2, R2 is selected for access.

8.4.4 Organizing Word Lengths (Different Memory Organization)

So far we have discussed the memories that can access a single storage cell or 1-bit at a time. In computer systems requirement is to access a group of bits at a time. In this section our prime concern is to discuss the organization that can access a group of bits. Moreover, this section also deals with different types of cell selection mechanism that can be used conveniently.

To begin with let us consider a 4 × 4-memory array organized as 16 × 1 memory i.e. 16 words each of length 1-bit, as shown in Fig. 8.8(*a*).

Here two 2 to 4 line decoders, namely decoder-'X' & decoder-'Y', are used to decode row and column addresses respectively. A cell in this array can be selected only when both *x* and *y* inputs are activated *i.e.,* at $x = y = 1$.

The same array can be organized as 8 × 2 memory *i.e.,* 8 words each of length 2-bits. Still the storage capacity remains 16-bits. This is shown in Fig. 8.8 (*b*). In this case decoder-Y is replaced by two 1 to 2 line decoders, Y1 and Y2. Both Y1 and Y2 are feeded by the same address line (A2) and work simultaneously. Thus, at any time two columns are activated as per the truth table given in table 1. Whenever any row is activated, *x* inputs of all the 4-cells in that row will be activated. Since two columns are activated at a time, two of the memory cells in the activated row will get their y input activated. Thus two memory cells have both *x* and *y* inputs activated i.e. two cells are selected. Alternately with this organization we can access a group of 2-bits at a time.

The array of Fig. 8.8 (*a*) can be organized as 4 × 4 memory i.e. 4 words each of length 4-bits. Still the storage capacity remains 16-bits. This is shown in Fig. 8.8 (*c*). To do so, decoder-Y is removed and cells are connected only to decoder-X. Thus to select memory cells only x inputs have to be activated. Whenever a row is activated *x* inputs of all the 4-cells are activated i.e. all the 4-cells of that row are selected. In other words, this organization can access a group of 4-bits at a time. Infact this addressing is an example of linear addressing because only rows have to be activated and columns are always selected.

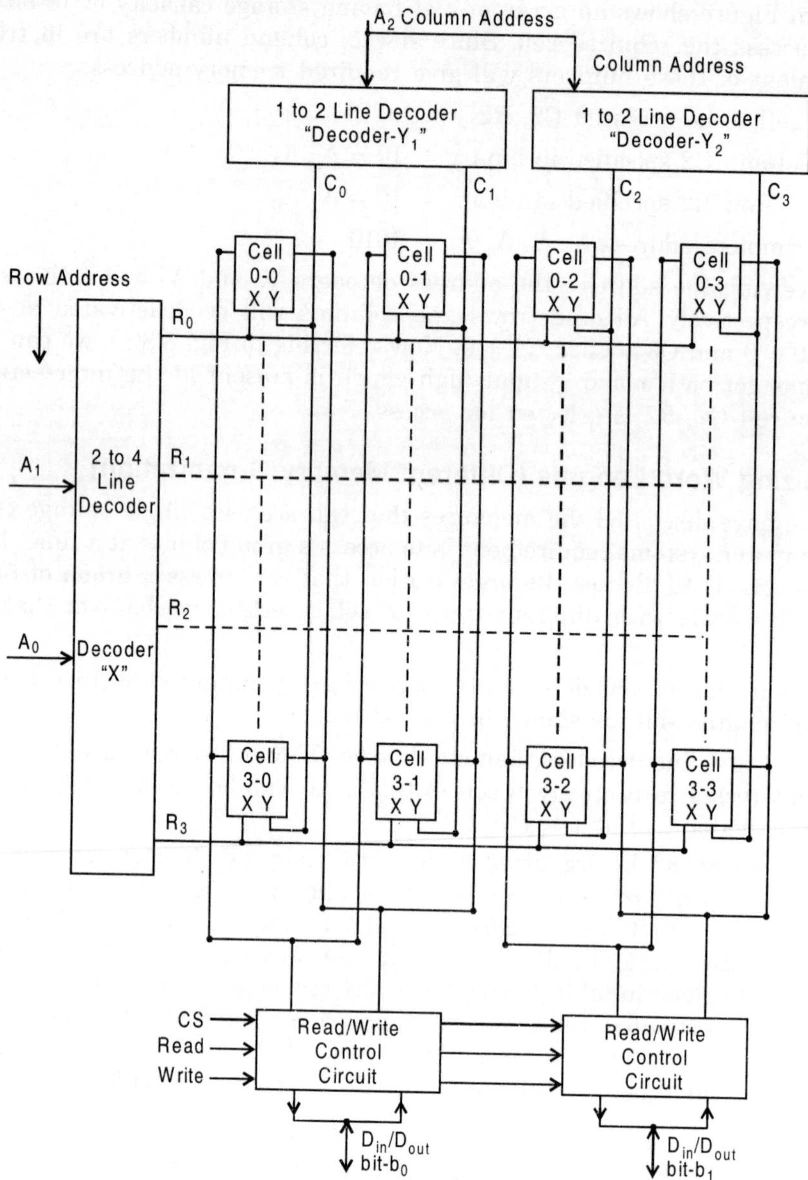

Fig. 8.8 (*b*) 4×4 array organized as 8×2 memory

Table 1 Truth Table for Fig. 8.8(b)

Column Address	*Selected*	
"A_2"	*Columns*	
0	C_0	C_2
1	C_1	C_3

Notice the reduction in number of address lines in each case. In Fig. 8.8 (*a*) 4-address lines were used whereas in Fig. 8.8 (*b*) 3-address lines and in Fig. 8.8 (*c*) only 2 address lines were used. It is because by increasing the word length the total numbers of words are reduced. When more than one cell (*i.e.*, group of cells) is accessed at a time than one storage location indicates a group of cells. Recall that a group of bits (or cells) is nothing but a word. It is a simple consequence of eqns. (1) and (2).

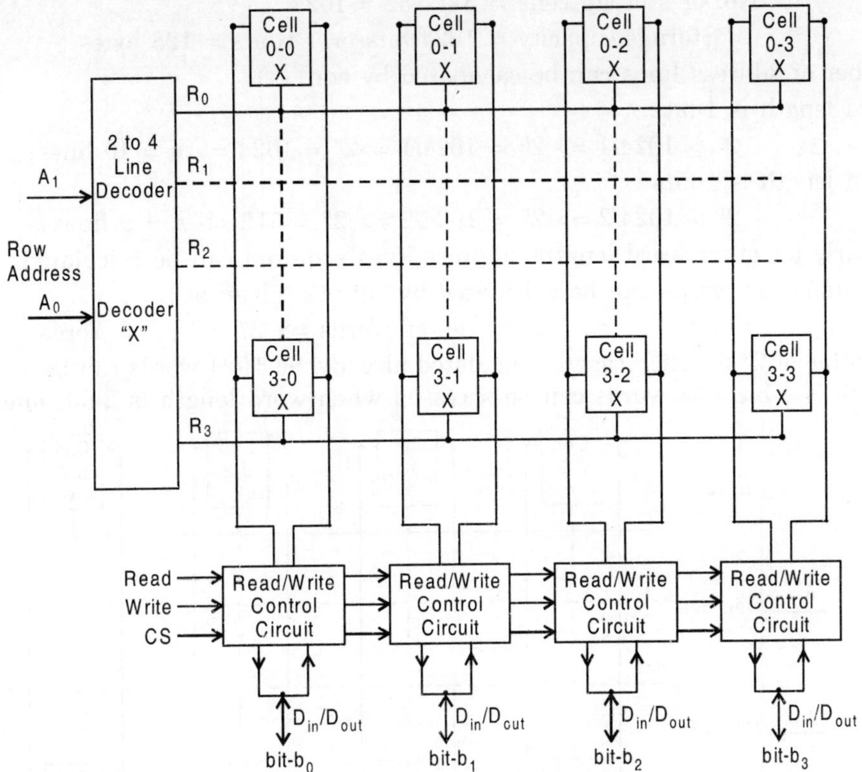

Fig. 8.8 (*c*) 4×4 array organized as 4×4 (4 words of length 4-bits) memory

by eqn. (1) we get

$$\text{No. of words} = \frac{\text{Storage Capacity (in bits)}}{\text{Word length (in bits)}} \qquad ...(3)$$

but by eqn. (2) we get $2^n \geq$ No. of Words

where n = no. of address lines

Putting it into eqn. (3) results

$$2^n \geq \frac{\text{Storage Capacity (in bits)}}{\text{Word length (in bits)}} \qquad ...(4)$$

Equation (4) clearly shows that number of address line 'n' is inversely proportional to word length. Thus, increasing word length will cause a reduction in no. of address lines.

Example. *A memory has cell array of 32 × 32. What would be the required number of address lines if the word length is*

(i) 1-bit (ii) 2-bits (iii) 4-bits (iv) 8-bits (v) 16-bits

Also find out the number of words in each case. What is the storage capacity of this memory in bits and in bytes?

Solution. We know that since a single cell stores 1-bit, total number of cells constitutes the storage capacity of this memory.

Given array is 32 × 32 so,

$$\text{No. of storage cells} = 32 \times 32 = 1024$$

$$\text{Storage capacity} = 1024 \text{ bits or } 1 \text{ Kbits} = 128 \text{ bytes}$$

Number of address lines can be calculated by eqn. (4)

when word length is 1-bit

$$2^n \geq 1024/1 \Rightarrow 2^n = 1024/1 \Rightarrow 2^n = 1024 \Rightarrow n = 10 \text{ lines}$$

when word length is 2-bits

$$2^n \geq 1024/2 \Rightarrow 2^n = 1024/2 \Rightarrow 2^n = 512 \Rightarrow n = 9 \text{ lines}$$

Similarly for other word lengths address lines required can be calculated.

The number of words can be calculated by eqn. (2) given as

$$2^n = \text{No. of Words} \Rightarrow 2^n = \text{No. of Words}$$

The value of 2^n for each case is calculated already. So 1024 words can be accessed when word length is 1-bit, 512 words can be accessed when word length is 2-bit, and so on.

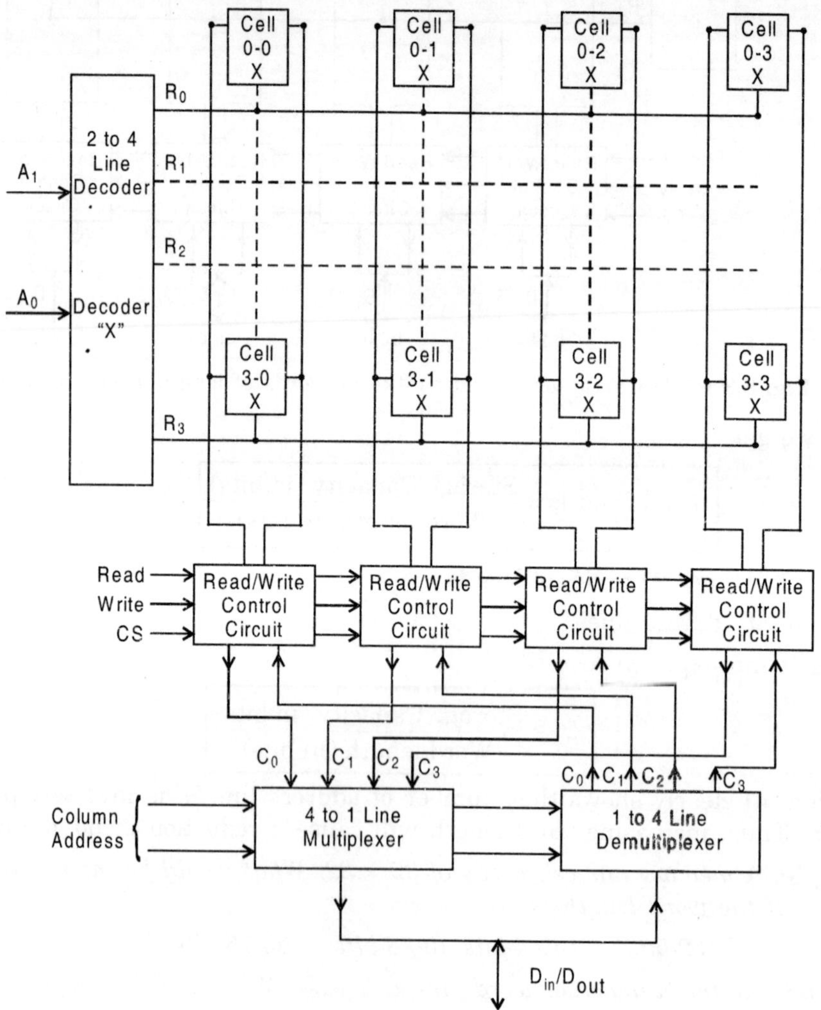

Fig. 8.9 As 16 × 1 memory organization

A slightly different organization of 4 × 4 memory cells as 16 × 1 memory is shown in Fig. 8.8. Compare the Fig. 8.9 with Fig. 8.8 (*a*). Here instead of using a column decoder, one 4 to 1 line MUX and a 1 to 4 line DeMUX is used. Whenever any row is activated all the 4-cells are selected. If read operation is requested, the multiplexer connects one of the selected cells to D_{in}/D_{out} output line. If write operation is requested, then demultiplexer connects its input line (i.e. D_{in}/D_{out} line) to one of its output line which is connected to one of the cells which were selected by activating the row. The select input of both MUX & DeMUX are feeded with the column address lines.

8.4.5 Classification of Semiconductor Memory

The semiconductor memory devices can be categorized in several ways according to their functional and architectural characteristics. Figure 8.10 shows the most general classification. As shown they fall in two general categories.

- Read Write Memory or Random Access Memory (RAM)
- Read Only Memory (ROM)

Read Write Memories or RAMs are those memories, which allows both read & write operation online. They are used in applications where data has to change continuously. They are also used for temporary storage of data. Since, each cell is a flip-flop a power off means loss of data. Thus, the RAMs are *Volatile Storage Devices*.

ROMs are those memory devices, which allows only read operation online and there is no write mode. ROMs are needed in applications where data does not change e.g. monitor programs, mathematical constants etc. Since the data is permanently stored in ROM, power failure does not result in loss of data. Hence, ROMs are *Non Volatile Storage Devices*. In fact, ROM is also a random access memory, but in practice saying RAM refers to read write memory.

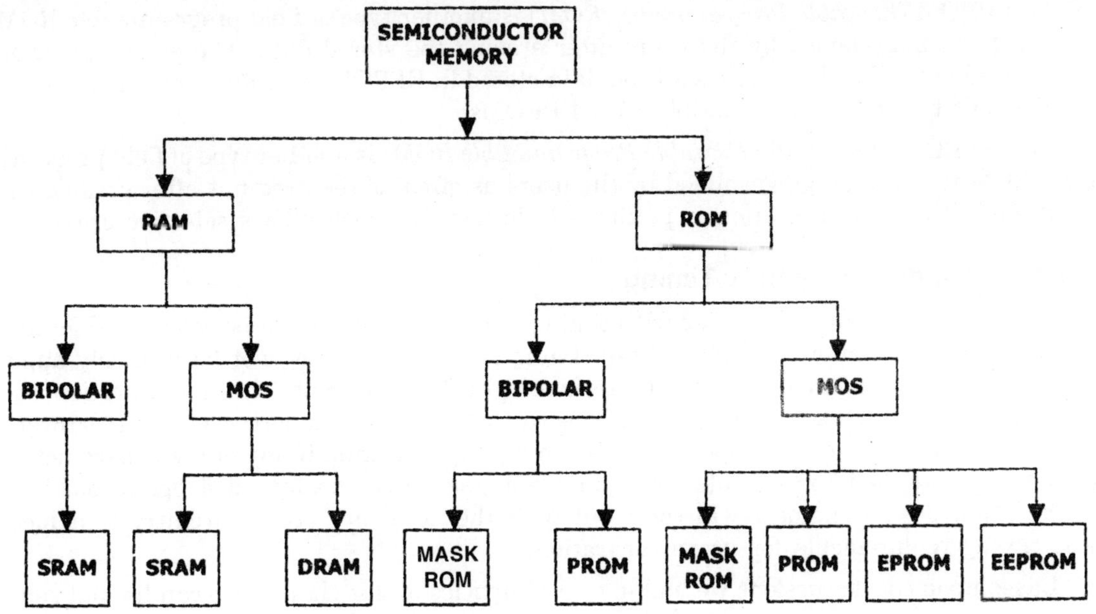

Fig. 8.10 Classification of semiconductor memories

Both the RAM & ROM are further divided into other subcategories, as shown in figure. We will define each of them in this section whereas detailed discussions proceed in the subsequent sections.

Primarily both RAMs & ROMs are classified as *Bipolar* and *MOS* memory depending upon the type of transistors used to construct the individual cell. If the cells are formed by using bipolar (or MOS) transistors the chip is called *bipolar (or MOS) memory chip*. High-speed operation is possible with bipolar chips but their storage capacity is lesser. However, economical MOS/CMOS chips have greater storage capacity, reduced size, and lesser power requirements.

Static RAM (SRAM) is a read write memory that uses two cross-coupled transistor (either bipolar or MOS) working as Flip-Flop to form the basic memory cell. SRAM holds the stored data indefinitely if power is on, hence the name static. SRAMs are very high speed but costly memories. A *Non Volatile RAM* (NVRAM) is formed by using battery backed up SRAM.

Dynamic RAM (DRAM) is a read write memory that uses a capacitor in conjunction with a MOS transistor to store 1-bit. Since the capacitors are leaky, they cannot hold the charges permanently. So, they must be recharged (refreshed) periodically to hold the stored data, hence the name dynamic. They are slower than SRAMs but their storage capacity is greater.

The ROMs are categorized according to the data storing process. The mechanism that stores the data into ROM is called as programming mechanism.

A simple ROM is programmed during the manufacturing process according to the data specified by the user. Such a ROM is referred as *Mask ROM* or *Mask Programmable ROM*.

A PROM (*Programmable ROM*) is a type of ROM that can be programmed by the users in the field. They are also referred as *Field Programmable ROM*. But once programmed they can not be altered.

An EPROM (*Erasable Programmable ROM*) is another type of field programmable ROM that can be reprogrammed by the users after erasing the stored data. The erase operation is performed optically which erases all the data stored in EPROM. Selective erase i.e. erasing just a segment of data is not possible with EPROMs.

An EEPROM (*Electrically Erasable Programmable ROM*) is another type of field programmable ROM that can be programmed by the users as often as required but after erasing the stored data. The erase operation is performed electrically which allows selective erase.

8.4.6 Semiconductor Memory Timing

In this section we are concerned with timing sequences which must be considered for any type of memory operation. Although these timing requirements may not be precisely applicable to all types of memories, but it is useful to grasp the concept. The time duration and sequence in which various signals have to be activated is of great importance. Manufacturer's data sheets of the memory ICs specify these timing requirements. If any of these parameters are not followed then the manufacturer does not guarantee the intended operation. It is expected that design engineer will come out with the necessary access circuitry that meet these timing requirements for proper operation.

Based upon the discussions we had in earlier articles of this chapter we can list out some of the operational requirements.

- A valid memory address must be applied to the address input.

- Before initiating any operation the chip must be enabled through–$\overline{CS}$ signal line.
- The appropriate control signal then is applied for a certain time.

More or less all type of memory operations is followed in above-mentioned sequence. For the sake of easiness we first consider the memory write operation and then read operation. The readers are advised to devote some time to understand these waveforms.

8.4.6.1 Memory Write Operation

Memory Write operation means setting the status of the selected memory cell either to 1 or to 0 according to the data supplied through the data bus. The timing waveforms for a typical memory write operation is shown in Fig. 8.11(a). A close inspection reveals that to write a data to a memory location one must–

- Apply a valid address to the address bus to select a location.
- Initiate chip select signal by making $\overline{CS} = 0$.
- Apply the write control signal by making $\overline{WR} = 0$ for a certain time.
- Apply the data to be written at the selected memory location to the data bus for a certain time.

In response the information present at data bus replaces the word at addressed memory location that might have been stored in it. The waveform also shows the different times that are typical and are explained as below:

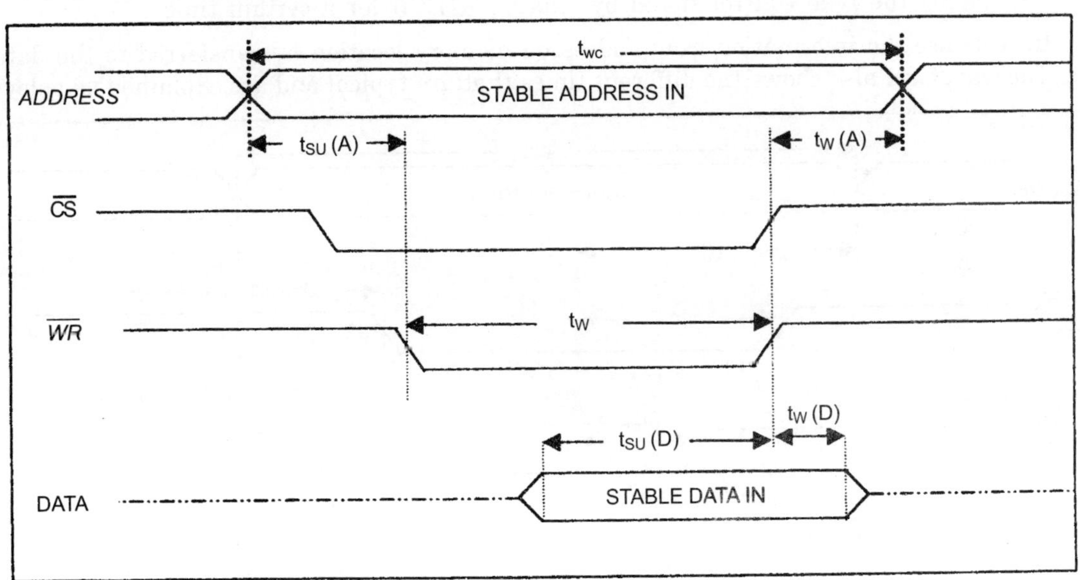

Fig. 8.11 (*a*) Simplified memory write cycle waveforms

Write Cycle Time (t_{WC}): This is defined as the minimum amount of time for which a valid address must be present at the address bus for writing a memory word. As earlier defined this is the minimum time that must lapse between two successive write cycles.

Address Setup Time t_{SU} *(A)*: It is defined as the minimum time for which the address must be stable at the address bus before $\overline{WR}$ goes low.

Address Hold Time t_H (A): This is defined as the minimum time for which the valid address must be stable at the address bus after $\overline{WR}$ rises.

Write Pulse Width (t_W): This is defined as the minimum amount of time for which $\overline{WR}$ must be low in order to store the data in the memory. It must be noted that t_W is measured from the later of write pulse or chip select going low to the write pulse going high. A write occurs during the overlap of low chip select and low write pulse i.e. when both are low. Also the write pulse should be high during all address transitions.

Data Setup Time t_{SU} (D): This is defined as the minimum time for which the data to be written at addressed location must be held stable at the data bus before $\overline{WR}$ rises.

Data Hold Time t_H (D): This is defined as the minimum time for which the data to be written at addressed location must be held stable at the data bus after $\overline{WR}$ rises.

8.4.6.2 Memory Read Operation

A *Read operation*, also called *sensing*, means detecting the status (i.e. information) of selected memory cell that whether it is 1 or 0. The timing waveforms for a typical memory read operation is shown in Fig. 8.11(b). A close inspection reveals that to read a data from a memory location one must

- Apply a valid address to the address bus to select a memory location.
- Initiate chip select signal by making $\overline{CS}$ = 0.
- Apply the read control signal by making $\overline{RD}$ = 0 for a certain time.

In response the information from addressed memory location is transferred to the data bus. The waveform also shows the different times that are typical and are explained as below.

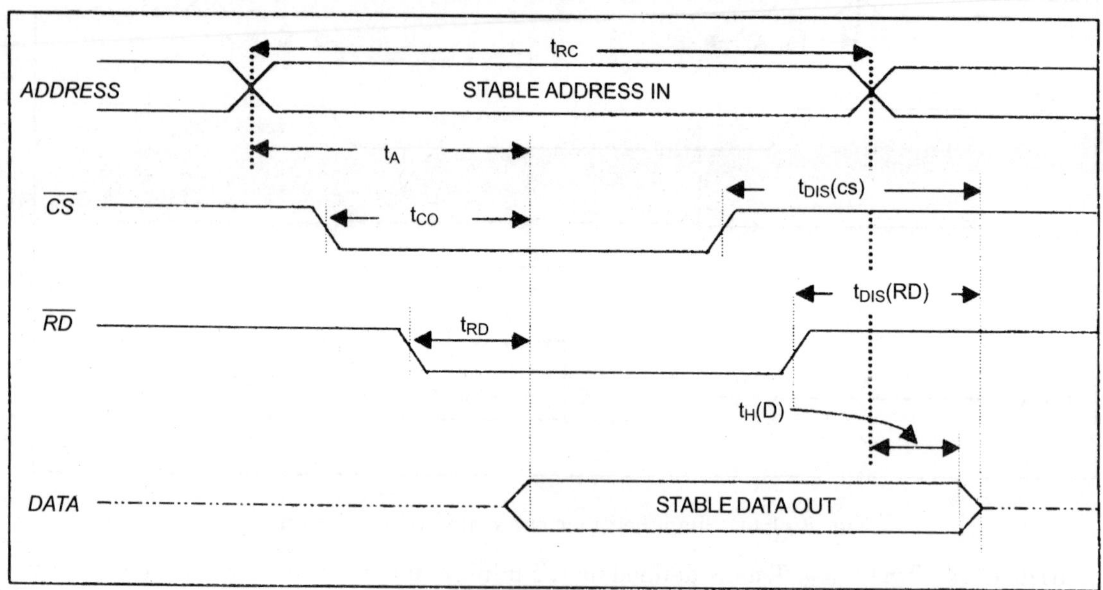

Fig. 8.11 (*b*) Simplified memory read cycle waveforms

Read Cycle Time (t_{RC}): This is defined as the minimum amount of time for which a valid address must be present at the address bus for reading a memory word. As earlier defined this is the minimum time that must lapse between two successive read cycles.

Read Access Time (t_A): Read Access Time or *Access time* is defined as the maximum delay between the appearance of stable address and appearance of stable data. This definition is based upon the assumption that chip select goes low after the application of address. Access time is at the most equal to the read cycle time i.e. $t_A \leq t_{RC}$.

Chip Select to Output Valid time (t_{CO}): This is the maximum time delay between the beginning of chip select signal and appearance of stable data at the data bus. It is some times also defined as the access time from chip select t_A(CS).

Read to Output Delay (t_{RD}): This is the maximum time delay between the beginning of read signal and appearance of stable data at the data bus. It is some times also defined as the access time from read t_A(RD) or read to output valid time.

Chip Select to Output Disable t_{DIS}*(CS)*: This is the maximum time delay after the end of chip select signal for data bus to go to high impedance state i.e. to get disabled. It is also defined as t_{OTD}, output 3-state from deselection.

Read to Output Disable t_{DIS}*(RD)*: This is the maximum time delay after the end of read signal for data bus to go to high impedance state i.e. to get disabled.

$$\text{Generally } t_{DIS}(CS) = t_{DIS}(RD) = t_{OTD}$$

Data Hold Time t_H*(D)*: This is defined as the minimum time for which the stable data is available at the data bus after the end of the valid address. It is also defined as t_V(A), output data valid after change in address or output hold from change in address t_{OHA}.

Example. *For a memory the read & write cycle times are 100 nsec. Find out the bandwidth for this memory. What will be the access rate if access time is 80 nsec?*

Solution. By section 8.2, we know that *Data Transfer Rate* or *Bandwidth* of a memory is given as

$$b_C = 1/t_C \text{ words/sec}$$

so bandwidth for given memory is $b_C = 1/(100 \times 10^{-9})$

$$b_C = 10 \times 10^9 \text{ words/sec.}$$

Again by section 8.2 we know that *Access Rate* of a memory is given as

$$b_A = 1/t_A \quad \text{words/sec}$$

so access rate of given memory is $b_A = 1/(80 \times 10^{-9})$

$$b_A = 12.5 \times 10^6 \text{ words/sec.}$$

or $$b_A = 125 \times 10^5 \text{ words/sec.}$$

8.4.7 Read Only Memory

A read only memory (ROM) is a semiconductor storage device that allows only read operation to be performed on it. It is used to store the information permanently. It is an essential part of computer system that stores *boot up* information required at the time of switch on by the operating systems. It has become an important part of many digital systems because of its high speed, large storage at low cost, and non-volatility. It offers great flexibility in system design and has got many applications including implementation of various complex combinational logic, lookup tables, character displays, embedded systems etc.

ROMs are mainly classified by the method that is used to store the information. The mechanism that stores the data into ROM is called as *programming mechanism*.

Basic organization of a ROM remains same as was discussed from 8.4.1 through 8.4.4 with the only difference that ROMs do not contain write signal. The actual consideration here would be the type of cells that a ROM device must use. Before going for discussion on the typical cells for different kind ROM cells we first consider some simpler ROM organization that can be worked out in labs to demonstrate the idea. The logic diagram of ROM unit is shown in Fig. 8.12.

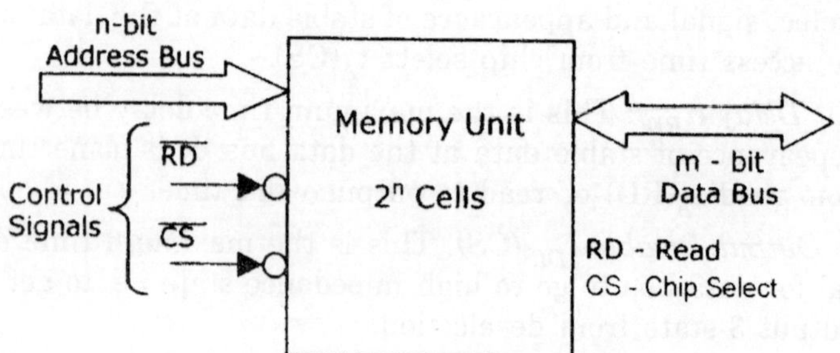

Fig. 8.12 Logic symbol of a ROM memory unit

8.4.7.1 Some Simple ROM Organizations

Diode Matrix ROM: They are the simplest types of ROM. They were discussed previously in chapter 5.

Multiplexer ROM: It is yet another simple type of ROM. In this case the multiplexer inputs are fixed to logic 1 or logic 0 as desired and the select inputs can be used as address lines to select the particular memory word. Figure 8.13 shows such a 4 × 1 ROM with the truth table of stored information.

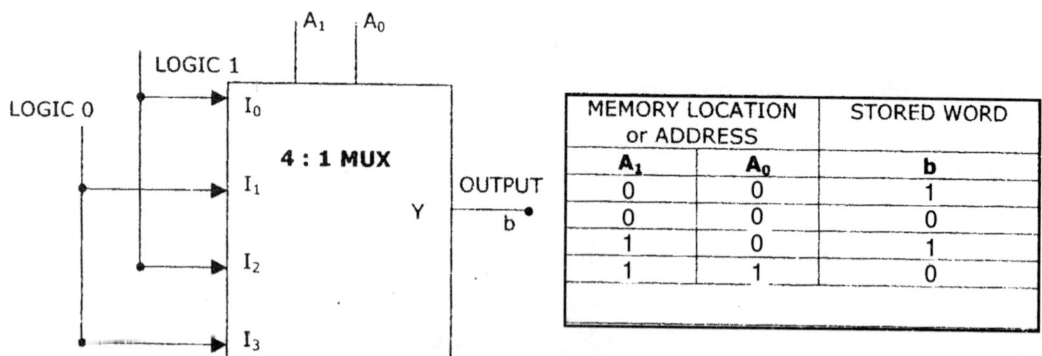

MEMORY LOCATION or ADDRESS		STORED WORD
A_1	A_0	b
0	0	1
0	0	0
1	0	1
1	1	0

Fig. 8.13 (*a*) A 4 to 1 line multiplexer connected as 4 × 1 ROM

Fig. 8.13 (*b*) Truth table for ROM in Fig. 8.13(*a*)

It is interesting to note that forming a multiplexer tree with such fixed data connections can make a ROM of larger word length thus increasing the storage capacity. Such a ROM is shown in Fig. 8.14. Figure 8.14 shows a 4 × 2 ROM with a storage capacity of 8-bits as opposite to the ROM of Fig. 8.13 which is a 4-bit ROM of word length of 1-bit. Note that the address lines are shorted together as usual to address the bits of same location from two MUX. Readers are encouraged to find out the words stored at different locations in this ROM.

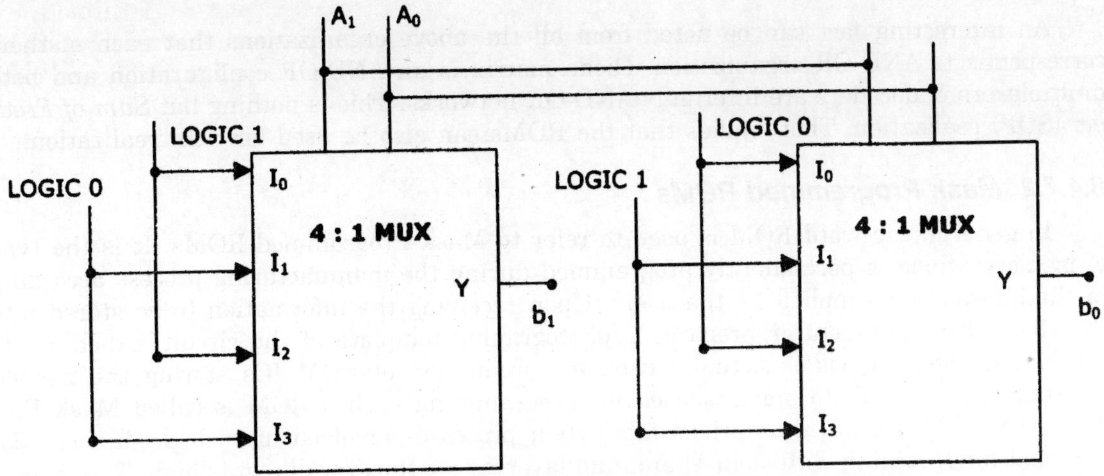

Fig. 8.14 Two 4 to 1 line multiplexer connected as 4 × 2 ROM

OR-Gate ROM: It is yet another simpler ROM but more versatile than the multiplexer ROM. In this case a decoder and n-input OR gates are used. The selection of decoder depends upon the number of words needed to store. The word length determines the number of OR-gates. The select inputs are used as address lines to select the particular memory word. Figure 8.15 shows such a 4 × 4 ROM.

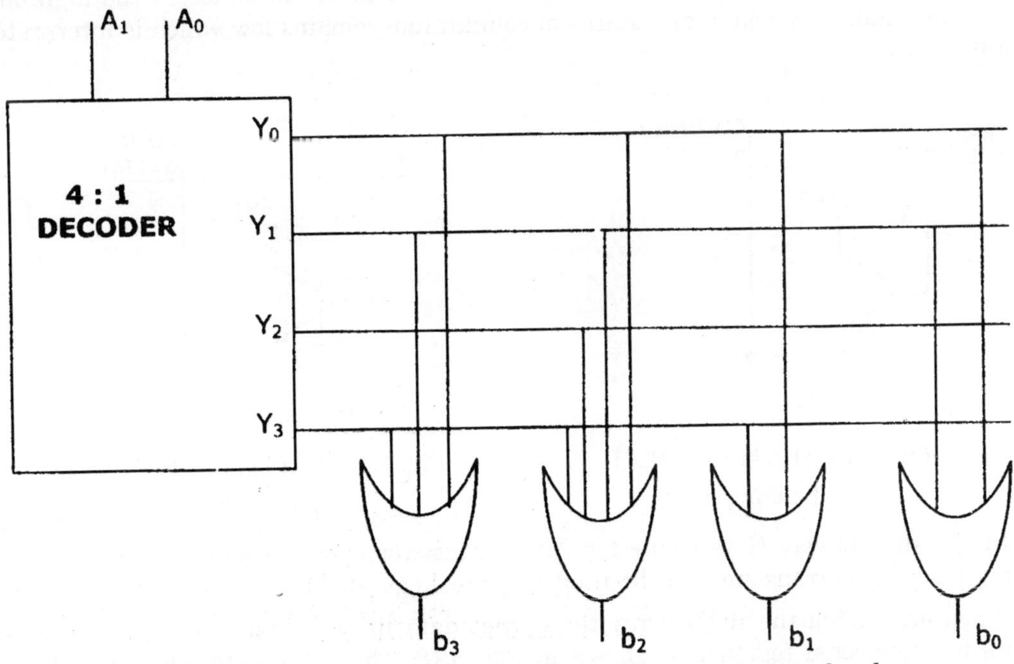

Fig. 8.15 16-bit, 4 × 4 ROM using OR-gates and 4 to 1 line decoder

Assuming that upon activation each of the decoder line becomes high we get that location 10 stores the word 0100. The readers are advised to find out the words stored at other locations.

An interesting fact can be noted from all the above organizations that each of them corresponds to AND-OR organization. Diode matrix is an AND-OR configuration and both multiplexers & decoders are internally AND-OR networks. This is nothing but *Sum of Product* (SOP) realization. This implies that the ROMs can also be used for SOP realization!!

8.4.7.2 Mask Programmed ROMs

In general, the word ROM is used to refer to Mask Programmed ROMs. It is the type of memory, which is permanently programmed during the manufacturing process according to the information supplied by the users. Upon receiving the information to be stored into the ROM the manufacturer prepares a photographic template of the circuit, called *Photo Mask*. This photo mask is actually used for production of ROM ICs storing the needed information. Since photo mask is used for programming such a ROM is called Mask Programmed ROM. Due to the various fabrication processes involved in such production like photomasking, etching, diffusion, Aluminium layering etc the overall cost is high. This makes them unsuitable when only a small number of ROMs are needed. For this reason, mask programmed ROMs are always used for high volume production.

Both the BJTs and MOS devices can be used as basic memory cell for this kind of ROM. Figure 8.16 shows the Bi-polar transistors connected to store logic 1 and logic 0. As shown the base connected transistor can be used to store logic 1. If row line is connected to the base of transistor then whenever ROW line goes high transistor turns ON and $+V_{CC}$ is connected to column line which is interpreted as Logic 1. But if base is unconnected the high on row line does not make it working and status of column line remains low which is interpreted as Logic 0.

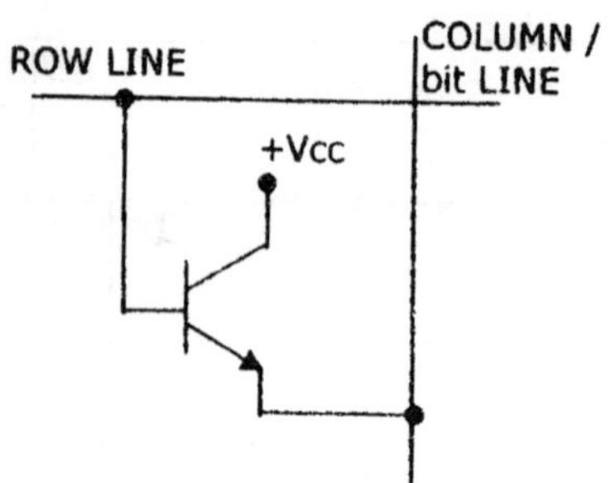

Fig. 8.16 (*a*) A bi-polar ROM cell storing '1'

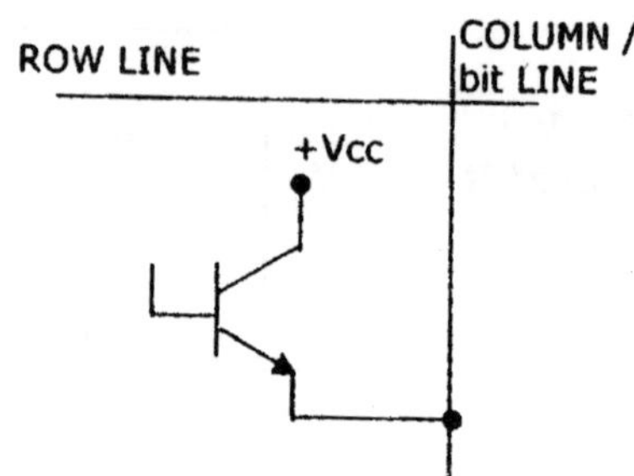

Fig. 8.16 (*b*) A bi-polar ROM cell storing '0'

In the similar way Gate connected MOS transistor can be used to store Logic 1 and floating Gate MOS transistor can be used to store Logic 0. This is shown in Figure 8.17.

When organizing the ROM array the output data lines of individual cells in the same column are connected together as shown in Fig. 8.18. This is done by assuming that such connections will form *Wired-ORing* and such assumption is valid for most of the semiconductor technology.

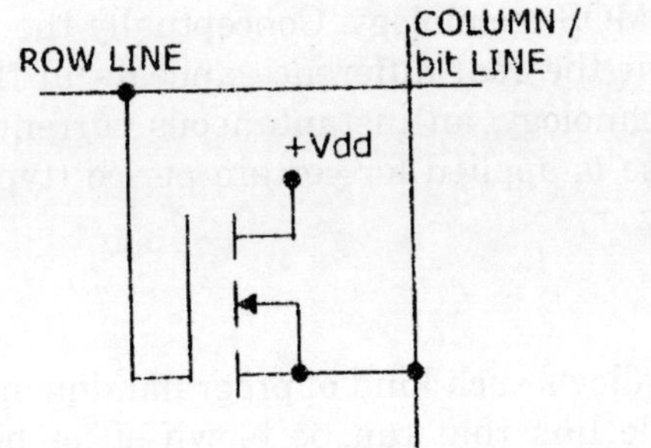

Fig. 8.17 (a) A MOS ROM cell storing '1'

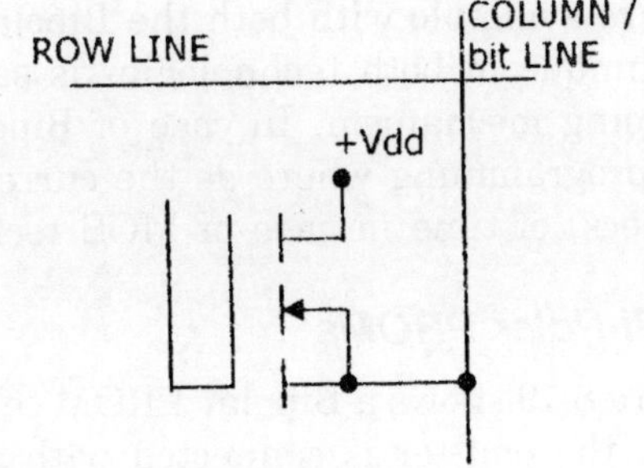

Fig. 8.17 (b) A MOS ROM cell storing '0'

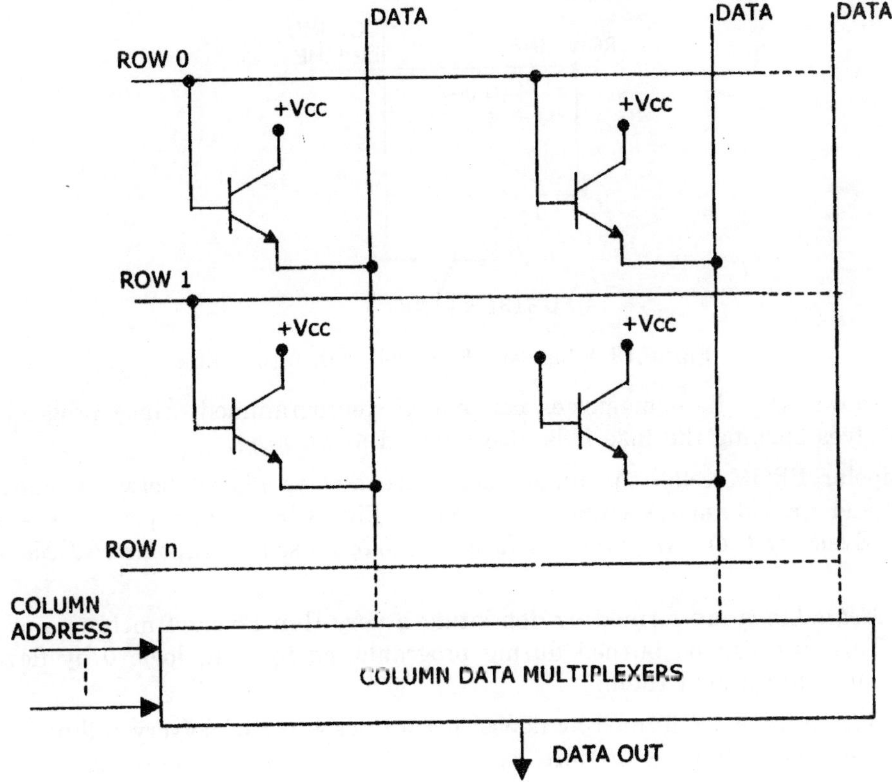

Fig. 8.18 An array of bipolar ROM cells

Finally column multiplexers may be used to select the data from a particular column only. This organization is identical to the one, shown in Fig. 8.8.

8.4.8 Programmable Read Only Memory (PROM)

The mask programmed ROMs have disadvantage that they are not flexible and with low volume production they become highly expensive. A Programmable ROM is a type of ROM which efforts to minimize these two disadvantages. These ROMs can be programmed by the users in the field, thus they are also called field programmable ROMs. These ROMs provide some flexibility to user but their disadvantage is that they can be programmed only once. PROMs are available with both the Bipolar and MOS technology. Conceptually the programming technique in both technologies is same, but the real difference appears in the actual programming mechanism. In case of Bipolar technology, an instantaneous current pulse is used for programming where as the current pulse is applied for certain period (typically for a few msecs) of time in case of MOS technology.

8.4.8.1 Bi-Polar PROMs

Figure 8.19 shows a Bipolar PROM cell that allows such kind of programming in PROMs. As shown the emitter is connected with a fusible link that can be blown off or burned by applying a high current to it. When the fuse remains intact it stores logic 1 and when it is burnt it stores logic 0. Initially all the cells have fuse intact i.e. they all stores logic 1 and at the of time of programming users can selectively burn them out to store logic 0.

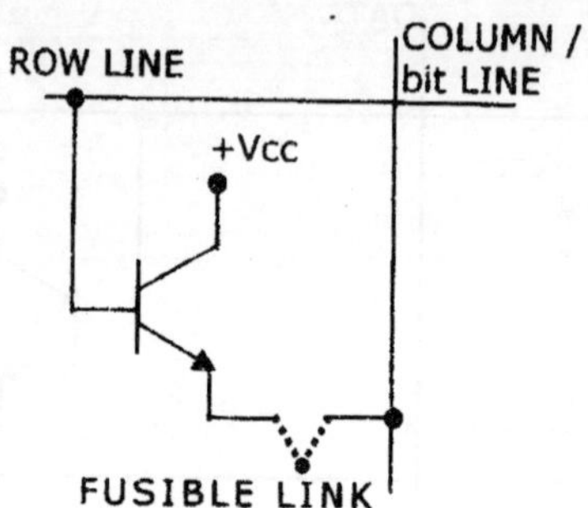

Fig. 8.19 A bipolar PROM cell with fusible link

This shows why these memories can not be reprogrammed. Since programming the PROM involves burning the fuse it is also referred as *burning.*

In Bipolar PROM cells, the fusible links are always placed between emitter of the transistor and the column/data line, as shown in Fig. 8.18. Such types of cells are called fuse cells. Basically there are types of fuse technology used in Bipolar PROMs as briefed below.

- *Metal Links* are formed by depositing a very thin layer of nichrome as material. This fuse can be burned during programming to store logic 0 by flowing large currents through them.
- *Silicon links* are formed by depositing a thick layer of polycrystalline silicon. This fuse can be burned to store logic 0 by flowing current pulse train of successive wider pulses. This programming process causes a temperature upto 1400°C at the fuse location, and oxidizing the silicon. This oxidization forms insulation around the opened link.
- *Avalanche Induced Migration (AIM) Process* this programming technique uses two PN junction diodes connected back to back as shown in Fig. 8.20. During the programming the diode D_1 is in reversed bias. There flows a large current in reverse direction due to avalanche of electrons. The heavy reverse current along with heat generated at the junction causes some aluminium ions to migrate. Due to this the base to emitter junction is shorted. Although this process requires higher currents and voltages than the earlier two, but this is faster.

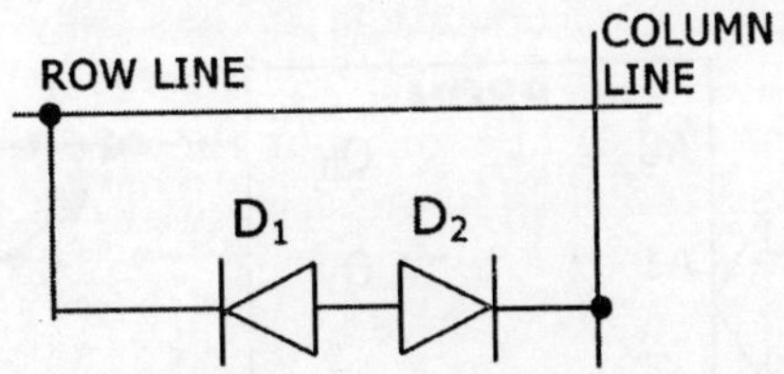

Fig. 8.20 A shorted junction cell

8.4.8.2 MOS PROMs

The fuse mechanism shown for Bipolar PROMs does not works for MOS technology where high current and resistance levels required for programming is incompatible with MOS impedance levels. Commonly used MOS fabrication techniques are

- Floating Gate Avalanche Injection MOS (FAMOS)
- MAOS

In FAMOS PROMs silicon gate MOSFET with no electrical connection to the gate is used as storage device. In this case gate is floating in an insulating layer of silicon dioxide. The operation of such device is based on charge transport on application of high voltage to the floating gate. This charge gets trapped in the gate when voltage is removed, as there is no electrical connection for gate. Logic symbol of this device is shown in Fig. 8.21.

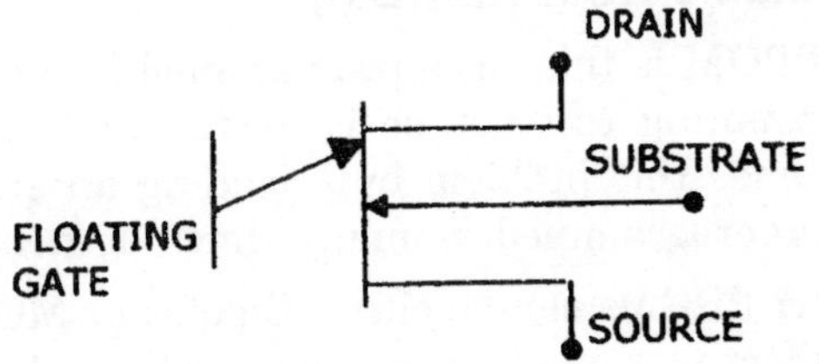

Fig. 8.21 Logic symbol of FAMOS device

It is interesting to note that the charged trapped in floating gate of FAMOS device can be removed if the cell is exposed to high energy UV radiation for some times.

A MOS memory cell that uses *Alumina* (Al_2O_3) and silicon nitride as gate dielectric for charge storage is called MAOS memory element. Applying negative and positive polarity voltages can program the MAOS elements. The MAOS elements can be erased by applying opposite polarity voltage at gate and thus provides reprogramming.

Note that the MOS techniques provide reprogramming feature as opposite to Bipolar technology.

8.4.8.3 PROM Programming

As explained above the programming of PROM requires varying amount of currents to be supplied to store the information. This is done through a special device called PROM Programmer that mainly contains programmable pulse generator to meet requirements of different devices. A simplified programming setup is shown in Fig. 8.22.

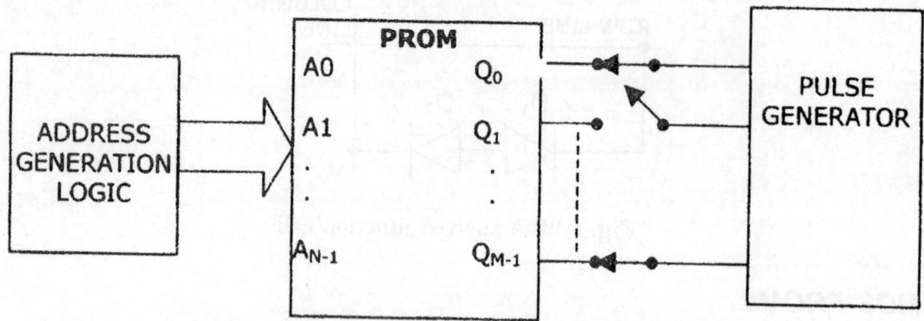

Fig. 8.22 Simplified PROM programming setup

As earlier discussed initially PROMs will contain all 1's or all 0's. Let initially all 1's are stored then to store information first of all find out the bit positions where 0s have to be stored. Connect the corresponding output bit line i.e. Q_i to the pulse generator. The address generation unit selects the location of the word to be programmed. After the selection of word pulse generator supply the appropriate amplitude pulse to burn out the fuse, thus storing 0 at connected bit positions. The process is repeated for all memory words and the PROM can be programmed completely. Although not shown in figure all the power supply and other connections are assumed to be present for PROM before programming.

8.4.9 Erasable Programmable ROM (EPROM)

One disadvantage of PROM is that once programmed it can not be reprogrammed. Any mistake during the programming can not be corrected and the whole chip may become useless. An EPROM overcomes this problem by providing an erase operation. EPROM is a PROM device that can be reprogrammed as many times as needed.

Note that both ROMs & PROMs can be either Bipolar or MOS but EPROMs can only be MOS. An indication of this fact was given in previous section when discussing the FAMOS & MAOS devices. In its simplest form an EPROM can be viewed as a FAMOS PROM with an additional gate, called control gate. Figure 8.23 shows the basic structure of such a device and its logic symbol. The programming mechanism is same as previous and to erase UV radiation is used.

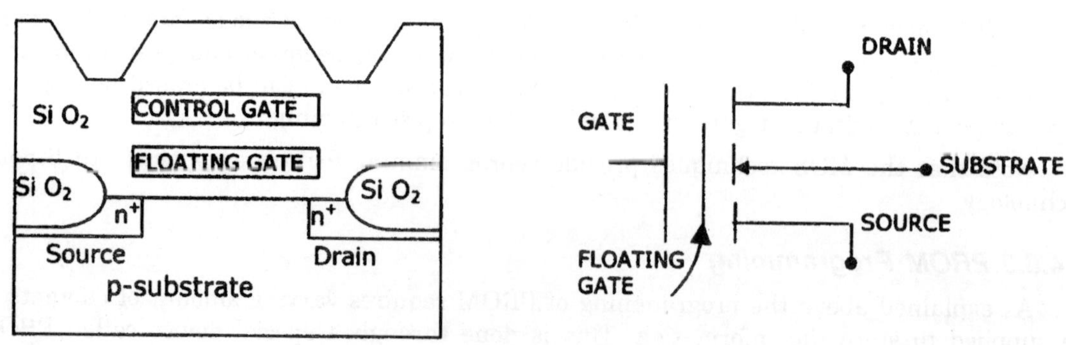

Fig. 8.23 A structure of EPROM cell

Shown in the figure is an n-channel enhancement mode MOSFET with insulated gate architecture. The additional gate, called control gate, is formed within the SiO_2 layer. The control gate is connected to the row address line. As usual presence or absence of stored charge represents the data.

8.4.9.1 EPROM Programming

To program the memory cell the floating gate must be charged i.e. trap the charge. To accomplish this a high positive voltage is applied between source & drain. A slightly higher +ve voltage is applied at the control gate. Thus high electric field strength is developed which highly energizes the electrons. These energetic electrons thus reach to floating gate and charge is accumulated. As the more electrons (i.e. negative charges) are accumulated the field strength reduces and further electron accumulation is inhibited. Since an insulating layer (SiO_2) surrounds this gate, no discharge path is available for accumulated electrons. Thus charge carriers remain trapped at gate. Thus, the programming requires charging the floating gates of individual memory cells. Programming of the EPROMs requires higher current levels for different times. For this purpose a special device, called EPROM Programmer, is used. All the EPROM chips must be physically removed from the memory boards and be connected to the EPROM programmer for reprogramming or programming. This is considered as the disadvantages of the EPROM.

To reprogram all cells of EPROMs must be erased. Illuminating EPROM to strong UV radiations carries out the erase operation. For this reason they are some times called as *UV erasable PROMs* or simply *UV EPROMs*. It typically requires a radiation of wavelength of about 254 nm to be present for about 20 minute. The photons of the incident radiation provide sufficient energy to the trapped electrons to escape to substrate through the SiO_2 layer.

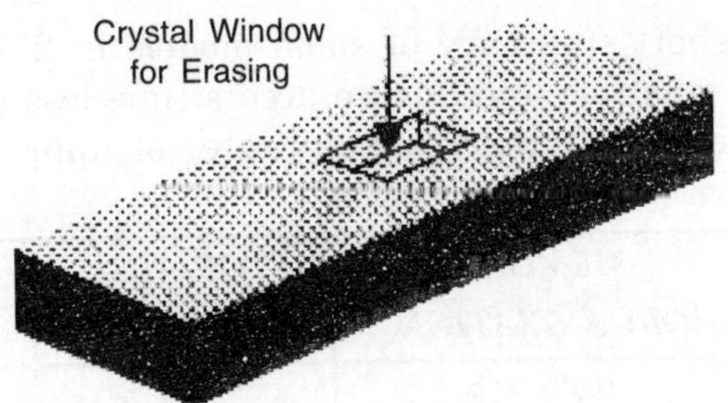

Fig. 8.24 Crystal window to erase data from EPROM

The EPROMs are provided with a transparent crystal window on the top of the IC, as shown in Fig. 8.24 to allow UV rays to enter to the chip to erase the data. Upon illumination contents of all the locations are erased, which is considered as the biggest disadvantage of EPROMs. At the time of working the EPROM window is covered with the opaque sticker to prevent it from unwanted exposure from sunlight or from other sources.

The EPROMs can be erased by direct sunlight in about one week or by room level fluorescent lighting in about one to three years.

8.4.9.2 The 27XXX EPROM Series

A popular *EPROM family* is 27XXX series and commercially available from many manufacturers. In the IC number XXX represents the storage capacity in Kbits. For example, in IC 2732A XXX = 32 means it can stores a total of 32 Kbits. However, all ICs of this series are organized to have a word length of a byte i.e. 8-bits. Hence, 2732A is a XXX/8 = 32 Kbits/ 8 = 4Kbyte or 4KB memory device. Since, length of each memory word is byte this IC contain 4 K (4 × 1024 = 4096) memory location. Thus, by equation 4 get that IC 2732A must have 12 address lines (as $2^{12} = 4096$) to select any of the 4096 words. In the similar way, we can work out for all the members of this family to find out the storage capacity and the width of address bus. Data bus width will always be same as 8-bit. Pin diagram of IC 2732A is shown in Fig. 8.24 (*a*). The signal is $\overline{OE}/V_{PP}$ a dual purpose signal. Under the normal operation a logic 0 on this pin allows the data from selected location to appear at data output lines. The 2732A is in programming mode when it is connected to 21V. When 2732 is erased all the memory cells are returned to logic 1 status. The programming is done by applying the required input data on the $O_0 - O_7$ pins.

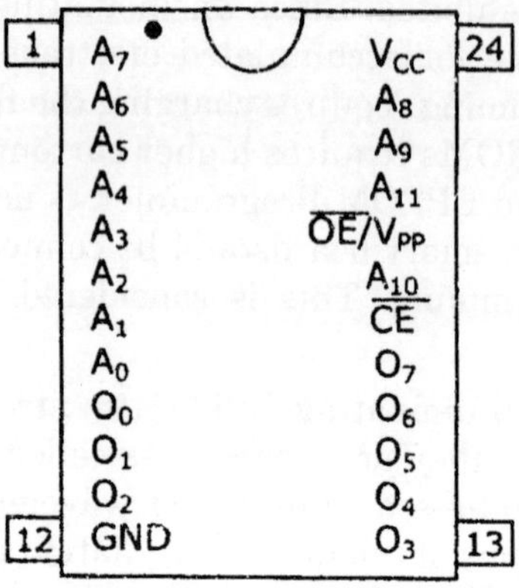

Pin			Pin
1	A_7	V_{CC}	24
	A_6	A_8	
	A_5	A_9	
	A_4	A_{11}	
	A_3	$\overline{OE}/V_{PP}$	
	A_2	A_{10}	
	A_1	$\overline{CE}$	
	A_0	O_7	
	O_0	O_6	
	O_1	O_5	
	O_2	O_4	
12	GND	O_3	13

A_0-A_{11}	Address Lines
$\overline{CE}$	Chip Enable
$\overline{OE}$ / V_{PP}	Out Enable/Programming
O_0-O_7	Data Output Lines

Fig. 8.24 (*a*) INTEL 2732A EPROM IC, 4K × 8 IC

Figure 8.25 list out a short summary of some models in 27XXX series. Also note that these chips are fully TTL compatible and has an access time less than 450 nsec that permits an access rate (most of the time Data Transfer Rate) of approximately 2 × 10^9 memory words/sec or 6 × 10^6 bits/sec.

EEPROM	*MEMORY*	*STORAGE*	*CAPACITY*
27XXX	*ORGANIZATION*	*KBYTES*	*BITS*
2708	1024 × 8	1 KB	8192
2716	2048 × 8	2 KB	32768
2732	4096 × 8	4 KB	32798
2764	8192 × 8	8 KB	95539
27128	16384 × 8	16 KB	131072
27256	32768 × 8	32 KB	262144
27512	65536 × 8	64 KB	524288

Fig. 8.25 27XXX series EPROMs

8.4.10 Electrically Erasable Programmable ROM (EEPROM)

One disadvantage of EPROM is that erase operation erase all the memory cells completely and reprogramming involves entering the complete data. There is no option of selective erase.

Another disadvantage is that for reprogramming the chip must be removed physically from the main boards. An EEPROM, also called as E^2PROM, overcomes these problems which allows selective erase operation and there is no need to remove the chip as it requires different voltage levels which can be provided on board. In EEPROMs the erase operation is performed electrically as opposite to EPROMs. For this reason, they are also called as *Electrically Altered PROM* i.e. the EAPROM. In these memories both the erase and programming can be done electrically by applying the controlled electrical pulses to the IC. Thus, changes can be made to selected memory locations only. The E^2PROMs can be programmed one byte at a time.

Recently a newer E^2PROM called *flash EEPROM* is introduced by Intel Corporation that has very simpler cells and thus have greater packing density. The *flash EEPROMs* devices can be programmed faster as compared to normal EEPROM and can be erased sector-by-sector. While the EEPROMs can be erased and programmed for parts of code, the entire *flash EEPROMs* memory must be erased & reprogrammed completely. The future seems to opt for *flash EEPROMs* because of their simpler cells, high speed, and high density which means lower cost per bit.

8.4.11 The Random Access Memory (RAM)

The Random Access Memory refers to the read write memory i.e. a memory that allows both reading and writing data online. In a computer system, these devices are used for temporary or scratchpad storage. Basic organization of a RAM remains same as was discussed from 8.4.1 through 8.4.4 with no changes.

RAMs are also available with both the Bipolar and MOS technology. Conceptually the operation in both technologies is same, but the real difference appears in the way they store the binary information in the memory cell. As shown in subsection 8.4.5 RAMs are classified as *Static* and *Dynamic*. The static RAM can be either Bipolar or MOS but dynamic RAM can only be MOS. Both types of RAMs are discussed in subsequent subsections.

8.4.12 Static Random Access Memory (SRAM)

Random Access Memories that consist of circuits that can retain the information as long as power is applied are referred as *Static Random Access Memories*. Flip-flop forms the basic memory element in a SRAM device. A static RAM contains an array of flip-flops, one for each bit. Logic diagram of SRAM cell is shown in Fig. 8.26. The figure shows an S-R latch with required signals. As shown the cell can only be selected by placing logic 1 to both the ROW & COL lines. Data can be entered to this cell only when WR=1 and RD=0 when both row and col are activated. At the time of writing the data the cell output is disabled and is available only when RD=1 and WR=0.

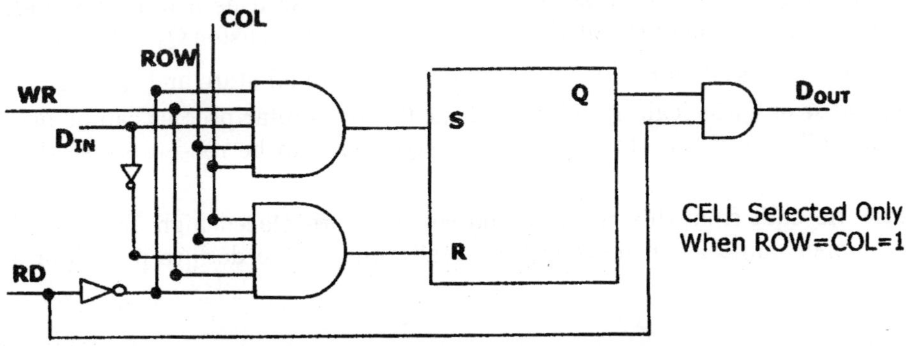

Fig. 8.26 Logic diagram of basic SRAM cell

Since SRAM contains array of flip-flops, a large number of flip-flops are needed to provide higher capacity memory. For this reason simpler flip-flop circuits using BJTs and MOS transistors are used for SRAM. This helps to save chip area and provides memory IC at relatively reduced cost, increased speed and reduces the powers dissipation as well. The static RAMs have very small access times typically less than 10 nsec. SRAM with battery backup is commonly used provide non-volatile RAM (NVRAM).

8.4.12.1 The Bi-Polar SRAM Cell

For the reasons stated above we use simpler flip-flop circuits to implement flip-flops in memory. The circuit shown in Fig. 8.27 (*a*) forms basic bipolar latch that can be used to store 1-bit of binary information. Note that it contains two cross-connected collector coupled transistors Q_1 and Q_2. At any time one of the transistor remains on and other remains off i.e. status of their collector are always complement to each other thus satisfying the basic definition of flip-flop. That's why the data lines are labeled as b and $\overline{b}$.

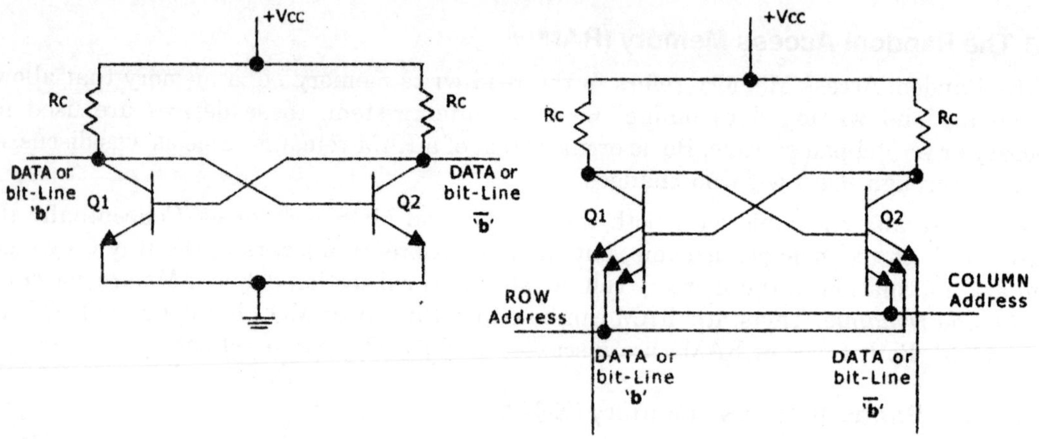

Fig. 8.27 (*a*) Basic bipolar latch (or storage cell) **Fig. 8.27** (*b*) Basic bipolar SRAM cell

Figure 8.27(*b*) shows the same latch but along with the necessary control and the data signals. Use of multi-emitter transistors accommodates the necessary signals of BJT SRAM cell. To read or write to this cell both the row and col lines must be high.

To ***write*** the two bit lines are placed with required status and row and col lines are placed high. This causes the two lower emitters of both the transistors to stop current conduction. If a logic 1 was to be store then bit line b placed as $b = 1$, this makes upper emitter of Q_1 to cease to conduct but the upper emitter of transistor Q_2 will remain conducting as its bit line is placed as $\overline{b} = 0$. This makes Q_1 to go to off state and Q_2 to go to on state, thus giving status of Q_1 as logic 1 and Q_2 as logic 0. The similar process can be used to store logic 0 to this cell. Every time the state of transistor Q_1 can be read to sense the status of this cell.

To ***read*** the data from this cell row and col lines are placed high and current flowing through the upper emitter can be sensed by read amplifier to declare it as 0 or 1.

8.4.12.2 The MOS SRAM Cell

A CMOS realization of SRAM cell is shown in Fig. 8.28. The cell comprises two cross-connected inverters, Q_1 and Q_2 as inverter 1 and Q_3 and Q_4 as inverter 2, thus forming a latch capable of storing 1-bit of information. When the cell stores logic 1 the voltage at point A is maintained high and that at B is maintained low by having Q_1 and Q_4 ON and Q_2 and Q_3 OFF. Thus if Q_5 and Q_6 are turned ON by activating the row address line, the bit lines will have their status respective to voltage at A and B. Finally upon activation of column address lines Q_7 and Q_8 turn ON and pass this information to read/sense amplifier. Note that the bit line is connected to all the cells of a column by forming wire-OR connection. The write operation for this cell is same as BJT cell in that bit lines are first fixed to required status and then the cell is forced to enter in that status.

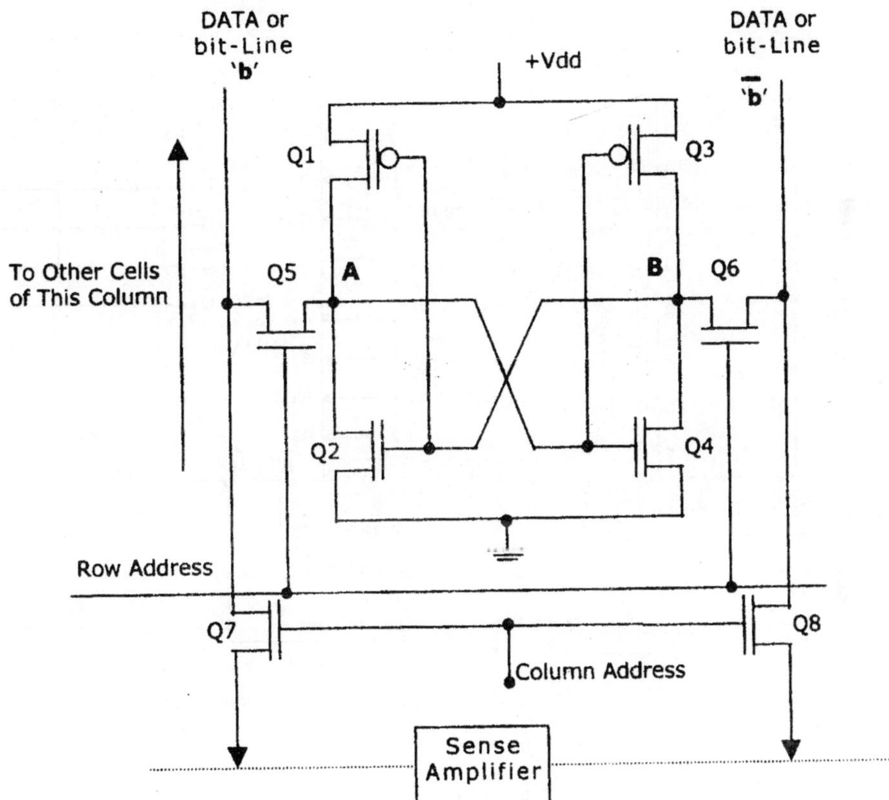

Fig. 8.28 CMOS SRAM storage cell

The power supply V_{dd} is 5 volts in standard CMOS SRAMs and 3.3 volts in low-voltage versions. The biggest advantage of using CMOS cell is very low power consumption because the current can flow in the cell only when they are accessed. Otherwise one of the transistor from each inverter remains OFF ensuring that there is no active path between the supply voltage V_{dd} and ground. Although CMOS is slower than the BJT but latest advances have reduced the difference in the Access time of the two types of SRAM. Also since CMOS devices have very large packing density, almost all the SRAM devices with capacity more than 1024 bits are MOS devices.

8.4.12.3 SRAM ICs

- ***IC 7489-94 bit TTL SRAM:*** Figure 8.29 shows the pin diagram and logic diagram along with truth table & signal description. This is a 94-bit SRAM organized as 91 × 4 memory. The data sheet for this memory device specify a maximum read data delay or access time of 50 nsec. Note that the device contains separate data busses for input and output. The output of this IC always the complement of bits stored in the memory. Moreover, this IC package provides open collector output thus it requires external pull up resistors to be used as shown in functional diagram. Note the third entry of the truth table with the mode of operation as inhibit storage. In this mode the device simply works as inverter in that the data present at the input complemented and presented at the output. In this mode neither the data can be stored nor the data read whether address is applied or not.

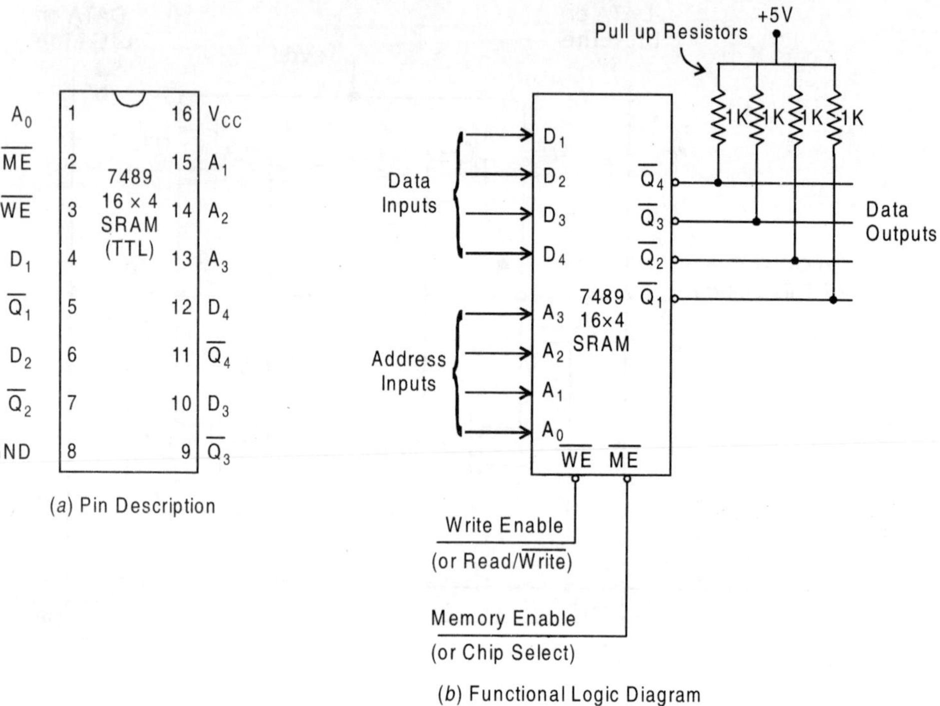

(*a*) Pin Description

(*b*) Functional Logic Diagram

Signal	Description
$\overline{ME}$	Selects or Enables the chip when $\overline{ME} = 0$
$\overline{WE}$	Selects write operation when $\overline{WE} = 0$ Selects read operation when $\overline{WE} = 1$
A_0 to A_3	Address lines to select any of the 16-words
D_1 to D_4	4-bit Data Input Bus
$\overline{Q}_1$ to $\overline{Q}_4$	4-bit Data Output Bus-Data output is complement of bits stored.

(*c*) Signal Description

Operating Mode	Inputs $\overline{ME}$	Inputs $\overline{WE}$	Output Condition
Write	0	0	Complement of Data Inputs
Read	0	1	Complement of Selected word
Inhibit Storage	1	0	Complement of Data Input. No read or write is done
Do Nothing	1	1	All Outputs High

(*d*) Truth Table

Fig. 8.29 64 bit (16×4) TTL SRAM IC

- **TMS 4016-2KB MOS SRAM:** Figure 8.30 shows the pin diagram of this MOS SRAM device. This device is representative of all other 2K × 8 SRAM MOS ICs such as Intel 2016, and most popular in this range is the 696 IC. The figure shows the signal description also, with the help of which truth table for this IC can be worked out.

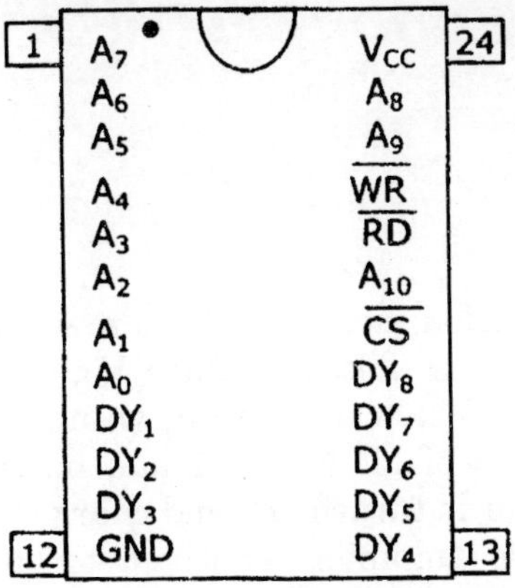

Pin	Signal	Signal	Pin
1	A_7	V_{CC}	24
	A_6	A_8	
	A_5	A_9	
	A_4	$\overline{WR}$	
	A_3	$\overline{RD}$	
	A_2	A_{10}	
	A_1	$\overline{CS}$	
	A_0	DY_8	
	DY_1	DY_7	
	DY_2	DY_6	
	DY_3	DY_5	
12	GND	DY_4	13

A_0-A_{10}	Address Lines
$\overline{CS}$	Chip Select when set to 0
$\overline{WR}$	Writes the word when set to 0
$\overline{RD}$	Reads the word when set to 0
DY_0-DY_7	Data input/output Lines i.e. bi-directional Data Bus

Fig. 8.30 TMS 4016 MOS SRAM IC, A 2K × 8 IC

An interesting point to note that many of the EPROM ICs and SRAM ICs (such as 27XXX & 62XXX family) are pin compatible. For example compare the Fig. 8.24 to the Fig. 8.30. One can find that few signal names are changed and WR signal is replaced by V_{PP}.

- Popular SRAM ICs are from 62XXX MOS family like IC-6264 a 8K × 8 memory, IC-62256 a 32K × 8 memory etc. Readers are encouraged to search & list out several other SRAM ICs from various references.

8.4.13 Dynamic Random Access Memory (DRAM)

SRAMs are faster but they come at high cost, as their cell requires several transistors. Less expensive RAMs can be obtained if simpler cells are used. A MOS storage cell based on dynamic charge storage is much simpler and can be used to replace the SRAM cells. Such a storage cell can not retain the charge (i.e. information) indefinitely and must be recharged again, hence these cells are called as dynamic storage cells. RAMs using these cells are called Dynamic RAMs or simply DRAMs.

8.4.13.1 Basic DRAM Cell

A basic DRAM cell is shown in Fig. 8.31, in which information is stored in the form of charge stored on a capacitor. The cell consists of a MOS transistor and a storage capacitor C. Information can be stored as presence or absence of charge on capacitor C. But the charge stored on the capacitor tend to decay with time. It is because of two facts- first even though the transistor is OFF there flows a small leakage current, and secondly the capacitor itself has a leakage resistance. The result is charge (i.e. information) can only be retained for a few milliseconds. Thus to retain the information for a longer period of time the capacitor must be recharged periodically before charge goes below a threshold value. The process of restoring charge is called as refreshing. Even though DRAM requires additional circuitry to deal periodic refresh and destructive read the advantage of greater storage capacity on a single chip make it worth, as it still offers a low cost per bit solution.

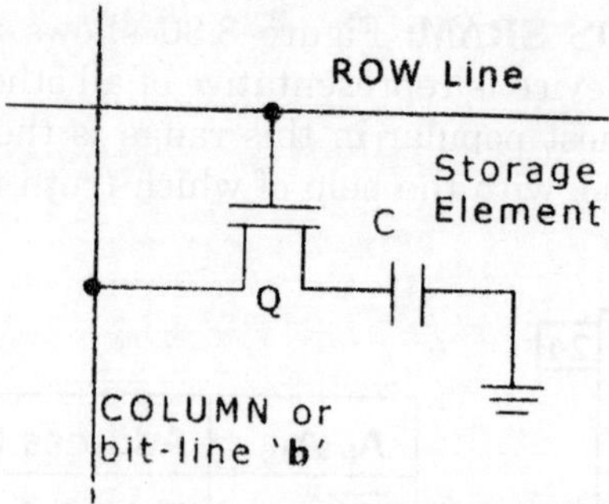

Fig. 8.31 Basic DRAM storage cell

- ***Read and Write Operation*** to store the information into the cell the row address line is activated to turn ON the transistor, and a voltage (either high or low corresponding to 1 or 0) is placed on the bit line. This causes a known amount of charge transfer to the storage capacitor if status of bit line is 1 and no charge transfer if bit line is 0. During read, the transistor is turned ON and charge on the capacitor is transferred to the bit line. A sense amplifier connected to bit line detects whether the stored charge is below or above a threshold value or not and accordingly interpret it 1 or 0. Since the read process discharges the capacitor, information is lost, it called as *Destructive Read Out (DRO)* process. For this reason the information that was read out is amplified and written back to the storage cell. A DRAM cell therefore is refreshed every time it is read. Alternately, the activation of row line refreshes the cell every time. This forms the basis of periodic refreshing of DRAM, as we will see later, to keep it from loosing the information. Also note that transistor itself acts as transmission gate so no additional transistors are needed as was required in MOS SRAM cell, see Fig. 8.28.

8.4.13.2 One MOS Transistor DRAM Cell

The circuit used for DRAM MOS cells may use one or more transistors. The simplest is one transistor cell as shown in Fig. 8.32(*a*). It consists of an NMOS transistor and a MOS capacitor, C_{MOS}, fabricated on a chip. The equivalent circuit is shown in Fig. 8.32(*b*), in which C_j represents the junction capacitance of transistor. If MOS capacitor plate lead & substrate leads are connected to ground, it then forms a parallel connection of C_j and CMOS. The storage capacitor is $C = C_j + C_{MOS}$ which can be used to store charge in dynamic cells.

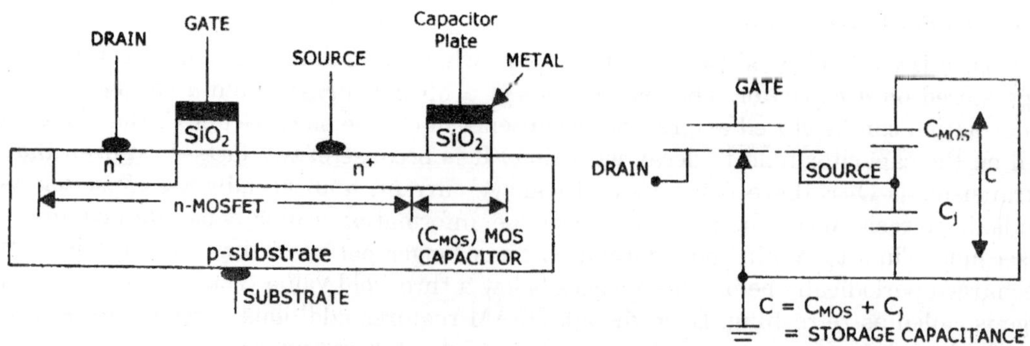

Fig. 8.32 (*a*) Fabrication of one transistor dynamic memory cell

Fig. 8.32 (*b*) Equivalent circuit diagram

8.4.13.3 DRAM Organization

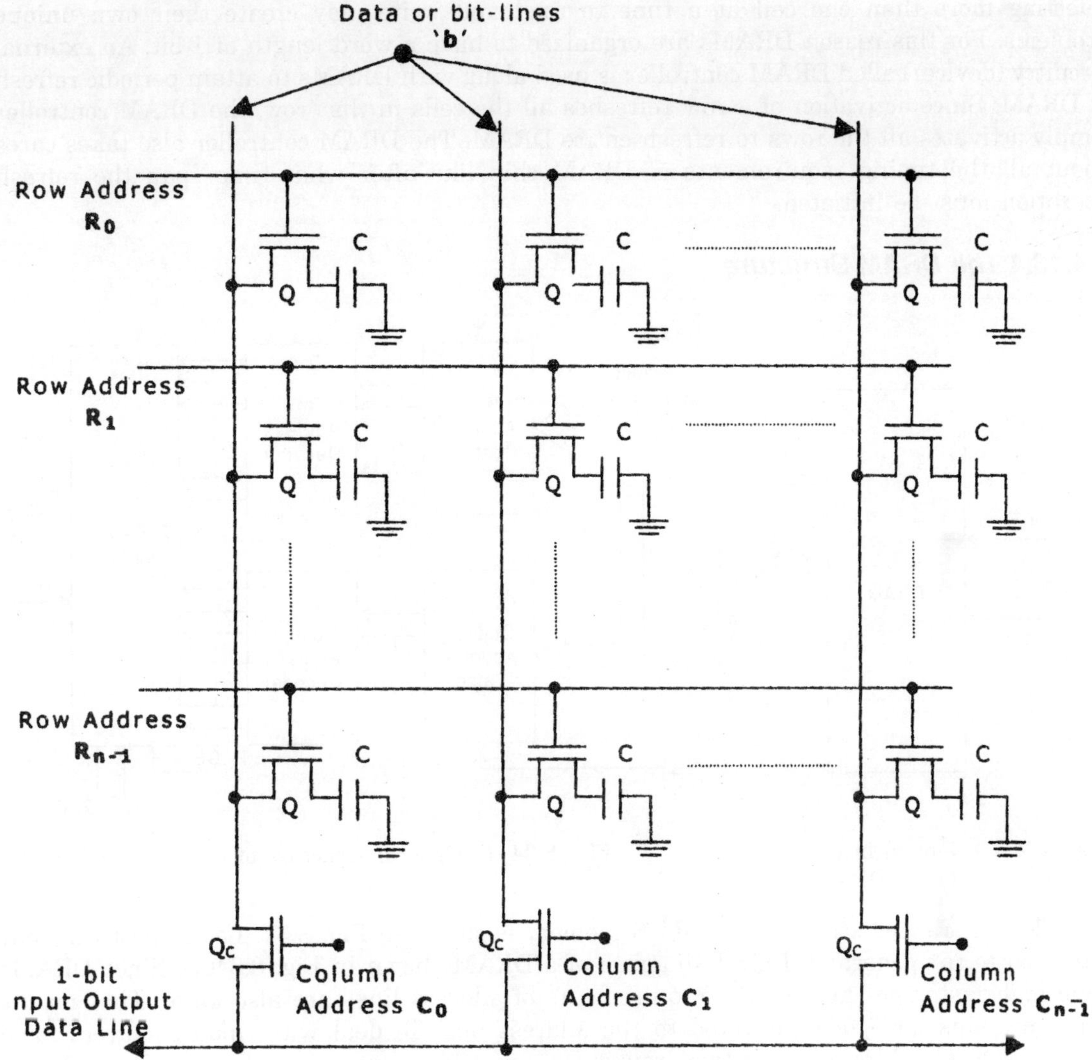

Fig. 8.33 Organization of n × 1 MOS DRAM memo

The organization of a DRAM allows many cells to be accessed together by a minimum amount of circuitry. A DRAM memory contains an array of basic dynamic storage cells as shown in Fig. 8.33. Readers are advised to study the organization carefully. The organization is such that upon activation of a row all the cells of that row are selected. Then cells transfer the stored information to their corresponding bit line. Thus, all the cells of the activated row must be subsequently written back i.e. refreshed, to restore the information. Application of column address will turn ON any one of the Q_C transistors connected to the bit line. Thus the data bit from desired cell is obtained which is then feeded to the sense amplifier before transferring to the output. Similarly for the write operation only one of the Q_C transistors will be turned ON and the data present at the input is transferred to the desired cell. From the above discussion it is clear that for any operation with the DRAM the row address must be applied first and then column address is applied. Moreover due to the fact that charge stored in DRAM cell is very small the actual read and write processes are too intricate. Thus selecting more than one cell at a time to read and write may create their own unique problems. For this reason DRAMs are organized to have a word length of 1-bit. An external circuitry (device) called DRAM controller is used along with DRAMs to attain periodic refresh of DRAM. Since activation of a row refreshes all the cells in that row, the DRAM controller simply activates all the rows to refresh entire DRAM. The DRAM controller also takes cares about all the timing requirements of DRAM chip like after what time span the refresh operation must be initiated.

8.4.13.4 The DRAM Structure

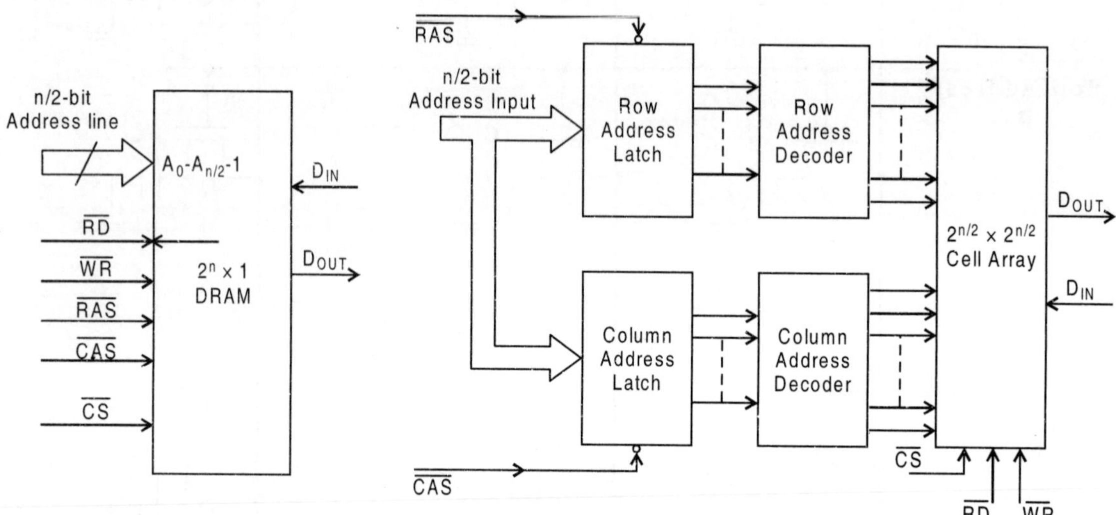

Fig. 8.34 (*a*) General block diagram of a $2^n \times 1$ DRAM

Fig. 8.34 (*b*) General structure of a $2^n \times 1$ DRAM

The generalized structure of DRAM memory is shown in Fig. 8.34. Let us first turn our attention to the generalized block diagram of a DRAM shown in Fig. 8.34(*a*). Since DRAMs have higher storage capacity the requirements of address lines are also large. That means too many pins have to be devoted to the address bus. To deal with, manufacturers have multiplexed the address bits in two groups; the row address and the column address. That's why the diagram shows n/2-bit address bus. The two signals $\overline{RAS}$ and $\overline{CAS}$ are used to tell when row address and when the column address respectively, is applied. RAS stands for *row address strobe* and CAS stands for *column address strobe*. A low on any of these two lines strobes the address present on address bus into an internal latch. The latch can be row address latch if RAS is activated or column address latch if CAS is activated, as shown in Fig. 8.34(*b*). Output of these two latches are used by the two decoders to address the required cell. Note that the $\overline{CAS}$ also performs the chip select function in most of the DRAMs. As previously stated for any operation first the row address followed by $\overline{RAS}$ is applied. After that the column address of desired cell followed by $\overline{CAS}$ is applied. The DRAM controller handles this sequence for the users and fulfills the required timing between the signals. In fact DRAM controllers are invisible for users from operational point of view, even though they actually communicate with them only. It is DRAM controller that actually communicates with DRAM

IC by means of various signals. A practically useful DRAM is IC-496 whose pin diagram is shown in Fig. 8.35. This IC employs architecture identical to the one shown in Fig. 8.34(*b*) but utilizes a 128 × 128 cell array. Note that the IC does not have any chip select signal, thus, it is performed by the $\overline{CAS}$ signal.

Signal	Pin		Pin	Signal
V_{BB} (–5V)	1	IC-4116 16K × 1	16	V_{SS} (0V)
D_{IN}	2		15	$\overline{CAS}$
$\overline{WE}$	3		14	D_{OUT}
$\overline{RAS}$	4		13	A_6
A_0	5		12	A_3
A_1	6		11	A_4
A_2	7		10	A_5
V_{DD} (+12V)	8		9	V_{CC} (+5V)

Fig. 8.35 Pin diagram of IC 499 a 91K × 1 DRAM

Refreshing DRAMs. To refresh, the DRAM controller utilizes the fact that activation of a row refreshes all the cells connected to that row. The DRAM controller simply outputs a row address and activates the $\overline{RAS}$ signal to refresh a row. Such a memory cycle is called as $\overline{RAS}$ cycle. The total number of such cycles needed to refresh all the rows constitute one refresh cycle that refreshes the entire DRAM. The DRAM controller initiates a refresh cycle after every predetermined time interval as specified by data sheet of DRAM. Yet another method is to interleave the refresh operation in which only one row is refreshed at a time and each successive row is refreshed after a fixed amount of time. Consider for example a DRAM having 5 rows and after 5 seconds datas are lost. Then it is require to refresh each successive row within 1 second so that all the rows are refreshed once in every 5 seconds. Readers are encouraged to figure out situations where interleaved refreshing is useful and compare this scheme with refreshing all the rows at once. In either the scheme when refresh operation is taking place the DRAM can not be accessed.

Simm Module. As shown in figure the DRAMs have a word length of 1-bit where as most modern computers requires to read or write an m-bit word at time. Therefore m identical DRAM ICs are connected in parallel to give m-bit word length. This is done by mounting the ICs on a single board called, *Single Inline Memory Module* or simply SIMM module. For example, IC 41256 is 256K × 1 DRAM where as 41256A8 is a SIMM module using 8 41256 ICs that offers a 256K × 8 DRAM memory.

Page Addressing Mode. A special addressing mode that is available with most DRAMs to support the transfer of large data block is page addressing mode. In this case the data bits are stored in successive cells of the same row. Thus, for the data transfer the row address has to be specified only once and by specifying the successive column address the data bits can be read. The data transfer rate becomes double in this case as compared to the normal case in which each time a cell is accessed both the row & column address must be issued.

Pseudostatic DRAMs. Newer DRAMs are available in which the refresh logic is incorporated in the memory chip itself. In this case, the dynamic nature of these memories are almost invisible. Such a dynamic memory is referred as pseudostatic, or simply PDRAMs. Note that when internal refreshing is going on the device can not be accessed.

8.4.14 SRAMs and DRAMs

Having been studied the SRAMs & DRAMs it is worth to devote some time in comparative study of the two memories. Here are the few, which must be appreciated for the two types.

- The SRAMs cells can either be Bi-polar or MOS but DRAM cells can only be MOS.
- The DRAM cells are much simpler than the SRAM cells.
- SRAMs can retain information as long as power is ON whereas DRAMs loose the retained information after a small time and thus requires refreshing.
- Data storage in DRAMs involves charge storage whereas in SRAMs it involves ON/OFF the transistors.
- DRAM cells can be constructed around a single transistor whereas SRAMs may require 9 transistors per cell. Thus, DRAMs offer higher storage density. Infact, the DRAMs are among the densest VLSI circuits in terms of transistor per chip. Due to this almost all the RAMs greater than 16KB are DRAMs.
- DRAMs are cost effective memory solution than the SRAMs.
- The actual read/write mechanism of DRAMs are much more tedious than that of SRAMs.
- DRAMs are always organized with the word length of 1-bit so that they are used with SIMM module whereas SRAMs are organized for many different word lengths.
- DRAMs require many hardware pins so that address lines have be multiplexed whereas it is seldom the case with SRAMs.
- Timing requirements of DRAMs are very complex as compared to the timing requirements of SRAMs.
- The DRAMs suffer from destructive read out so that each read must be followed by a write whereas there is no such problem with SRAMs.
- The DRAM requires extra hardware circuitry, called DRAM Controllers, for its proper operation whereas SRAM does not need such supports.
- DRAMs are slower than the SRAMs due to destructive read out, address multiplexing and mainly due to refresh requirements.
- Future expansion of DRAM is easier as they are used with SIMM module in which case expansion involves replacement of board on the slot of main board. Needless to mention that using SIMM further increases the storage capacity than achieved by using a single DRAM chip.
- The DRAM cells offer lesser power consumption compared to either BJT or MOS SRAM cell.

At this point, we strongly recommend the readers to find out similar comparison between other memory devices like BJT & MOS SRAMs, BJT & MOS ROMs & PROMs etc.

8.4.15 Memory System Design

An important aspect of study of memories to design a memory system that suits for a particular application. A design engineer is suppose to design a memory system that meets number of required storage words and the word length with the available resources. The memory system design mainly involves the two things

- The Address Space Allocation or Address decoding.
- Obtaining the required storage capacity with a given word length.

The first point requires the understanding of address range available with memory IC and that with the available processing system. The second problem mainly requires different types of connection that can be made with the memory ICs to achieve the required capacity and word length. We access these steps one by one in following subsections to elaborate the idea. It has been our experience that students are often feared about how to determine number of address lines, how to find out address ranges and how it is written. Thus, we first address this issue in order to make the newcomers feel easy.

8.4.15.1 Determining Address Lines and Address Range

To determine the number of address lines in a memory IC we first see its organization (e.g. 1K × 8) to find out the number of storage locations or words in that memory. For example, in 1K × 8 memory 1K = 1024 storage locations or words are present. Now, by using equation 2, we determine the number of address lines in a memory. We have equation 2 as

$$2^n \geq \textit{Number of Storage Locations}$$

Where n = number of address lines and is always integer.

Address bits →	A_9	A_8	A_7	A_6	A_5	A_4	A_3	A_2	A_1	A_0	*Address In Hex*
Starting Address	0	0	0	0	0	0	0	0	0	0	000_H
Ending Address	1	1	1	1	1	1	1	1	1	1	$3FF_H$

Fig. 8.36 Address bit map or address range for 1K × 8 memory

So, for 1K × 8 memory $2^n \geq 1K = 1024$ results in $n = 10$ i.e. it has 10 address lines. The address bits are written as A_0, $A_1 \ldots A_{n-1}$ i.e., for $n = 10$ they are labeled as A_0, $A_1 \ldots A_8$. To determine the address range we use the fact that addresses are specified as binary bit and each binary bit can have max change from 0 to 1. Thus, to determine range we consider this max change for address bits at a time as shown in Fig. 8.36 for 1K × 8 memory. While writing the memory addresses the usual way is to write in hexadecimal instead of binary for the sake of easiness.

Thus, the address ranges from 000_H to $3FF_H$ for a 1K × 8 memory. The bit wise representation of memory address as shown in Fig. 8.36 is called as *Address Bit Map*. We will use such type of bit maps in designing problems. From the above, *we can conclude that adding* $3FF_H$ *to the initial address will give the last address of 1K memory locations*. The same can be said alternately, that by changing the 10 address bits we traverse through any of the 1K address location.

8.4.15.2 Parallel and Series Connections of Memory

In this subsection we are concerned with how to connect more than one memory ICs to get the desired word length and desired number of address location. In general, we can connect the memory in three different configurations. They are

- Parallel Connection
- Series Connection
- Series-Parallel Connection

The *parallel connection* of memory ICs are used when we need to increase the word length while keeping the address location same. In this case, the address lines and control signals are tied together by forming wired-OR and data lines from all the ICs are taken at the output to provide desired word length. Let we have two ICs organized as 1K × 4 and we need a memory of 1K × 8. A Parallel connection shown in Fig. 8.37 can solve this issue.

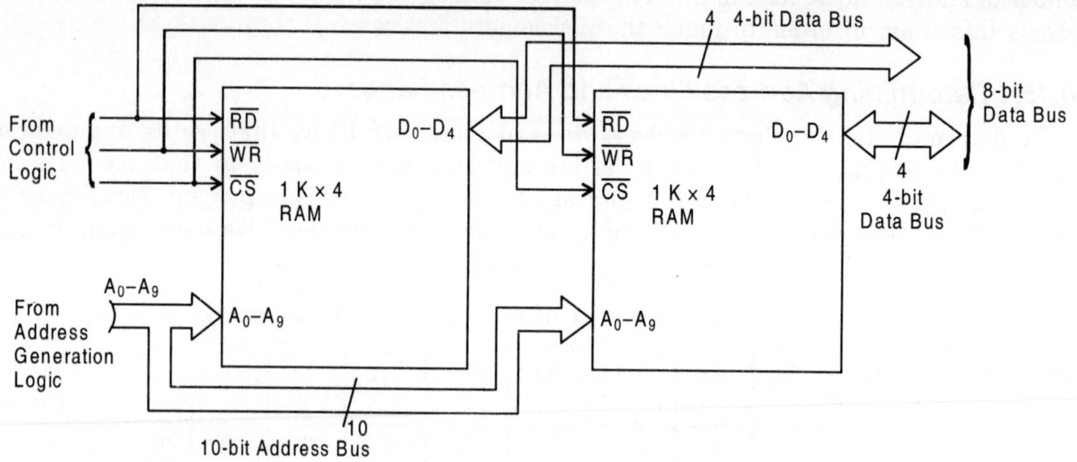

Fig. 8.37 Parallel connection of 1024 × 4 RAMs to obtain 1024 × 8 RAM

Observe the figure carefully and notice that 4-bit data bus from two ICs are taken at the output in parallel giving a total of 8-bit data bus, that's why it is called parallel connection. Any group of the four bits can be defined as least significant or most significant.

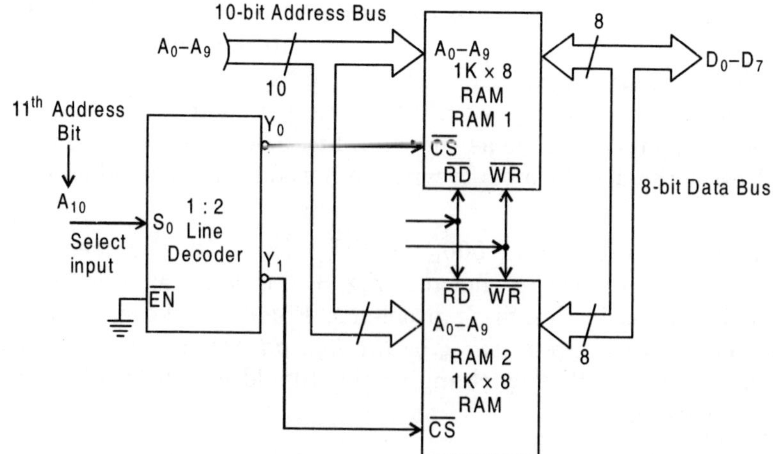

Fig. 8.38 (*a*) Series connection of two 1K × 8 RAMs to obtain 2K × 8 RAM

The *series connection* of memory ICs are used when we need to increase the address location while keeping the word length same. In this case the address lines, data lines and control signals are all tied together by forming wired-OR. A decoder can then be used to select one of the ICs. This provides the desired numbers of address locations. Let we have two ICs organized as 1K × 8 and we need a memory of 2K × 8. A series connection shown in Fig. 8.38(*a*) can solve this issue.

Since there are two ICs of 1K locations each, when combined together they can give 2K locations. But having 2K locations requires 9-address bits. The select input of 2:1 line decoder provides this 9th address bit as labeled in figure. When this bit is 0 the upper RAM is selected and when 1, lower RAM is selected. Thus, 2K locations can be accessed. The address bit map for this connection is shown in Fig. 8.38(*b*). Observe how the address is changing for the two values of A_{10} which is connected to the select input of decoder. Thus, we can say that we have address bank of 2K locations. As clearly seen by the bit map 1st 1K addresses of this bank are assigned to the upper RAM and 2nd 1K are assigned to the lower RAM. Since, the two RAMs are connected one after another they are called to be series connected.

Address bits →	A_{10}	A_9	A_8	A_7	A_6	A_5	A_4	A_3	A_2	A_1	A_0	*Address In Hex*
Starting Address	0	0	0	0	0	0	0	0	0	0	0	000_H
Ending Address	0	1	1	1	1	1	1	1	1	1	1	$3FF_H$
Starting Address	1	0	0	0	0	0	0	0	0	0	0	400_H
Ending Address	1	1	1	1	1	1	1	1	1	1	1	$7FF_H$

Fig. 8.38 (*b*) Address bit map or address range for 1K × 8 memory

The *series-parallel connection* of memory ICs are used when we need to increase both the word length and the number of address locations. This is the combination of first two connections. If readers are through with the first two configurations this can easily be understood. For this type of connection, we can consider the connection shown in Fig. 8.38(*a*) in which each of the 1K × 8 ICs are derived from Fig. 8.37. Readers are instructed to draw the connections to obtain a memory of 2K × 8 by using the 4-ICs of 1K × 4 on the basis of Fig. 8.37 and 8.38(*a*). In this fashion, the desired word length and desired number of storage locations can be obtained. In fact, it is the connection often required.

8.4.15.3 Address Space Allocation—The Address Decoding

The address decoding problem is associated with what addresses to be assigned to which memory devices from entire address space available, and what exactly should be the scheme to assign it. In nutshell, it is the problem of memory address allocation. In a general microprocessor system the number of address lines are large allowing it to access large amount of memory. Whereas the memory ICs available may be of smaller storage capacities i.e. they can process smaller number of address lines. Now the problem could be which of the ICs are to be given which address. Infact, the problem can be solved by careful analysis of requirement and applying the appropriate decoding scheme.

There are two types of decoding schemes, *Exhaustive Decoding* and *Partial Decoding*. In the exhaustive or fully decoded system, all the address lines of processor must be decoded. In case of partial decoding scheme, all the address lines of processor are not decoded. But in this case, the addresses may not be unique and more than one address can be assigned to same memory word.

To understand it let us consider a residential colony in which house numbers can be a 4-digit numbers and any of the house can be allotted a number between 0000 to 9998. Now suppose only the numbers from 1000 to 2000 are allotted to houses. In this case saying house numbers 500 or even 3500 or 4500 is same as 1500 as we know that only house numbers are one from 1000 to 2000. Saying just 500 as house number means *Partial Decoding* in case of memory address decoding. Where as saying house number 1500 is *Exhaustive Decoding*. Now consider the case when allotted house numbers are 1000 to 3000 then saying just 500 to mention same house creates trouble that whether it refers to 1500 or 2500 as both are present. In memories it means two memory ICs are assigned same address. So both the ICs will start working at a time causing *BUS Contention* to occur. *BUS Contention* is a situation in which more than one device attempts to drive the system bus simultaneously in which case buses may be damaged. To avoid all such problems, we must give complete house number i.e. 1500 to refer to that house. Thus before saying 500 we must ensure that there is no other house number whose last 3 digits are 500. The same is ensured in memory systems too, to avoid bus contention. Thus, irrespective of the decoding schemes the goal of design is to assign unambiguous (or unique) addresses to memory devices. The way two schemes are realized are explained in next subsection.

8.4.15.4 Formation of Memory System

In this section, we are concerned with how to deal with the requirements of different systems. Here, we must analyze the different requirement given to us and check the resources available and then take the appropriate steps to solve the problem. Infact, a design engineer is supposed to come with the appropriate access circuitry for the memory modules given to him/her in order to obtain the required system. It has our best observation and reading that it is far better if we consider the problems to illustrate the ideas to students. So, we start with the example in this subsection to consider the different situation and its fulfillment. *Note that we illustrate many important facts and make the important conclusions through these examples only.* We first start with the assumption that processing system can be a microprocessor or any other system capable of providing the required number address bits and control signals required for memory. Our job is mainly to find out the decoding circuitry for the circuits.

Example 1. *It is required to obtain a memory system of 2K × 8 bits for a certain application. Given that memory ICs available are 1K × 8. Obtain the desired system.*

Solution. By reading the problem it is revealed that we have already solved this problem even before saying any thing!!!!. If surprised see Fig. 8.38 (a), it is same as required by problem. Sit comfortably now because we are already through with it. For the reference, we reproduce the Fig. 8.38 (a) and 8.38 (b) again.

It's now time to list out the different steps in order to solve these types of problems. Carefully read all steps we take to solve this problem again.

1. > *Determine the total number of address lines required for the memory system to be designed.*

Given total requirement is 2K × 8 means we need a *total of 9 address lines* because $2^9 = 2048 = 2K$.

2. > *Determine the number of address bits available with the given memory ICs.*

Given ICs are 1K × 8 means we have 10 address lines available with the individual ICs because $2^{10} = 1024 = 1K$.

3. > *Determine the number of bits available for the chip decoder.*

Since total required address bits are 9 and individual memories can handle 10-bits of address, we are left with 1 address bit. This extra bit can be feeded to the decoder as shown in Fig. 8.39 (*a*).

4. > *Draw the Address Bit Map for Entire Memory Space and indicate change in bits for individual memory ICs.* This is needed to assign particular address to individual ICs. The same is shown in Fig. 8.39 (*b*).

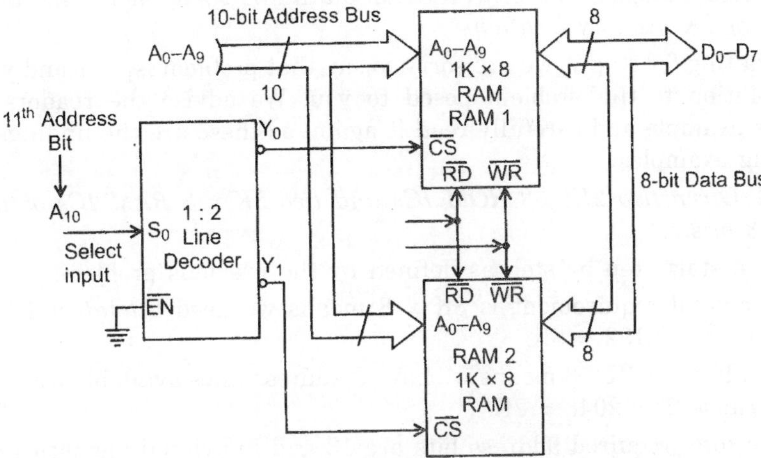

Fig. 8.39 (*a*) Series connection of two 1K × 8 RAMs to obtain 2K × 8 RAM

Decoder bit | Address bits handled by memory ICs

	A_{10}	A_9	A_8	A_7	A_6	A_5	A_4	A_3	A_2	A_1	A_0	HEX Addr	IC
Start Addr	0	0	0	0	0	0	0	0	0	0	0	000_H	RAM 1
End Addr	0	1	1	1	1	1	1	1	1	1	1	$3FF_H$	
Start Addr	1	0	0	0	0	0	0	0	0	0	0	400_H	RAM 2
End Addr	1	1	1	1	1	1	1	1	1	1	1	$7FF_H$	

Fig. 8.39 (*b*) Address bit map for Fig. 8.39(*a*) of example 1

Observe that bit A_{10} does not change anywhere within the address range for RAM1 and remains 0 always. The same is true for RAM2 where bit A_{10} is 1 always. This justifies the step 3 to take the left over bits for decoder. As one can *see* that *the decoder bit selects how much memory address (1K in this case) can be selected by each output of the decoder*. Infact, this is the main purpose of drawing address bit maps.

5. > *Draw the Connection diagram for the memory ICs as per Address Bit Map.* This is the one we have done already in Fig. 8.39(*a*). But this is done for this example only. We will draw the connection diagram in this step only for the other examples.

6. > *Draw the Address Map*. In this step we diagrammatically represents the entire memory address space and show which address is assigned to which IC as shown in Fig. 8.39(*c*).

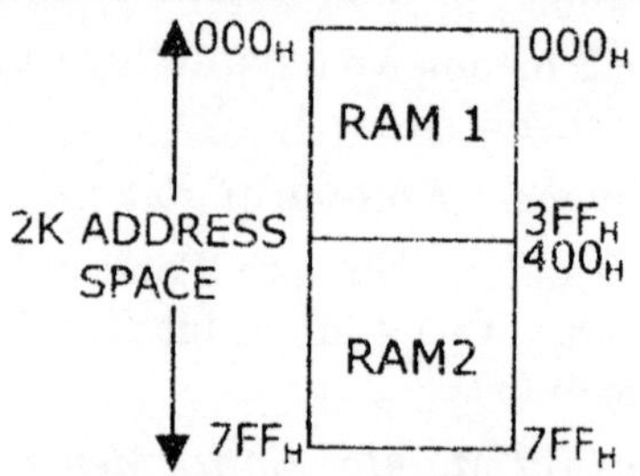

Fig. 8.39 (*c*) Address map for Fig. 8.39 (a)

From this address map *we also conclude that adding* $7FF_H$ *to the initial address will give the last address of 2K memory locations.*

If you are waiting for step 7, we are sorry to say that problem is over and you have given the complete solution to the problem posed to you. We advise the readers to go to the beginning of this example and carefully read it again, as these are the findings we would be using in following examples.

Example 2. *Given two 2K × 8 ROM ICs and two 2K × 8 RAM ICs obtain a memory system of 8K × 8 bits.*

Solution. We start step by step as defined by the previous problem.

1. > Given total requirement is 8K × 8 means we need a *total of 13 address lines* because 2^{13} = 8192 = 8K.
2. > Given ICs are 2K × 8 means we have 9 address lines available with the individual ICs because 2^9 = 2048 = 2K.
3. > Since total required address bits are 13 and individual memories can handle 9-bits of address, we are left with 2-address bit, which can be feeded to the decoder.
4. > *Draw the Address Bit Map for Entire Memory Space and indicate change in bits for individual memory ICs.* This is needed to assign particular address to individual ICs. The same is shown in Fig. 8.40(*a*).

Decoder bit | ← Address bits handled by memory ICs → |

	A_{12}	A_{11}	A_{10}	A_9	A_8	A_7	A_6	A_5	A_4	A_3	A_2	A_1	A_0	HEX Addr	IC
Start Addr	0	0	0	0	0	0	0	0	0	0	0	0	0	0000_H	ROM 1
End Addr	0	0	1	1	1	1	1	1	1	1	1	1	1	$07FF_H$	
Start Addr	0	1	0	0	0	0	0	0	0	0	0	0	0	0800_H	ROM 2
End Addr	0	1	1	1	1	1	1	1	1	1	1	1	1	$0FFF_H$	
Start Addr	1	0	0	0	0	0	0	0	0	0	0	0	0	1000_H	RAM 1
End Addr	1	0	1	1	1	1	1	1	1	1	1	1	1	$17FF_H$	
Start Addr	1	1	0	0	0	0	0	0	0	0	0	0	0	1800_H	RAM 2
End Addr	1	1	1	1	1	1	1	1	1	1	1	1	1	$1FFF_H$	

Fig. 8.40 (*a*) Address bit-map for 8K × 8 memory system of example 2

It is clear from the address bit map that each decoder output (selected by A_{12} and A_9 bits) selects one 2K-memory address space.

5. > Connection diagram for the memory ICs is shown in Fig. 8.40 (*b*).

6. > The Address Map is shown in Fig. 8.40 (*c*).

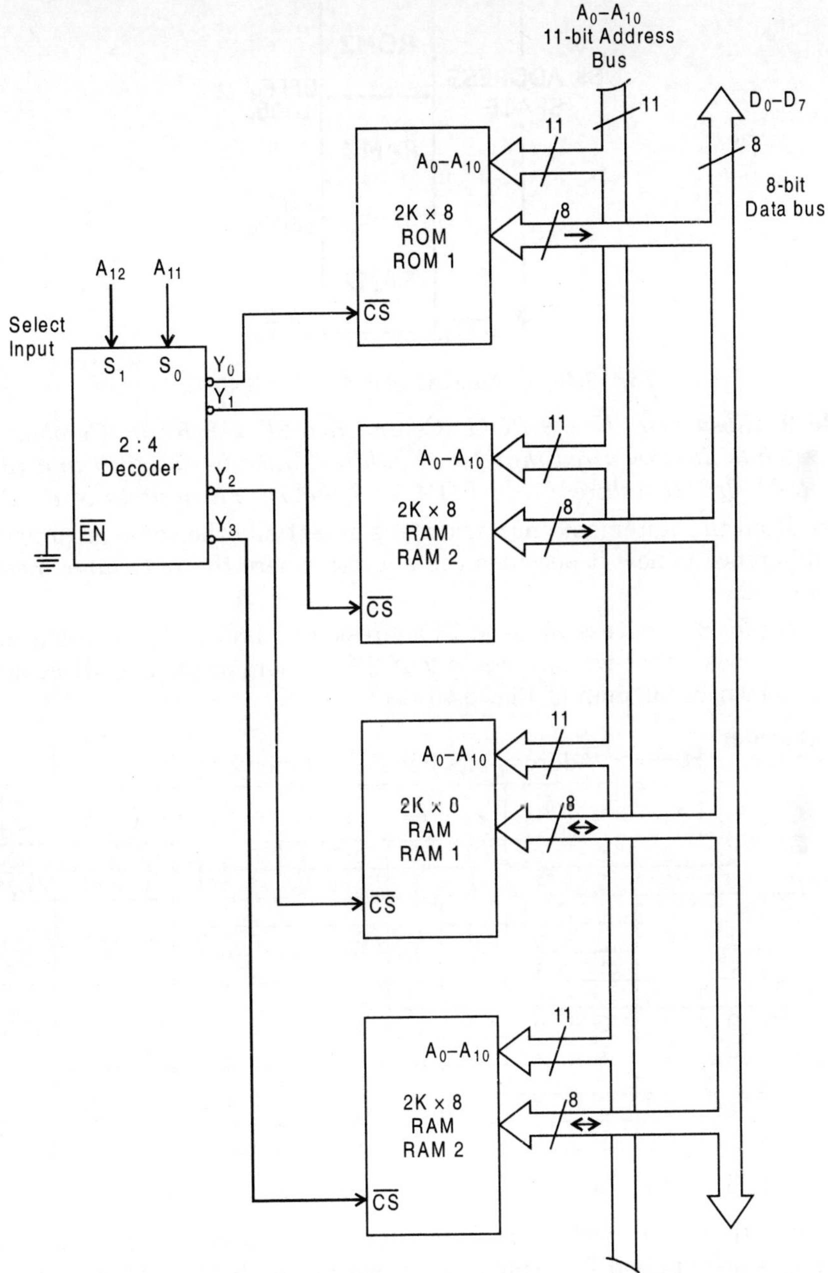

Fig. 8.40 (*b*) Formation of 8KB memory system using RAMs and ROMs

From this address map *we also conclude that adding $1FFF_H$ to the initial address will give the last address of 8K memory locations.* Alternately, we can say that we must change 13 address bits at a time to traverse through 8K memory locations.

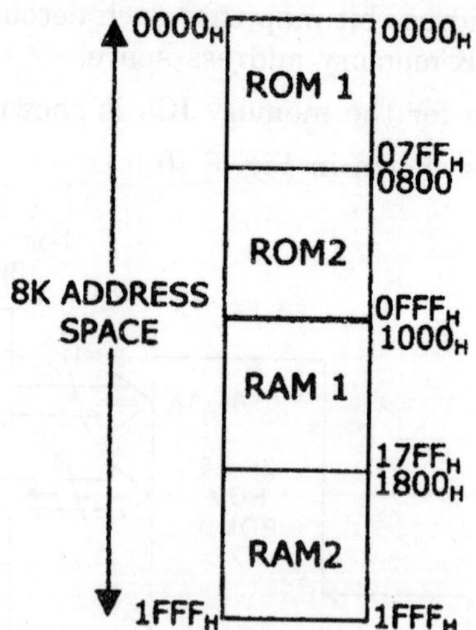

Fig. 8.40 (*c*) Address map for Fig. 8.40(*b*)

Example 3. *Given two 2K × 8 ROM ICs and two 2K × 8 RAM ICs obtain a memory system of 8K × 8 bits. It is required that 1st 2K address must be assigned to a ROM, 2nd 2K address to a, RAM, 3rd 2K address to a ROM, and the 4th 2K address to a RAM.*

Solution. Read the statement once again it is essentially the same as previous problem, but the only difference is now it specifies the address where the particular memory devices are to be connected.

We know that for 8K address we need 13 address lines and with the arrangement shown in Fig. 8.40 (b) it reveals that we can access four 2K memory segments whose addresses can be obtained as shown in bit map of Fig. 8.40 (*a*).

	Decoder bit		← Address bits handled by memory ICs →												
	A_{12}	A_{11}	A_{10}	A_9	A_8	A_7	A_6	A_5	A_4	A_3	A_2	A_1	A_0	HEX Addr	IC
Start Addr	0	0	0	0	0	0	0	0	0	0	0	0	0	0000_H	ROM 1
End Addr	0	0	1	1	1	1	1	1	1	1	1	1	1	$07FF_H$	
Start Addr	0	1	0	0	0	0	0	0	0	0	0	0	0	0800_H	RAM 1
End Addr	0	1	1	1	1	1	1	1	1	1	1	1	1	$0FFF_H$	
Start Addr	1	0	0	0	0	0	0	0	0	0	0	0	0	1000_H	ROM 2
End Addr	1	0	1	1	1	1	1	1	1	1	1	1	1	$17FF_H$	
Start Addr	1	1	0	0	0	0	0	0	0	0	0	0	0	1800_H	RAM 2
End Addr	1	1	1	1	1	1	1	1	1	1	1	1	1	$1FFF_H$	

Fig. 8.41 (*a*) Address bit-map for 8K × 8 memory system of example 3

Now the only thing we need to do is to connect the device at correct outputs of the decoder, which can be obtained from bit map shown in Fig. 8.41(*a*). Go through the details of this map carefully.

The similar connection diagram and memory address map would also appear but with the ICs changed to positions required by problem. Figure 8.41(*b*) and 8.41(*c*) shows this.

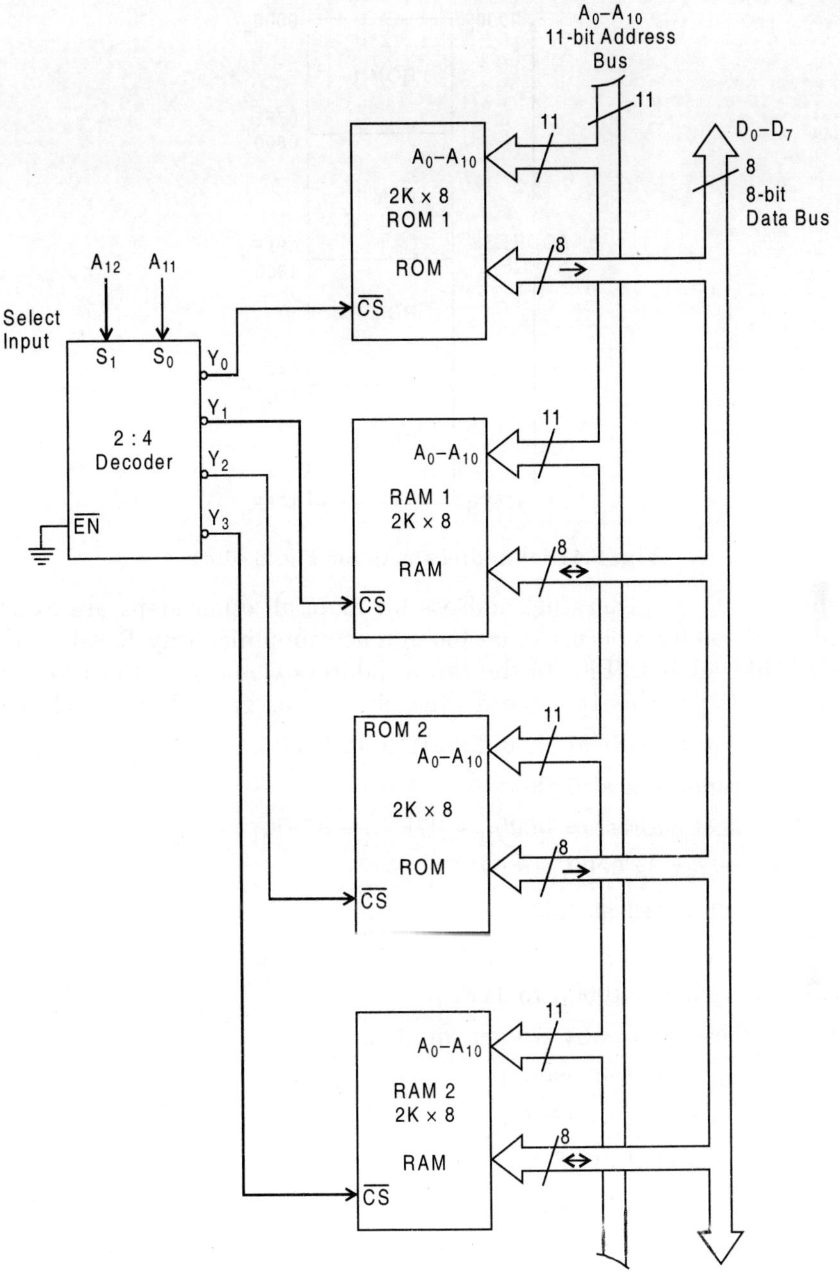

Fig. 8.41 (*b*) Formation of 8KB memory system using RAMs and ROMs

Example 4. *Given two 2K × 8 ROM ICs and two 2K × 8 RAM ICs obtain a memory system of 8K × 8. It is required that two ROMs must be connected to addresses beginning from 0000H and 1000H. Also the RAMs should be connected to begin from* 0800_H *and* 1800_H.

Solution. Example 4 is just another way or giving statement for example 3. But we will consider it separately to understand that what should be the way to solve these types of issues.

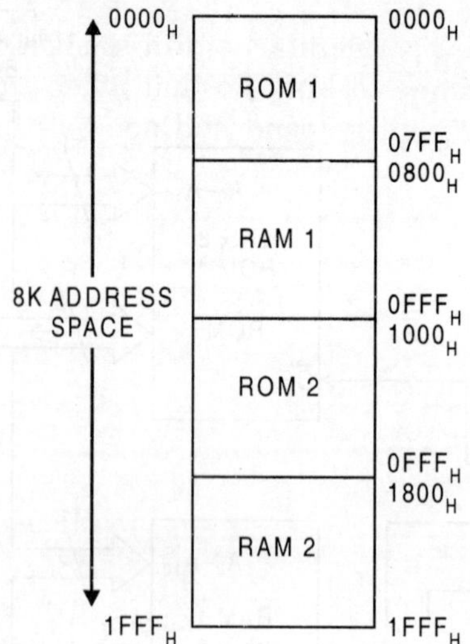

Fig. 8.41 (*c*) Address map for Fig. 8.41(*b*)

Except the step 4 i.e., generating address bit map, all other steps are exactly same so that we consider only address bit map that too in much simplified way. Recall that in example 1, we concluded that adding $7FF_H$ to the initial address would give the last address of 2K memory locations. We use this to generate the bit map as given ICs are 2K × 8.

For ROMs: given that a ROM should start from 0000_H and 1000_H

First when connected at 0000_H:

$$\text{Last address} = 0000_H + 07FF_H = 07FF_H$$

So, the address range is 0000_H to $07FF_H$

Second when connected at 1000_H:

$$\text{Last address} = 1000_H + 07FF_H = 17FF_H$$

So, the address range is 1000_H to $17FF_H$

For RAMs: in the similar way we get for RAMs

First when connected at 0800_H:

Address range is 0800_H to $0FFF_H$

Second when connected at 1800_H:

Address range is 1800_H to $1FFF_H$

Compare these addresses with those shown in the bit map of Fig. 8.41 (*a*), you reached the same map. Thus, the entire Fig. 8.41 can redrawn for example 4.

It is interesting to see that instead of adding 07FF if we would have changed the least 9 significant address bits (as 2^9 = 2K) of initial address to 1 even in that case also we could have reached the same point. Infact this is what we have done till example 3. Either the way we proceed it is same because we found the number 07FFH from the fact that the highest hexadecimal number that can be written by 9 address bits is 07FFH. Many a times you will find that adding a number to initial address is easier because it is less time consuming than case of drawing complete bit map. But this short hand is useful only when you are through with concept and saying a hexadecimal number displays its bit pattern in your mind.

Shadowing or Address Overlapping means assigning more than one address space to the memory device. It is the situation similar with a student who has his roll number and university enrolment number unique to him. So that when called by enrolment number or by roll number the same student will respond and none of the other students respond. An example will illustrate this best.

Example 5. *A system needs a 4K × 8 memory system. But the requirement is such that a single 2K × 8 RAM IC be used and the two consecutive 2K memory addresses should be overlapped to this device only.*

Solution. We know that for 4K system we will have 12 address lines. The 2K devices can process only 9 address bits thus the left over is 1-bit that can be used to access two 2K segments. By address overlapping we mean that both the segments would be referenced for same device. This is shown in Fig. 8.42.

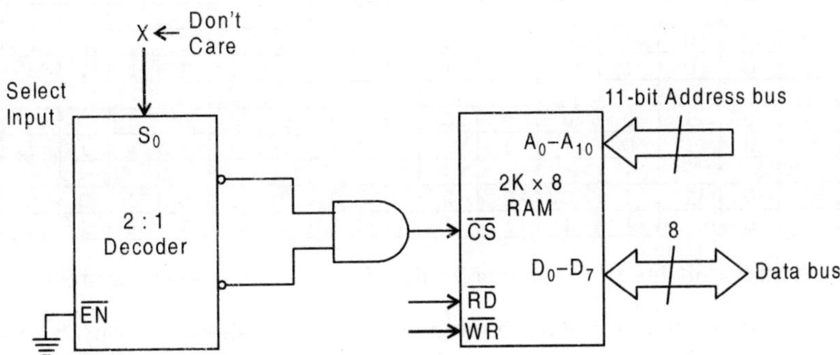

Fig. 8.42 Illustration of memory address overlapping with 2K × 8 IC

Notice the use of AND gate used to generate chip select for memory device. As shown, upon activation decoder outputs goes low so whether activated output is Y_0 or Y_1 AND gate output is always low. This means memory is selected irrespective of the status of select input of decoder.

If we draw (readers are instructed to draw) the bit map for 4K as two 2K address block we see that first 2K block is 000 to $7FF_H$ and second 2K block is 800_H to FFF_H. With the configuration shown in Fig. 8.42 both the address blocks refers to the same IC thus address blocks are overlapped to each other and hence the name address overlapping. In the similar way, we can overlap any two equal sized address blocks present in entire address map (or space).

We now turn our attention to exhaustive and partially decoded memory system. For illustration purpose we again consider the problem posed in example 2. Because write now we are very much through with it and once we understand the idea we can apply it to design. Read the problem statement carefully as it is going to create the difference.

Example 6. *It is required to obtain an 8K × 8 memory system for 8085 microprocessor system that has an addressing capability of 64K locations. Given memories are 2K × 8 ROM ICs and 2K × 8 RAM ICs. Obtain the exhaustive or fully decoded as well as partially decoded system.*

Solution. We first determine the number of address lines this processor has. 64K means 2^{16} *i.e.*, the processor has 16 address bits where as the requirement for memory is just

13-bits, see step 1 in example 2. This means 3 address bits of processor will remain unused (or unchanged). This is shown by the address bit maps in Fig. 8.43, which now contains 16 bits instead of 13.

For exhaustive decoding: As stated for this scheme we must decode all the address bits of microprocessor. The address bit map for this scheme is shown in Fig. 8.43(*a*). It shows that bit A_{13} to bit A_{15} are unchanged for entire 8K-address space. Since in this scheme we must decode all address bits, we can use these 3-bits to generate $\overline{EN}$ signal for Decoder which can also be used enable or disable the memory system by enabling/disabling decoder.

← Unchanged bit → | Decoder bit | ← Address bits handled by memory ICs →

	A_{15}	A_{14}	A_{13}	A_{12}	A_{11}	A_{10}	A_9	A_8	A_7	A_6	A_5	A_4	A_3	A_2	A_1	A_0	HEX Addr	IC
Start Addr	0	0	0	0	0	0	0	0	0	0	0	0	0	0	0	0	0000_H	ROM
End Addr	0	0	0	0	0	1	1	1	1	1	1	1	1	1	1	1	$07FF_H$	1
Start Addr	0	0	0	0	1	0	0	0	0	0	0	0	0	0	0	0	0800_H	RAM
End Addr	0	0	0	0	1	1	1	1	1	1	1	1	1	1	1	1	$0FFF_H$	1
Start Addr	0	0	0	1	0	0	0	0	0	0	0	0	0	0	0	0	1000_H	ROM
End Addr	0	0	0	1	0	1	1	1	1	1	1	1	1	1	1	1	$17FF_H$	2
Start Addr	0	0	0	1	1	0	0	0	0	0	0	0	0	0	0	0	1800_H	RAM
End Addr	0	0	0	1	1	1	1	1	1	1	1	1	1	1	1	1	$1FFF_H$	2

Fig. 8.43 (*a*) Address bit-map for exhaustive decoded 8K × 8 memory system of example 6

The diagram for decoder along with enable generation is shown in Fig. 8.43 (*b*). The figure indicates which decoder output to be connected to which IC. Infact connections will be same as Fig. 8.40 (*b*) and the decoder of Fig. 8.43 (*b*) will replace decoder of Fig. 8.40 (*b*).

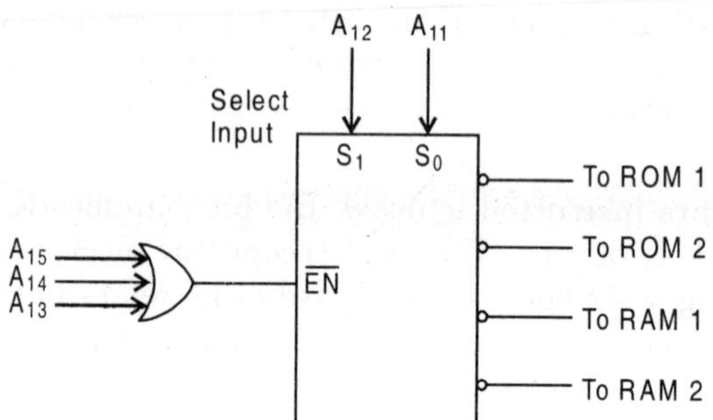

Fig. 8.43 (*b*) Exhaustive decoded system for example 6

Notice the use of OR gate. Output of OR gate is low only when all its inputs are low otherwise high. Thus, saying an address higher than $1FFF_H$ will not be responded by this system. If we say 8000H it means bit A_{15} goes high and output of OR gate goes high thus disabling the decoder and consequently the entire memory system.

For partial decoding: Here the idea is not to use all the address bits of processor for address decoding. As the 3 most significant address bits A_{15}, A_{14}, A_{13} are unchanged we can say that they are not used i.e. don't care. So, it hardly matters if we use them for decoding or not. The address bit map shown in Fig. 8.44 (*a*) shows this.

← Unused bits → | Decoder bit | ← Address bits handled by memory ICs → |

	A_{15}	A_{14}	A_{13}	A_{12}	A_{11}	A_{10}	A_9	A_8	A_7	A_6	A_5	A_4	A_3	A_2	A_1	A_0	HEX Addr	IC
Start Addr	X	X	X	0	0	0	0	0	0	0	0	0	0	0	0	0	0000_H	ROM 1
End Addr	X	X	X	0	0	1	1	1	1	1	1	1	1	1	1	1	$07FF_H$	
Start Addr	X	X	X	0	1	0	0	0	0	0	0	0	0	0	0	0	0800_H	RAM 1
End Addr	X	X	X	0	1	1	1	1	1	1	1	1	1	1	1	1	$0FFF_H$	
Start Addr	X	X	X	1	0	0	0	0	0	0	0	0	0	0	0	0	1000_H	ROM 2
End Addr	X	X	X	1	0	1	1	1	1	1	1	1	1	1	1	1	$17FF_H$	
Start Addr	X	X	X	1	1	0	0	0	0	0	0	0	0	0	0	0	1800_H	RAM 2
End Addr	X	X	X	1	1	1	1	1	1	1	1	1	1	1	1	1	$1FFF_H$	

Fig. 8.44 (*a*) Address bit-map for partially decoded 8K × 8 memory system of example 6

This means irrespective of the status of A_{15}, A_{14}, and A_{13} the memory system can be used. Thus, Fig. 8.40 (*b*) is direct implementation of partially decoded system, as it does not include these bits.

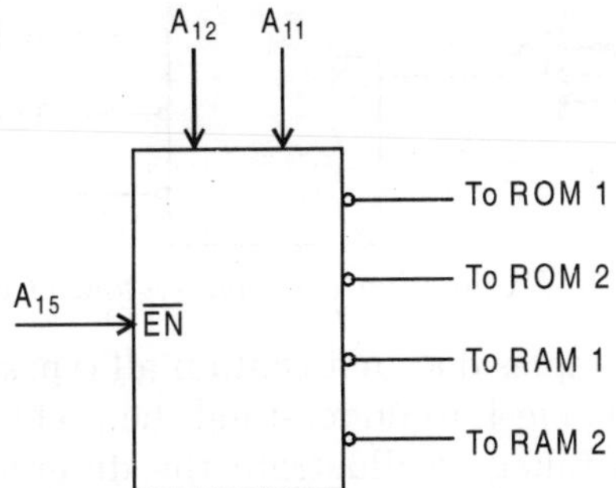

Fig. 8.44 (*b*) Partially decoded system for example 6.

A variation of implementation is shown in Fig. 8.44 (b), which uses A_{15} to enable the address decoder. Since A_{14} and A_{13} are not used it is still a partially decoded system.

Example 7. *It is required to obtain an 8K × 8 memory system for 8085 microprocessor system that has an addressing capability of 64K locations. Given memories are 2K × 8 ROM ICs and 2K × 8 RAM ICs. Obtain the exhaustive decoded system, which maps the 8K-memory system to begin from* 8000_H.

Solution. It is essentially the same problem as was done in example 6. The change now is to assign the particular address. For this, already we have seen example 4. So we do not go to complete design rather we just draw the bit-map and decoder circuit required by stating the net outcomes of detailed design.

The only change in bit map of Fig. 8.45 (*a*) as compared to that of Fig. 8.43 (*a*) is just the status of A_{15}, which is set to 1 now. It is because 8K system has to begin from 8000_H. No other change can be observed in this bit map. So we connect A_{15} through NOT gate to the input of OR gate to generate the enable signal for this address range. This is shown in Fig. 8.44 (*b*).

The readers are directed draw an address map for this design as was drawn in Fig. 8.40(*c*) of example 2.

← Unchanged bit → | Decoder bit | ← Address bits handled by memory ICs → |

	A_{15}	A_{14}	A_{13}	A_{12}	A_{11}	A_{10}	A_9	A_8	A_7	A_6	A_5	A_4	A_3	A_2	A_1	A_0	HEX Addr	IC
Start Addr	1	0	0	0	0	0	0	0	0	0	0	0	0	0	0	0	8000_H	ROM 1
End Addr	1	0	0	0	0	1	1	1	1	1	1	1	1	1	1	1	$87FF_H$	
Start Addr	1	0	0	0	1	0	0	0	0	0	0	0	0	0	0	0	8800_H	RAM 1
End Addr	1	0	0	0	1	1	1	1	1	1	1	1	1	1	1	1	$8FFF_H$	
Start Addr	1	0	0	1	0	0	0	0	0	0	0	0	0	0	0	0	9000_H	ROM 2
End Addr	1	0	0	1	0	1	1	1	1	1	1	1	1	1	1	1	$97FF_H$	
Start Addr	1	0	0	1	1	0	0	0	0	0	0	0	0	0	0	0	9800_H	RAM 2
End Addr	1	0	0	1	1	1	1	1	1	1	1	1	1	1	1	1	$9FFF_H$	

Fig. 8.45 (*a*) Address bit-map for exhaustibe decoded 8K × 8 memory system of example 7

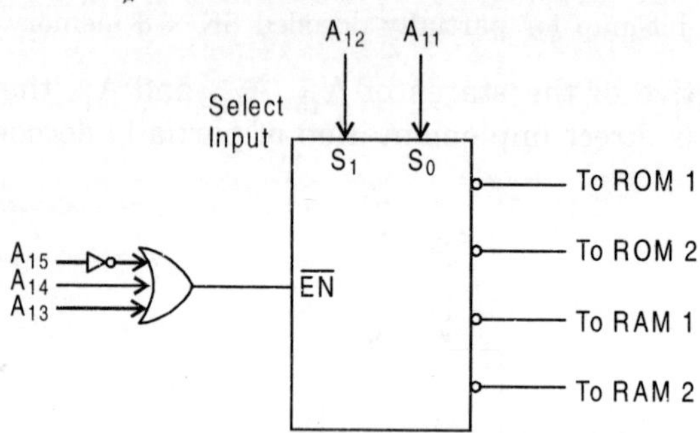

Fig. 8.45 (*b*) Exhaustive decoded system of example 7

Although all the above examples does not contain all types design variation we hope that it will serve as the best building block to understand the problem of memory bank formation. Notice that same system was taken to illustrate the different ideas because once we are familiar with a system then it is easy to view it in different situations. In general, the problems define IC numbers not the organization (e.g., 2K × 8) of memories for which user must know the address bus width and data bus width. For this reason readers are advised to prepare a list of ICs and memorize the bus width and control signals.

8.5 EXERCISE

1. What is meant by primary memory, secondary memory and mass storage devices?
2. Define the terms storage capacity, memory word, storage location, and word length.
3. Determine the storage capacity of the memories shown in Fig. 8.46.

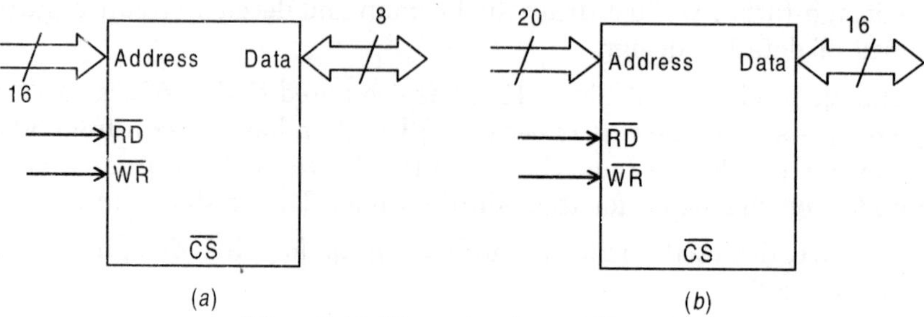

Fig. 8.46 Memories for problem 3

4. Define the terms (*i*) Access Time (*ii*) Cycle time (*iii*) Bandwidth (*iv*) Volatile (*v*) Dynamic (*vi*) Static
5. Explain organization of tracks in magnetic memories with the help of figure.
6. What are optical memories?
7. Explain basic semiconductor memory organization with the help of figure.
8. With the help of figure explain the memory cell addressing.
9. In the Fig. 8-b how memory cell "Cell 3-2" is accessed?
10. Explain the memory cell organization shown in Fig. 8.5. How a particular cell can be selected in this case?
11. A memory has cell array of 94 × 94, what will be the required number of lines of word length (*a*) 2-bits (*b*) 4-bits (*c*) 8-bits.
12. Explain the sequence of operation to be performed to write a data into memory?
13. A memory has access time of 50 n sec. It has an additional delay called precharge time of 20 n sec. Find out the access rate and bandwidth of this unit?
14. Explain how a multiplexer can be used as ROM?
15. What are Mask Programmed ROM? How logic 1s and logic 0s are stored in such ROMs.
16. Explain the MOS PROMS. What are their advantages over Bipolar PROMs?
17. Explain the basic structure of an EPROM cell. Why they are so popular?
18. How an EPROM can be reprogrammed?
19. Explaining the basic logic structure of SRAM?
20. With the help of figure explain the operation of a bipolar SRAM cell and MOS SRAM cell?
21. What is Dynamic RAM? Why they are most widely used?
22. With the help of figure explain the basic DRAM storage cell.
23. Draw the organization of a DRAM. How a particular cell can be read or write in this organization?
24. What are DRAM controllers? Why they are needed for operation of DRAM?
25. What is partial decoding and exhausitive decoding?
26. For a 16 Address-bit processor systems it is required to form 32 KB memory system using one 2764, four 4016 and two 6264 memory ICs. Obtain the decoding circuitry and draw the address map. Use exhaustive decoding.
27. A system needs to create 16KB of memory using 6224 ICs. It is required that the 2nd and 3rd 8K address should overlap. Obtain the decoding circuitry and draw the address maps.
28. For a 16 address bit processor it is required to obtain 16KB memory system using IC 6232. Use partial decoding schemes and obtain the decoding circuitry and address maps.

9

ASYNCHRONOUS SEQUENTIAL LOGIC

9.0 INTRODUCTION

Much of today's logic design is based on two major assumptions: all signals are binary, and time is discrete. Both of these assumptions are made in order to simplify logic design. By assuming binary values on signals, simple Boolean logic can be used to describe and manipulate logic constructs. By assuming time is discrete, hazards and feedback can largely be ignored. However, as with many simplifying assumptions, a system that can operate without these assumptions has the potential to generate better results.

An asynchronous sequential circuit is a sequential circuit whose behaviour depends only on the order in which its input signal changes and can be affected at any instant of time.

These circuits have state that is not synchronized with a clock *i.e.*, state variables of such circuits may change at any point of time.

The memory elements are either **unclocked flip-flops or time-delay elements**. The memory capability of time-delay device depends on the finite amount of time it takes for the signal to propagate through digital gates.

Asynchronous sequential circuits are useful in applications where the input signals to the system may change at any time, independent of the internal clock.

Resemble as combinational circuits with feedback.

General Structure

An asynchronous sequential circuit (Fig. 9.0) consists of

- A combinational circuit and delay elements connected to form a feedback loop.
- n input variables, m output variables and k internal states.
- Present state and next state variables, called **secondary variables** and **excitation variables** respectively.

State transition occurs when there is an input change. Changes in the input variables do not instantaneously change the secondary variables y. It takes certain amount of time.

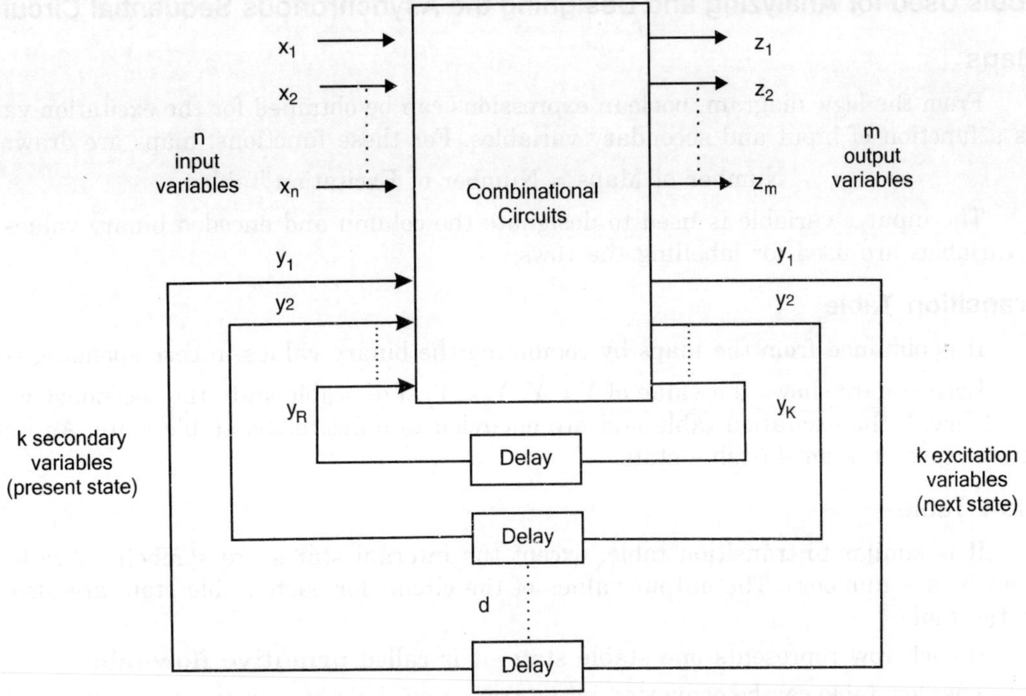

Fig. 9.0 Block diagram of an asynchronous sequential circuit.

The signal from the input terminals propagate through the combinational circuit to the excitation variables, which generates new values for the next state. These values propagate through the delay elements and become the new present state for the secondary variables.

Steady state condition,

$$\forall\ i - 1, 2, \ldots, k$$

$$y_i\ (t) = Y_i\ (t)$$

i.e., when secondary variables are equal to the corresponding excitation variables, the condition is known as steady-state. These variables are not same during transition.

Stable and Unstable

If the next state is the same as the current state, it is **stable** state. Updating the current state with the next state doesn't change anything. On the other hand, if the next state is different than the current state, it is **transient** state and as the current state is updated with the next state, the circuit will change states.

More formally, the system is **stable** if the circuit reaches a steady-state condition with $y_i = Y_i$. Otherwise, the circuit is in a continuous transition and is called **unstable.**

Fundamental Mode

To ensure proper operation, simultaneous changes of two or more input variables are usually prohibited. Since only one input variable can change at any time and the time between two input changes must be longer than the time it takes the circuit to reach a stable state. So, in **fundamental mode** operation, the input signals change one at a time and only when the circuit is in a stable condition.

Tools Used for Analyzing and Designing the Asyhchronous Sequential Circuits

Maps

From the logic diagram, boolean expressions can be obtained for the excitation variables as a function of input and secondary variables. For these functions, maps are drawn.

Number of Maps = Number of Excitation Tables.

The input x variable is used to designate the column and encoded binary values of the y variables are used for labelling the rows.

Transition Table

It is obtained from the maps by combining the binary values in corresponding squares.

Each square shows the value of $Y = Y_1 Y_2 ... Y_k$. For stable state the secondary variables must match the excitation table and are encircled to indicate the stable state. An uncircled entry represents an unstable state.

Flow Table

It is similar to transition table, except the internal states are symbolized with letters than binary numbers. The output values of the circuit for each stable state are also shown in the table.

If each row represents one stable state, it is called **primitive flow-table**.

The flow table can be converted into a transition table by assigning each state a distinct value.

Asynchronous circuits keep the assumption that signals are binary, but remove the assumption that time is discrete. This has several possible benefits:

No Clock Skew

Clock skew is the difference in arrival times of the clock signal at different parts of the circuit. Since asynchronous circuits by definition have no globally distributed clock, there is no need to worry about clock skew. In contrast, synchronous systems often slow down their circuits to accommodate the skew. As feature sizes decrease, clock skew becomes a much greater concern.

Lower Power

Standard synchronous circuits have to toggle clock lines, and possibly precharge and discharge signals, in portions of a circuit unused in the current computation. For example, even though a floating point unit on a processor might not be used in a given instruction stream, the unit still must be operated by the clock. Although asynchronous circuits often require more transitions on the computation path than synchronous circuits, they generally have transitions only in areas involved in the current computation.

Note: that there are techniques being used in synchronous designs to address this issue as well.

Average-Case Instead of Worst-Case Performance

Synchronous circuits must wait until all possible computations have completed before latching the results, yielding worst-case performance. Many asynchronous systems sense when a computation has completed, allowing them to exhibit average-case performance. For circuits such as ripple-carry adders where the worst-case delay is significantly worse than the average-case delay, this can result in a substantial savings.

Easing of Global Timing Issues

In systems such as a synchronous microprocessor, the system clock, and thus system performance, is dictated by the slowest (*critical*) path. Thus, most portions of a circuit must be carefully optimized to achieve the highest clock rate, including rarely used portions of the system. Since many asynchronous systems operate at the speed of the circuit path currently in operation, rarely used portions of the circuit can be left unoptimized without adversely affecting system performance.

Better Technology Migration Potential

Integrated circuits will often be implemented in several different technologies during their lifetime. Early systems may be implemented with gate arrays, while later production runs may migrate to semi-custom or custom ICs. Greater performance for synchronous systems can often only be achieved by migrating all system components to a new technology, since again the overall system performance is based on the longest path. In many asynchronous systems, migration of only the more critical system components can improve system performance on average, since performance is dependent on only the currently active path. Also, since many asynchronous systems sense computation completion, components with different delays may often be substituted into a system without altering other elements or structures.

Automatic Adaptation to Physical Properties

The delay through a circuit can change with variations in fabrication, temperature, and power-supply voltage. Synchronous circuits must assume that the worst possible combination of factors is present and clock the system accordingly. Many asynchronous circuits sense computation completion, and will run as quickly as the current physical properties allow.

Robust Mutual Exclusion and External Input Handling

Elements that guarantee correct mutual exclusion of independent signals and synchronization of external signals to a clock are subject to *metastability*. A metastable state is an unstable equilibrium state, such as a pair of cross-coupled CMOS inverters at 2.5V, which a system can remain in for an unbounded amount of time. Synchronous circuits require all elements to exhibit bounded response time. Thus, there is some chance that mutual exclusion circuits will fail in a synchronous system. Most asynchronous systems can wait an arbitrarily long time for such an element to complete, allowing robust mutual exclusion. Also, since there is no clock with which signals must be synchronized, asynchronous circuits more gracefully accommodate inputs from the outside world, which are by nature asynchronous.

With all of the potential advantages of asynchronous circuits, one might wonder why synchronous systems predominate. The reason is that asynchronous circuits have several problems as well. Primarily, asynchronous circuits are more difficult to design in an ad hoc fashion than synchronous circuits. In a synchronous system, a designer can simply define the combinational logic necessary to compute the given functions, and surround it with latches. By setting the clock rate to a long enough period, all worries about hazards (undesired signal transitions) and the dynamic state of the circuit are removed. In contrast, designers of asynchronous systems must pay a great deal of attention to the dynamic state of the circuit. Hazards must also be removed from the circuit, or not introduced in the first place, to avoid incorrect results. The ordering of operations, which was fixed by the placement of latches in a synchronous system, must be carefully ensured by the asynchronous control logic. For complex systems, these issues become too difficult to handle by hand.

Finally, even though most of the advantages of asynchronous circuits are towards higher performance, it isn't clear that asynchronous circuits are actually any faster in practice. Asynchronous circuits generally require extra time due to their signaling policies, thus increasing average-case delay. Whether this cost is greater or less than the benefits listed previously is unclear, and more research in this area is necessary.

Even with all of the problems listed above, asynchronous design is an important research area. Regardless of how successful synchronous systems are, there will always be a need for asynchronous systems. Asynchronous logic may be used simply for the interfacing of a synchronous system to its environment and other synchronous systems, or possibly for more complete applications.

9.1 DIFFERENCE BETWEEN SYNCHRONOUS AND ASYNCHRONOUS

Sequential circuits are divided into two main types: *synchronous* and *asynchronous*. Their classification depends on the timing of their signals.

Synchronous sequential circuits change their states and output values at discrete instants of time, which are specified by the rising and falling edge of a free-running *clock signal*. The clock signal is generally some form of square wave as shown in Fig. 9.1 below.

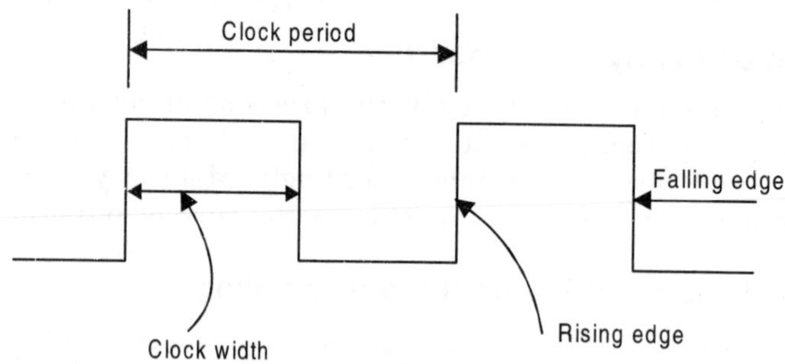

Fig. 9.1 Clock signal

From the diagram you can see that the *clock period* is the time between successive transitions in the same direction, that is, between two rising or two falling edges. State transitions in synchronous sequential circuits are made to take place at times when the clock is making a transition from 0 to 1 (rising edge) or from 1 to 0 (falling edge). Between successive clock pulses there is no change in the information stored in memory.

The reciprocal of the clock period is referred to as the *clock frequency*. The *clock width* is defined as the time during which the value of the clock signal is equal to 1. The ratio of the clock width and clock period is referred to as the duty cycle. A clock signal is said to be *active high* if the state changes occur at the clock's rising edge or during the clock width. Otherwise, the clock is said to be *active low*. Synchronous sequential circuits are also known as *clocked sequential circuits*.

The memory elements used in synchronous sequential circuits are usually flip-flops. These circuits are binary cells capable of storing one bit of information. A flip-flop circuit has two outputs, one for the normal value and one for the complement value of the bit stored in it. Binary information can enter a flip-flop in a variety of ways, a fact which give rise to the different types of flip-flops.

In *asynchronous* sequential circuits, the transition from one state to another is initiated by the change in the primary inputs; there is no external synchronization. The memory commonly used in asynchronous sequential circuits are time-delayed devices, usually implemented by feedback among logic gates. Thus, asynchronous sequential circuits may be regarded as combinational circuits with feedback. Because of the feedback among logic gates, asynchronous sequential circuits may, at times, become unstable due to transient conditions.

The differences between synchronous and asynchronous sequential circuits are:

- In a clocked sequential circuit, a change of state occurs only in response to a synchronizing clock pulse. All the flip-flops are clocked simultaneously by a common clock pulse. In an asynchronous sequential circuit, the state of the circuit can change immediately when an input change occurs. It does not use a clock.
- In clocked sequential circuits, input changes are assumed to occur between clock pulses. The circuit must be in the stable state before next clock pulse arrives. In asynchronous sequential circuits input changes should occur only when the circuit is in a stable state.
- In clocked sequential circuits, the speed of operation depends on the maximum allowed clock frequency. Asynchronous sequential circuits do not require clock pulses and they can change state with the input change. Therefore, in general the asynchronous sequential circuits are faster than the synchronous sequential circuits.
- In clocked sequential circuits, the memory elements are clocked flip-flops. In asynchronous sequential circuits, the memory elements are either unclocked flip-flops (latches) or gate circuits with feedback producing the effect of latch operation.

In clocked sequential circuits, any number of inputs can change simultaneously (during the absence of the clock). In asynchronous sequential circuits only one input is allowed to change at a time in the case of the level inputs and only one pulse input is allowed to be present in the case of the pulse inputs. If more than one level inputs change simultaneously or more than one pulse input is present, the circuit makes erroneous state transitions due to different delay paths for each input variable.

9.2 MODES OF OPERATION

Asynchronous sequential circuits can be classified into two types:

- Fundamental mode asynchronous sequential circuit
- Pulse mode asynchronous sequential circuit

Fundamental Mode

In fundamental mode, the inputs and outputs are represented by levels rather than pulses. In fundamental mode asynchronous sequential circuit, it is also assumed that the time difference between two successive input changes is larger than the duration of internal changes. Fundamental mode operation assumes that the input signals will be changed only when the circuit is in a stable state and that only one variable can change at a given time.

Pulse Mode

In pulse mode, the inputs and outputs are represented by pulses. In this mode of operation the width of the input pulses is critical to the circuit operation. The input pulse must be long enough for the circuit to respond to the input but it must not be so long as to be present even after new state is reached. In such a situation the state of the circuit may make another transition.

The minimum pulse width requirement is based on the propagation delay through the next state logic .The maximum pulse width is determined by the total propagation delay through the next state logic and the memory elements.

In pulse-mode operation, only one input is allowed to have pulse present at any time. This means that when pulse occurs on any one input, while the circuit is in stable state, pulse must not arrive at any other input. Figure 9.2 illustrates unacceptable and acceptable input pulse change. X_1 and X_2 are the two inputs to a pulse mode circuit. In Fig. 9.2 (*a*) at time t_3 pulse at input X_2 arrives.

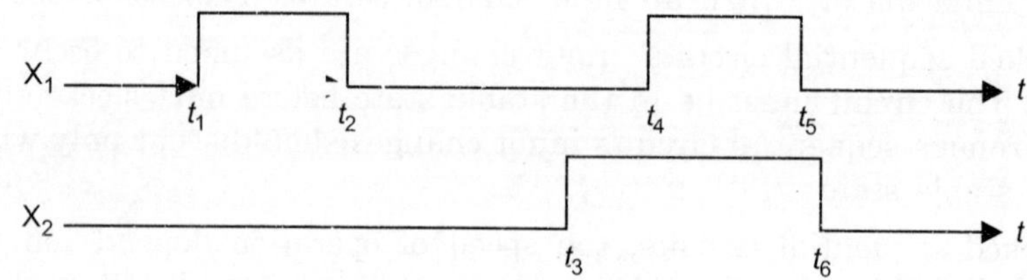

Fig. 9.2 (*a*) Unacceptable pulse mode input changes

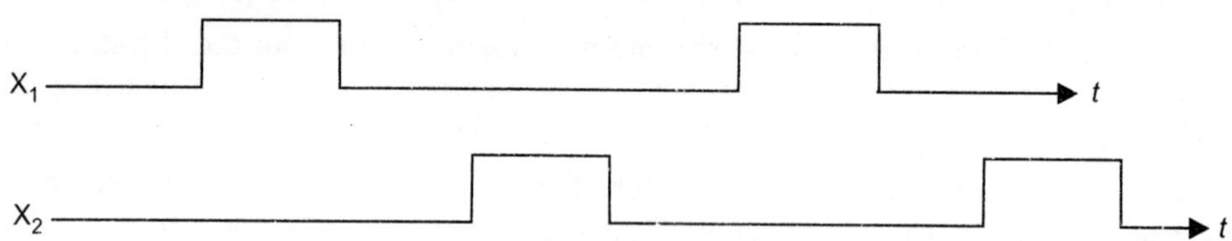

Fig. 9.2 (*b*) Acceptable pulse mode input changes

While this pulse is still present, another pulse at X_1 input arrives at t_4. Therefore, this kind of the presence of pulse inputs is not allowed.

Both fundamental and pulse mode asynchronous sequential circuits use unclocked S-R flip-flops or latches. In the design of both types of circuits, it is assumed that a change occurs in only one input and no change occurs in any other inputs until the circuit enters a stable state.

9.3 ANALYSIS OF ASYNCHRONOUS SEQUENTIAL MACHINES

Analysis of asynchronous sequential circuits operation in fundamental mode and pulse mode will help in clearly understanding the asynchronous sequential circuits.

9.3.1 Fundamental Mode Circuits

Fundamental mode circuits are of two types:

- Circuits without latches
- Circuits with latches

9.3.2 Circuits without Latches

Consider a fundamental mode circuit shown in Fig. 9.3.

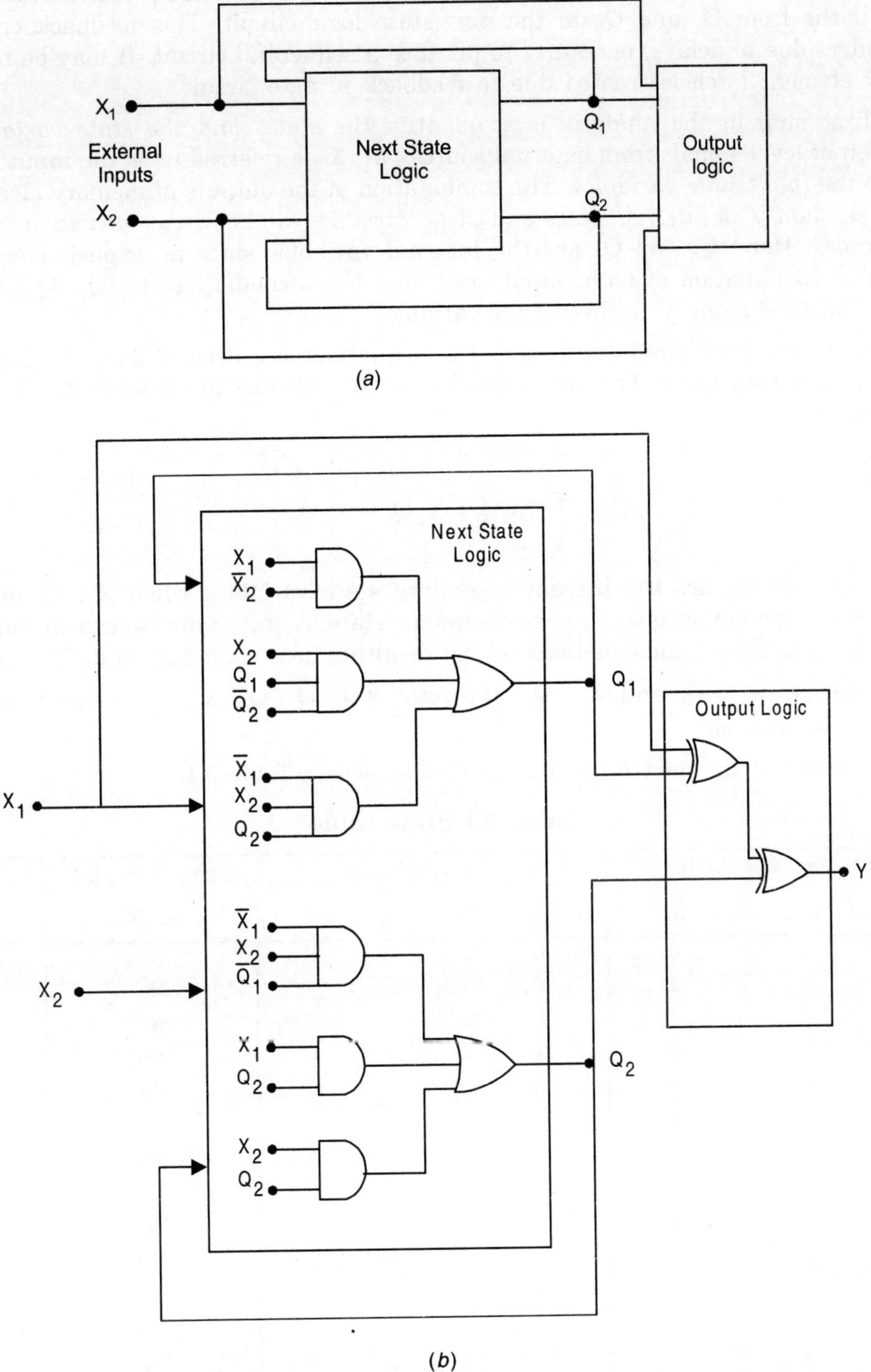

Fig. 9.3 Fundamental mode asynchronous sequential circuit without latch (*a*) block diagram (*b*) circuit diagram

This circuit has only gates and no explicit memory elements are present. There are two feedback paths from Q_1 and Q_2 to the next-state logic circuit. This feedback creates the latching effect due to delays, necessary to produce a sequential circuit. It may be noted that a memory element latch is created due to feedback in gate circuit.

The first step in the analysis is to identify the *states* and the *state variables*. The combination of level signals from external sources X_1, X_2 is referred to as the input state and X_1, X_1 are the *input state variables*. The combination of the outputs of memory elements are known as *secondary*, or *internal states* and these variables are known as *internal* or *secondary state variables*. Here, Q_1 and Q_2 are the internal variables since no explicit elements are present. The combination of both, input state and the secondary state (Q_1, Q_2, X_1, X_2) is known as the *total state*. Y is the output variable.

The next secondary state and output logic equations are derived from the logic circuit in the next-state logic block. The next-secondary state variables are denoted by Q_1^+ and Q_2^+ these are given by

$$Q_1^+ = X_1 \overline{X}_2 + \overline{X}_1 X_2 Q_2 + X_2 Q_1 \overline{Q}_2$$

$$Q_2^+ = \overline{X}_1 X_2 \overline{Q}_1 + X_1 Q_2 + X_2 Q_2$$

$$Y = X_1 \oplus Q_1 \oplus Q_2$$

Here, Q_1 and Q_2 are the present secondary state variables when X_1, X_2 input-state variables occur, the circuit goes to next secondary state. A state table shown in Table 9.1 is constructed using these logic equations. If the resulting next secondary state is same as the present state, i.e. $Q_1^+ = Q_1$ and $Q_2^+ = Q_2$, the total state Q_1, Q_2, X_1, X_2 is said to be stable. Otherwise it is unstable.

The stability of the next total state is also shown in Table 9.1.

Table 9.1 State table

Present total state				*Next total state*				*Stable total state*	*Output*
Q_1	Q_2	X_1	X_2	Q_1^+	Q_2^+	X_1	X_2	*Yes/No*	Y
0	0	0	0	0	0	0	0	Yes	0
0	0	0	1	0	1	0	1	No	0
0	0	1	1	0	0	1	1	Yes	1
0	0	1	0	1	0	1	0	No	1
0	1	0	0	0	0	0	0	No	1
0	1	0	1	1	1	0	1	No	1
0	1	1	1	0	1	1	1	Yes	0
0	1	1	0	1	1	1	0	No	0
1	1	0	0	0	0	0	0	No	0
1	1	0	1	1	1	0	1	Yes	0
1	1	1	1	0	1	1	1	No	1
1	1	1	0	1	1	1	0	Yes	1
1	0	0	0	0	0	0	0	No	1
1	0	0	1	1	0	0	1	Yes	1
1	0	1	1	1	0	1	1	Yes	0
1	0	1	0	1	0	1	0	Yes	0

9.3.3 Transition Table

A state table can be represented in another form known as *transition table*. The transition table for the state table of Table 9.1 is shown in Fig. 9.4.

In a transition table, columns represent input states (one column for each input state) and rows represent secondary states (one row for each secondary state). The next secondary state values are written into the squares, each indicating a total state. The stable states are circled. For any given present secondary state (Q_1 Q_2), the next secondary state is located in the square corresponding to row for the present secondary state and the column for the input state (X_1 X_2).

For example, for Q_1 Q_2 = 11 and X_1 X_2 = 00, the next secondary state is 00 (third row, first column) which is an unstable state.

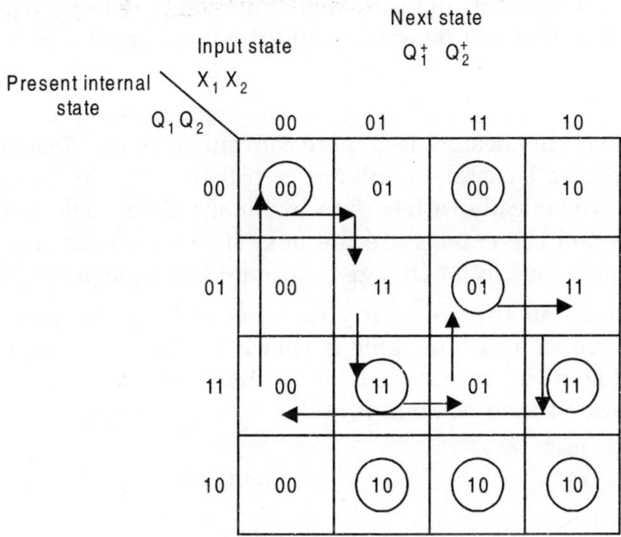

Fig. 9.4 Transition table for table 9.1

For a given input sequence, the total state sequence can be determined from the transition table.

Example. *For the transition table shown in Fig 9.4, the initial total state is Q_1 Q_2 X_1 X_2 = 0000. Find the total state sequence for an input sequence X_1 X_2 = 00, 01, 11, 10, 00.*

Solution. For a given internal state of the circuit, a change in the value of the circuit input causes a horizontal move in the transition table to the column corresponding to the new input value. A change in the internal state of the circuit is reflected by a vertical move. Since a change in the input can occur only when the circuit is in a stable state, a horizontal move can start only from a circled entry.

The initial total state is 0000 (first row, first column) which is a stable state. When the input state changes from 00 to 01, the circuit makes a transition (horizontal move) from the present total state to the next total state 0101 (first row, second column) which is unstable. Next, the circuit makes another transition from 0101 to 1101 (vertical move) (second row, second column) which is also an unstable state. Finally in the next transition (vertical move) it comes to stable state 1101 (third row, second column). All these transitions are indicated by arrows. Thus we see that a single input change produces two secondary state changes before a stable total state is reached. If the input is next changed to 11 the circuit goes to total state 0111 (horizontal move) which is unstable and then to stable total state 0111 (vertical move). Similarly, the next input change to 10 will take the circuit to unstable total state 1110 (horizontal move) and finally to stable total state 1110 (vertical move). A change in input state from 10 to 00 causes a transition to unstable total state 0000 (horizontal move) and then to stable total state 0000 (vertical move), completing the state transitions for the input sequence. All the state transitions are indicated by arrows.

The total state sequence is

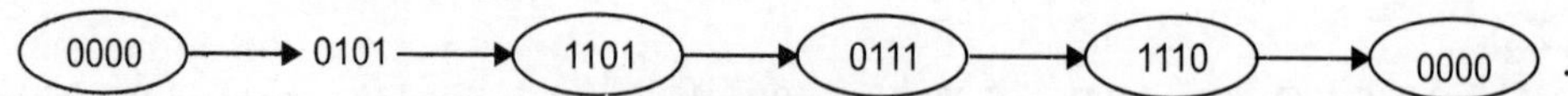

From the preceding discussions we see that from the logic diagram of an asynchronous sequential circuit, logic equations, state table, and transition table can be determined. Similarly, from the transition table, logic equations can be written and the logic circuit can be designed.

9.3.4 Flow Table

In asynchronous sequential circuits design, it is more convenient to use *flow table* rather than transition table. A flow table is basically similar to a transition table except that the internal states are represented symbolically rather than by binary states. The column headings are the input combinations and the entries are the next states, and outputs. The state changes occur with change of inputs (one input change at a time) and logic propagation delay.

The flow of states from one to another is clearly understood from the flow table. The transition table of Fig. 9.4 constructed as a flow table is shown in Fig. 9.5. Here, *a*, *b*, *c*, and *d* are the states. The binary value of the output variable is indicated inside the square next to the state symbol and is separated by a comma. A stable state is circled.

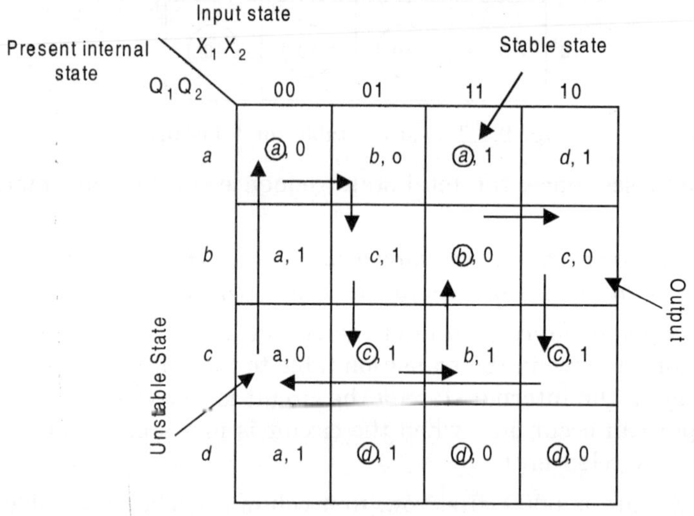

Fig. 9.5 Flow table

From the flow table, we observe the following behavior of the circuit.

When X_1 X_2 = 00, the circuit is in state ⓐ. It is a stable state. If X_2 changes to 1 while X_1 = 0, the circuit goes to state *b* (horizontal move) which is an unstable state. Since *b* is an unstable state, the circuit goes to *c* (vertical move), which is again an unstable state. This causes another vertical move and finally the circuit reaches a stable state ⓒ. Now consider X_1 changing to 1 while X_2 = 1, there is a horizontal movement to the next column. Here *b* is an unstable state and therefore, there is a vertical move and the circuit comes to a stable state ⓑ. Next change in X_2 from 1 to 0 while X_1 remaining 1 will cause horizontal move to state *c* (unstable state) and finally to stable state ⓒ due to the vertical move. Similarly changing X_1 from 1 to 0 while X_2 = 0 will cause the circuit to go to the unstable state *a* and finally to stable state ⓐ. The flow of circuit states are shown by arrows.

In the flow table of Fig. 9.5, there are more than one stable states in rows. For example, the first row contains stable states in two columns. If every row in a flow table has only one stable state, the flow table is known as a *primitive flow table.*

From a flow table, transition table can be constructed by assigning binary values to each state and from the transition table logic circuit can be designed by constructing K-maps for Q_1^+ and Q_2^+.

9.3.5 Circuits with Latches

Analysis Procedure (Circuits using SR Latches)

Asynchronous sequential circuits constructed with SR latches may or may not have external feedback paths. (A feedback loop is always there within the latch itself). The procedure for analyzing an asynchronous sequential circuit with SR latches is as follows:

1. Label each latch output with Y_i and its external feedback path (if any) with y_i for $i = 1, 2, \ldots, k$.
2. Derive the Boolean functions for the S_i and R_i inputs in each latch.
3. Check whether $SR = 0$ for each NOR latch or whether $\bar{S}\bar{R} = 0$, for each NAND latch. If this condition is not satisfied, there is a possibility that the circuit may not operate properly.
4. For each NOR latch, evaluate $Y = S + \bar{R}y$

   or, for each NAND latch, evaluate $Y = \bar{S} + Ry$
5. Construct a map with y's representing the rows and inputs x representing the columns.
6. Plot the transition table for $Y = Y_1 Y_2 \ldots\ldots Y_k$ in the map.
7. Encircle all the stable states where $Y = y$.

Circuit Implementation with *SR* Latches

An asynchronous sequential circuit can be implemented with *SR* latches. It is a procedure of obtaining the logic diagram from a given transition table. The inputs *S* and *R* are to be derived from the given transition table and excitation table.

The procedure for circuit implementation with *SR* latches is as follows:

1. From the given transition table (which specifies the excitation function $Y = Y_1 Y_2 \ldots Y_k$), derive a pair of maps for inputs S_i and R_i for each $i = 1, 2, \ldots k$, by using excitation table for *SR* latch.

2. Derive S_i and R_i in simplified form. Here, S_i and R_i cannot be made equal to 1 in the same minterm square.
3. Draw the logic digram with k latches along with additional required gates to generate the S and R Boolean functions. For NOR latches, S and R Boolean functions are used in the usual form but for NAND latches, they are used in complemented form.

In chapter 6 latches were introduced. Latch circuits using NAND and NOR gates are shown in Fig. 9.6.

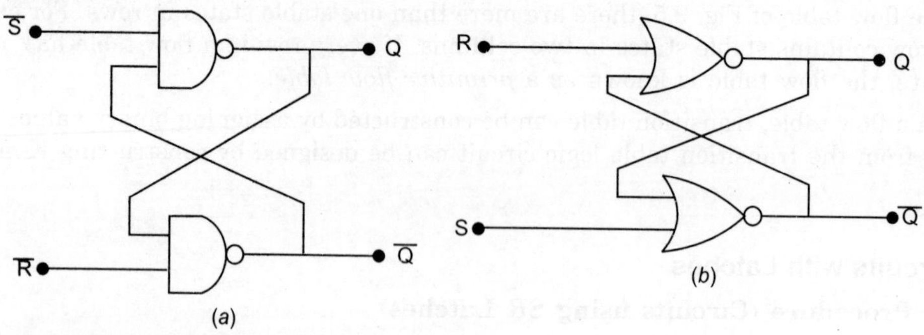

Fig. 9.6 (*a*) $\overline{S}$-$\overline{R}$ latch using NAND gates (*b*) S-R latch using NOR gates

For the circuit of Fig. 9.6(*a*), the next-state equation is

$$Q^+ = \overline{\overline{S}.\overline{Q}} = \overline{\overline{S}.\overline{(Q\overline{R})}}$$

$$= S + \overline{R}Q$$

Similarly, for the circuit of Fig. 9.6(*b*), the next-state equation is

$$Q^+ = \overline{R + \overline{Q}} = \overline{R + \overline{(S + Q)}}$$

$$= \overline{R}\,.\,(S + Q)$$

$$= S\overline{R} + \overline{R}Q$$

Since, S = R = 1 is not allowed, which means SR = 0, therefore,

$$S\overline{R} = S\overline{R} + SR = S(\overline{R} + R) = S$$

which gives,

$Q^+ = S + \overline{R}Q$ It is same as the next-state equation for the circuit of Fig. 9.6(*a*)

The transition table of S-R latch is shown in Fig. 9.7.

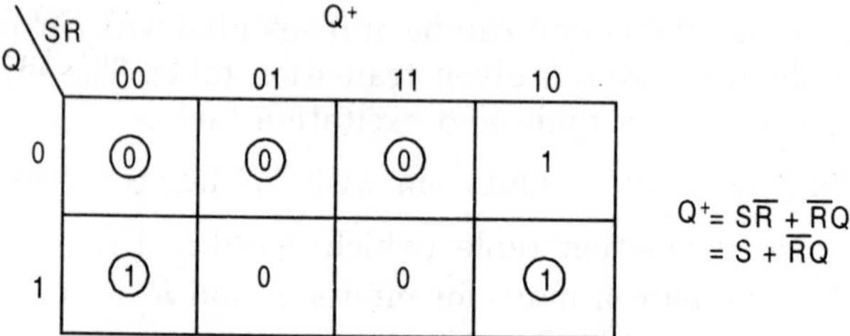

Fig. 9.7 Transition table of S-R latch

From the transition table of S-R FLIP-FLOP, we observe that when SR changes from 11 to 00 the circuit will attain either the stable state ⓪ (first row, first column) or ① (second row, first column) depending upon whether S goes to 0 first or R goes to 0 first respectively. Therefore, S = R = 1 must not be applied.

Consider an asynchronous sequential circuit with latches shown in Fig. 9.8.

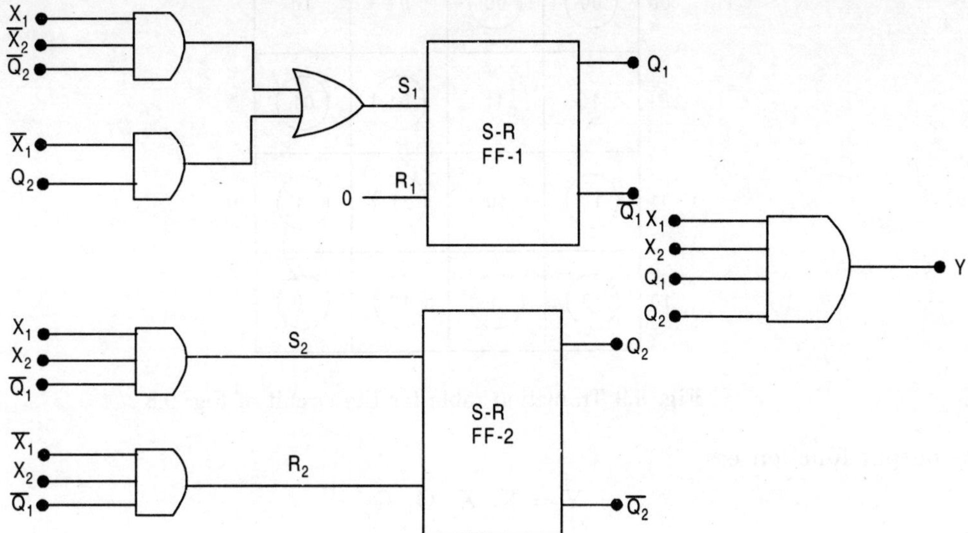

Fig. 9.8 Asynchronous sequential circuit with latches

For FF-1, $R_1 = 0$ and the excitation equation for S_1 is

$$S_1 = X_1 \overline{X}_2 \overline{Q}_2 + \overline{X}_1 Q_2$$

The next-state equation is

$$Q_1^+ = S_1 + \overline{R}_1 Q_1$$

Substituting the value of S_1 we obtain,

$$Q_1^+ = X_1 \overline{X}_2 \overline{Q}_2 + \overline{X}_1 Q_2 + Q_1$$

Similarly, the excitation equations for FF-2 are

$$S_2 = X_1 X_2 \overline{Q}_1,\ R_2 = \overline{X}_1 X_2 \overline{Q}_1$$

The next-state equation is

$$Q_2^+ = S_2 + \overline{R}_2 Q_2$$

$$= X_1 X_2 \overline{Q}_1 + \overline{\overline{X}_1 X_2 \overline{Q}_1} . Q_2$$

Using next-state equation for FF-1 and FF-2, transition table is obtained as shown in Fig. 9.9.

Q_1Q_2 \ X_1X_2	00	01	11	10
00	(00)	(00)	01	10
01	11	11	(01)	(01)
11	(11)	10	(11)	(11)
10	(10)	(10)	(10)	(10)

$Q_1^+ Q_2^+$

Fig. 9.9 Transition table for the circuit of Fig. 9.8

The output function is

$$Y = X_1 X_2 Q_1 Q_2$$

Its flow table is shown in Fig. 9.10.

Q_1Q_2 \ X_1X_2	00	01	11	10
a	(a), 0	(a), 0	b, 0	d, 0
b	c, 0	c, 0	(b), 0	(b), 0
c	(c), 0	d, 0	(c), 1	(c), 0
d	(d), 0	(d), 0	(d), 0	(d), 0

$Q_1^+ Q_2^+$

Fig. 9.10 Flow table for the circuit of Fig. 9.8

From a flow table, transition table can be obtained by assigning binary values to the states. From the transition table, logic equations can be obtained by constructing K-maps for S and R inputs of every latch. For this, the excitation table of S-R latch will be used. Logic circuit can then be designed using the logic equation for S, R inputs of every latch.

Example. *Design logic circuit using S-R latches for the transition table of Fig. 9.4.*

Solution. Since, there are two internal states Q_1 and Q_2, therefore, two S-R latches are required for the design of logic circuit. Let the two latches be L_1 and L_2. The inputs and outputs of these latches are given as

Latch	*Inputs*	*Outputs*
L_1	S_1, R_1	Q_1, $\overline{Q}_1$
L_2	S_2, R_2	Q_2, $\overline{Q}_2$

The excitation table of an S-R latch is given in Table 9.2. This is same as for S-R flip-flop.

Table 9.2 Excitation table of *S-R* latch

Present state	*Next state*	*Inputs*	
Q	Q^+	S	R
0	0	0	×
0	1	1	0
1	0	0	1
1	1	×	0

To determine S_1 and R_1 for different values of X_1 X_2, we make use of Q_1 and Q_1^+ values for every square of transition table. For example, this square in the first row and first column gives $Q_1 = 0$ and $Q_1^+ = 0$. This means, for the present state 0 the circuit gives next state as 0 for Q_1. Corresponding to this we find the value of S_1 and R_1 using the Table 9.2, which are $S_1 = 0$ and $R_1 = X$.

Thus the entry in the cell corresponding to X_1 X_2 = 00 and Q_1 Q_2 = 00 for K-map of S_1 will be 0 and for K-map of R_1 it will be X. Similarly, K-map entries are determined for S_1 and R_1.

Following similar procedure, K-maps for S_2 and R_2 are constructed. The K-maps are given in Fig. 9.11.

From the K-map of Fig. 9.11, we obtain logic equations for S_1, R_1, S_2, and R_2.

$$S_1 = X_1\overline{X}_2 + \overline{X}_1 X_2 Q_2$$

$$R_1 = \overline{X}_1 + X_1 X_2 Q_2$$

$$S_2 = \overline{X}_1 X_2 \overline{Q}_2$$

$$R_2 = \overline{X}_1\overline{X}_2$$

The logic circuit is shown in Fig. 9.12.

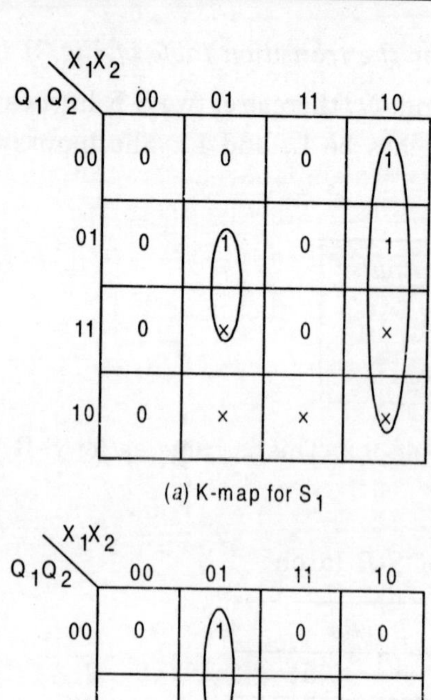

(*a*) K-map for S_1

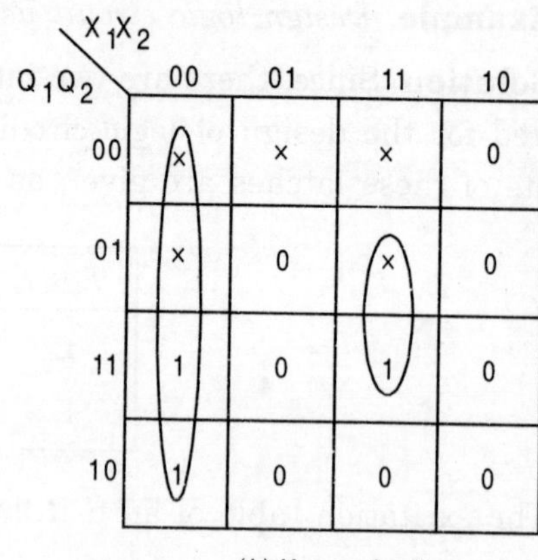

(*b*) K-map for R_1

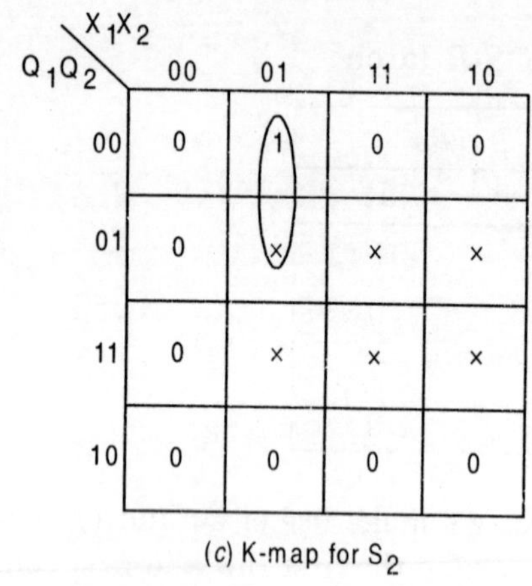

(*c*) K-map for S_2

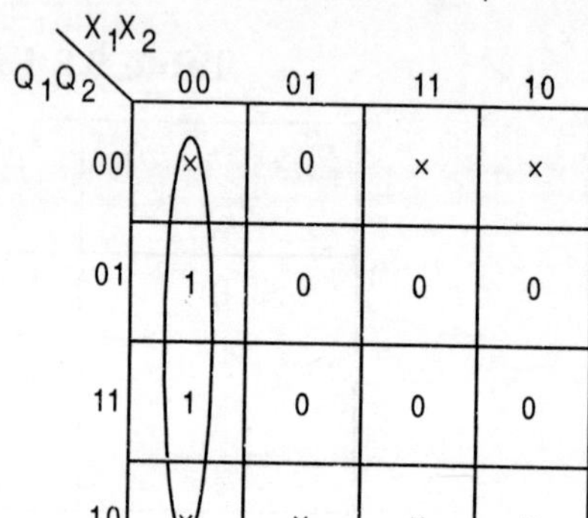

(*d*) K-map for R_2

Fig. 9.11

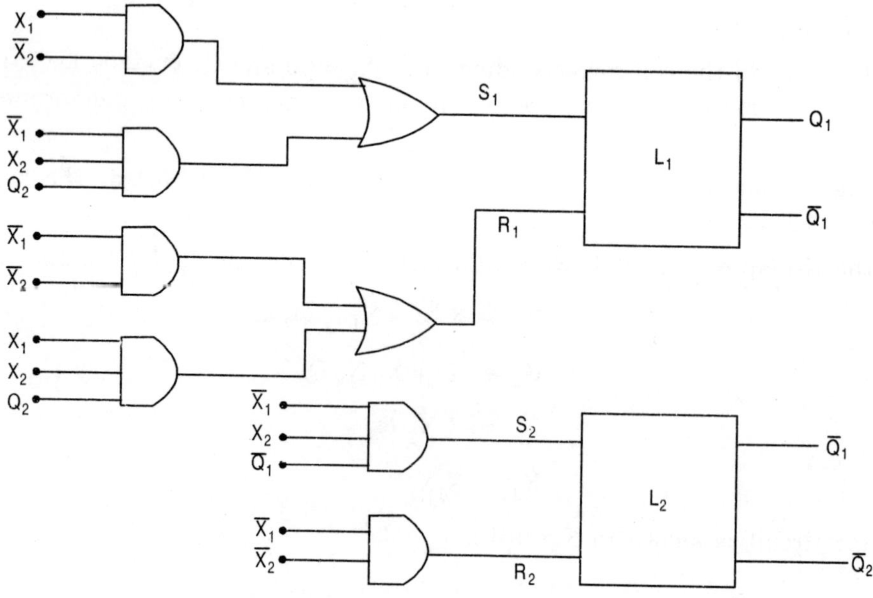

Fig. 9.12

9.3.6 Races and Cycles

A *race* condition exists in an asynchronous sequential circuit when more than one state variable change value in response to a change in an input variable. This is caused because of unequal propagation delays in the path of different secondary variables in any practical electronic circuit. Consider a transition table shown in Fig. 9.13. When both the inputs X_1 and X_2 are 0 and the present state is $Q_1\,Q_2$ = 00, the resulting next state $Q_1^+\,Q_2^+$ will have Q_1^+ = 1 and Q_2^+ = 1 simultaneously if the propagation delays in the paths of Q_1 and Q_2 are equal.

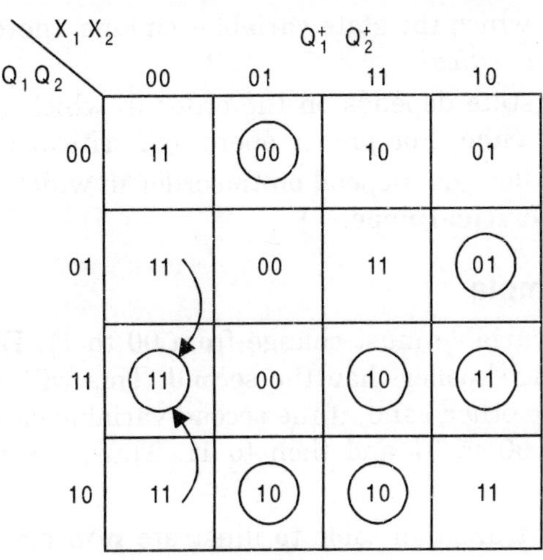

Fig. 9.13

Since Q_1 and Q_2 both are to change and in general the propagation delays in the paths of Q_1 and Q_2 are not same, therefore, either Q_1 or Q_2 may change first instead of both changing simultaneously. As a consequence of this the circuit will go to either state 01 or to state 10.

If Q_2^+ changes faster than Q_1^+, the next state will be 01, then 11 (first column, second row) and then to the stable state (11) (first column, third row) will be reached. On the other hand, if Q_1^+ changes faster than Q_2^+, the next-state will be 10, then 11 (first column, fourth row) and then to the stable state (first column, third row) will be reached. In both the situations, the circuit goes to the same final stable state (11). This situation, where a change of more than one secondary variable is required is known as a *race*.

There are two types of races: *noncritical race* and *critical race*.

In the case of noncritical race, the final stable state in which the circuit goes does not depend on the sequence in which the variables change. The race discussed above is a noncritical race. In the case of critical race, the final stable state reached by the circuit depends on the sequence in which the secondary variables change. Since the critical race results in different stable states depending on the sequence in which the secondary states change, therefore, it must be avoided.

Race Conditions

This is the problem of changing multiple state variables simultaneously in response to the input variable. The situation where multiple state variables can change at the same time, is called a **race**. The state variables are **racing** to see which one can change first.

A race condition exists in an asynchronous circuit when two or more binary state variables change value in response to a change in an input variable. This may cause the state variable to change in an unpredictable manner, when unequal delays are encountered.

Categories of Race Conditions

Based on the order by which the state variables change, there are two categories of race conditions : *critical* and *non-critical.*

When the final stable state depends on the order in which the state variables change, this is the case of **critical race**. For proper operation, critical races must be avoided.

If the final stable state does not depend on the order in which the state variables change, the race is known as **non-critical race**.

Explanation with an Example

Consider, if the state variable must change from 00 to 11. Due to the delay difference, the first variable may change sooner than the second. This will result in sequence from 00 to 10 and then to 11. On the other hand, if the second variable changes sooner than the first, the sequence will be from 00 to 01 and then to 11. Thus, the order of change cannot be predicted in advance.

Consider the following transition table to illustrate **non-critical** races:

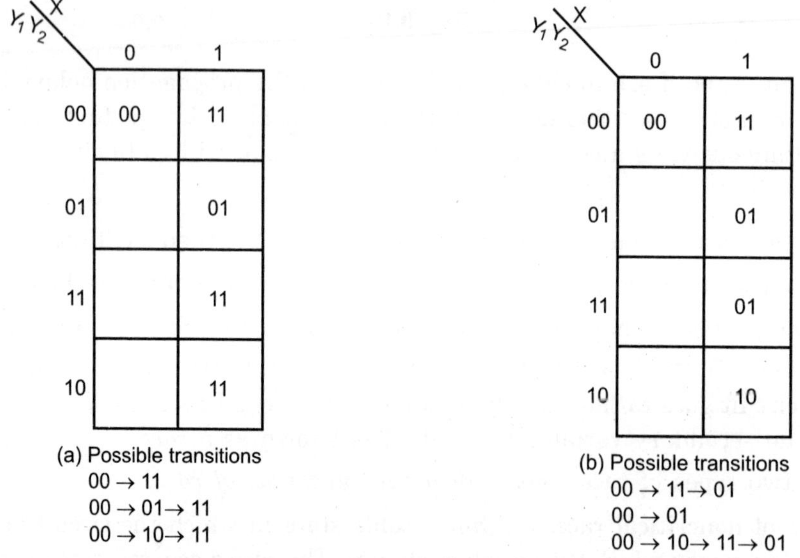

Fig. 9.13 (*a*) Example of non-critical races

Starting, with the total stable state $y_1y_2\ x = 000$ and the input changes from 0 to 1, the state variables must change from 00 to 11. This is the race condition. Following are the three possible ways in which the state variable changes:

- Simultaneously changing both variables from 00 to 11 or
- Changes in sequence from 00 to 01 then to 11 or
- Changes in sequence from 00 to 10 then to 11.

Since in all cases, the final state is the same, so race is non-critical.

To illustrate the critical races, following are the examples:

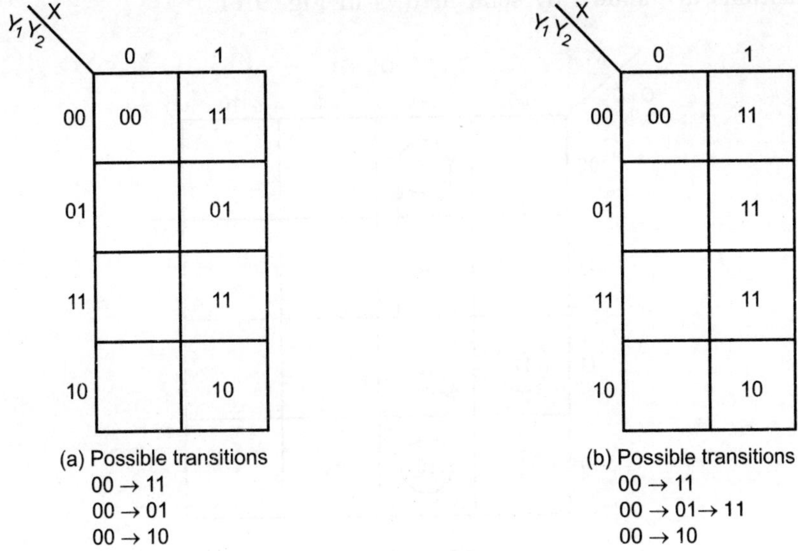

Fig. 9.13 (*b*) Examples of critical races

Starting with the stable state y_1y_2 = 00 and input changes from 0 to 1, the state must change from 00 to 11 (for race condition). Both variables may change simultaneously. From table (*a*), due to unequal propagation delay, if Y_2 changes to 1 before Y_1 does, the circuit goes to state 01. If Y_1 changes before Y_2 then the circuit goes to state 10. Hence the race is critical because circuit goes to different stable states.

How to Avoid Race Condition

Races may be avoided by

- Making a proper binary assignment to the state variables. Assignment should be done in a way that only one state variable changes at one time during transition.
- Directing the circuit through intermediate unstable states with a **unique** state variable change.

Stability Considerations

This is another difficulty that arises in an asynchronous sequential circuits due to the feedback connections. This causes the circuit to become unstable i.e., circuit oscillates between unstable states.

The transition table is a tool used for detecting the instability occurrence.

Example. *In the transition table of Fig. 9.13, consider the circuit in stable total state 1100. Will there be any race, if the input state changes to 01? If yes, find the type of race.*

Solution. When the circuit is in stable total state, $X_1 X_2 = 00$. Now X_2 changes to 1 while $X_1 = 0$. From Fig. 9.13 we see that the required transition is to state 00. If Q_1^+ and Q_2^+ become 00 simultaneously, then the transition will be

$$\textcircled{11} \rightarrow 00 \rightarrow \textcircled{00}$$

These transitions are shown by solid arrows in Fig. 9.14.

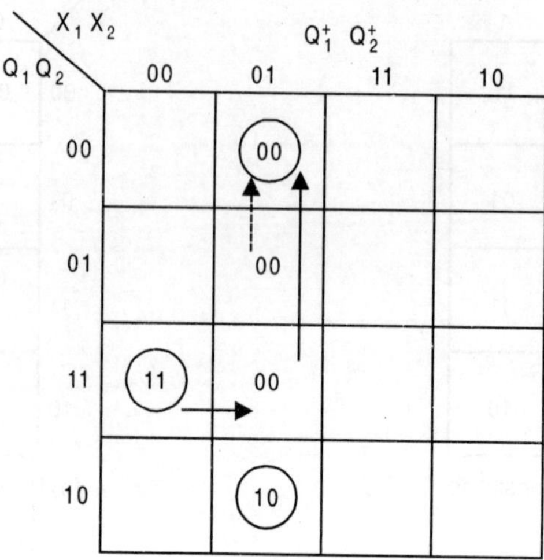

Fig. 9.14

If Q_2^+ becomes 0 faster than Q_1^+, the circuit will go to the state 10 and then to $\textcircled{10}$, which is a stable state. The transition is

$$\textcircled{11} \rightarrow 10 \rightarrow \textcircled{10}$$

On the other hand, if Q_1^+ becomes 0 faster than Q_2^+, the transition will be

$$\textcircled{11} \rightarrow 01 \rightarrow 00 \rightarrow \textcircled{00}$$

It is shown by dotted arrow in Fig. 9.13. Thus, we see that the circuit attains different stable states $\textcircled{00}$ or $\textcircled{10}$ depending upon the sequence in which the secondary variables change.

Therefore, the race condition exists in this circuit and it is critical race.

Races can be avoided by making a proper binary assignment to the state variables in a flow table. The state variables must be assigned binary numbers in such a way so that only one state variable can change at any one time when a state transition occurs in the flow table. The state transition is directed through a unique sequence of unstable state variable change. This is referred to as a *cycle*. This unique sequence must terminate in a stable state, otherwise the circuit will go from one unstable state to another unstable state making the entire circuit unstable.

9.3.7 Pulse-Mode Circuits

In a pulse-mode asynchronous sequential circuit, an input pulse is permitted to occur only when the circuit is in stable state and there is no pulse present on any other input.

When an input pulse arrives, it triggers the circuit and causes a transition from one stable state to another stable state so as to enable the circuit to receive another input pulse. In this mode of operation critical race can not occur. To keep the circuit stable between two pulses, flip-flops whose outputs are levels, must be used as memory elements.

For the analysis of pulse-mode circuits, the model used for the fundamental-mode circuits is not valid since the circuit is stable when there are no inputs and the absence of a pulse conveys no information. For this a model similar to the one used for synchronous sequential circuits will be convenient to use.

In pulse-mode asynchronous circuits the number of columns in the next-state table is equal to the number of input terminals.

Consider a pulse-mode circuit logic diagram shown in Fig. 9.15. In this circuit there are four input variables X_1, X_2, X_3, and X_4, and Y is the output variable. It has two states Q_1 and Q_2.

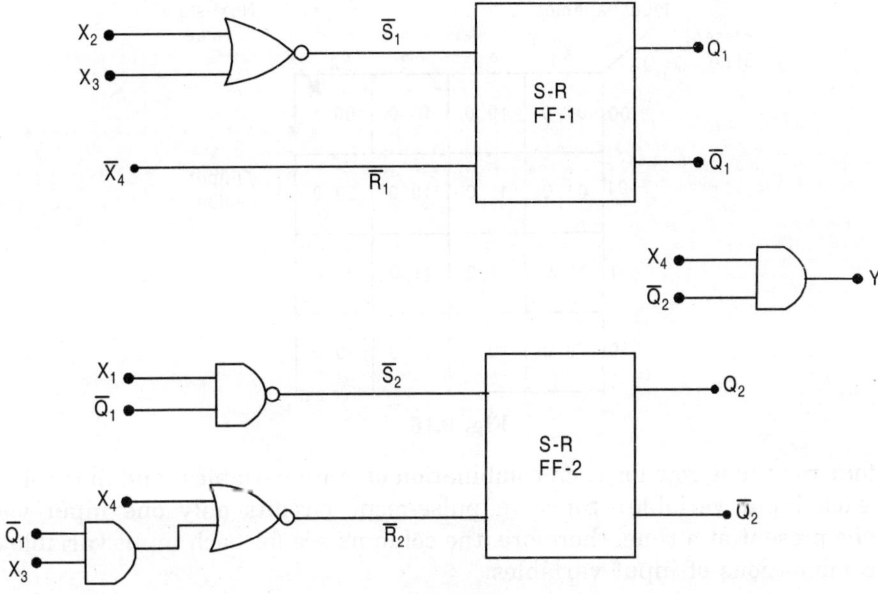

Fig. 9.15

The excitation equations are:

$$\overline{S}_1 = \overline{(X_2 + X_3)} \quad \text{or} \quad S_1 = X_2 + X_3$$

$$\overline{R}_1 = \overline{X}_4 \quad \text{or} \quad R_1 = X_4$$

$$\overline{S}_2 = \overline{\overline{Q}_1 X_1} \quad \text{or} \quad S_2 = \overline{Q}_1 X_1$$

$$\overline{R}_2 = \overline{(X_4 + \overline{Q}_1 X_3)} \quad \text{or} \quad R_2 = X_4 + \overline{Q}_1 X_3$$

The output equation is: $$Y = X_4 \overline{Q}_2$$

The next-state equations are obtained by using the excitation equations and the characteristic equation of latch.

These are:

$$Q_1^+ = S_1 + \bar{R}_1 Q_1$$
$$= X_2 + X_3 + \bar{X}_4 Q_1$$

and

$$Q_2^+ = S_2 + \bar{R}_2 Q_2$$
$$= \bar{Q}_1 X_1 + \overline{(X_4 + \bar{Q}_1 X_3)} \,.\, Q_2$$
$$= \bar{Q}_1 X_1 + \bar{X}_4 \,.\, (\overline{\bar{Q}_1 X_3}) \,.\, Q_2$$
$$= \bar{Q}_1 X_1 + \bar{X}_4 \,.\, (Q_1 + \bar{X}_3) \,.\, Q_2$$
$$= \bar{Q}_1 X_1 + Q_1 Q_2 \bar{X}_4 + Q_2 \bar{X}_3 \bar{X}_4$$

The transition table is constructed by evaluating the next-state and output for each present state and input value using next-state equations and output equation. The transition table is shown in Fig. 9.16.

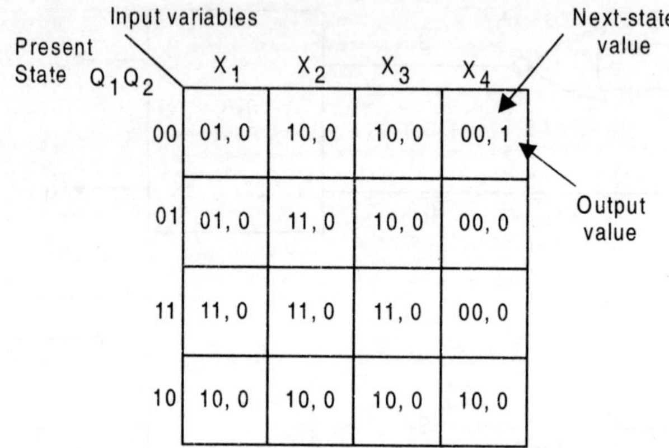

Present State Q_1Q_2	X_1	X_2	X_3	X_4
00	01, 0	10, 0	10, 0	00, 1
01	01, 0	11, 0	10, 0	00, 0
11	11, 0	11, 0	11, 0	00, 0
10	10, 0	10, 0	10, 0	10, 0

Fig. 9.16

It has four rows (one row for each combination of state variables) and four columns (one column for each input variable). Since in pulse-mode circuits only one input variable is permitted to be present at a time, therefore, the columns are for each input variable only and not for the combinations of input variables.

Flow table can be constructed from the transition table and is shown in Fig. 9.17. Here, S_0, S_1, S_2, and S_3 are the four state variables.

Present State Q_1Q_2 \ Input variables	X_1	X_2	X_3	X_4
S_0	S_1, 0	S_3, 0	S_3, 0	S_0, 1
S_1	S_1, 0	S_2, 0	S_3, 0	S_0, 0
S_2	S_2, 0	S_2, 0	S_2, 0	S_0, 0
S_3	S_3, 0	S_3, 0	S_3, 0	S_0, 1

Fig. 9.17

From a flow table, a transition table can be constructed by assigning binary values to the states. From a transition table next-state equations can be obtained and the logic diagram can then be obtained.

9.4 ASYNCHRONOUS SEQUENTIAL CIRCUIT DESIGN

Design of asynchronous sequential circuits is more difficult than that of synchronous sequential circuits because of the timing problems involved in these circuits. Designing an asynchronous sequential circuit requires obtaining logic diagram for the given design specifications. Usually the design problem is specified in the form of statements of the desired circuit performance precisely specifying the circuit operation for every applicable input sequence.

9.4.1 Design Steps

1. Primitive flow table is obtained from the design specifications. When setting up a primitive flow table it is not necessary to be concerned about adding states which may ultimately turn out to be redundant. A sufficient number of states are to be included to completely specify the circuit performance for every allowable input sequence. Outputs are specified only for stable states.
2. Reduce the primitive flow table by eliminating the redundant states, which are likely to be present. These redundant states are eliminated by merging the states. *Merger diagram* is used for this purpose.
3. Binary numbers are assigned to the states in the reduced flow table. The binary state assignment must be made to ensure that the circuit will be free of critical races. The output values are to be chosen for the unstable states with unspecified output entries. These must be chosen in such a way so that momentary false outputs do not occur when the circuit switches from one stable state to another stable state.
4. Transition table is obtained next.
5. From the transition table logic diagram is designed by using the combinational design methods. The logic circuit may be a combinational circuit with feedback or a circuit with *S-R* latches.

The above design steps is illustrated through an example.

Example. *The output (Y) of an asynchronous sequential circuit must remain 0 as long as one of its two inputs X_1 is 0. While $X_1 = 1$, the occurrence of first change in another input X_2, should give Y = 1 as long as $X_1 = 1$ and becomes 0 where X_1 returns to 0. Construct a primitive flow table.*

Solution. This circuit has two inputs X_1, X_2 and one output Y. For the construction of flow table, the next-state and output are required to be obtained. The flow table is shown in Fig. 9.18.

For $X_1 X_2 = 00$, let us take state a. When the circuit has $X_1 X_2 = 00$ the output is 0 (since $X_1 = 0$) and the circuit is in stable state ⓐ. The next-state and output are shown in the first column, first row of Fig. 9.18. Since only one input is allowed to change at a time, therefore, the next input may be $X_1 X_2 = 01$ or 10.

If $X_1 X_2 = 01$, let us take another state b, correspondingly the second row of the flow table corresponds to state b. when the inputs change from $X_1 X_2 = 00$ to 01, the circuit is required to go to stable state ⓑ and output is 0 (since $X_1 = 0$). Therefore, the entry in the second column, first row will be b, 0 and in the second column, second row will be ⓑ, 0. The output corresponding to unstable state b is taken as 0 so that no momentary false outputs occur when the circuit switches between stable states. On the other hand if $X_1\ X_2 = 10$, the circuit is required to go to another stable state ⓒ with output 0. Therefore, the entries in the fourth column, first row and fourth column, third row will be respectively c, 0 and ⓒ, 0.

Present-state \ $X_1 X_2$	00	01	11	10
a	ⓐ, 0	b, 0	–, –	c, 0
b	a, 0	ⓑ, 0	d, 0	–, –
c	a, 0	–, –	e, –	ⓒ, 0
d	–, –	b, 0	ⓓ, 0	f, –
e	–, –	b, –	ⓔ, 1	f, 1
f	a, –	–, –	e, 1	ⓕ, 1

Fig. 9.18 Flow table

Since, both the inputs cannot change simultaneously, therefore, from stable state ⓐ, the circuit cannot go to any specific state corresponding to $X_1\ X_2 = 11$ and accordingly the entry in the third column, first row will be –, –. The dashes represent the unspecified state, output.

Now consider the stable state ⓑ. The inputs $X_1\ X_2$ can change to 00 or 11. If $X_1\ X_2 = 00$, the circuit will go to state a. Therefore, the entry in the first column, second row will be a, 0. From this unstable state the circuit goes to stable state ⓐ. On the other hand if $X_1\ X_2 = 11$, then the circuit goes to a new state d. The output corresponding to $X_1\ X_2 = 11$ will be 0 since, there is no change in X_2, which is already 1. Therefore, the entry in the third column, second row will be d, 0. The fourth row corresponds to state d, and the entry in the third column, fourth row, will be ⓓ, 0. From ⓑ, the circuit is not required to go to any specific state and therefore, the entry in the fourth column, second row will be –,–.

Similarly, now consider stable state ⓒ. The inputs can change to $X_1\ X_2 = 11$ or 00. If $X_1\ X_2 = 11$, the circuit goes to a new stable state ⓔ and the output will be 1, since X_2 changes from 0 to 1 while $X_1 = 1$. The entry in the third column, third row will be c, –. Output has to change from 0 to 1 from stable state ⓒ to stable state ⓔ, which may or may not change to 1 for unstable e. The entry in the third column, fifth row will be ⓔ, 1. The entry in the second column third row will be –, – and the entry in the first column, third row will be a, 0 (for $X_1\ X_2 = 00$).

In the same manner, we consider the stable state ⓓ and obtain the entries f, – (fourth column, fourth row); ⓕ, 1 (fourth column, sixth row); b, 0 (second column, fourth row) and –, – (first column, fourth row).

Similar procedure applied to ⓔ and ⓕ, yields the remaining entries of the flow table.

Since, every row in the flow table of Fig. 9.18 contains only one stable state, therefore, this flow table is a primitive flow table.

9.4.2 Reduction of States

The necessity of reducing the number of states has been discussed in chapter 6 and the equivalent states have been defined. When asynchronous sequential circuits are designed, the design process starts from the construction of primitive flow table. A primitive flow table is never completely specified. Some states and outputs are not specified in it as shown in Fig. 9.18 by dashes. Therefore, the concept of equivalent states cannot be used for reduction of states. However, incompletely specified states can be combined to reduce the number of states in the flow table. Two incompletely specified states can be combined if they are *compatible.*

Two states are *compatible* if and only if, for every possible input sequence both produce the same output sequence whenever both outputs are specified and their next states are compatible whenever they are specified. The unspecified outputs and states shown as dashes in the flow table have no effect for compatible states.

Example 1. *In the primitive flow table of Fig. 9.18, find whether the states a and b are compatible or not. If compatible, find out the merged state.*

Solution. The rows corresponding to the states a and b are shown in Fig. 9.19. Each column of these two states is examined.

X_1X_2	00	01	11	10
a	ⓐ, 0	b, 0	–, –	c, 0
b	a, 0	ⓑ, 0	d, 0	–, –

Fig. 9.19

Column-1. Both the rows have the same state a and the same output 0. a in first row is stable state and in the second row is unstable state.

Since for the same input both the states a and b have the same specified next-state a and the same specified output 0. Therefore, this input condition satisfies the requirements of compatibility.

Column-2. The input condition X_1 X_2 = 01 satisfies the requirements of compatibility as discussed for column-1.

Column-3. The first row has unspecified next-state and output and the second row has specified state and output. The unspecified state and output may be assigned any desired state and output and therefore, for this input condition also the requirements of compatibility are satisfied.

Column-4. The requirements of compatibility are satisfied for the reasons same as applicable to column-3.

Therefore, we conclude that since the next-states and the outputs for all the input combinations are compatible for the two states a and b, the two states are compatible.The merged state will be as shown in Fig. 9.20.

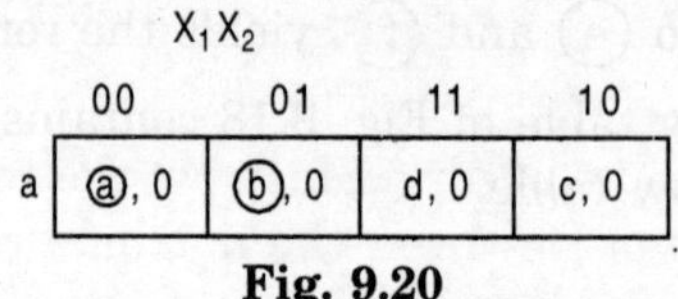

X_1X_2	00	01	11	10
a	ⓐ, 0	ⓑ, 0	d, 0	c, 0

Fig. 9.20

When the merged state entries are determined a circled entry and an uncircled entry results in a circled entry, since the corresponding state must be stable as shown in Fig. 9.20.

Example 2. *In the primitive flow table of Fig. 9.18 find whether the states a and e are compatible or not. Examine their compatibility if the entries in the fourth column for the states a and e have same output.*

Solution. The partial flow table for states a and e of Fig. 9.18 is shown in Fig. 9.21.

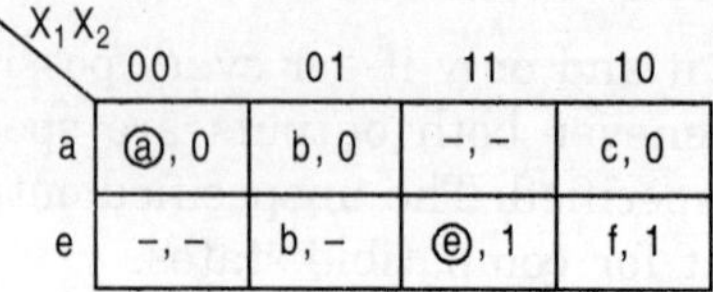

X_1X_2	00	01	11	10
a	ⓐ, 0	b, 0	–, –	c, 0
e	–, –	b, –	ⓔ, 1	f, 1

Fig. 9.21

From this we observe the following

Column-1 compatible

Column-2 compatible

Column-3 compatible

Column-4 not compatible, since the outputs are different.

Therefore, the states a and e are not compatible.

In case of same output in column-4, the outputs are said to be not conflicting and the states a and e are compatible if and only if the states c and f are compatible. This is referred to as c, f is implied by a, b or a, b implies c, f.

9.4.3 Merger Diagram

A *merger diagram (or graph)* is prepared for a primitive flow table to determine all the possible compatible states (maximal compatible states) and from this a minimal collection of compatibles covering all the states.

A merger graph is constructed following the steps outlined below:

- Each state is represented by a vertex, which means it consists of n vertices, each of which corresponds to a state of the circuit for an n- state primitive flow table. Each vertex is labelled with the state name.
- For each pair of compatible states an undirected arc is drawn between the vertices of the two states. No arc is drawn for incompatible states.
- For compatible states implied by other states a broken arc is drawn between the states and the implied pairs are entered in the broken space.

The flow table is required to be examined for all the possible pairs of states. All the pairs are checked and the merger graph is obtained. Thus, we see that the merger graph displays all possible pairs of compatible states and their implied pairs. Next, it is necessary to check whether the incompatible pair (s) does not invalidate any other implied pair. If any implied pair is invalidated it is neglected. All the remaining valid compatible pairs form a group of *maximal* compatibles.

The maximal compatible set can be used to construct the reduced flow table by assigning one row to each member of the group. However, the maximal compatibles do not necessarily constitute the set of *minimal* compatibles. The set of minimal compatibles is a smaller collection of compatibles that will satisfy the row merging.

The conditions that must be satisfied for row merging are:

- the set of chosen compatibles must *cover* all the states, and
- the set of chosen compatibles must be *closed*.

The condition of covering requires inclusion of all the states of the primitive flow graph in the set of chosen compatibles. This condition only defines a *lower bound* on the number of states in the minimal set. However, if none of their implied pairs are contained in the set, the set is not sufficient and this is referred to as *closed* condition not being satisfied. Therefore, condition of *closed covering* is essentially required for row merging.

9.5 ESSENTIAL HAZARDS

Similar to static and dynamic hazards in a combinational circuits, essential hazards occur in sequential circuits. Essential hazard is a type of hazard that exists only in asynchronous sequential circuits with two or more feedbacks. Essential hazard occurs normally in toggling type circuits. It is an error generally caused by an excessive delay to a feedback variable in response to an input change, leading to a transition to an improper state. For example, an excessive delay through an inverter circuit in comparison to the delay associated with the feedback path many cause essential hazard. Such hazards cannot be eliminated by adding redundant gates as in static hazards. To avoid essential hazard, each feedback loop must be designed with extra care to ensure that the delay in the feedback path is long enough compared to the delay of other signals that originate from the input terminals.

Even though an asynchronous sequential circuit (network) is free of critical races and the combinational part of the network is free of static and dynamic hazards, timing problems due to propagation delays may still cause the network to malfunction and go to the wrong state. To better understand, consider for example the network of Fig. 9.22.

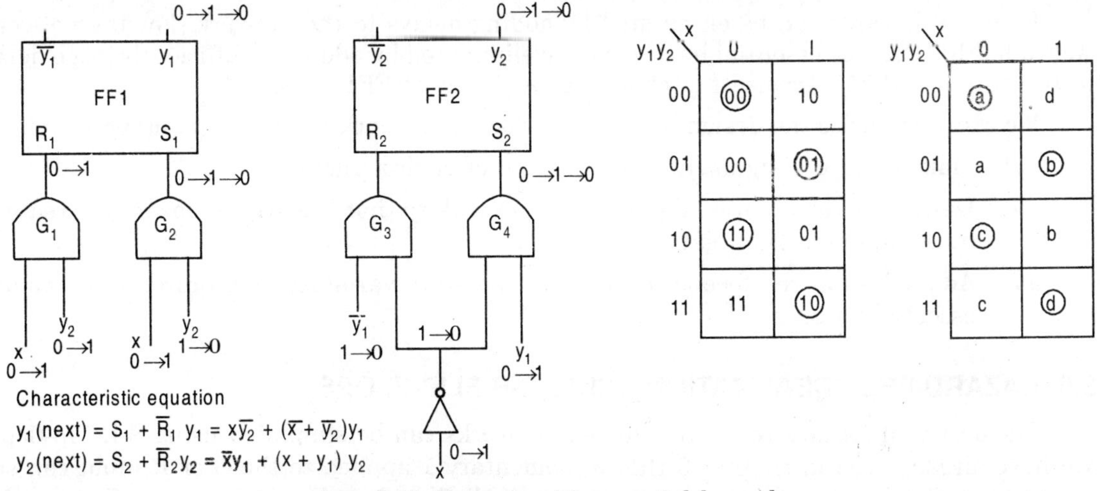

Fig. 9.22 Network with essential hazards

There are no hazards in the combinational part of the network, and flow table inspection shows that there are no critical races. If we start in state ⓐ and change x to 1, the network should go to state ⓓ. Let us consider the following possible sequence of events.

(*i*) x changes 0 to 1.

(*ii*) Gate 2 (G2) output changes 0 to 1.

(*iii*) Flip-flop (FF1) output y_1 changes 0 to 1.

(*iv*) G4 output changes 0 to 1.

(*v*) FF2 output changes 0 to 1.

(*vi*) Inverter output $\bar{x}$ changes 1 to 0.

(*vii*) G1 output changes 0 to 1, G2 output changes back to 0, and G4 output changes back to 0.

(*viii*) Flip-flop output y_1 changes back to 0.

Though the network should go to stage ⓓ when change x to 1 but the final state of the network is ⓑ instead of ⓓ. The malfunction illustrated in example network of Fig. 9.22 is referred to as an essential hazard. This came about because the delay in inverter was large than the other delays in the network, so that part of the network having value $x = 1$ while other part have value $x = 0$. The final result was that the network acted as if the input x had changed three times instead of once so that the network went through the sequence of states $y_1y_2 = 00, 10, 11, 01$. Essential hazards can be located by inspection of the flow table. An essential hazard can be defined as follows:

A flow table has an essential hazard starting in stable total state ⓢ for input variable x_i if and only if the stable total state reached after one change in x_i different from the stable total state reached after three changes in x_i.

If an essential hazard exists in the flow table for total stable state ⓢ and input x_i, then due to some combination of propagation delays network go to the wrong state when x_i is changed starting in ⓢ on realization. This occurs because the change in x_i reaches different parts of the network at different times.

In order to test a flow table for essential hazards it is necessary to test each stable total for each possible input change using the definition of essential hazard given.

Essential hazards can be eliminated by adding delays to the network. For the network shown in Fig. 9.22, the essential hazard can be eliminated by adding a sufficiently large delay to the output of FF1, because the change in x output of FF1 does.

We can summarize the design of an asynchronous network is free of timing problems as:

(*i*) Make a state assignment which is free of critical races.

(*ii*) Design the combinational part of the network so that it is free of hazards (if require by adding redundant gates).

(*iii*) Add delays in the feedback paths for the state variables as required to eliminate essential hazards.

9.6 HAZARD-FREE REALIZATION USING S-R FLIP-FLOPS

The design of hazard-free asynchronous networks can be simplified using S-R flip-flops. We have already seen in chapter 6 that a momentary 1 applied to the S or R input can set or reset the flip-flop, however a momentary 0 applied to S or R will have no effect on the flip-flop state. Since a 0-hazard can produce a momentary false 1, the networks realizing S and R must be free of 0-hazards but the S and R networks may contain 1-hazards. A minimum two-level sum of products expression is free of 0-hazards but it may contain 1-hazards. For this reason, the minimum sum of products can be used as a starting point for realizing the S-R flip-flop input equations. Simple factoring or transformation which do not introduce 0-hazards can be applied to the minimum sum-of-products expressions, in the process of realizing S and R.

A typical network structure with the S-R flip-flop driven by 2-level AND-OR networks constructed from cross-coupled NOR gates is shown in Fig. 9.23(a). Figure 9.23(b) shows equivalent network structure with multiple input NOR gates. The two structure are equivalent since in both cases.

$$Q = \overline{(\bar{Q} + R_1 + R_2 + \ldots)}$$

$$\bar{Q} = \overline{(Q + S_1 + S_2 + \ldots\ldots.)}$$

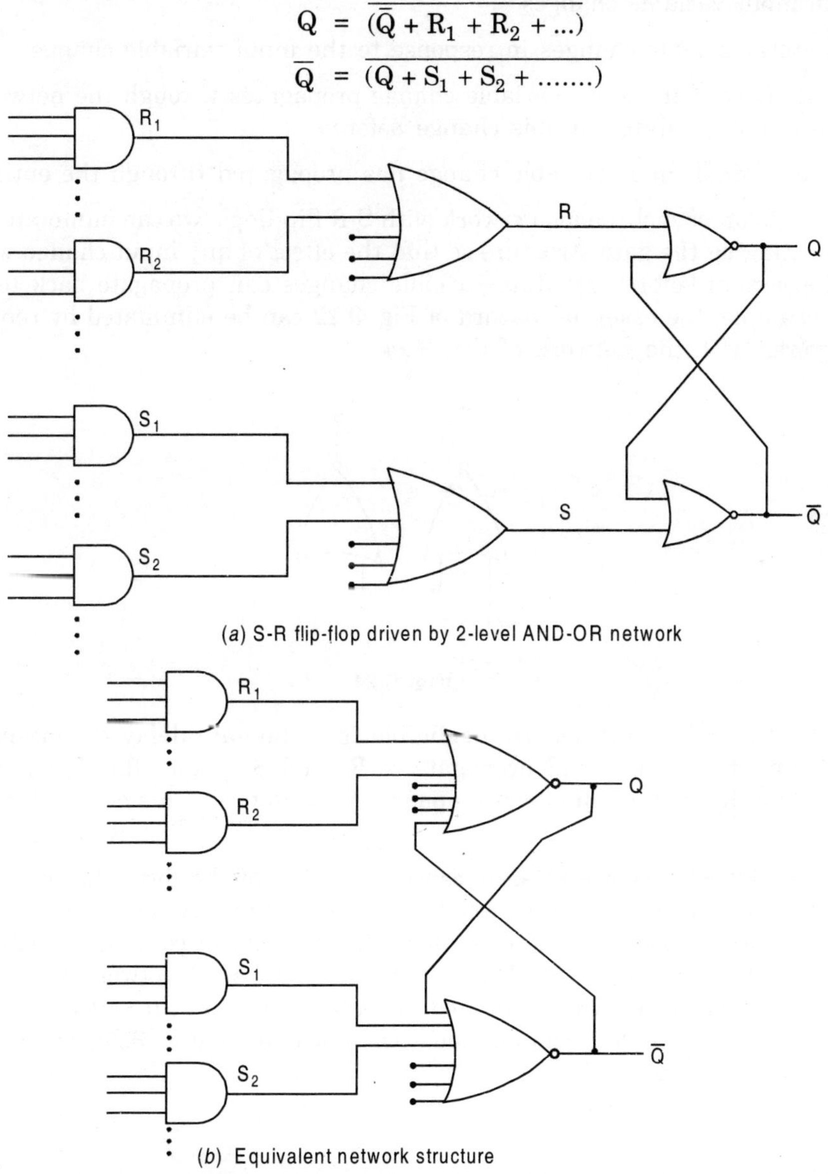

(*a*) S-R flip-flop driven by 2-level AND-OR network

(*b*) Equivalent network structure

Fig. 9.23 Gate structures for S-K flop-flip realization of flow table

Even if an asynchronous network is realized using S-R flip-flops and S and R networks are free of 0-hazards, essential hazards may still be present. Such essential hazards may be eliminated as discussed previously by adding delays in the feedback paths for the state variables.

An alternative method for eliminating essential hazards involves changing the gate structure of the network. This method can be applied only if wiring delays are negligible and all the gate delays are concentrated at the gate outputs.

As illustrated in previous section, the following sequence of events is needed for an essential hazard to cause a network of malfunction.

(*i*) An input variable changes.

(*ii*) A state variable changes in response to the input variable change.

(*iii*) The effect of the state variable change propagates through the network and initiates another state variable change before.

(*iv*) The original input variable change has propagated through the entire network.

Therefore, in an asynchronous network with S-R flip-flops, we can eliminate the essential hazards by arranging the gate structure so that the effect of any input change will propagate to all flip-flop inputs before any state variable changes can propagate back to the flip-flop inputs. For example, the essential hazard of Fig. 9.22 can be eliminated by replacing the R_2 and S_2 networks with the network of Fig. 9.24.

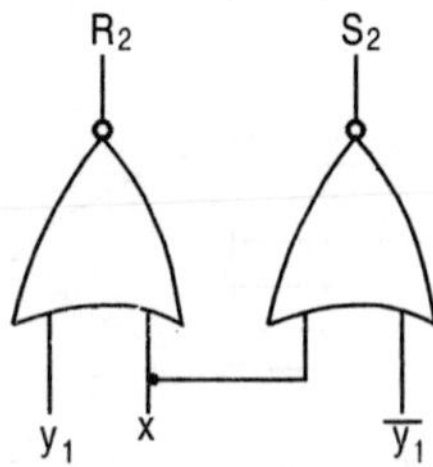

Fig. 9.24

Assuming that wiring delays are negligible that the gate delay is concentrated at the gate output any change in x will propagate to R_2 and S_2 before flip-flop 1 output y_1 can change state and this change in y_1 can propagate to R_2 and S_2. This eliminates the essential hazard.

In the Fig. 9.23 (*b*), each AND gate can have inputs of the form shown in Fig. 9.25 (*a*), where x's are external inputs to the circuit, and the y's are feedback from flip-flop outputs. If there are essential hazards in the flow table, then the circuit could malfunction due to the inverter delays. By replacing the AND gate with the NOR-AND network of Fig. 9.25 (*b*), the inverters on the x variables are eliminated. Therefore by replacing all of the AND gate in Fig. 9.23 with the NOR-AND combinations as indicated in Fig. 9.25, all of the essential hazards will be eliminated.

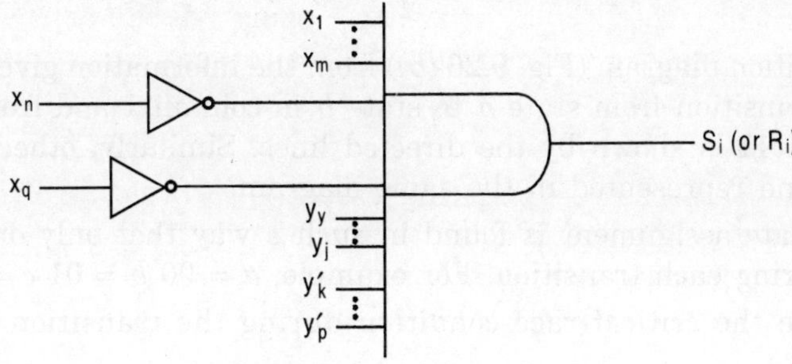

(*a*) AND gate with general inputs

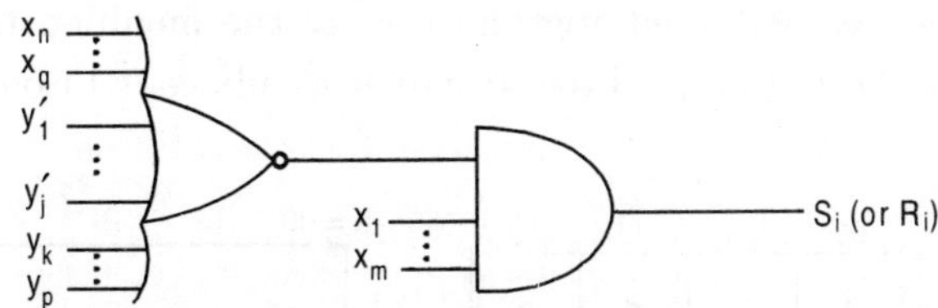

(*b*) Replacement for (*a*)

Fig. 9.25 A gate transformation for elimination of essential hazards

9.7 RACE FREE ASSIGNMENTS

There are *two* methods for making race-free assignments:

1. By adding extra rows in the flow table
2. Multiple-row method

1. Race-Free Assignment by Adding Extra Rows in the Flow Table (or Shared-row Method)

This method is also referred as shared-row method. No critical races can occur in a two-row flow table.

Three-Row Flow-Table Example

A three-row flow table requires two binary variables for state assignment. To avoid critical races, binary assignment should be done by drawing a **transition diagram** which represents all the transitions between any two states.

Example. *Consider a reduced flow table given in Fig. 9.26 (a).*

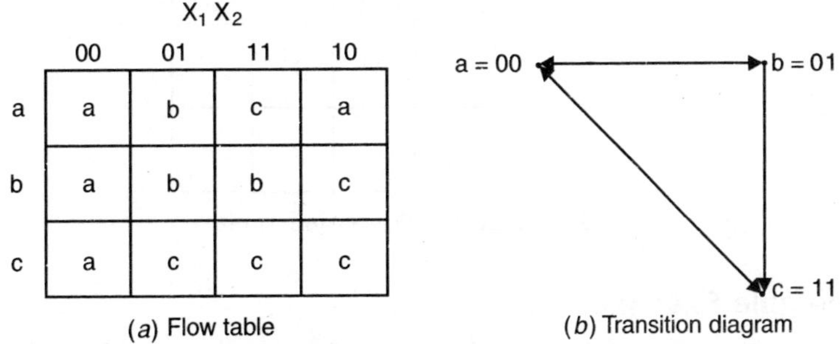

$x_1 x_2$	00	01	11	10
a	a	b	c	a
b	a	b	b	c
c	a	c	c	c

(*a*) Flow table

(*b*) Transition diagram

Fig. 9.26 (*a*) Three-row flow table example

Procedure:

I. Draw transition diagram (Fig. 9.26 (*b*)) from the information given in the table. For example, transition from state *a* to state *b* in column 1 and from state *a* to state *c* in column 11 is shown by the directed lines. Similarly, other rows can also be inspected and represented in the same diagram.

II. A binary state assignment is found in such a way that only one binary variable changes during each transition. For example, $a = 00$ $b = 01$ $c = 11$

It will raise the critical race condition during the transition from state *a* to *c* or vice-versa.

III. To obtain race-free assignment, add one extra row to the flow table which does not increase the number of binary state variables but allows the cycle formation between two stable states. After adding one row '*d*' the modified table is shown in Fig. 9.26 (b). '*d*' is assigned as 10 which is adjacent to both *a* and *c*.

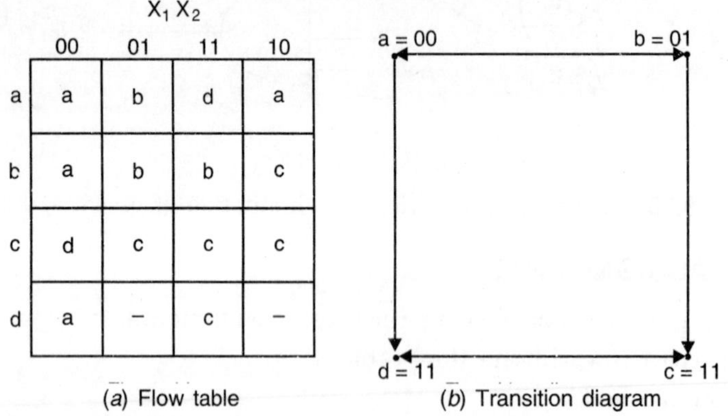

(*a*) Flow table (*b*) Transition diagram

Fig. 9.26 (*b*) Flow table with an extra row

Due to the cycle, the transition from *a* to *c* must go through *d* *i.e.*, $a = 00$, $d = 10$, $c = 11$ which avoids a critical race. Corresponding to the flow table Fig. 9.26 (*b*), the transition table is shown in Fig. 9.26 (c).

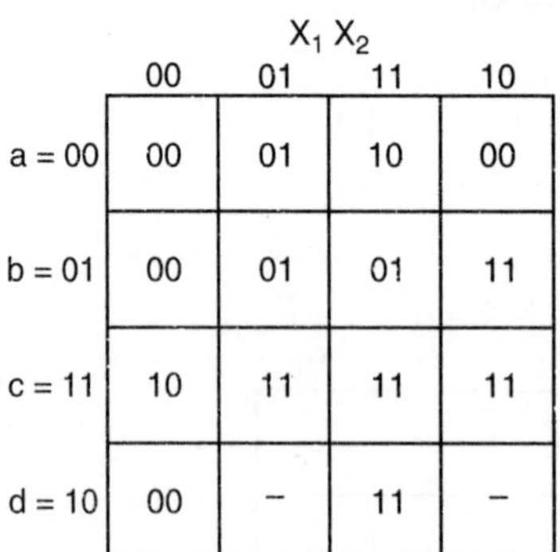

	00	01	11	10
a = 00	00	01	10	00
b = 01	00	01	01	11
c = 11	10	11	11	11
d = 10	00	–	11	–

Fig. 9.26 (*c*) Transition table

Four-row Flow-table Example

A table with four rows requires a minimum of two state variables. In some cases, race-free assignment is possible with only two binary state variables, other variables correspond to extra rows play role in avoiding critical races.

Example. *Consider the following flow table and its corresponding transition diagram:*

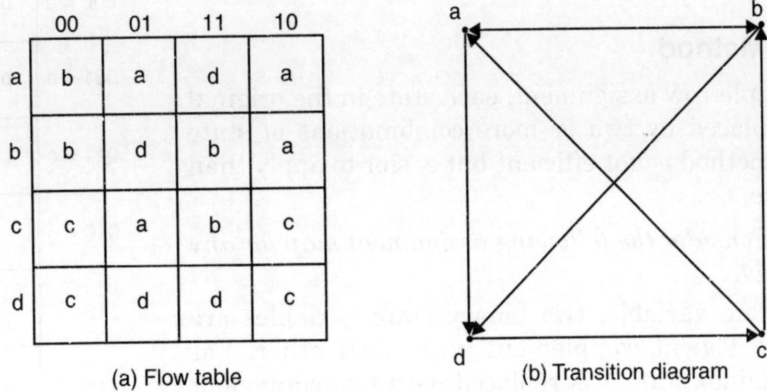

	00	01	11	10
a	b	a	d	a
b	b	d	b	a
c	c	a	b	c
d	c	d	d	c

(a) Flow table (b) Transition diagram

Fig. 9.27 Four-row flow-table example

Transition diagram shows that if there were no transition in the diagonal direction (from b to *d* or from *c* to *a*) then adjacent assignment would be found easily. Diagonal transition restricts the adjacency requirement of binary state variables. Therefore, at least three binary state variables are needed.

Method

For any four-row state table, a **state assignment** map is used for assigning binary values to each state. The assignment ensures that a **cycle** is produced so that only one binary variable changes at a time.

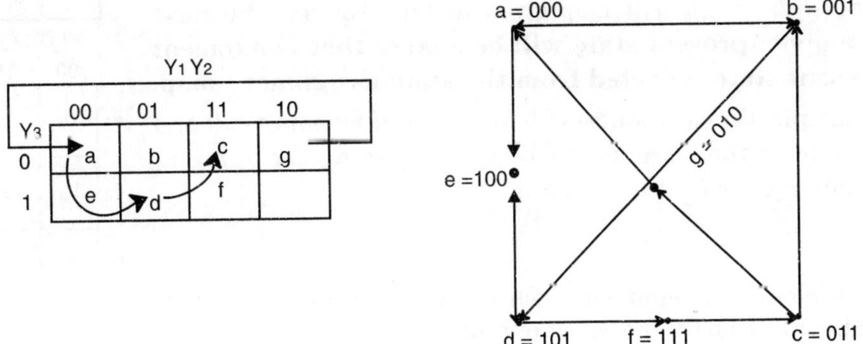

Fig. 9.28 Choosing extra rows for the flow table

Original states *a*, *b* and *c* are assigned as 000, 001, 011 *i.e.*, they have adjacent assignment. Extra states are *e*, *f* and *g*. (Here $d \neq 100$ because $c = 011$). The transition from *a* to *d* must be directed through the extra state *e* to produce cycle. Similarly, the transition from *c* to *a* is directed through *g* and transition from *d* to *c* goes through *f*.

Expanded Flow Table

Now, the four-row flow-table can be expanded to a seven-row flow-table that is free of critical races (Fig. 9.29).

The uncircled states in three extra rows provide a race-time transition between the stable states.

	00	01	11	10
000 = a	b	a	e	a
001 = b	b	d	b	a
011 = c	c	g	b	c
010 = g	–	a	–	–
110 –	–	–	–	–
111 = f	c	–	–	c
101 = d	f	d	d	f
100 – e	–	–	d	–

Fig. 9.29 State assignment of modified flow table

Multiple-Row Method

In the multiple-row assignment, each state in the original flow table is replaced by two or more combinations of state variables. This method is not efficient but easier to apply than the previous one.

Example. *Consider the following assignment map for any four-row flow table:*

For each state variable, two binary state variables are used. These are logical complement with each other. For example, the original state a is replaced with two equivalent states a_1 = 000 and a_2 = 111. The output values also be same in a_1 and a_2 (not shown here).

State a_1 is adjacent to b_1, c_2 and d_1, a_2 is adjacent to c_1, b_2 and d_2. Similarly, each state is adjacent to three states of different letter designation. This is shown in Fig. 9.30(a).

The expanded table of multiple row assignment for the original four-row flow table (Fig. 9.27 (a)) is shown in the following Fig. 9.30(b). It is formed by replacing each row with two rows. For example, a is replaced by rows a_1 and a_2.

First, enter all the stable states. Now, unstable states are filled in by using *state-assignment* map given in Fig. 9.30(a). **The next state for a given present state will be a state that is adjacent to the present state, selected from the state-assignment map.**

For example, the next states of b are a and d for inputs 10 and 01 respectively. In the expanded table, the next states for b_1 are a_1 and d_2. Similarly, for b_2 are a_2 and d_1.

Y_1 \ Y_1Y_2	00	01	11	10
0	a_1	b_1	c_1	d_1
1	c_2	d_2	a_2	b_2

(a) Binary assignment

	00	01	11	10
a_1	b_1	a_1	d_1	a_1
a_2	b_2	a_2	d_2	a_2
b_1	b_1	d_2	b_1	a_1
b_2	b_2	d_1	b_2	a_2
c_1	c_1	a_2	b_1	c_1
c_2	c_2	a_1	b_2	c_2
d_1	c_1	d_1	d_1	c_1
d_2	c_2	d_2	d_2	c_2

(b) Flow Table

Fig. 9.30 Multiple-row assignment

Remark

In multiple-row assignment, each stable state has two binary assignments with exactly the same output.

9.8 SOLVED EXAMPLES

Example 1. *Construct merger diagram for the primitive flow table of Fig. 9.18. Determine maximal compatibles and the minimal set of compatibles.*

Solution. For construction of merger diagram, every row of the primitive flow table is checked with every other row to determine compatibility of states.

Consider row-1 (state a)

a, b are compatible

a, c are compatible
a, d are compatible if c, f are compatible
a, e are compatible if c, f are compatible
a, f are compatible if c, f are compatible

row-2 (state b)
b, c are compatible if d, e are compatible
b, d are compatible
b, e are not compatible (outputs are conflicting)
b, f are not compatible (outputs are conflicting)

row-3 (state c)
c, d are compatible if e, d and c, f are compatible
c, e are not compatible (outputs are conflicting)
c, f are not compatible (outputs are conflicting)

row-4 (state d)
d, e are not compatible (outputs are conflicting)
d, f are not compatible (outputs are conflicting)

row-5 (state e)
e, f are compatible

The primitive flow table has six states therefore, there are six vertices in the merger diagram as shown in Fig. 9.31.

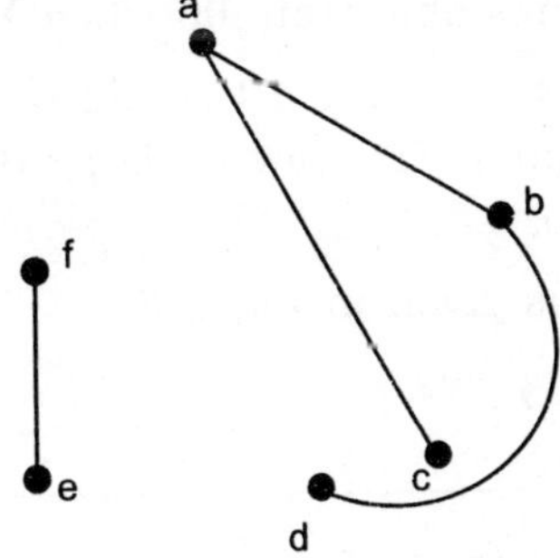

Fig. 9.31 Merger diagram

Solid arcs are drawn between (a, b), (a, c), (b, d) and (f, e) vertices. Corresponding to these states being compatibles. Since (c, f) and (d, e) are not compatible, therefore, there are no implied pairs available.

From the merger diagram, we get the maximal compatibles:

(a, b), (a, c), (b, d), (e, f)

Since (a, b) is covered by (a, c) and (b, d), therefore, the minimal set is (a, c), (b, d), (e, f).

Example 2. *Determine the reduced flow table of Fig. 9.19.*

Solution. From the merger diagram, we have obtained three pairs of compatible states: These compatibles are merged and are represented by

a, c : S_0

b, d : S_1

e, f : S_2

The reduced flow table is shown in Fig. 9.32.

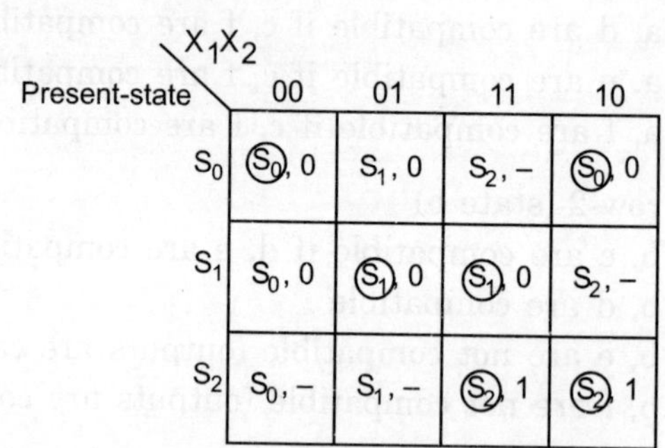

Present-state \ X_1X_2	00	01	11	10
S_0	(S_0), 0	S_1, 0	S_2, –	(S_0), 0
S_1	S_0, 0	(S_1), 0	(S_1), 0	S_2, –
S_2	S_0, –	S_1, –	(S_2), 1	(S_2), 1

Fig. 9.32 Reduced flow table

Example 3. *Assign binary states to the reduced flow table of Fig. 9.32. Avoid critical race.*

Solution. Let us assign the following binary states to S0, S1, and S2 for the reduced flow table of Fig. 9.32.

S0 → 00

S1 → 01

S2 → 11

The transition table will be as shown in Fig. 9.33.

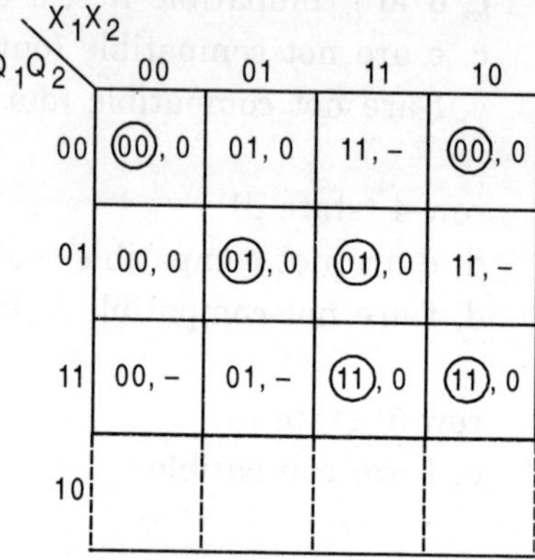

Q_1Q_2 \ X_1X_2	00	01	11	10
00	(00), 0	01, 0	11, –	(00), 0
01	00, 0	(01), 0	(01), 0	11, –
11	00, –	01, –	(11), 0	(11), 0
10				

Fig. 9.33 Transition table

In the transition table of Fig. 9.33, we observe that race condition occurs in the following cases:

(*i*) From stable state 00 to unstable state 11 when X_1 X_2 changes from 10 to 11.

(*ii*) From stable state 11 to unstable state 00 when X_1 X_2 changes from 10 to 00.

To avoid critical race, one unstable state 10 is added with the entries 00, –; –, –; 11, –; –, – and the entries in third column, first row is changed from 11, – to 10, – and in first column, third row from 00, – to 10, –.

The modified transition table is given in Fig. 9.34.

Q_1Q_2 \ X_1X_2	00	01	11	10
00	(00), 0	01, 0	10, –	(00), 0
01	00, 0	(01), 0	(01), 0	11, –
11	10, –	01, –	(11), 1	(11), 1
10	00, –	–, –	11, –	–, –

Fig. 9.34 Modified transition table

Example 4. *Design logic circuit with feedback for the transition table of Fig. 9.34.*

Solution. The K-maps for Q_1^+, Q_2^+, and Y determined from the transition table are given in Fig. 9.35.

From the K-maps, we obtain,

$$Q_1^+ = X_1 X_2 \overline{Q}_2 + \overline{X}_2 Q_1 Q_2 + X_1 \overline{X}_2 Q_2 + X_1 Q_1$$

$$Q_2^+ = \overline{X}_1 X_2 + X_1 Q_2 + X_1 Q_1$$

$$Y = Q_1$$

Logic circuit using gates can be obtained from the above logic equations.

Thus, we see that the design steps outlined above can be used to design an asynchronous sequential circuit.

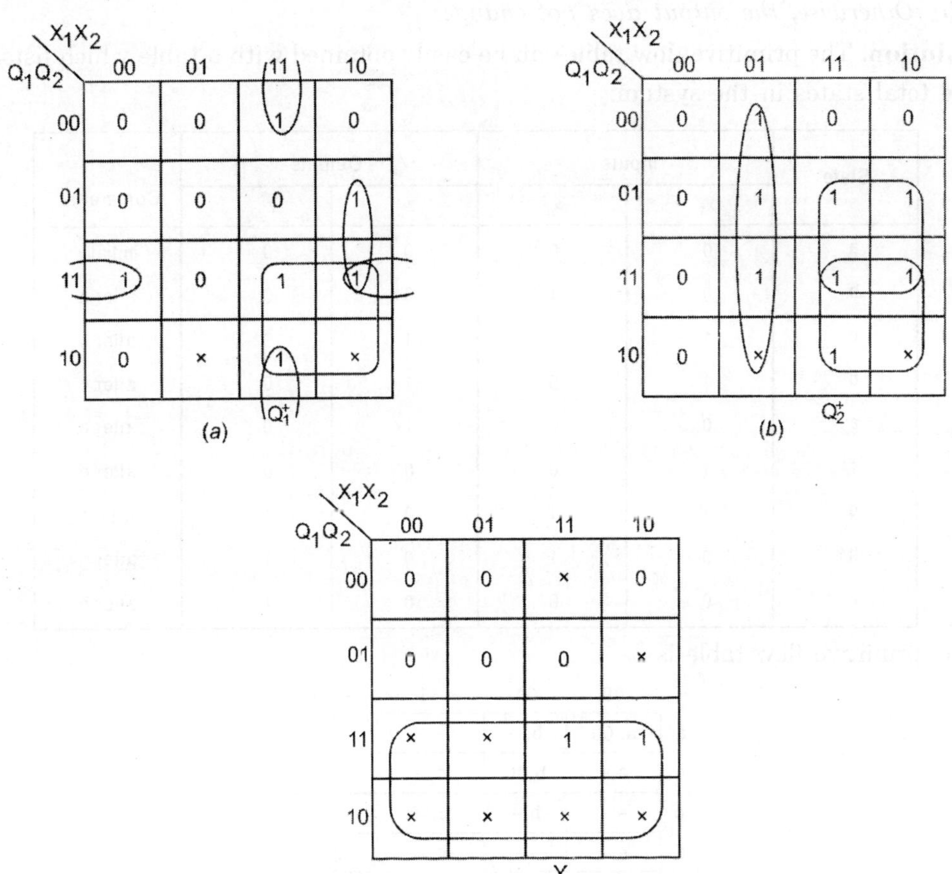

Fig. 9.35 K-Maps for (*a*) Q_1^+ (*b*) Q_2^+ (*c*) Y

Example 5. *In the state transition table of Fig. 9.13, if $X_1 X_2$ = 10 and the circuit is in stable state* (01), *find the cycle when X_2 is changed to 1 while X_1 remaining 1.*

Solution: The circuit is in stable state (01) (fourth column, second row). When X_2 changes to 1, the circuit will go to the state 11 (third column, second row), then to state 10 (third column, third row) and finally to the stable state (10) (third column, fourth row). Thus, the cycle is

$$(01) \rightarrow 11 \rightarrow 10 \rightarrow (10)$$

Example 6. *Obtain a primitive flow table for a circuit with two inputs x_1 and x_2 and two outputs z_1 and z_2, that satisfy the following four conditions:*

(a) When $x_1x_2 = 00$, the output is $z_1z_2 = 00$.

(b) When $x_1 = 1$ and x_2 changes from 0 to 1, the output is $z_1z_2 = 01$.

(c) When $x_2 = 1$ and x_1 changes from 0 to 1, the output is $z_1z_2 = 10$.

(d) Otherwise, the output does not change.

Solution. The primitive flow table can be easily obtained with a table which lists all the possible total states in the system:

State	Inputs		Outputs		Comments
	x_1	x_2	z_1	z_2	
a	0	0	0	0	initially
b	0	1	0	0	after a
c	1	1	1	0	after b
d	1	0	1	0	after c
e	0	0	1	0	after d
f	1	0	0	0	after e
g	1	1	0	1	after f
h	0	1	0	1	after g
i	0	0	0	1	after h

The primitive flow table is

	00	01	11	10
a	**a**, 00	b, –	–, –	f, –
b	a,–	**b**,00	c, –	–,–
c	–,–	b,–	**c**,10	d,–
d	e,–	–,–	c, –	**d**,10
e	**e**,10	–,–	–, –	–, –
f	e, –	–, –	g, –	**f**, 00
g	–, –	h, –	**g**, 01	f, –
h	i,–	**h**, 01	g, –	–, –
i	**i**, 01	–, –	–, –	–, –

Example 7. *Investigate the transition table given in the figure and determine all race conditions and whether they are critical or non-critical. Also determine whether there are any cycle.*

$y_1y_2 \backslash x_1x_2$	00	01	11	10
00	10	00	11	00
01	01	01	10	01
11	01	00	11	11
10	11	00	10	10

Transition table

Solution. Let's first consider the transition when input **changes from 00 to 01.** The transition is 10 → 00. Here, only one state value is changed. Hence there is **no race condition.** When input changes from 01 to 11, the transition is 00 → 11 → 10 → 11 or 00→11 → 10. Here, different stable states are achieved, which indicate the possibility of **critical** race condition.

Similarly, there is no race condition when input changes from 11 to 10, since the transitions are 11 → 10 → 11 or 11 → 10. In any case, only one state variable has changed. For cycle, there is a unique sequence of transitions which is shown in first and second column i.e., transition 10→00.

Example 8. *Merge each of the primitive flow tables shown below. Proceed as follows:*

(a) *Find all compatible pairs by means of an implication table.*
(b) *Find the maximal compatibles by means of a merger diagram.*
(c) *Find a minimal set of compatibles that covers all the states and is closed.*

	00	01	11	10
a	a, 0	b, –	–, –	e, –
b	a, –	b, 0	c, –	–, –
o	–, –	d, –	c, 0	h, –
d	a, –	d, 1	–, –	–, –
e	a, –	–, –	f, –	e, 0
f	–, –	g, –	f, 0	h, –
g	a, –	g, 0	–, –	–,
h	a, –	–, –	–, –	h, 0

(i)

a	a, 1	f, –	–, –	e, –
b	c, –	–, –	i, –	b, 0
c	c, 0	d, –	–, –	b, –
d	c, –	d, 0	g, –	–, –
e	a, –	–, –	g, –	e, 1
f	a, –	f, 1	g, –	–, –
g	–, –	d, –	g, 0	j, –
h	h, 0	d, –	–, –	j, –
i	–, –	f, –	i, 1	b, –
j	a, –	–, –	i, 1	k, 0

(ii)

Solution. (I) For table I:

(a) To find the compatible pairs by means of an *implication table:*

Compatible pairs are:

(*a*, *b*) (*a*, *c*) (*b*, *h*) (*c*, *d*) (*c*, *h*) (*d*,*e*) (*d*, *h*) (*e*, *g*) (*e*, *h*) (*f*, *g*) (*f*, *h*) (*g*, *h*)

(*b*) To find maximal compatibles by *merger diagram :* By observing the merger diagram, maximal compatible pairs are:

(*a*, *b*) (*b*, *h*) (*a*, *e*) (*c*, *d*, *h*) (*d*, *e*, *h*) (*e*, *g*, *h*) (*f*, *g*, *h*)

(c) To find minimal set of compatibles which satisfy the *closed-covering condition:*

If we remove (*a*, *e*), (*b*, *h*) and (*d*, *e*, *h*) the set is (*a*, *b*) (*c*, *d*, *h*), (*e*, *g*, *h*) (*f*, *g*, *h*).

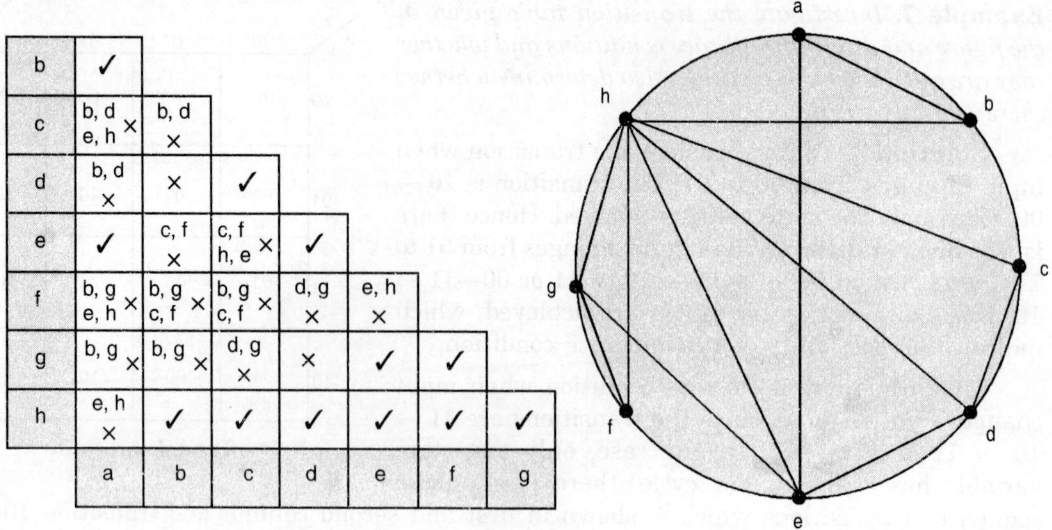

	a	b	c	d	e	f	g
b	✓						
c	b, d e, h ×	b, d ×					
d	b, d ×	×	✓				
e	✓	c, f ×	c, f h, e ×	✓			
f	b, g e, h ×	b, g c, f ×	b, g c, f ×	d, g ×	e, h ×		
g	b, g ×	b, g ×	d, g ×	×	✓	✓	
h	e, h ×	✓	✓	✓	✓	✓	✓

Table-I

All the 8 states from the given flow table are included in the set. Thus, the set satisfies the covering condition.

There are no implied states from (*a, b*); (*a, e*); (*b, h*); (*c, d*); (*c, h*); (*d, e*); (*d, h*); (*e, g*); (*e, h*); (*e, h*); (*f, g*); (*f, h*); (*g, h*) as seen from the implication table (I), so the **closure** condition is also satisfied.

Therefore, minimal set of compatibles, satisfying closed covering condition:

(*a, b*) (*c, d, h*) (*e, g, h*) (*f, g, h*)

II. For table (II):

(a) To find all compatible pairs from the *implication table:*

	a	b	c	d	e	f	g	h	i
b	a, c e, b ×								
c	×	✓							
d	a, c i, g ×	i, g ×	✓						
e	✓	×	c, a b, e ×	c, a ×					
f	✓	a, c i, g ×	c, a d, f ×	×	✓				
g	f, d e, k ×	i, g b, k ×	b, k ×	✓	e, k ×	f, d ×			
h	×	c, h b, k ×	c, h b, k ×	c, h ×	a, h e, k ×	a, h f, d ×	✓		
i	b, e ×	✓	d, f ×	d, f g, i ×	g, i e, b ×	g, i ×	×	d, f k, b ×	
j	e, k ×	a, c b, k ×	c, a b, k ×	c, a g, i ×	g, i e, k ×	g, i ×	×	h, d ×	b, k ×

Table-II

Compatible pairs are :

(a, e) (a, f) (b, c) (b, i) (c, d), (d, g) (e, f) (g, h)

(b) To find maximal compatibles from the *merger diagram:*

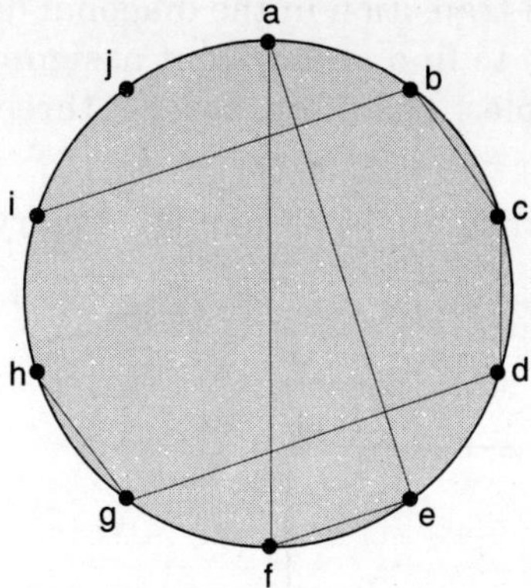

There is only one triangle corresponds to the maximal compatible set (a, e, f)

∴ Maximal compatibles are : (a, e, f) (b, c) (b, i) (c, d) (d, g) (e, f) (g, h) (j)

Here j is isolated *i.e.*, not compatible with other states.

(c) To find minimal set of compatibles which satisfy the *closed-covering condition:*

If we left (b, c) (d, g) and (e, f) remaining sets are : (a, e, f) (b, i) (c, d) (g, h) (j).

All the 10 states are included in this set. Thus, the set satisfies the **covering** condition. There are no implied states for compatible states (obtained in part (a)), so the **closure** condition is also satisfied.

Hence the following is the maximal compatible set which satisfies the closed-covering condition:

(a, e, f) (b, i) (c, d) (g, h) (j)

Example 9. *Obtain a binary state assignment for the reduced table shown below, to avoid critical-race conditions:*

y_1y_2 \ x_1x_2	00	01	11	10
a	a, 0	a, 1	b, –	d, –
b	a,–	b, 0	b, 0	c, –
c	a,–	–, –	d, –	c, 0
d	a,–	a,–	d, 1	d, 1

Solution. From the given reduced flow table, draw the transition diagram as shown alongside.

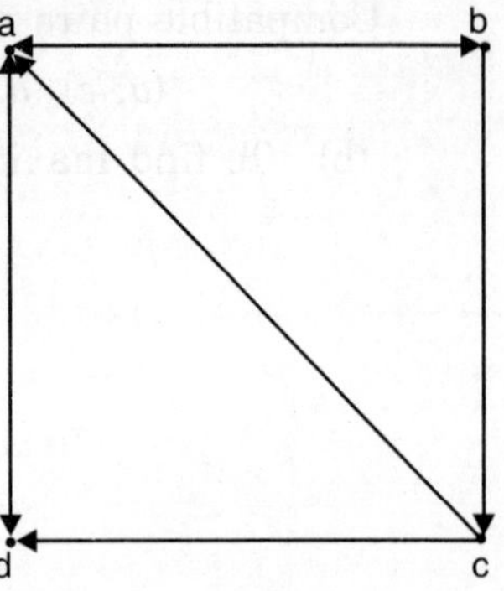

A four-row flow table requires a minimum of two state variables. But, here, there is a transition in the diagonal direction (from *c* to *a*), it is not possible to find an adjacent assignment for the state with two state variables. Therefore, at least three binary state variables are needed.

Redrawing the transition diagram by assigning 3-binary values for each state:

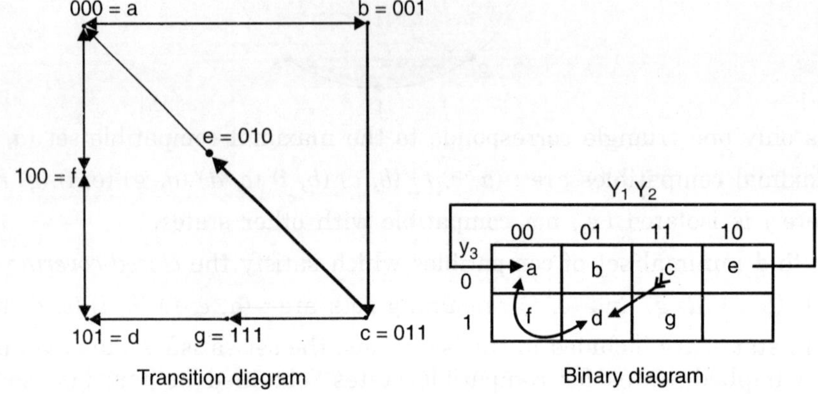

Transition diagram

Binary diagram

By using 3-variable binary assignment to each state, the four-row table can be expanded to a seven-row table that is free of critical races. In the expanded table, only 4 rows have a stable states out of seven rows. The uncircled states are responsible to provide race-free transition between the stable states.

	00	01	11	10
000 = a	(a),0	(a),1	b, –	f, –
001 = b	a, –	(b),0	(b), 0	c, –
011 = c	e, –	–, –	g, –	(c),–
010 = e	a, –	–	–	–
110 = –	–	–	–	–
111 = g	–	–	d , –	–
101 = d	f, –	f, –	(d),1	(d),1
100 = f	a, –	a, –	–	d,1

State assignment to modified table

Example 10. *Consider the reduced flow table given below:*

(*a*) *Obtain the transition diagram and show that at least three state variables are needed for a race-free binary state assignment.*

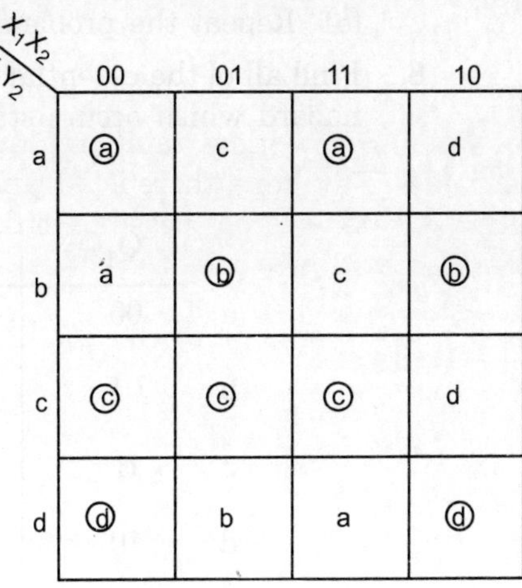

Y_1Y_2 \ X_1X_2	00	01	11	10
a	(a)	c	(a)	d
b	a	(b)	c	(b)
c	(c)	(c)	(c)	d
d	(d)	b	a	(d)

Solution. (*a*) *Transition diagram for a four-row table is:*

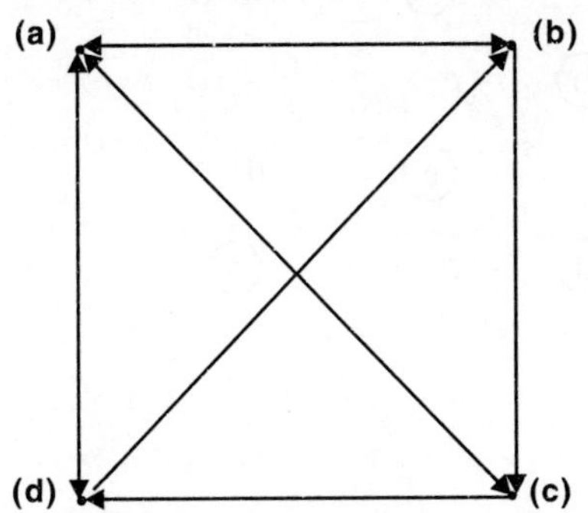

From the transition diagram, it is clear that at least three binary state variables due to the transitions along the diagonals (*i.e.*, from *a* to *c* and from *d* to *b*).

9.9 EXERCISE

1. (*a*) Explain the difference between asynchronous and synchronous sequential circuits.

(*b*) Define fundamental mode of operation.

(*c*) Define pulse mode of operation.

(*d*) Explain the difference between stable and unstable states.

(*e*) What is the difference between internal state and total state?

2. Describe the design procedure for asynchronous sequential circuits.

3. What do you mean by critical and non-critical races? How can they be avoided?

4. Describe cycles in asynchronous sequential circuits.

5. Design a JK flip-flop asynchronous sequential circuit that has two inputs and single output. The circuit is required to give an output equal to 1 if and only if the same input variable changes two or more times consecutively.

6. Design an asynchronous circuit that has two inputs and single output. The circuit is required to give an output whenever the input sequence 00, 10, 11 and 01 are received but only in that order.

7. (*a*) Design an asynchronous binary counter with one pulse input and two outputs, capable of counting from zero to three. When the circuit is pulsed after the count has reached three, it should return to zero. The output should provide continuously the count modulo 4.

(*b*) Repeat the problem for level inputs and outputs.

8. Find all of the essential hazards in the following flow table. How can table essential hazard which occurs starting in b be eliminated?

	Q_1Q_2	X_1X_2 = 00	01	11	10
a	00	(a)	b	(a)	d
b	01	a	(b)	c	–
c	11	–	d	(c)	d
d	10	a	(d)	a	(d)

10

ALGORITHMIC STATE MACHINE

10.0 INTRODUCTION

Finite state machines are a powerful tool for designing sequential circuits, but they are lacking in that they do not explicitly represent the algorithms that compute the transition or output functions, nor is timing information explicitly represented. We can recast the idea of a state machine to include a representation of the algorithms. The result is an *algorithmic state machine*, or ASM. "The ASM chart separates the conceptual phase of a design from the actual circuit implementation." An algorithmic state machine diagram is similar to a flowchart but with some differences. Square boxes represent the states, diamonds represent decisions, and ovals represent outputs. States can also have outputs, and the outputs associated with a state are listed in the state box. State boxes are labelled with the state name and possibly with a binary code for the state. The basic unit of an ASM is the ASM block. An ASM block contains a single state box, a single entry path, and one or more exit paths to other ASM blocks. Algorithmic state machines capture the timing of state transitions as well.

10.1 DESIGN OF DIGITAL SYSTEM

In the earlier chapters, we have presented the analysis and design of various types of Digital System for specified task. A close look on all such systems reveal that these systems can be viewed as collection of two subsystems:

(*i*) The Data Processing or manipulating subsystem which include the operation such as shifting, adding, counting, dividing etc.

(*ii*) The control subsystem or simply control. This subsystem has to initiate, superwise and sequence the operation in data-processing unit.

Usually design of data processor is a fair and simple design. But design of control logic with available resources is a complex and challenging part, perhaps because of timing relations between the event. And in this chapter we are majorily concerned with design of control.

The control subsystem is a sequential circuit whose internal states dictate the control command to sequence the operations in data processing unit. The digital circuit used as control subsystem is responsible to generate a time sequence of control signals that initiates operation in data processor, and to determine the next state of control subsystem itself. The task of data processing and control sequence are specified by means of **a hardware algorithm.**

An algorithm is a collection of **produces** that tells how to obtain the solution. A **flow chart** is a simple way to represent the sequence of procedures and decision paths for algorithm.

A **hardware algorithm** is a procedure to implement the problem with the available hardware or resource. A flowchart for hardware algorithm translates the word statements to an information of diagram, that enumerates the sequence of operations along with the necessary condition for their execution.

A special flowchart, developed specifically to define the "Digital Hardware Algorithm" is called as an **Algorithmic State Machine** (ASM) chart. In fact, a sequential circuit is alternately called as *state machine* and forms the basic structure of a digital system. A conventional flow-chart describes the sequence of procedural steps and decision paths for an algorithm without concern for their timing relationship. The ASM chart describes the sequence of events as well as the timing relationship between the states of sequential controllers and the events that occur while going from one state to next state. The ASM chart is specifically adapted to accurately specify the control sequence and data processing in digital systems, while considering the constraints of available hardware.

10.2 THE ELEMENTS AND STRUCTURE OF THE ASM CHART

An ASM chart is composed of four elements. These are the "state box", the "decision box", the "conditional output box" and "edges".

State Boxes

State boxes describe a state of the ASM. In general an ASM is a sequential system. The state box represents the condition of the system. The symbol for a state box is as follows:

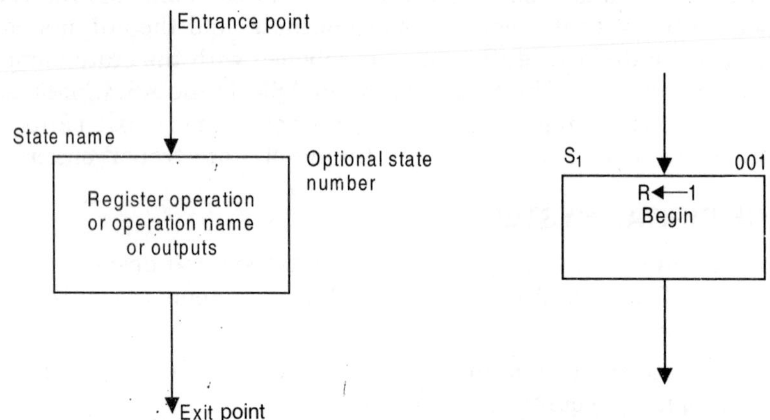

Fig. 10.1 (*a*) State box **Fig. 10.1** (*b*) Example of state box

The state box has exactly one entrance point and one exit point. The state box also has a name and is often assigned a number for clarity. Inside the state box we place the names of the system outputs that must be asserted while the system is in that state. We can also place variable assignments in the state box, in order to "remember" the fact that the system has been in that state. This is useful in program design.

Note: Sequential systems are systems with memory; their output depend on their input as well as their history. The historical information that the system stores is called a 'state'. Combinational systems are the opposite, having no memory. Combinational systems output depends only on present inputs.

Decision Boxes

Decision boxes are used to show the examination of a variable and the outcomes of that examination. In this model, the outcome of a decision is always either true or false. This means that there is always exactly one input to a decision box and exactly two exits from the box. The exit points are always labelled "true" or "false" for clarity. When input condition is assigned a binary value, the two exit paths are labelled 1 and 0. 1 in place of 'True' and 0 in place of 'False'. The condition which is being examined is written inside the decision box. The symbol for a decision box is shown here:

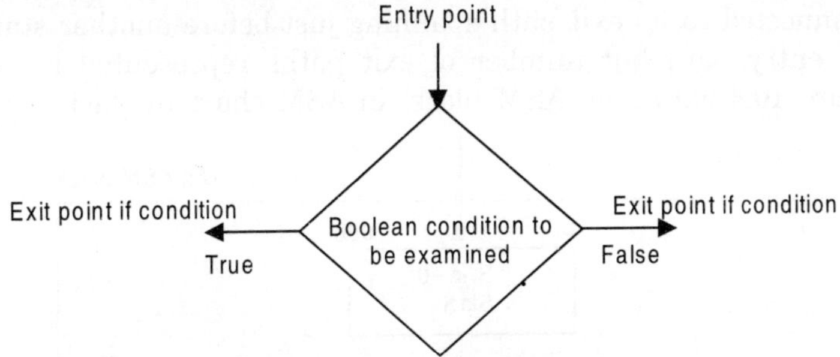

Fig. 10.2 Decision Box

Note that the decision box does not imply a system state. When a decision box is executed the decision is made and the system proceeds immediately to the next item in the ASM chart.

Conditional Output Boxes

Conditional output boxes are used to show outputs which are asserted by the system "on the way" from one state to another. The symbol for the conditional output box is shown here in Fig. 10.3(a).

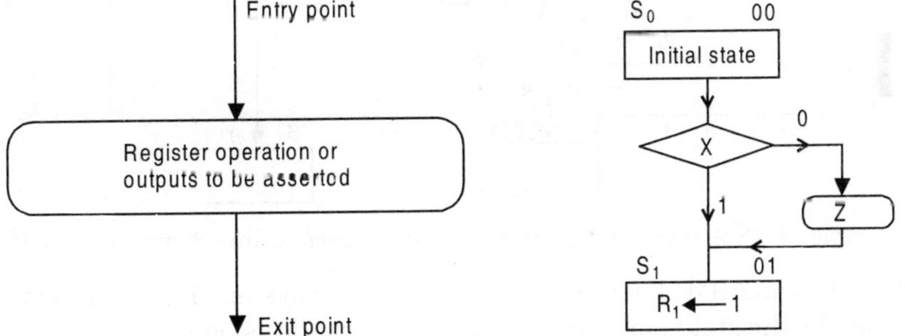

Fig. 10.3 (*a*) Conditional box **Fig. 10.3** (*b*) Use of conditional box.

There is always one entry and one exit point from a conditional output box. Inside the box we write the name of the signal that must be asserted by the system as it passes from one state to another. Input path to conditional box must come from an exit path of decision box.

Conditional output boxes do not imply a system state. We can put variable assignments in the state box. Figure 10.3(*b*) shows an example using conditional box. The first one in diagram is initial state (Labled S_0) system which attains certain conditions fulfilled before starting the actual process. The control then checks the input X. If X = 0, then control generates the Z output signal and go to state S_1 otherwise it moves to next state without generating Z. $R_1 \leftarrow 1$ in state box S_1 is a register operation that loads R_1 by 1.

Edges are used to connect other ASM chart elements together. They indicate the flow of control within the system. The symbol for an edge is as follows:

$\uparrow \downarrow \rightarrow \leftarrow$

Note that the edge must always indicate which direction is being taken, by using one or more arrow heads.

10.2.1 ASM Block

An ASM block is a structure that contains one state box and all decision boxes and conditional boxes connected to its exit path spanning just before another state box. An ASM block has only one entry path but number of exit paths represented by the structure of decision boxes. Figure 10.4 shows an ASM block, in ASM chart by dashed lines around it.

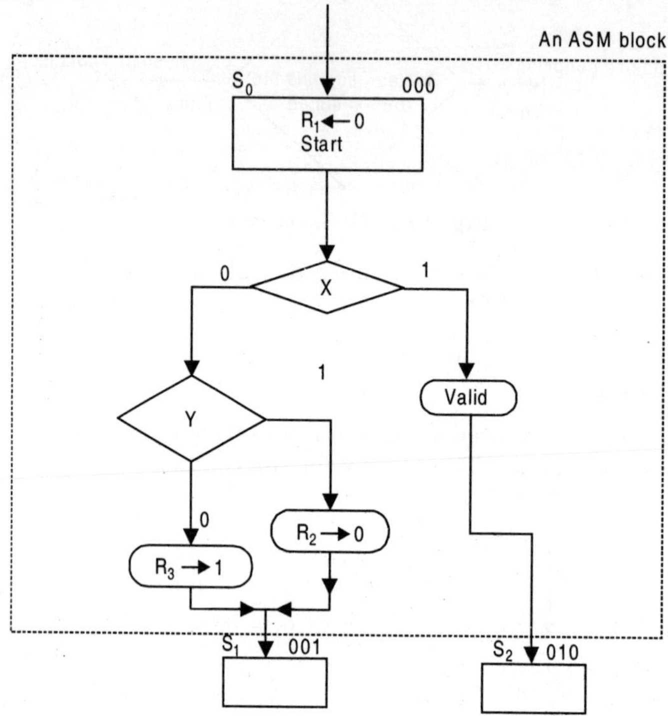

Fig. 10.4 Example of ASM block-structure enclosed by dashed line represent an ASM block

Each ASM block is an ASM chart represents the state of system during one clock pulse. The operations specified by the state box, conditional boxes, and decision boxes are executed during a common clock pulse while the system is in S_0 state. The same clock pulse is also responsible to move the controller to one of the next states, S_1 or S_2 determined by binary status of X and Y. A state box without any decision or conditional boxes constitutes a simple block.

10.2.2 Register Operation

A digital system consists of one or more registers for data storage. Thus, operation to be performed on data is actually performed on the register that stores the data. A register is a general designation which includes shift registers, counters, storage registers, and single flip-flops. A single flip-flop is identified as 1-bit register. A register is designated by use of one or more capital letters such as A, B, RA, RB, R_1, R_2. In practice, most convenient designation is letter R along with numbers such as R_1, R_2, ..., R_n.

Register operations can be increment, decrement, shift, rotate, addition, cleaning, copying, data transfer, etc. The data transfer to a register from another register or from the result of mathematical operations etc. are shown (or symbolized) by directed arrow whose head is towards target register and tail is towards source register. Figure 10.5 summarizes the symbolic notations for some of register operations.

Symbolic Notation of Operation	*Initial Value of Target Register*	*Initial Value of Source Register*	*Value of Target Register after Operation*	*Description of Operation*
$A \leftarrow B$	A = 01010	B = 00000	A = 00000	Copy the content of register B into register A
$R_1 \leftarrow 0$	R_1 = 11111	–	R_1 = 00000	Clear register R_1
$A \leftarrow A + B$	A = 01010	B = 00101	A = 01111	Add the contents of register B to register A and put result in A.
$R \leftarrow R - 1$	R = 01101	–	R = 01100	Decrement register R by 1.
$R \leftarrow R + 1$	R = 00101	–	R = 00110	Increment register R by 1.
"Shift Left A"	A = 10111	–	A = 01110	Shift content of A to left by 1-bit
"Rotate Right A"	A = 10111	–	A = 11011	Rotate content of A to right by 1-bit
$R \leftarrow 1$	R = 01010	–	R = 11111	Set content R to 1

Fig. 10.5 Symbolic representation or register operation

Assume 5-bit register to understand the operations. Note that shift and rotate operation are not same. Shift left means MSB ← MSB-1, MSB-1 ← MSB-2, ..., LSB + 1 ← LSB, LSB ← 0. If rotate right operation then MSB ← LSB, MSB-1 ← MSB, ... LSB ← LSB + 1. It is clear that in shift operation loose MSB if left-shift or LSB if right-shift because as above explained MSB was overwritten by content of MSB-1, and prior to this value of MSB was not saved. And that's why a 0 is inserted at LSB. In rotate operation, we don't loose the status of bits. If we rotate left then status of MSB is transferred to LSB and then it is overwritten by the value of MSB-1.

Equipped with this many knowledge and understanding we are able to draw and understand the simple ASM charts and with analysis and synthesis we can figure out the similarity of ASM charts with that of state diagram.

In the next section, we present some simple examples to give you a feel of ASM chart and its representation.

10.2.3 Example of ASM Charts

Example 1. ***1-Bit Half Adder:*** *The half adder take the two data bits if START input is activated otherwise it remains in initial state. The data bits are read into register R_1 and R_2 and sum and carry bits are maintained in register R_3 and R_4, respectively ASM chart for this task is shown in Fig. 10.6.*

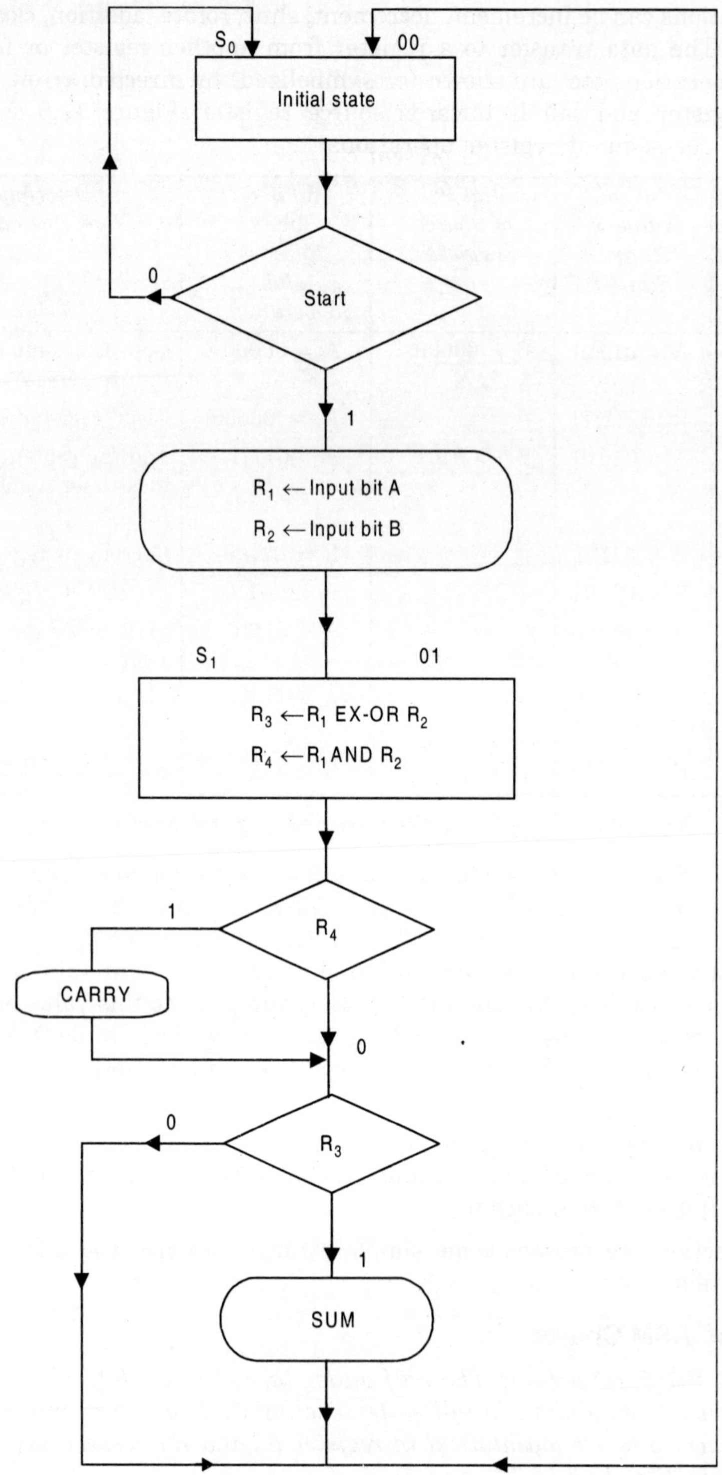

Fig. 10.6 ASM chart for 1-bit half adder

By observing the ASM chart of Fig. 10.6 we see that there are three state boxes which contributes to three ASM blocks. And we know that the state box and conditional blocks are executed by one common clock pulse corresponding to the state defined by state box in the particular ASM block.

Example 2. *MOD-5 Counter:* *We present counter having one input X and one output Z. Counter will have five states, state 0 (i.e., S_0) to state 4 (i.e., S_4) and it moves to next state only and only if input X = 1 at the time of arrival of clock pulse. If X = 0 at this time counter does not move to next state and maintains its current state. Also when in state S_4 then X = 1 at clock pulse moves the system to next state S_0 i.e., to initial state so that counting can be restarted from 000. The output Z produces a pulse when X = 1 at 5 clock pulses or when state changes from S_4 to S_0.*

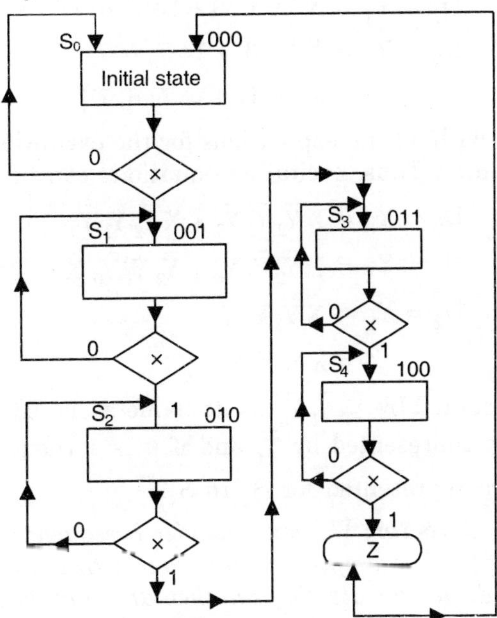

Fig. 10.7 ASM chart for MOD-5 counter

Note that in the ASM chart shown in Fig. 10.7 state boxes are blank and does not specify any operation. But the blocks corresponding to these states contain decision boxes which means that only operation to be done in these states is to test the states of input.

Now let us consider the synthesis and we wish to use D flip-flop. The 3D flip-flops together are used to assign the 3-bit binary name to states. We know that for D flip-flops excitation input D_i should be same as next state variable Y_i. By simply observing the assigned state on ASM chart of Fig. 10.7, we carry out the task.

Let the three outputs of flip-flops are $Y_2Y_1Y_0$ *i.e.,* the three bit binary name for state.

(1) First finding the changes in bit Y_0 in ASM chart. When present state is 000 or 010 then the next value of Y_0 has to become 1. Thus

$$D_0 = Y_0 = X\left[\Sigma(0, 2) + \Sigma\phi(5, 6, 7)\right]$$

Similarly, for Y_1 and Y_2.

(2) The next value of Y_1 has to become 1 only for the present states 001 and 010. So

$$D_1 = Y_1 = X\left[\Sigma(1, 2) + \Sigma\phi(5, 6, 7)\right]$$

(3) Similarly, next value of Y_2 has to become 1 only for the present state 011. So

$$D_2 = Y_2 = X[\Sigma(3) + \Sigma\phi(5, 6, 7)]$$

(4) Similarly, next value of Z has to become 1 only and only for present state 100. So

$$Z = X[\Sigma(4) + \Sigma\phi(5, 6, 7)]$$

Note that state 5, 6, 7 *i.e.*, 101, 110, 111 never occurs and that's why these three states are written $\Sigma\phi(5, 6, 7)$.

Thus, the synthesis equations can be summarized as

$$D_0 = Y_0 = X.[\Sigma(0, 2) + \Sigma\phi(5, 6, 7)]$$

$$D_1 = Y_1 = X.[\Sigma(1, 2) + \Sigma\phi(5, 6, 7)]$$

$$D_2 = Y_2 = X.[\Sigma(3) + \Sigma\phi(5, 6, 7)]$$

and

$$Z = X.[\Sigma(4) + \Sigma\phi(5, 6, 7)]$$

Here input X is ANDed with all the expressions for the excitations. In this system, input X is used to enable the counter. Thus, excitation equations can be given as–

$$D_0 = Y_0 = X\overline{Y}_2\overline{Y}_1\overline{Y}_0 + X\overline{Y}_2 Y_1 \overline{Y}_0$$

$$D_1 = Y_1 = X\overline{Y}_2\overline{Y}_1 Y_0 + \overline{Y}_2 Y_1 \overline{Y}_0 X$$

$$D_2 = Y_2 = X\overline{Y}_2 Y_1 Y_0$$

and

$$Z = X Y_2 \overline{Y}_1 \overline{Y}_0$$

Where state S_0 is represented by $\overline{Y}_2\overline{Y}_1\overline{Y}_0$ as its name assigned was 000. In fact if value of state value is 0 then it is represented by $\overline{Y}_i$ and if it is 1 then use Y_i.

Similarly, the states are represented for S_1 to S_4.

Example 3. *Sequence Detector:* *We now consider a sequence detector to detect "0101" and allowing the overlapping. This example is chosen to illustrate the similarity between state diagram. If we have already drawn a state diagram then drawing the ASM chart is a very easy job. The state diagram to detect 0101 is shown in Fig. 10.8.*

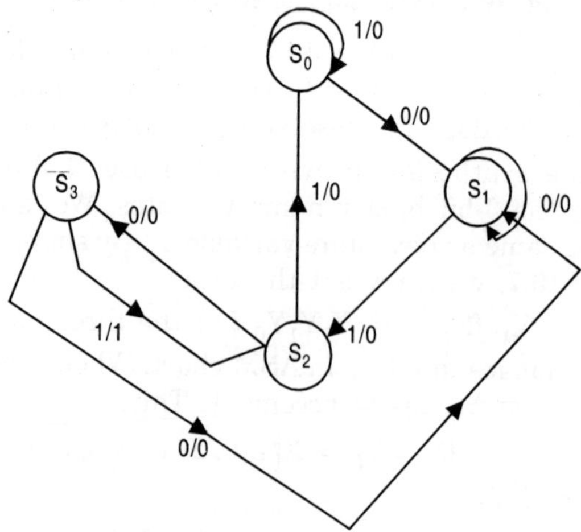

Fig. 10.8 State diagram of 0101 sequence detector with overlapping

Number marked with segments connecting two states is value of input and output and written as Input/output. If its 1/0 means Input-1 then Output is 0, while in this state.

It is evident from the state diagram that there are four states S_0, S_1, S_2, S_3. So, 2-bit binary code can be associated to these states to identify the particular state. Thus we use two D flip-flops to assign binary number Y_1Y_0 to the states. As earlier indicated use of D flip-flop makes the excitation table same as transition table because for D flip-flop $D_i = Y_i$. In fact we carry out these exercise after drawing ASM chart, but here we did it earlier to reflect the similarity between ASM chart and state graph as major difference between the two is indication of timing relation in the drawing. Below is the ASM chart *i.e.*, Fig. 10.9 for the problem.

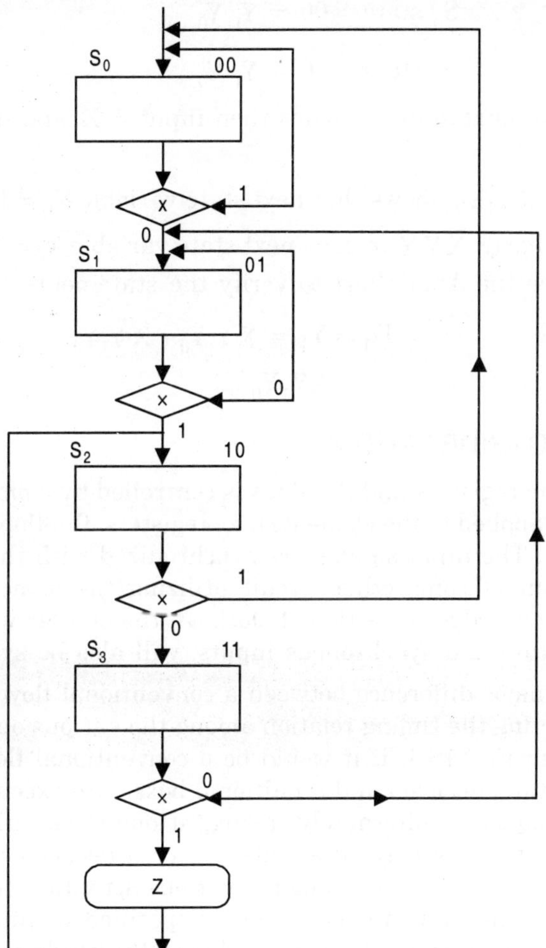

Fig. 10.9 ASM chart for 0101 sequence detector

Note that, as in earlier examples, here also X is input through which the sequence is applied to the system. Z is output which goes high when the intended sequence is detected. Note the similarity between the state diagram and ASM chart. A close inspection of two graphs, shows that for every state of state diagram, there exist one ASM block and there is one state box per ASM block in the ASM chart. Thus, there are four ASM blocks in Fig. 10.9 each of which contains a decision box and last one contains a conditional box also in addition to state box. We again assert the fact that all the operations owing to an ASM block are to be completed in the same clock period. We now consider synthesis to find out the equations for excitations.

Here also we use observation (as was done in example 2) to determine that when next state variables Y_0 and Y_1 become 1. Thus

$$D_0 = Y_0 = \overline{X}[\Sigma(0, 2)]$$

$$\therefore \quad D_0 = Y_0 = \overline{X}\,\overline{Y}_1\,\overline{Y}_0 + \overline{X}\,Y_1\,\overline{Y}_0$$

as the $\quad S_0 \text{ state} = 00 = \overline{Y}_1\,\overline{Y}_0$

$$S_2 \text{ state} = 10 = Y_1\,\overline{Y}_0$$

If input X = 0 make next state to comes then input = $\overline{X}$ and if input X = 1 causes the next state then input = X.

In equation for D_0, $\overline{X}\,\overline{Y}_1\overline{Y}_0$ shows that next state variable $Y_0 = 1$ when X = 0 and present state is S_0 (*i.e.,* 00). Similarly, $\overline{X}\,Y_1\overline{Y}_0$ means next state variable $Y_0 = 1$ if the input X = 0 while in state S_2 (*i.e.,* 10). See the ASM chart to verify the statements.

Similarly, $\quad D_1 = Y_1 = X\,\overline{Y}_1 Y_0 + \overline{X}\,Y_1\overline{Y}_0$

and $\quad Z = XY_1Y_0$

10.3 ASM TIMING CONSIDERATIONS

The timing of all the registers and flip-flops is controlled by a *master clock generator*. The clock pulses are equally applied to the elements (*i.e.,* registers, flip-flops) of both data processing and control subsystems. The input signals are synchronized with the clock as normally they happen to be the output of some other circuit utilizing the same clock. Thus the inputs change the state during an edge transition of clock. In the similar way, the outputs, that are a function of present state and synchronous inputs, will also be synchronous.

We re-insert that major difference between a conventional flow chart and ASM chart is in defining and interpreting the timing relation among the various operations. Let us consider the ASM chart shown in Fig. 10.4. If it would be a conventional flow chart, then the listed operations within the state, decision and conditional boxes are executed sequentially *i.e.,* one after another in time sequence. Alternately saying, at one clock pulse only one of the boxes will be executed, where the box may be a state box or a decision box or a conditional box. Thus, a total denial of timing relation among the various activities. In contrast to it, an entire ASM block is treated as one unit. All the activities specified within the block must happen in synchronism with the transition of positive edge of the clock, while the system changes from current state to next state. Here it is assumed that all the flip-flops are positive edge triggered. For illustration purpose consider the ASM chart shown in Figs. 10.4 and 10.10 shows the transition of control logic between the states.

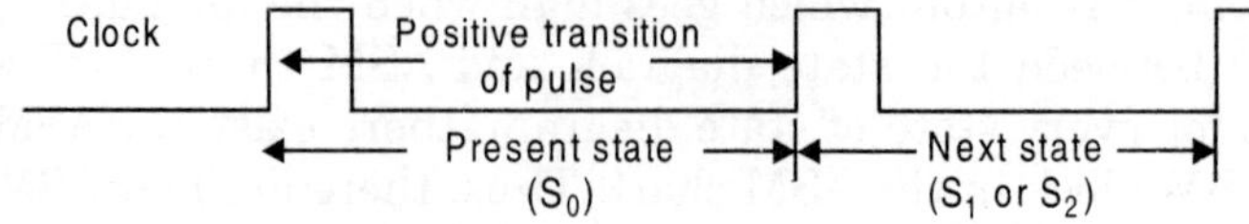

Fig. 10.10 Transition between states

In order to understand the state transition at the positive edge of clock refer the Figs. 10.4 and 10.10 simultaneously along with the following discussion.

The arrival of first positive transition of clock, transfers the control subsystem into S_0 state. The activities listed in various boxes of ASM block, corresponding to S_0 state can now be executed, as soon as the positive edge of second-clock pulse arrives. At the same time depending upon values of inputs X and Y the control is transferred to next state which may be either state S_1 or state S_2. Referring to the ASM block indicated by dashed line in Fig. 10.4 we can list out operation that occur simultaneously when the positive edge of second clock pulse appears. They are–

Recall that system is in S_0 state before second-clock pulse

(1) Register R_1 is cleared.

(2) If input X is 1, the output signal VALID is generated and the control enters in S_2 state.

(3) If input X is 0, then the control tests the input Y.

(4) If input Y is 0 register R_3 is set to one. If input Y is 1 register R_2 is cleared. In either case next state will be S_1 state.

Observe the ASM chart closely (in Fig. 10.4), and we find that next state is decided by the status of input X only. If X = 1 then next is S_2 state and when X = 0 then whether input Y = 0 or 1 the next state will always be S_1. Also note that the operation in the data processing subsect and change in state of control subsystem occur at the same time, during the positive transition of same clock pulse. We now consider a design example to demonstrate timing relation between the components of ASM chart.

Example 1. *Design a digital system having one 4-bit binary counter 'C' whose internal bits are labelled C_4 C_3 C_2 C_1 with C_4 MSB and C_1 as LSB. It has two flip-flops named 'X' and 'Y'. A start signal incrementing the counter 'C' by 1 on arrival of next clock pulse and continues to increment until the operation stops. Given that the counter bits C_3 and C_4 determines the sequence of operation. The system must satisfy following–*

(1) Initiate the operation when start signals = 1 by clearing counter 'C' and flip-flop "Y", i.e., C = 0000 and Y = 0.

(2) If counter bit C_3 = 0, it causes E to be cleared to 0 i.e., E = 0 and the operation proceeds.

(3) If counter bit C_3 = 1, E is set to 1 i.e., E = 1 and

(a) if C_4 = 0, count proceeds.

(b) if C_4 = 1, F is set to 1 i.e., F = 1 on next clock pulse and system stops counting.

Solution. ASM chart for the given problem is shown in Fig. 10.11. A close inspection reveals that:

When, no operation, system is in initial state S_0, and keep waiting for start signals 'S'.

When S = 1, counter C = 0000 and Y = 0 and simultaneously control goes to S_1 state. It means clearing of counter 'C' and flip-flop 'Y' occurs during S_0 state.

The counter is incremented by 1 during state S_1, on the arrival of every clock pulse. During each clock pulse simultaneously with increment during same transition of clock, one of the three possibilities is tested to determine the next state:

(1) Either X is cleared and control stays at S_1 (C_3 = 0).

or

(2) X is set (X = 1) and control maintains S_1 state (A_4A_3 = 10).

or

(3) X is set and control advanced to state S_2 (A_4A_3 = 11).

When in S_2 state flip-flop 'Y' is set to 1 and control move back to its initial state S_0. The ASM chart consists of three blocks, one external input S, and two status inputs S_4 and S_3.

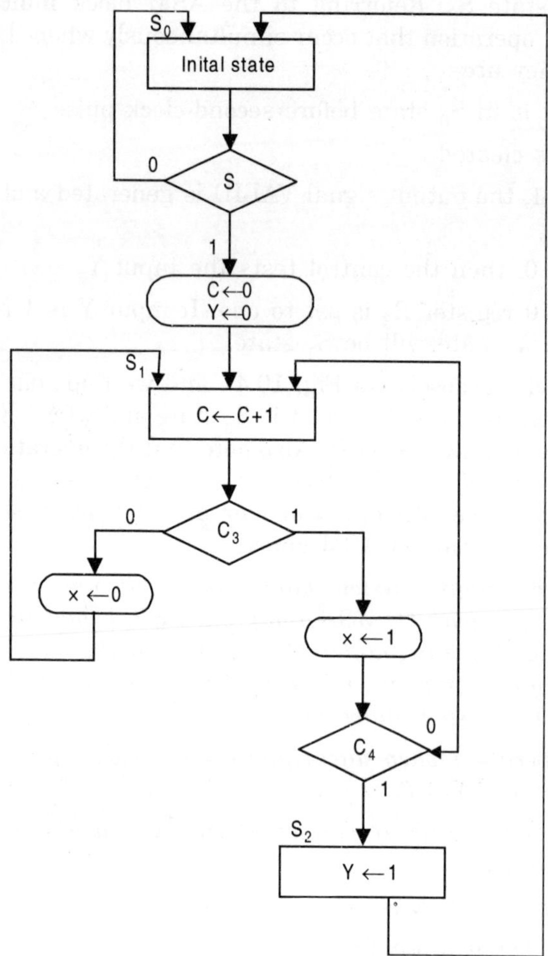

Fig. 10.11 ASM chart for cxample 10.4

Example 2. *Design a digital system for weight computation in a given binary word.*

Solution. The weight of a binary number is defined as the number of 1's contained in binary representation. To solve the problem, let the digital system have

1. R – A register where binary work is stored.
2. W – A register that counts number of 1's in binary number stored in R.
3. F – A flip-flop.

The operation of the system is to shift a single bit of R into F. Then check the output of the F. If it is 1 increment count in W by 1. If it is 0 no increment in W. The moment all the bits are shifted and tested operation stops and W contains the weight of the given word.

The ASM chart for this problem is shown in Fig. 10.12. Note that the system have 3 inputs S, Z and F. Z is used to sense whether all the bits in register R are 0 or not. Z = 1 indicates that register R contains all zeros, and the operation must stop.

Initially machine is in state S_0 and remains in state S_0 until the switch S (*i.e.*, start signal) is made 1. If S = 1 then in S_0 state, the clock pulse causes, input word to be loaded into R, counter W to have all 1's and machine transferred to state S_1.

In state S_1 a clock pulse causes two works simultaneously. First it increments W by 1. If W is incremented for the first time, then the count in W becomes all 0's as initially it was all 1's second it tests Z. If Z = 0 machine goes to state S_2. If Z = 1 machine goes to state S_0 and count in W is weight of the binary word.

In state S_2, a bit of register R is shifted into flip-flop F, and machine goes to state S_3.

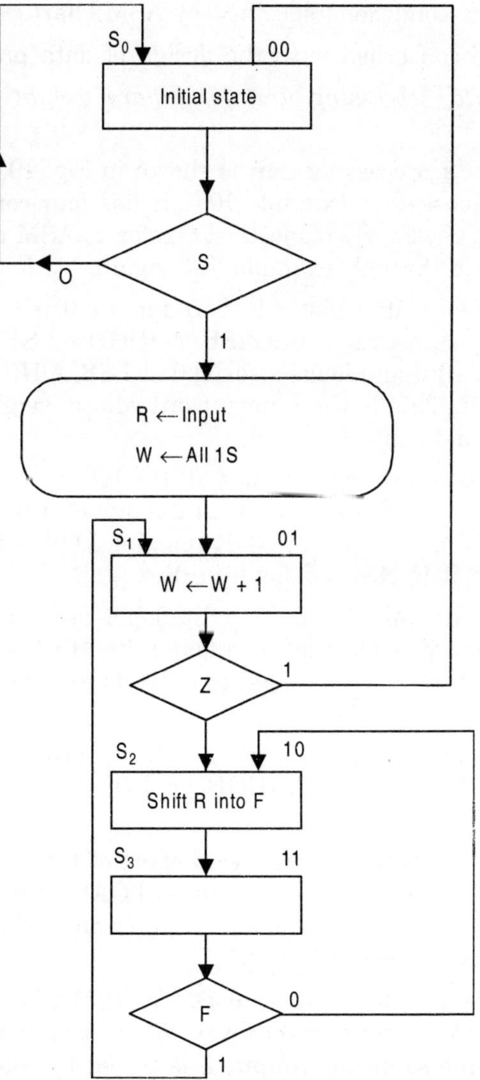

Fig. 10.12 ASM chart for weight computation

In state S_3, shifted bit in F is tested. If F = 0 the machine goes to state S_2 to shift next bit into F. If F = 1 it goes to state S_1 to increment to count in W.

10.4 DATA PROCESSING UNIT

Once the ASM chart is prepared, the system (or machine) can be designed. The design is splitted in two parts.

a. Data Processing Unit

b. Control Unit

The data processing unit contains the element that performs the operations like increment the count, shift a bit etc.

The control unit is the subsystem that is responsible to move the machine from one state to another, according to the conditions specified by ASM chart.

In this section we are concerned with the design of data processing unit only.

Example. *Design Data processing unit for binary weight computation, discussed in example 10.5.*

Solution. Proposed data processing unit is shown in Fig. 10.13. The control sub system has 3 inputs S, Z, F as discussed in example 10.5. It has four control signals C_0, C_1, C_2, C_3 corresponding to states S_0, S_1, S_2, S_3 respectively (refer to ASM chart shown in Fig. 10.12). We advise the readers to go through example 10.5 again.

Next shown in figure is a shift register R. Serial input '0' is a data input. Each time data in R is shifted left this input inserts a 0 at LSB. A HIGH on SHIFT LEFT input shifts the data present in R to left by 1 bit and loads a serial 0 at LSB. A HIGH on LOAD INPUT DATA loads the input data into R. This is the binary word whose weight is to be calculated. This word is loaded as parallel data into R.

A NOR gate is used to determine whether all the bits of R are 0 or not. All the bits of R are brought to the input of NOR. As soon as all the bits of R become 0 output of NOR goes high. If any of the bit of R is 1 output of NOR remains LOW. Output of NOR is feeded as input Z to controls, where it is checked for 0 or 1.

A flip-flop F is connected at MSB of R. This flip-flop is used to collect each shifted bit from register R. Every time R receives shift left command, its MSB is shifted out and is received in flip-flop F. The output of flip-flop is feeded to controls as input F, where it is checked for 1 or 0.

Last in figure is counter W which is yet another register acting as counter. A HIGH on LOAD INPUT loads all 1s into W. A HIGH on INCREMENT increments the count in W by 1.

Initially the system is in state S_0 (refer ASM chart of Fig. 10.12. along with Fig. 10.13). As soon as START = 1, C_0 is activated. This causes LOAD INPUT DATA signals of both R and W to go HIGH. Thus, binary word is loaded into R and initial count is loaded into W. At the same time the machine moves to state S_1.

In state S_1 signal C_1 is activated. This causes INCREMENT signal of W to go HIGH and consequently the count in W is incremented by 1. When it is incremented for the first time, All 1's become all 0's. At the same time input Z is tested by controls. If Z = 1 control goes back to state S_0. If Z = 0 control goes to state S_2.

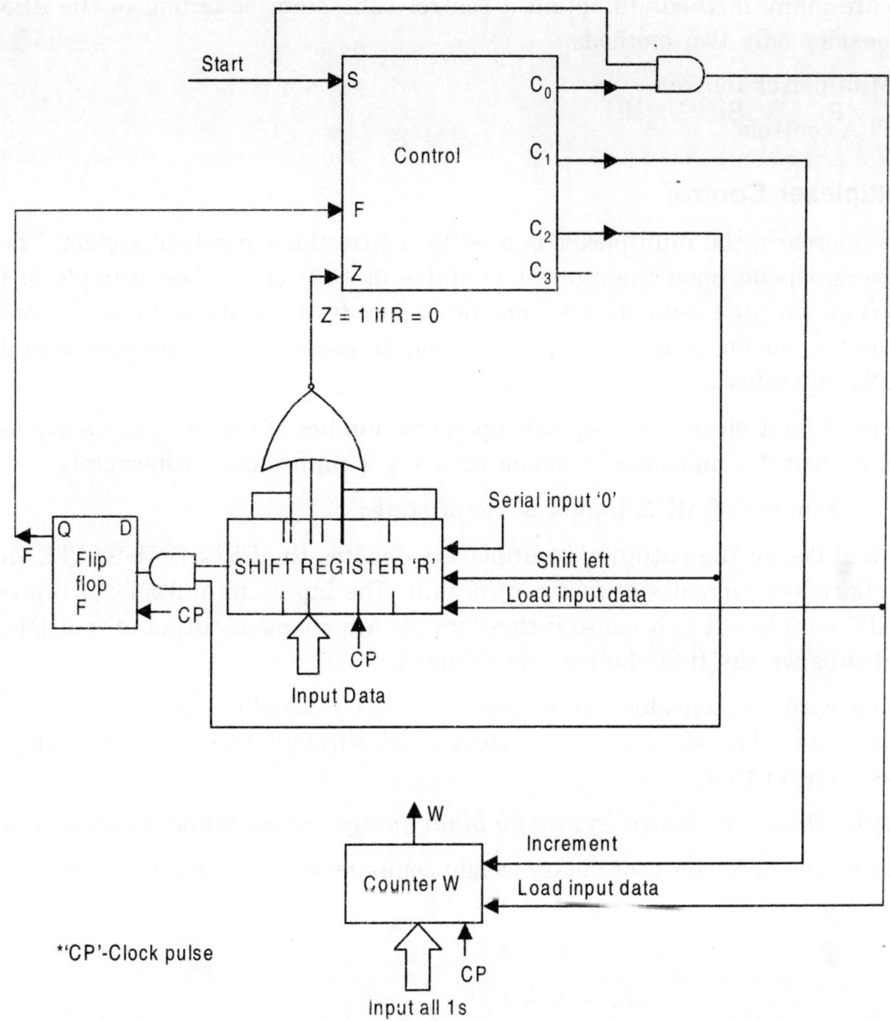

Fig. 10.13 Data processing unit for binary weight computation

In state S_2, the signal C_2 is activated. The C_2 causes shift left of R to go high and enables the flip-flop F. Thus, the content of R is shifted to left by 1-bit. Hence, MSB of R is shifted out and is collected by F and at the same time a 0 is inserted at the LSB through serial input. Now the machine moves on to state S_3.

In state S_3 the output of F is tested by control subsystem. If F = 1 machine should go to state S_1 *i.e.*, C_1 to be activated next. If F = 0 machine should go to state S_2 *i.e.*, C_2 to be next. Since all these activities are internal to control subsystem, C_3 is not connected to any element of data processing unit. In fact C_3 is used to activate the signals C_1 or C_2 and is processed internally by control unit.

10.5 CONTROL DESIGN

As stated earlier the job of the control subsystem is to move the machine from one state to other state according to inputs given to it. In general, variables defined by the "decision boxes" in an ASM chart are treated as inputs to the control subsystem.

There are many methods to obtain a control subsystem according to the ASM charts. Here we consider only two methods.

(*i*) Multiplexer controls

(*ii*) PLA controls

10.5.1 Multiplexer Control

In this approach, the multiplexer is used to realize the control subsystem. The number of multiplexers depend upon the number of states in ASM chart. For example, if there are four-states then we need 2-bit binary number to specify these states uniquely. So, we take two multiplexers, one for each bit of representation. In general if '*n*' is number of multiplexers then $2^n \geq$ No. of states.

The type of multiplexer also depends upon the number of states. If there are four states in ASM chart then the multiplexer should be a 4 × 1 multiplexer. Alternately

$$\text{Number of MUX inputs} \geq \text{No. of states}$$

In general design the output of multiplexers denotes the PRESENT STATE variable as these outputs reflect current status of control unit. The inputs to multiplexer represents the NEXT STATE variable. It is because if these inputs are changed output of multiplexers may change and thus we say that the state is changed.

To being with we consider our example of binary weight computation illustrated in example 10.5 and 10.6. We urge the readers to go through these two examples carefully before proceeding further.

Example. *Design the control system for binary weight computation, by using multiplexers.*

Solution. The ASM chart for binary weight computation is drawn in Fig. 10.12. Referring to the chart we find that there are 4 states. So,

$$2^n \geq 4$$

or

$$n \geq 2$$

So, we take 2 multiplexers (n = 2) MUX 1 and MUX 0. Since there are 4 states we select 4 input multiplexers *i.e.*, 4 × 1 multiplexers.

After selecting the multiplexers next step is to draw state table, as shown in Fig. 10.14(*a*). The first 3 columns of the table shows present states, next state and the inputs that causes the next state. Last column of the table is multiplexer input. As earlier stated multiplexer inputs are next-state variables. Thus, entries in these columns are made by making observations on inputs and next state. For example, if present state is S_0 *i.e.*, multiplexer output $Y_1 = 0$ and $Y_0 = 0$, then status of switch S decides the next state. If S = 0 the next state is S_0 *i.e.*, $Y_1 = 0$ and $Y_0 = 0$. If S = 1 the next state is S_1 *i.e.*, $Y_1 = 0$ and $Y_0 = 1$. Hence when S = 0 $Y_0 = 0$ and when S = 1 $Y_0 = 1$ so we say $Y_0 = S$. Since $Y_1 = 0$ always the first entry in MUX inputs column is 0 S. Consequently input I_0 of MUX1 must be connected to 0 and input I_0 of MUX0 must be connected to S. The same is shown in Fig. 10.14(*b*). Readers are advised to verify all the rows of state table in similar way.

Present State			*Inputs*			*Next State*		*MUX Inputs*	
	Y_1	Y_0	S	Z	F	Y_1	Y_0	$D_1 = Y_1$ MUX1	$D_0 = Y_0$ MUX0
S_0	0	0	0	×	×	0	0	0	S
			1	×	×	0	1		
S_1			×	1	×	0	0		
	0	1	×	0	×	1	0	$\overline{Z}$	0
S_2	1	0	×	×	×	1	1	1	1
S_3	1	1	×	×	0	1	0		
			×	×	1	0	1	$\overline{F}$	F

(a) State Table

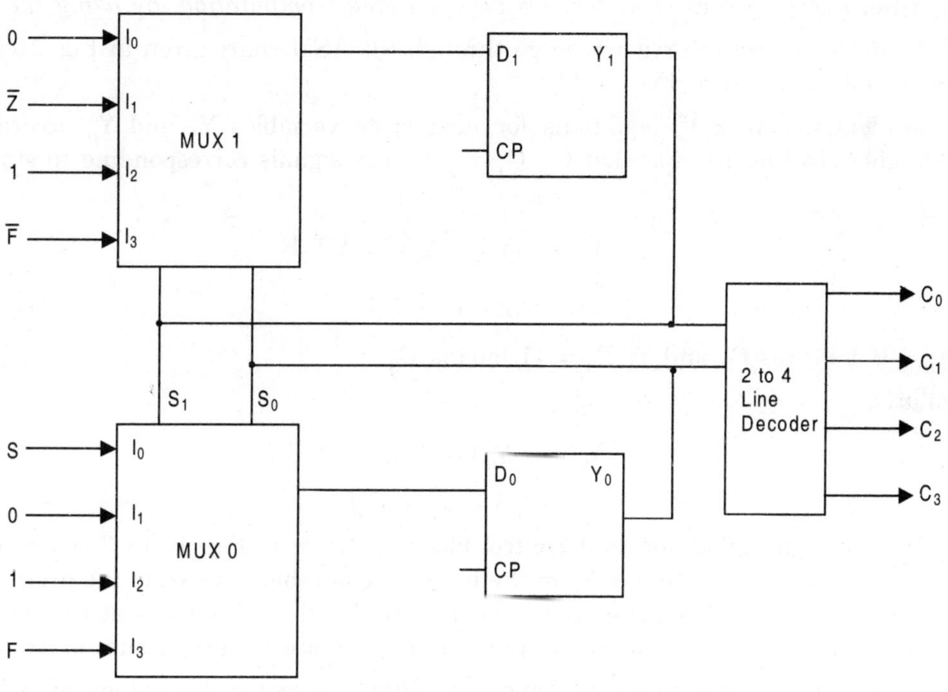

(b) Logic Diagram

Fig. 10.14 Control subsystem for binary weight computation

Figure 10.14(*b*) shows the complete control design for weight computation. The outputs of multiplexers are feeded to D flip-flops, whose outputs Y_1 and Y_0 are brought back to select lines S_0 and S_1 of multiplexers. Y_1 and Y_0 are decoded further by using 2 to 4 line decoder to generate the control signals C_0, C_1, C_2, C_3 corresponding to states S_0, S_1, S_2, S_3, respectively.

To understand the operation let us consider that control is in state S_0 so $Y_1 = 0$ and $Y_0 = 0$ *i.e.*, $S_1 = S_0 = 0$. Since $S_1S_2 = 00$, input I_0 of both the multiplexers are selected. As long as $S = 0$, both $Y_1 = Y_0 = 0$ and machine remains in state S_0. As soon as $S = 1$ output of MUX0 becomes 1 and consequently $Y_1 = 0$ and $Y_0 = 1$. Thus signal C_1 is activated and select inputs become $S_1 = 0$ and $S_0 = 1$. Hence inputs I_1 of both multiplexers selected. Note that by activation of C_1 state S_1 has arrived. At the input I_1 of MUX1, $\overline{Z}$ is connected whose value is responsible to make $Y_1 = 0$, $Y_0 = 0$ or $Y_1 = 1$, $Y_0 = 0$. Thus input Z is tested in state 1, which was to be done in S_1 according to the ASM chart shown in Fig. 10.12. Likewise the complete operation can be verified.

10.5.2 PLA Control

Use of PLA to realize, control subsystem makes the system more compact and efficient. PLAs have internal AND-OR array *i.e.*, the outputs of PLA represent sum of product. Thus, overall strategy is to prepare an SOP equation for each bit of state representation. For example, if 4-states are there we need two bits of representation. Thus, we need two SOP equations. After getting the SOP equations next step is to prepare PLA program table. Such a table is an input-output table according to which PLAs are programmed.

Example. *Design a control system for binary weight computation, by using the PLA.*

Solution. We advise the readers to go through the ASM chart given in Fig. 10.12 and multiplexer control shown in Fig. 10.14.

We now obtain two SOP equations for next state variables Y_1 and Y_0 according to state table given in Fig. 10.14(*a*). Let C_0, C_1, C_2, C_3 are signals corresponding to states S_0, S_1, S_2, S_3.

$$Y_1 = \overline{Y}_1 Y_0 \overline{Z} + Y_1 \overline{Y}_0 + Y_1 Y_0 \overline{F}$$

or

$$Y_1 = C_1 \overline{Z} + C_2 + C_3 \overline{F}$$

as (Y_1 Y_0 = 0 1 means C_1 and Y_1 Y_0 = 11 means C_3)

Similarly,

$$Y_0 = \overline{Y}_1 \overline{Y}_0 S + Y_1 \overline{Y}_0 + Y_1 Y_0 F$$

$$= C_0 S + C_2 + C_3 F$$

The PLA program table and PLA control block is shown in Fig. 10.15. Let us examine the PLA program table. Note that $Y_1 Y_0$ in the input side of table represents the present state and $Y_1 Y_0$ in the output side represents the next state. Further all entries at the input side is made for product terms and at the output side entries are results of sum of products.

First four rows in the program table are simply showing the values of $Y_1 Y_0$ and corresponding state to be excited. For example, if present state is $Y_1 = 0$ and $Y_0 = 0$ then it is state S_0 and signal C_0 is activated. This is shown in first row. At the output side $Y_1 Y_0$ shows next state. Now observe the third row, which shows that machine is in state S_2 so C_2 is activated. But according to the ASM chart shown in Fig. 10.12, if the machine is in state S_2 it goes to state S_3 without testing any input. Hence, at the output side of table we marked $Y_1 = 1$ and $Y_0 = 1$. Note that Y_1 and Y_0 on the output side are filled up according to the two SOP equations obtained in the beginning. Infact the first four rows are used to show what will be the control signal to be activated when machine is in a state.

Product Terms	*Inputs*					*Outputs*					
	Y_1	Y_0	S	Z	F	Y_1	Y_0	C_0	C_1	C_2	C_3
$C_0 = \overline{Y}_1 \overline{Y}_0$	0	0						1	0	0	0
$C_1 = \overline{Y}_1 Y_0$	0	1						0	1	0	0
$C_2 = Y_1 \overline{Y}_0$	1	0				1	1	0	0	1	0
$C_3 = Y_1 Y_0$	1	1						0	0	0	1
$C_1\overline{Z} = \overline{Y}_1 Y_0 \overline{Z}$	0	1		0		1	0	0	1	0	0
$C_3\overline{F} = Y_1 Y_0 \overline{F}$	1	1			0	1	0	0	0	0	1
$C_0 S = \overline{Y}_1 \overline{Y}_0 S$	0	0	1			0	1	1	0	0	0
$C_3 F = Y_1 Y_0 F$	1	1			1	0	1	0	0	0	1

(*a*) PLA Program Table

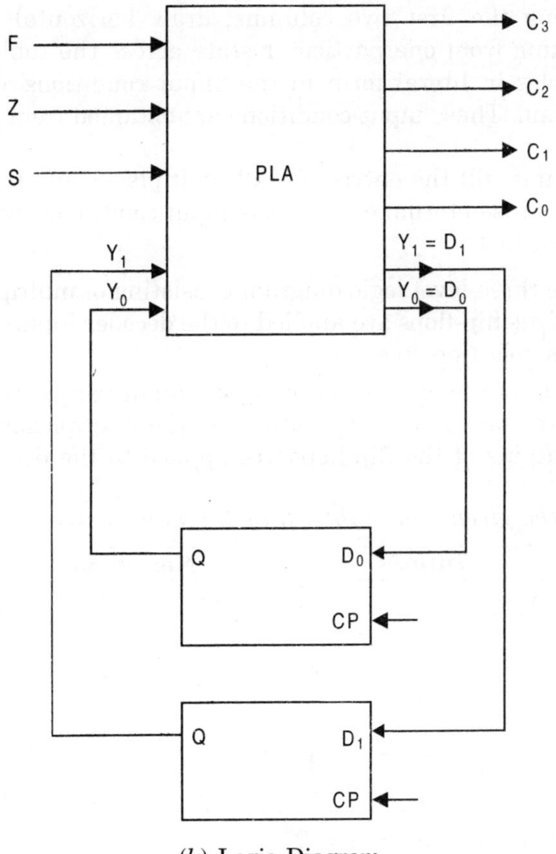

(*b*) Logic Diagram

Fig. 10.15 Control subsystem for weight computation using PLA

The rest of the four rows in PLA program table shows the input to be tested when machine is in a state and what should be the next state if testing is true. Consider the 7th row having product terms entry $C_0S = \overline{Y}_1\overline{Y}_0S$. This tests the input S when machine is in state S_0. At the input side $Y_1 = Y_0 = 0$ to show state S_0 and entry S = 1 is status of input S. At the output side in this row $C_0 = 1$ as machine is in state S_0. Next $Y_1 = 0$ and $Y_0 = 1$ at the output side indicates that since input S = 1, the machine must go to state S_1 at next clock pulse.

10.5.3 Design Examples with Multiplexers

Steps of Designing with Multiplexers:

Step 1. Obtain ASM chart.

Step 2. Obtain state diagram.

Step 3. Obtain state table.

Step 4. Obtain the table for multiplexer input conditions as follows:

No. of multiplexers required = number of flip-flops required.

- Formulate a table containing the columns for 'present state', 'next state', 'input conditions' and 'multiplexer inputs'.
- After completing the first two columns, draw horizontal lines indicating all the transitions taking from one particular state across the table.
- Put the variables in literal form in the 'input conditions' column according to the next-state column. These input conditions are obtained from the decision boxes in the ASM chart.
- In the last column, fill the entries for all multiplexer inputs. For each present state, multiplexer input is determined from the input conditions, when the next state of the flip-flop is equal to 1.

Step 5. Obtain the three-level logic diagram consisting of multiplexer, flip-flop and then decoder. The outputs of the flip-flops are applied to the decoder inputs and also to select inputs for the multiplexers as selection lines.

The purpose of each multiplexer is to produce an input to its corresponding flip-flop equal to the binary value of the next state. The outputs of the multiplexers are applied to the flip-flops as inputs. The outputs of the flip-flops are applied to the decoder as inputs.

Example 1. *For the given state table, find the multiplexer required for control design.*

Present State		Inputs			Next-state		Outputs		
G_1	G_0	S	A_3	A_4	G_1	G_0	T_0	T_1	T_2
0	0	0	×	×	0	0	1	0	0
0	0	1	×	×	0	1	1	0	0
0	1	×	0	×	0	1	0	1	0
0	1	×	1	×	0	1	0	1	0
0	1	×	1	1	1	1	0	1	0
1	1	×	×	×	0	0	0	0	1

Solution: ∵ Number of state = 4 = 2^2

∵ Number of flip-flops = 2

= number of multiplexers required.

Design with Multiplexers

The sequence register and decoder control consist of following *three* components:

1. Flip-flops (hold the present binary state value)
2. Decoder (generates the control outputs) and
3. Gates (to determine the next state and output signals).

A combinational circuit can be implemented with multiplexers instead of individual gates. Replacing the gates with multiplexers results a *three level* components:

1. *First level,* consists of multiplexers that determine the next state of the register.
2. *Second level,* consists of a register (made up of flip-flops) that holds the present binary state,
3. *Third level,* has a decoder to provide a separate output for each control state.

Example 2. *Design the control circuit using (i) sequence register and decoder (ii) one flip-flop per state (iii) multiplexer, for the following state diagram:*

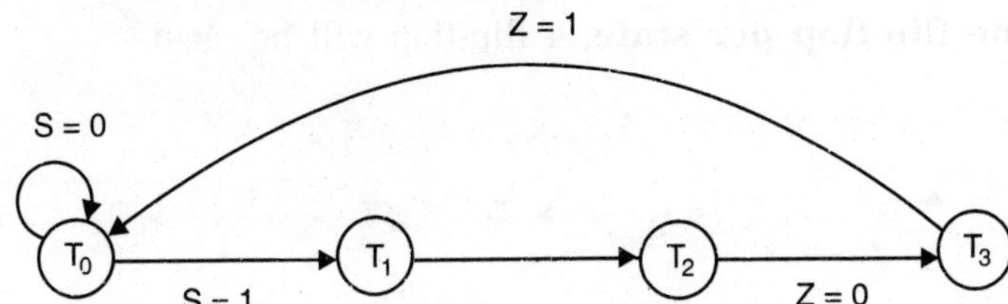

Solution. First we formulate the corresponding state table as follows:

State	*Present state*		*Inputs*		*Next state*		*Outputs*			
name	*A*	*B*	*S*	*Z*	*A*	*B*	T_0	T_1	T_2	T_3
T_0	0	0	0	×	0	0	1	0	0	0
T_0	0	0	1	×	0	1	1	0	0	0
T_1	0	1	×	×	1	0	0	1	0	0
T_2	1	0	×	0	1	1	0	0	1	0
T_3	1	1	×	1	0	0	0	0	0	1

(i) **Number of flip-flops required = 2**

Flip-flop input equations from the next-state column:

$$D_A = T_1 + T_2\bar{Z}$$

$$D_B = T_0 S + T_2\bar{Z}$$

Since 2 flip-flops are used here, so at 2-to-4-line decoder is required.

Output lines of the decoder represents the following states:

$$T_0 = \bar{A}\,\bar{B}$$
$$T_1 = \bar{A}\,B$$
$$T_2 = A\,\bar{B}$$
$$T_3 = A\,B$$

The logic diagram is as follows:

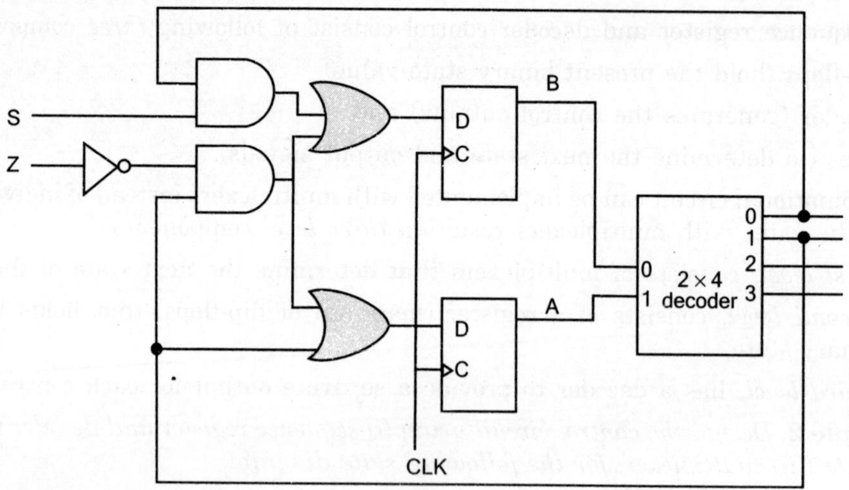

(ii) **Using one flip-flop per state**, 4 flip-flop will be used.

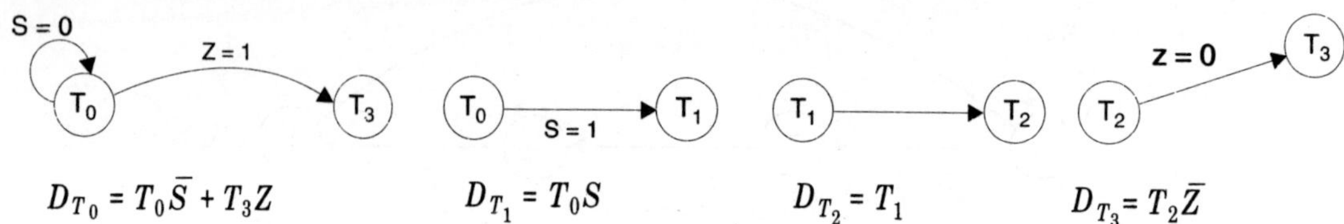

$D_{T_0} = T_0\bar{S} + T_3 Z$ $D_{T_1} = T_0 S$ $D_{T_2} = T_1$ $D_{T_3} = T_2\bar{Z}$

The logic diagram is as follows:

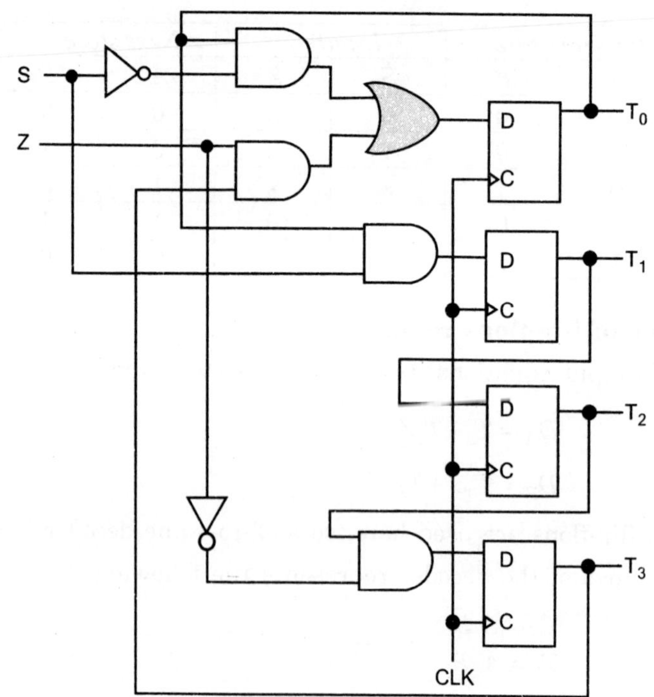

(iii) **Using multiplexers:**

Number of states = 4 = 2^2

∴ Number of flip-flops required = 2

= number of multiplexers required.

∴ 2 multiplexers of 4-to-1 line are required.

Table for multiplexer inputs is:

Present State		***Next State***		***Input Conditions***	***MUX Inputs***	
A	***B***	***A***	***B***	***A***	***MUX1***	***MUX2***
0	0	0	0	$\bar{S}$	0	S
0	0	0	1	S		
0	1	1	0	ϕ	1	1
1	0	1	1	$\bar{Z}$	$\bar{Z}$	$\bar{Z}$
1	1	0	0	Z	1	1

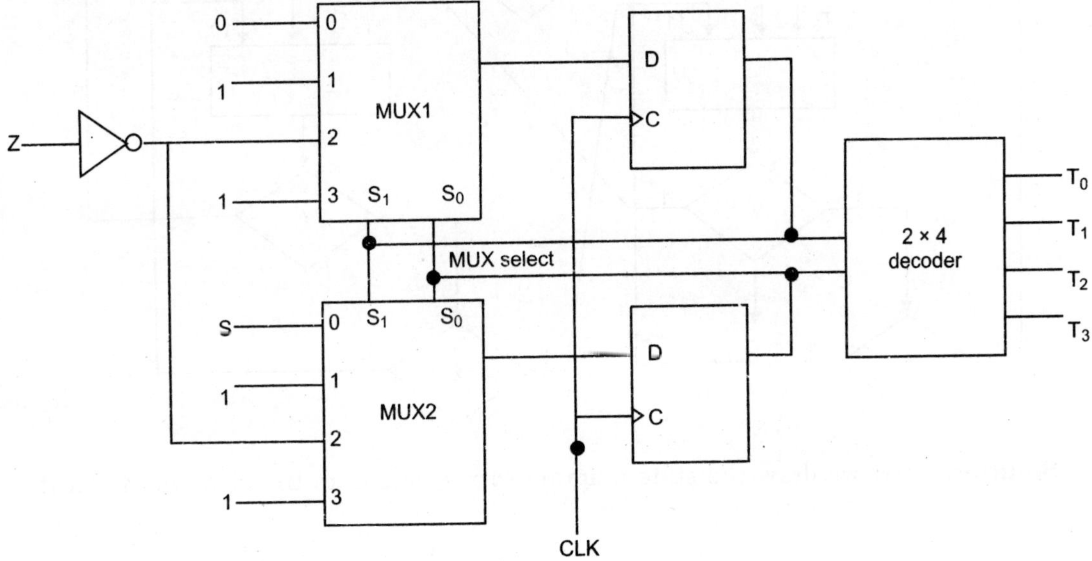

Example 3. *Following ASM chart consists of four states and four control inputs w, x, y and z. The binary assignment for each state is indicated at the upper right corner of the state boxes. Implement the control circuit using:*

(i) *Classical method*

(ii) *Sequence register and decoder*

(iii) *Multiplexer.*

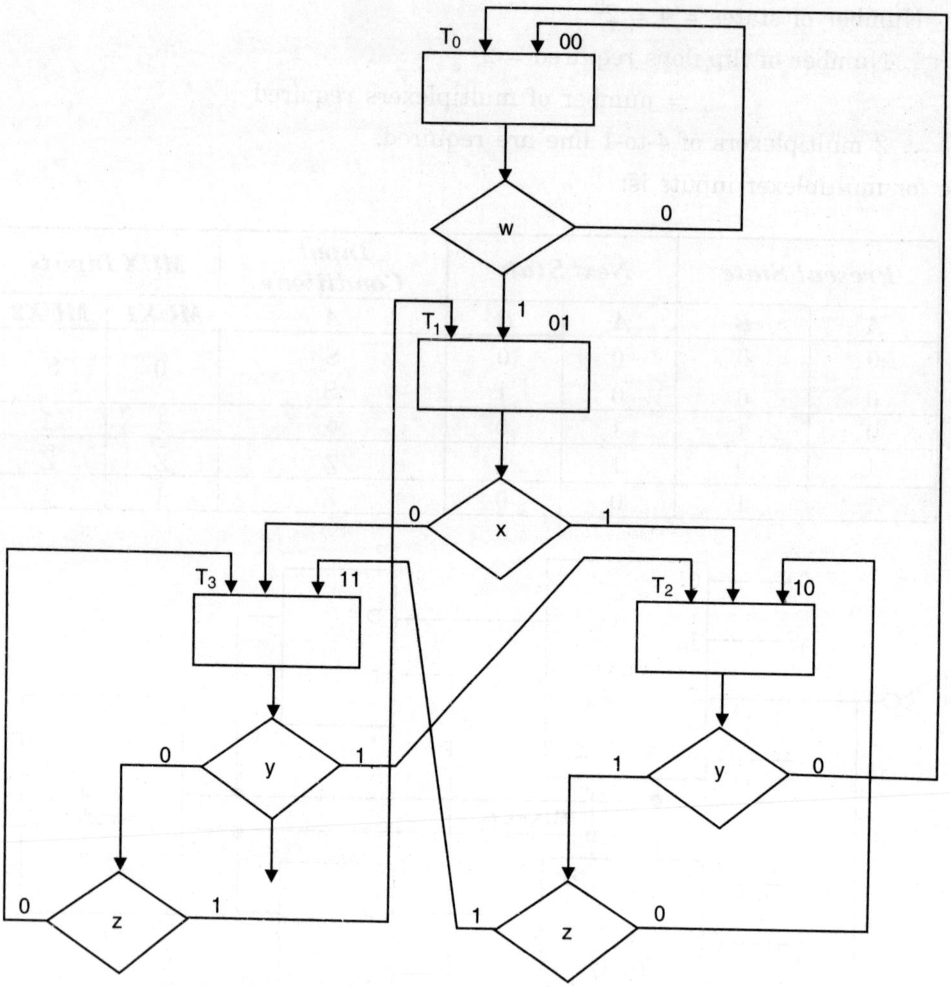

Solution. First we draw the state diagram corresponding to the given ASM chart.

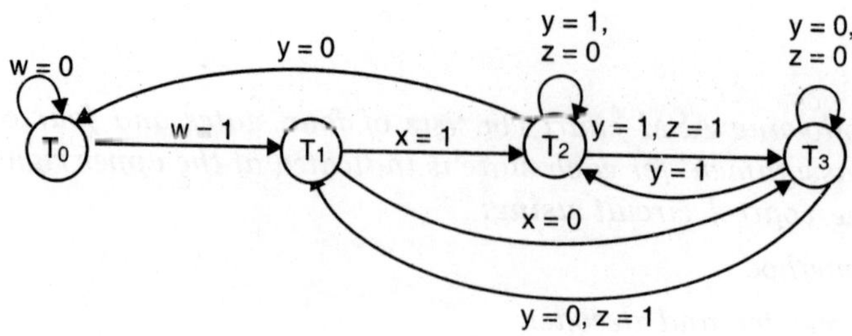

Corresponding state table is:

Present State Symbol	*Present State*		*Inputs*				*Next State*		*Outputs*			
	A	*B*	*w*	*x*	*y*	*z*	*A*	*B*	T_1	T_2	T_3	T_4
T_0	0	0	0	×	×	×	0	0	1	0	0	0
T_0	0	0	1	×	×	×	0	1	0	1	0	0
T_1	0	1	×	1	×	×	1	0	0	0	1	0
T_1	0	1	×	0	×	×	1	1	0	0	0	1
T_2	1	0	×	×	1	0	1	0	0	0	1	0
T_2	1	0	×	×	0	×	0	0	1	0	0	0
T_2	1	0	×	×	1	1	1	1	0	0	0	1
T_3	1	1	×	×	0	0	1	1	0	0	0	1
T_3	1	1	×	×	1	×	1	0	0	0	1	0
T_3	1	1	×	×	0	1	0	1	0	1	0	0

(i) Using classical method:

Here, a 6-variable K-map is required to simplify the input equation which is not an easy task. So instead of using maps, we can obtain the equations directly from the state table by inspection. These equations may or may not be in simplified form.

To design a circuit with D flip-flops, equations are derived from the next-state column:

$$\begin{aligned} D_A &= T_1 x + T_1 \bar{x} + T_2 y + T_3 \bar{y}\,\bar{z} + T_3 y \\ &= T_1 (x + \bar{x}) + T_2 y + T_3 \bar{y}\,\bar{z} + T_3 y \\ &= T_1 + T_2 y + T_3 \bar{y}\,\bar{z} + T_3 y \end{aligned}$$

$$\begin{aligned} D_B &= T_0 \omega + T_1 \bar{x} + T_2 yz + T_3 \bar{y}\,\bar{z} + T_3 \bar{y} z \\ &= T_0 \omega + T_1 \bar{x} + T_2 yz + T_3 \bar{y} \end{aligned}$$

Four output functions are:

$$T_0 = \bar{A}\bar{B} + A\bar{B} = \bar{B}$$

$$T_1 = \bar{A}\bar{B} + AB = (\overline{A \oplus B})$$

$$T_2 = \bar{A}B + A\bar{B} + AB = (A \oplus B)$$

$$T_3 = \bar{A}B + A\bar{B} + AB = (A \oplus B) + AB$$

Logic diagram using classical method

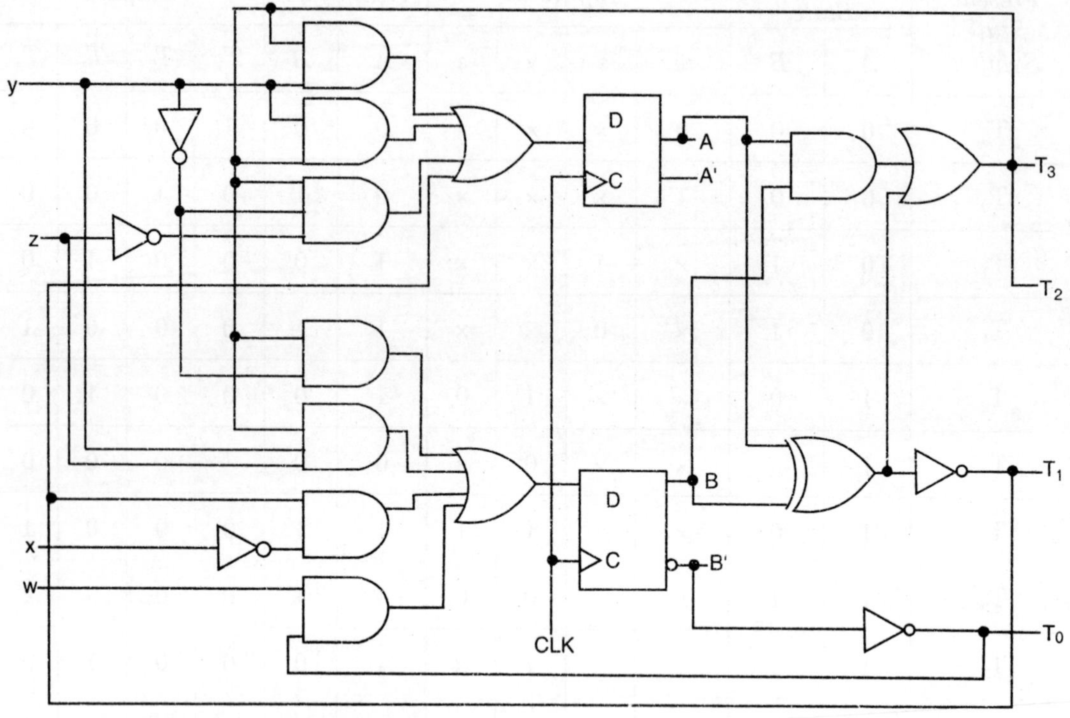

(ii) Using sequence register and decoder:

Number of flip-flops required = 2

∴ a 2-to-4-line decoder is required.

Flip-flop input equations are:

$$D_A = T_1 + T_2 y + T_3 \bar{y}\bar{z} + T_3 y$$
$$D_B = T_0 \omega + T_1 \bar{x} + T_2\ y\ z + T_3 \bar{y}$$

Outputs for decoder,

$$T_0 = \bar{A}\bar{B}; \quad T_1 = \bar{A}B; \quad T_2 = A\bar{B}; \qquad T_3 = \bar{A}\bar{B}$$

Logic diagram is similar to as in part (i) except the outputs from D. Flip-flops are treated as inputs to 2-to-4-line decoder and 4 output lines of the decoder are four states T_0, T_1, T_2 and T_3.

(iii) Using multiplexers:

Number of flip-flops required = 2 = number of multiplexers.

∴ Two 4-to-1 multiplexer is required.

Now formulating multiplexer input table as follows:

Present state		*Next State*		*Inputs*	*MUX Inputs*	
A	*B*	*A*	*B*	*Conditions*	*MUX1*	*MUX2*
0	0	0	0	$\overline{\omega}$		
0	0	0	1	ω	0	$\overline{\omega}$
0	1	1	0	x	1	$\overline{x}$
0	1	1	1	$\overline{x}$		
1	0	1	0	$y\overline{z}$		
1	0	0	0	$\overline{y}$	$y\overline{z} + yz = y$	yz
1	0	1	1	yz		
1	1	1	1	$\overline{y}\,\overline{z}$		
1	1	1	0	y	$y + \overline{y}\,\overline{z}$	$\overline{y}\,\overline{z} + \overline{y}z = \overline{y}$
1	1	0	1	$\overline{y}z$		

Logic diagram is as follows:

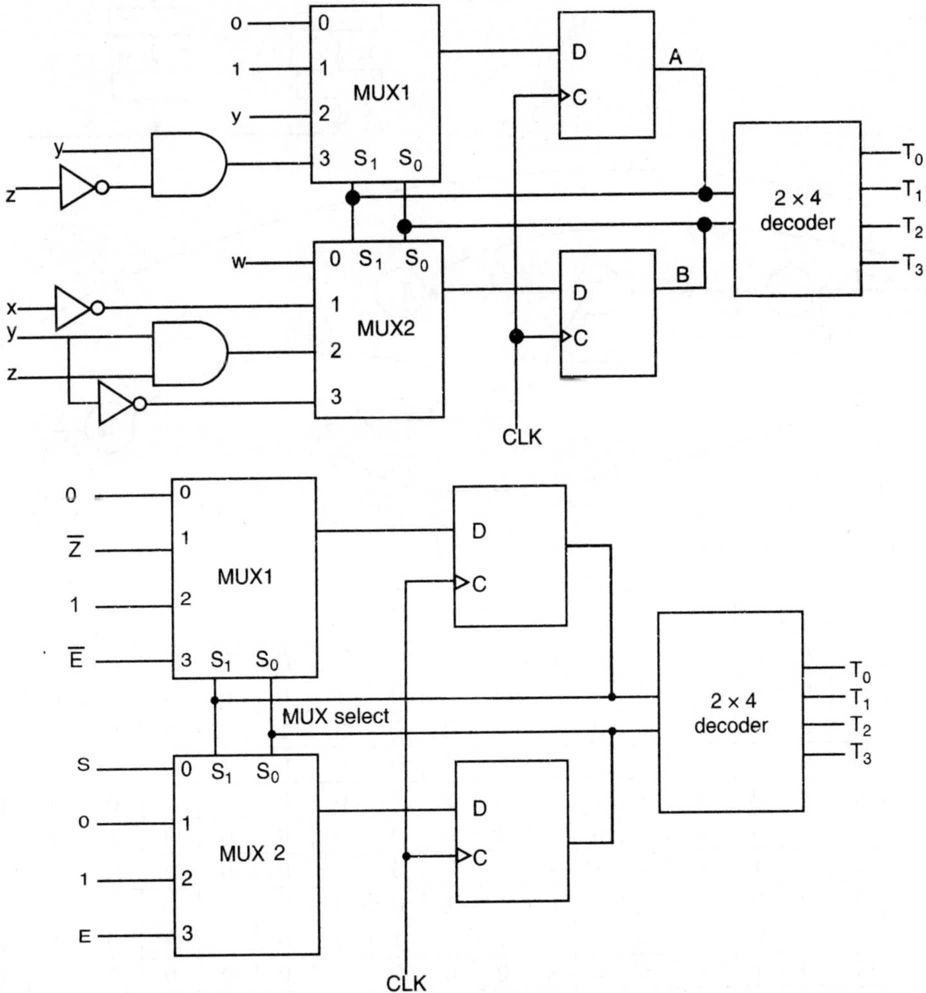

Example 4. *Consider the adajcent ASM chart. The register operations are not specified, because we are interested only in designing the control logic.*

(a) Draw the equivalvent state diagram.

(b) List the state table.

(c) Design the control with three D flip-flops, a decoder and gates.

(d) Design the control with one flip-flop per state.

(e) Derive a table showing the multiplexer input conditions for the control.

(f) Design the control using the multiplexer.

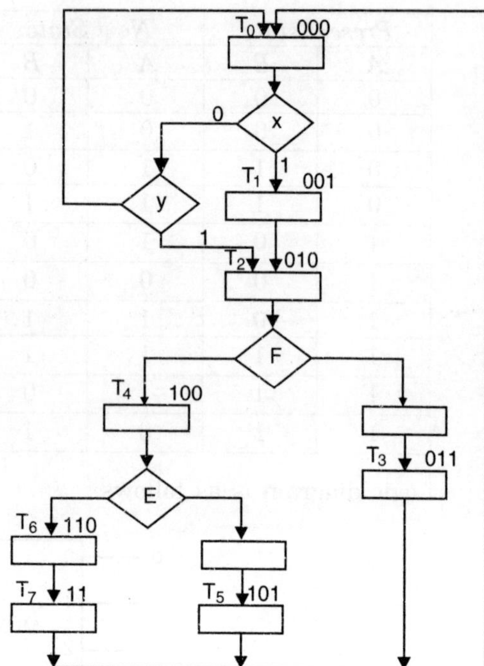

Solution. (a) The equivalent state diagram is as follows:

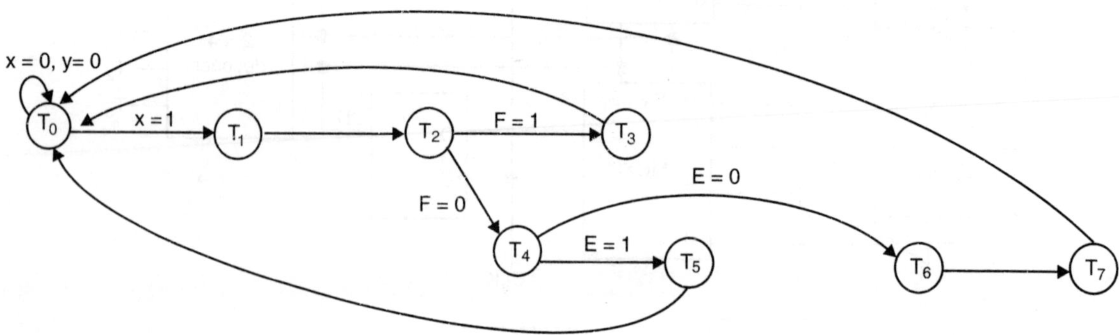

(b) State table is as follows:

Present State Symbol	*Present State*			*Inputs*				*Next State*			*Outputs*							
	A	*B*	*C*	*x*	*y*	*F*	*E*	*A*	*B*	*C*	T_0	T_1	T_2	T_3	T_4	T_5	T_6	T_7
T_0	0	0	0	0	0	×	×	0	0	0	1	0	0	0	0	0	0	0
T_0	0	0	0	1	×	×	×	0	0	1	0	1	0	0	0	0	0	0
T_1	0	0	1	×	×	×	×	0	1	0	0	0	1	0	0	0	0	0
T_2	0	1	0	×	×	1	×	0	1	1	0	0	0	1	0	0	0	0
T_2	0	1	0	×	×	0	×	1	0	0	0	0	0	0	1	0	0	0
T_3	0	1	1	×	×	×	×	0	0	0	1	0	0	0	0	0	0	0
T_4	1	0	0	×	×	×	0	1	1	0	0	0	0	0	0	0	1	0
T_4	1	0	0	×	×	×	1	1	0	1	0	0	1	0	0	1	0	0
T_5	1	0	1	×	×	×	×	0	0	0	1	0	0	0	0	0	0	0
T_6	1	1	0	×	×	×	×	1	1	1	0	0	0	0	0	0	0	1
T_7	1	1	1	×	×	×	×	0	0	0	1	0	0	0	0	0	0	0

(a) To design the control using D flip-flops, decoder and gates:

Number of flip-flops required = 3

∴ a 3-to-8-line decoder is required.

Flip flop input equations are :

$$D_A = T_2\bar{F} + T_4\bar{E} + T_4E + T_6$$

$$D_C = T_2\bar{F} + T_4 + T_6$$

$$D_B = T_1 + T_2F + T_4\bar{E} + T_6$$

$$= T_0x + T_2F + T_4E + T_6$$

Logic diagram is :

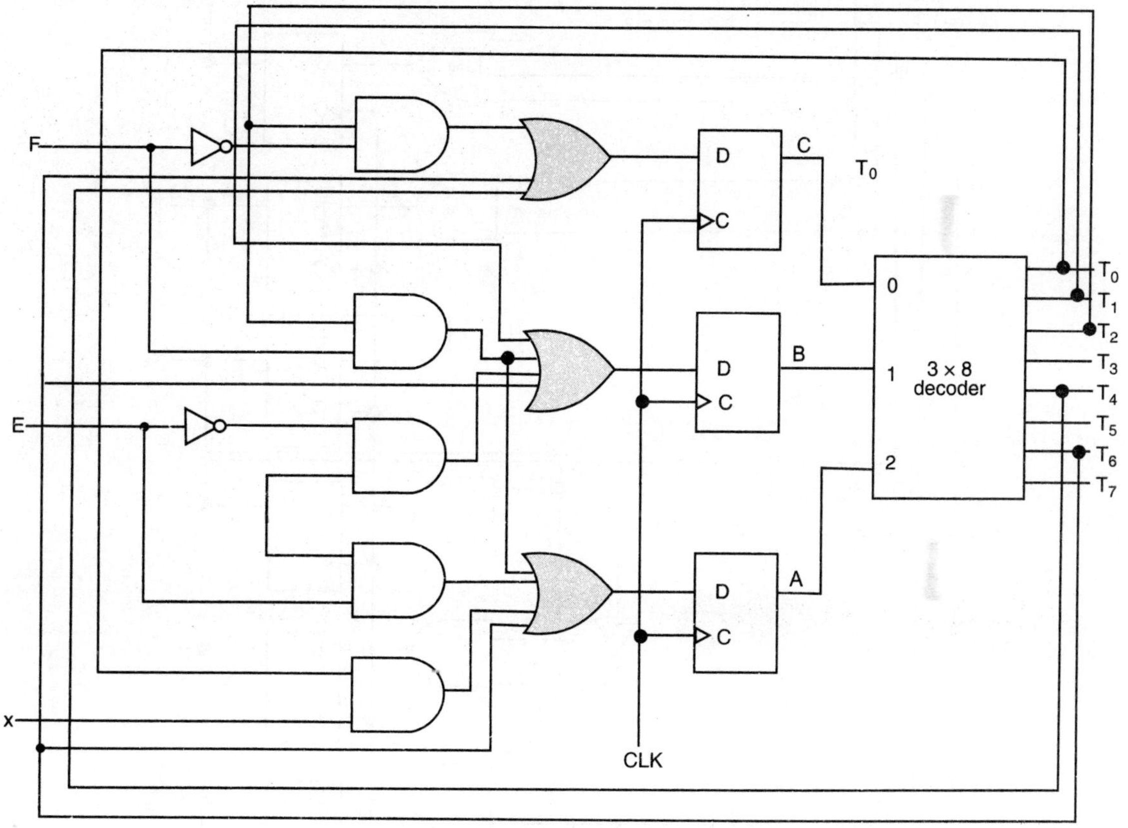

(d) To design the control using one flip-flop per state:

First, we will find out the flip-flop input equations:

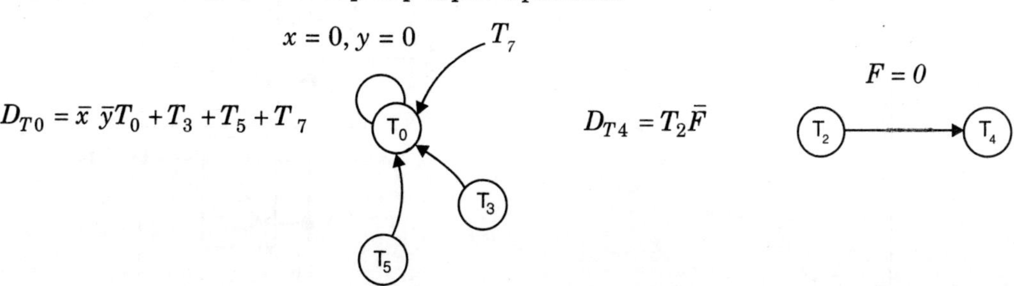

$$D_{T0} = \bar{x}\,\bar{y}T_0 + T_3 + T_5 + T_7$$

$$D_{T4} = T_2\bar{F}$$

$D_{T1} = T_0 x$ $T_0 \xrightarrow{x = 1} T_1$

$D_{T2} = T_1$ $T_1 \rightarrow T_2$

$D_{T3} = T_2 F$ $T_2 \xrightarrow{F = 1} T_3$

$D_{T5} = T_4 E$ $T_4 \xrightarrow{E = 1} T_5$

$D_{T6} = T_4 \bar{E}$ $T_4 \xrightarrow{E = 0} T_6$

$D_{T7} = T_6$ $T_6 \xrightarrow{E = 0} T_7$

Logic diagram is as follows

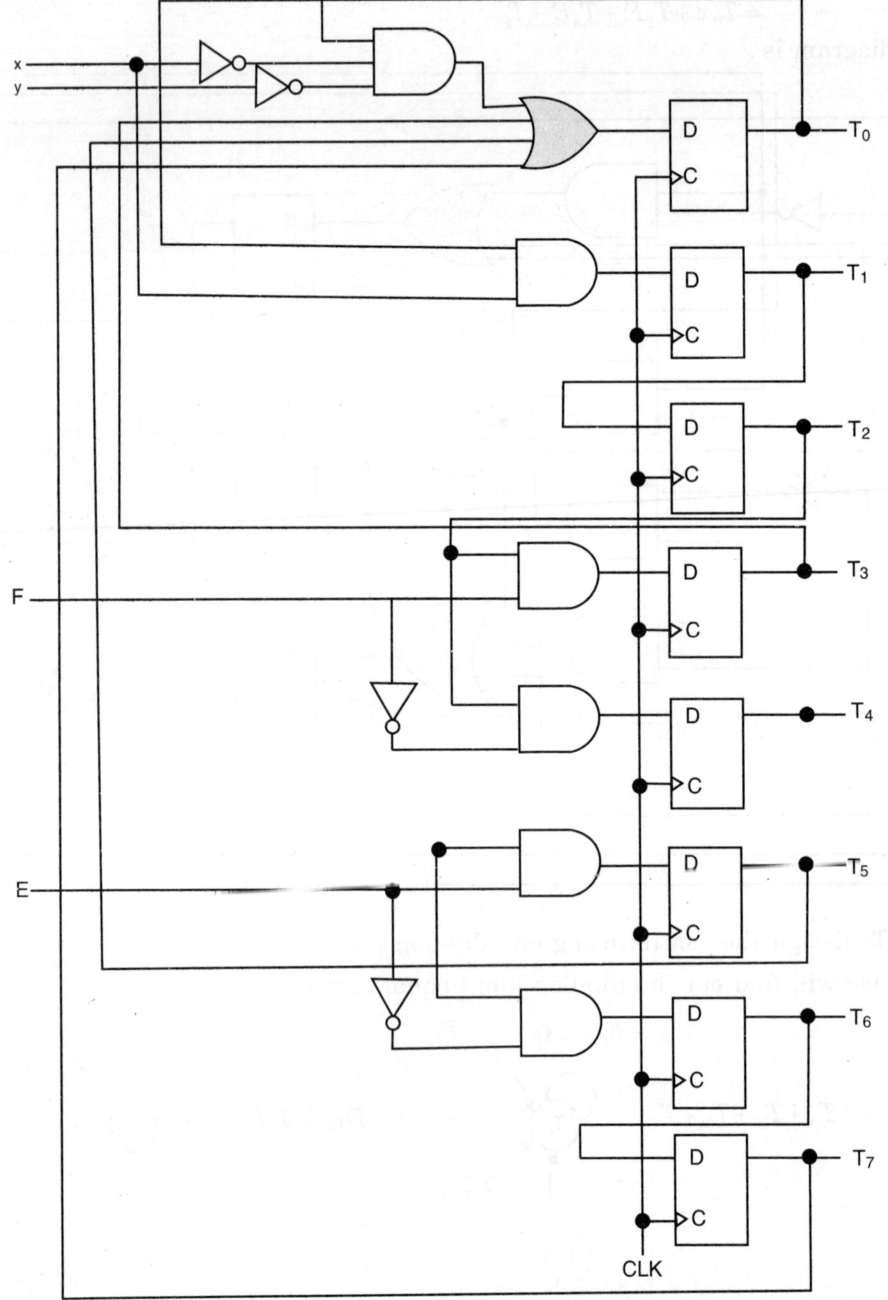

(e) A table shows the multiplexer imput conditions:

$\because$ No of state = 8 = 2^3

$\therefore$ No of flip–flops required = 3 = number of multiplexers required.

$\therefore$ Three 8–to–1 multiplexers are required to implement along with a 3–to–8 line decoder

Present state			Next-state			Inputs	Multiplexer		Inputs
A	B	C	A	B	C		MUX1	MUX2	MUX2
0	0	0	0	0	0	$\bar{x}\,\bar{y}$			
0	0	0	0	0	1	x	0	0	x
0	0	1	0	1	0	None	0	1	0
0	1	0	0	1	1	F			
0	1	0	1	0	0	$\bar{F}$	$\bar{F}$	F	F
0	1	1	0	0	0	None	0	0	0
1	0	0	1	1	0	$\bar{E}$			
1	0	0	1	0	1	E	1	$\bar{E}$	E
1	0	1	0	0	0	None	0	0	0
1	1	0	1	1	1	None	1	1	1
1	1	1	0	0	0	None	0	0	0

(f) Designing the control using multiplexers

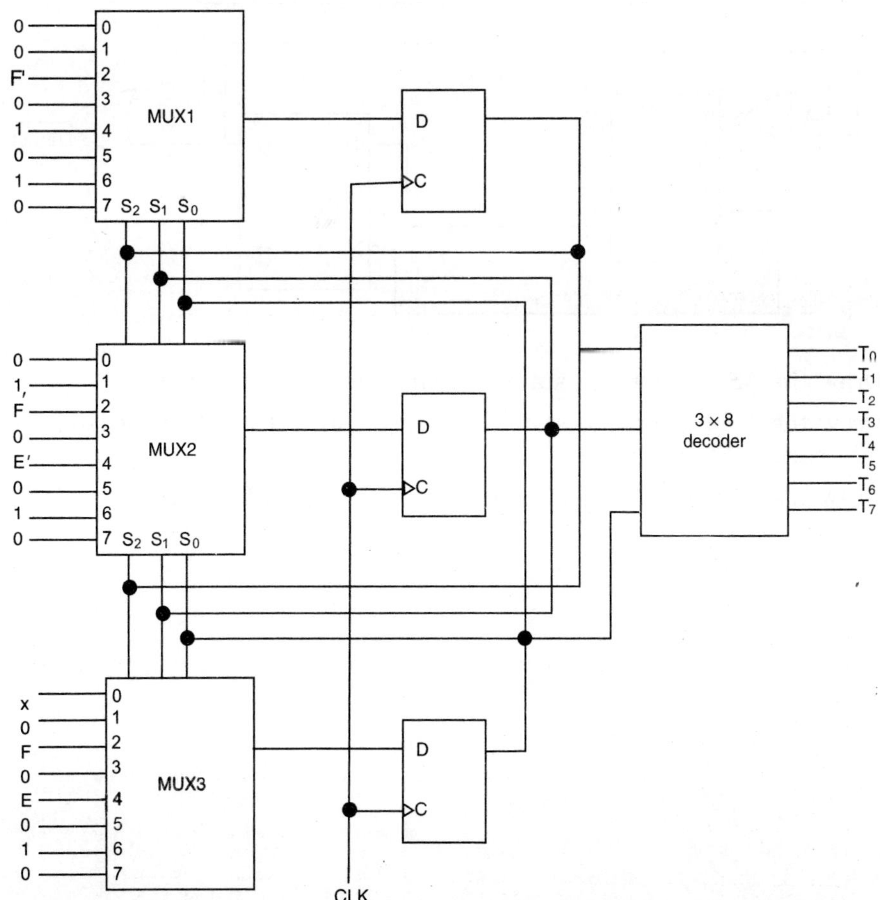

10.6 EXERCISE

1. Draw the ASM chart for a binary multiplier.
2. A binary stream is arriving serially. Stream is such that LSB arrives first and MSB arrives last. System requirement is such that the system must output the 2's complement of each incomming bit serially. Draw the ASM chart and design control subsystem and data processing subsystem for this system.
3. Draw the ASM chart to compare two 4-bits binary data.
4. Draw the ASM chart for 1-bit full adder.
5. Draw the ASM chart for 2-bit binary counter having one enable input.
6. Design a synchronous state machine to generate following sequence of states.

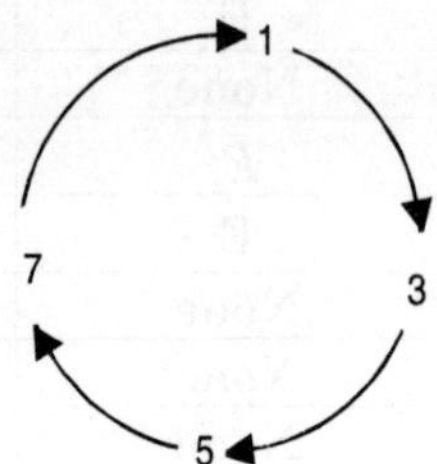

7. Draw the ASM chart and state diagram for the circuit shown

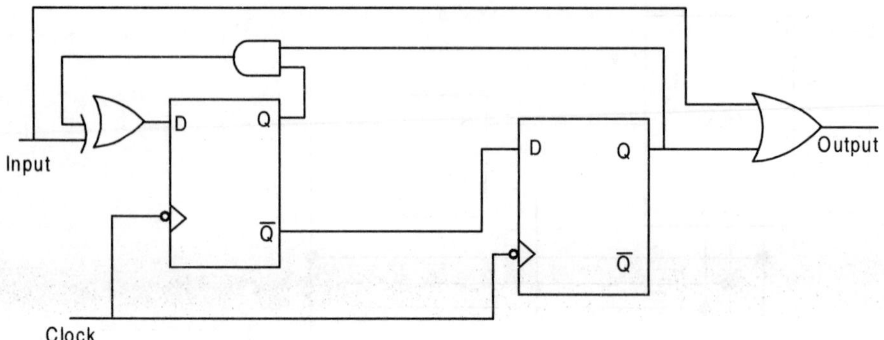

8. Draw the ASM chart and state diagram for decade counter.
9. Draw the ASM chart and state diagram to convert two-digit hexadecimal number into packed BCD number.
10. Draw the ASM chart and state diagram for 1-bit full subtractor.

APPENDIX-A: THRESHOLD LOGIC

A.0 INTRODUCTION

In this book we have been concerned with the logic design of switching circuits constructed of gates and/or bilateral devices. There also exists entirely different type of switching device called the *threshold element*. By the use of threshold logic we get many important theoretical results regarding switching function.

The circuits implemented with threshold elements have usually considerable reduction in the number of gates, inputs and components as well as in the size of the final circuit than the corresponding circuit implemented with conventional gate.

Then question arises why we are not using threshold element in place of conventional gate? The answer of this question is that presently threshold elements are not easy to manufacture as conventional gates and also threshold elements are not as fast to operate and therefore are of very limited usefulness.

Other limitation of threshold logic is its sensitivity to variations in circuit parameters. Also while the input-output relations of circuits constructed of conventional gates can be specified by switching algebra, different algebraic means must be developed for threshold circuits.

A.1 THE THRESHOLD ELEMENT OR T GATE

The symbol of a threshold element is shown in Fig. A.1. A threshold element or T gate has n two-valued inputs say $x_1, x_2, ..., x_n$ and a single two-valued output y. The T-gates internal parameters are threshold T and weights $w_1, w_2, ..., w_n$. For each input variable x_i there is corresponding associated weight w_i.

The $\sum_{i=1}^{n} w_i x_i$ is called the weighted sum of the element.

The output y is a function of input variables $x_1, x_2, ..., x_n$ and depends on weighted sum of the element. If the value of threshold T is less than weighted sum, then output will be 0 otherwise output will be 1. Therefore, the input-output relation of a T-gate is defined as follows:

$$y = 1 \text{ if and only if } \sum_{i=1}^{n} w_i x_i \geq \text{T} \qquad \text{...(A.1)}$$

$$= 0 \text{ if and only if } \sum_{i=1}^{n} w_i x_i < \text{T}$$

The values of the threshold T and the weights $w_i (i = 1, 2,..., n)$ may be real, finite, positive or negative numbers.

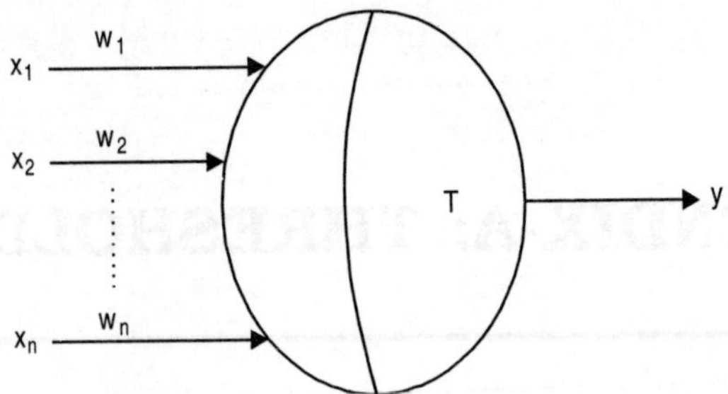

Fig. A.1 Symbol of a threshold element

Example. *The input-output of the threshold element is shown in Fig. A.2. Find the switching function implemented by threshold element.*

Solution.

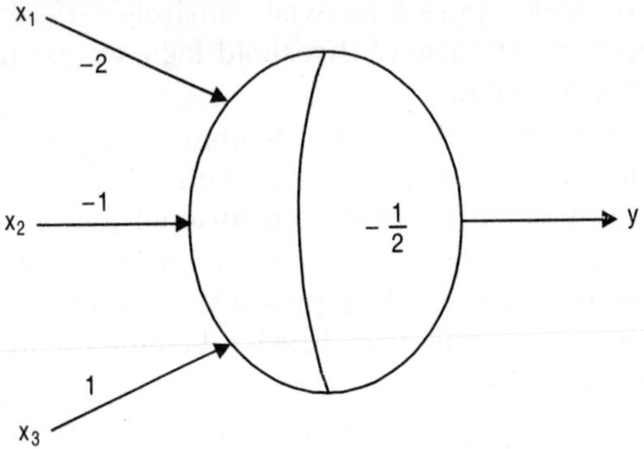

Fig. A.2 A threshold element

The input-output relation of the threshold element shown in Fig. A.2 is given in Table A.1.

Table A.1 Input-output relation of the T-gate shown in Fig. A.2.

Input variables			*Weighted sum*	*Output*
x_1	x_2	x_3	$-2x_1 - x_2 + x_3$	y
0	0	0	0	1
0	0	1	1	1
0	1	0	−1	0
0	1	1	0	1
1	0	0	−2	0
1	0	1	−1	0
1	1	0	−3	0
1	1	1	−2	0

The weighted sum is computed in the centre column for every input combination. For weighted sum greater than or equal to $-\frac{1}{2}$ output is 1 and 0 for weighted sum less than $-\frac{1}{2}$.

We can find the switching function by using K-map.

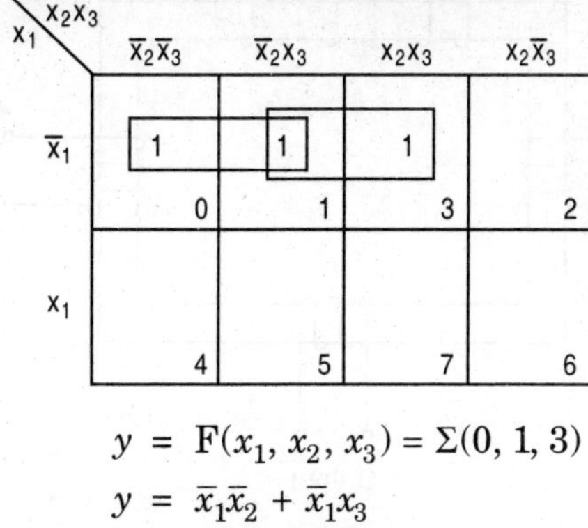

$$y = F(x_1, x_2, x_3) = \Sigma(0, 1, 3)$$

$$y = \bar{x}_1\bar{x}_2 + \bar{x}_1x_3$$

Therefore, this T-gate realizes the switching function

$$y = \bar{x}_1\bar{x}_2 + \bar{x}_1x_3 .$$

A.2 PHYSICAL REALIZATION OF THRESHOLD GATE

A threshold element can be realized in many ways. Since input associated weights may be positive or negative numbers. In this section, we have shown two methods: one which only provides positive weights (resistor-transistor T-gate) and other which provides both positive and negative weights (magnetic core T-gate).

The resistor-transistor T-gate is shown in Fig. A.3.

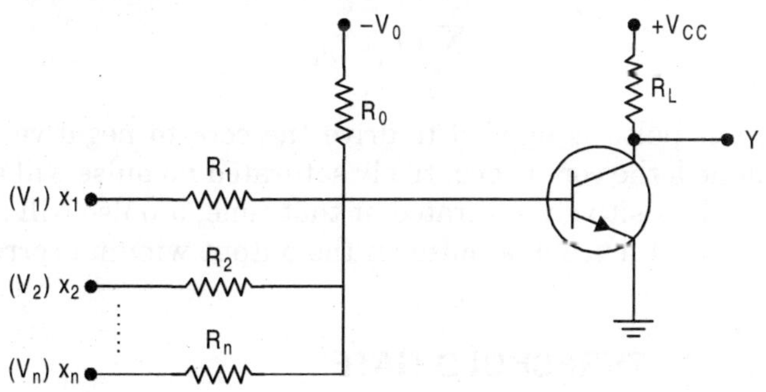

Fig. A.3 Resistor-transistor threshold element.

The resistors R_1, R_2, ..., R_n forms a linear summer which can compute weighted sum of input voltages ($x_1, x_2, ..., x_n$), provided values of resistors are not same. Whenever inputs are at logic 0 level, transistor is in cut-off and output goes to high (1 level) : When one or more inputs are logic 1 then the weighted sum of these input voltages are applied at the base of transistor. If this sum is greater than the threshold value determined by R_0 then the transistor goes to saturation and output goes low. The threshold in the above circuit is determined by resistor R_0 and the voltage source V_0. Because all resistors have positive values, this gate is capable of providing only positive weights.

The magnetic core T-gate is shown in Fig. A.4.

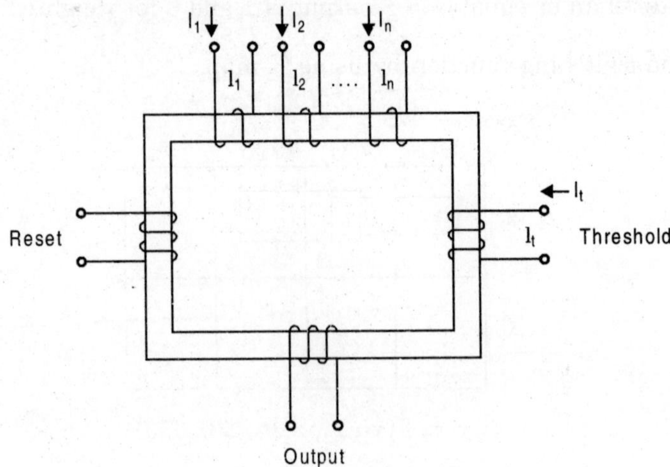

Fig. A.4 Magnetic core T-gate

The magnetic core T-gate provides both negative and positive weights. For the n-input T-gate, a magnetic core of $n + 3$ windings are used. The presence or absence of current in each input winding determined the value of each input variable x_i and the corresponding weights w_i is a function of the number of turns l_i. The threshold T is determined by the turns l_t, the constant current I_t and their direction. The value of the output is determined by the magnetization state of the core. Depending upon the direction of the windings, both positive and negative weights can be obtained. Initially core is in negative saturation.

To determine the direction of saturation which depends on the value of the sum;

$$\sum_{i=1}^{n} I_i l_i + I_t l_t \tag{A.2}$$

a sufficiently large reset pulse is applied to drive the core to negative saturation. If at the time of reset pulse applied, the core is negatively saturated no pulse will appear on the output winding and if the core is positively saturated at that time, a pulse will appear on the output winding. The presence or absence of a pulse on the output winding corresponds respectively, to output values 1 or 0.

A.3 CAPABILITIES OF THRESHOLD GATE

A threshold element can be considered a generalization of the conventional gates, because any of them can be realized by a single threshold element. The threshold elements are more powerful and capable than conventional diode and transistor gates because of single T-gates ability to realize a larger class of functions than is realizable by any one conventional gate. The following example shows the NOR gate implementation by single T-gate.

Example. *A two-input NOR gate can be realized by a single threshold element with weights –1, –1 and threshold* $T = -\frac{1}{2}$ *as shown in Fig. A.5.*

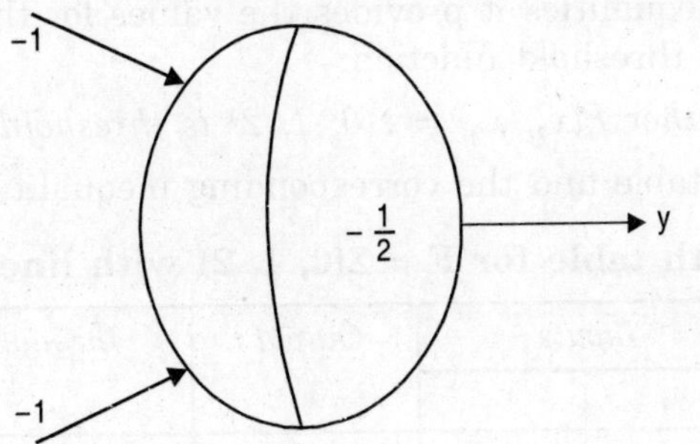

Fig. A.5 A T-gate realization of NOR operation.

Since NOR is a functionally complete operation any switching function can be realized by threshold element alone. The input-output relation of the threshold element shown in Fig. A.5 is given in Table A.2.

Table A.2 Input-Output relation of the gate shown in Fig. A.5

Input variables		*Weighted sum*	*Output*
x_1	x_2	$-x_1 - x_2$	y
0	0	0	1
0	1	–1	0
1	0	–1	0
1	1	–2	0

Though large class of switching functions can be realized by single threshold elements. But not every switching function is realizable by only one threshold element. The following example support the above statement.

Example. *Is it possible to implement EX-OR operation with a single threshold element.*

Solution. For two input EX-OR operation

$$F(x_1, x_2) = x_1\bar{x}_2 + \bar{x}_1 x_2$$

That is, output is 1 for dissimilar inputs.

If w_1 and w_2 are associated weights

$$\left.\begin{aligned} w_1 &\geq T \\ w_2 &\geq T \end{aligned}\right\} \qquad \text{(A)}$$

and $$w_1 + w_2 < T \qquad \text{(B)}$$

The inequalities (A) and (B) have conflicting requirements so no threshold value can satisfy them. Therefore, EX-OR cannot be realized by a single threshold element.

A switching function which can be realized by a single T-gate is known as threshold function.

To check whether a switching function $F(x_1, x_2, ..., x_n)$ is a threshold function or not, we have to derive 2^n linear, simultaneous inequalities for which weighted sum must exceed or equal to threshold T if F = 1 and for which weighted sum must be less than T if F = 0. If solution exists to above inequalities it provides the values for threshold and weights. If, no solution exists, F is not a threshold function.

Example. *Check whether $F(x_1, x_2) = \Sigma(0, 1, 2)$ is threshold function or not.*

Solution. The truth table and the corresponding inequalities are given in Table A.3.

Table A.3 Truth table for F = Σ(0, 1, 2) with linear inequalities

Inputs		*Output*	*Inequality*
x_1	x_2	F	
0	0	1	$0 \geq T$
0	1	1	$w_2 \geq T$
1	0	1	$w_1 \geq T$
1	1	0	$w_1 + w_2 < T$

From the inequality corresponds to combination 0 we deduce that T must be negative. And the inequality corresponds to combination 1 shows that both w_1 and w_2 must also negative. For smallest integer value, we obtain

$$w_1 = -1$$

$$w_2 = -1$$

$$T = -1\frac{1}{2}$$

Since all the inequalities are satisfied, hence F is a threshold function.

Example. *Show that a function having inputs $x_1, x_2, ..., x_j, ..., x_n$ is realizable by a single T-gate having weights $w_1, w_2, ..., w_j, ..., w_n$ and threshold T, after complementing one of the inputs, say x_j, the same function can be realized by a single T-gate with weights $w_1, w_2, ..., -w_j, ..., w_n$ and threshold $T - w_j$.*

Solution. We can specify *a* threshold element by a weight threshold vector V = {T; w_1, w_2, ..., w_n} and its input variables.

Let the function $F_1(x_1, x_2, ..., x_j, ..., x_n)$ be realizable by single T-gate V_1 = {T; w_1, w_2, ..., w_j, ..,. w_n}. Then from the definition of T-gate we can write

if $$w_jx_j + \sum_{i \neq j} w_ix_i \geq T \quad \text{then } F_1 = 1 \qquad ...(A.3)$$

and $$< T \quad \text{then } F_1 = 0$$

When input x_j is complemented we assume that same function F_1 can be realized by a single T-gate V_2 = {T – w_j; w_1, w_2, ..., w_j, ..,. w_n} for the inputs $x_1, x_2, ..., \bar{x}_j, ..., x_n$.

Then this T-gate must satisfy the following inequalities:

if $$-w_j\bar{x}_j + \sum_{i \neq j} w_ix_i \geq T - w_j \quad \text{then } F_2 = 1 \qquad ...(A.4)$$

and $$< T - w_j \quad \text{then } F_2 = 0$$

where F_2 is the function realized by the T-gate V_2.

To prove that F_1 and F_2 are identical functions, suppose $x_j = 0$ so that $\bar{x}_j = 1$. Hence, equations (A.3) and (A.4) becomes identical and if suppose $x_j = 1$ so that $\bar{x}_j = 0$. Again equations (A.3) and (A.4) becomes identical.

Since both F_1 and F_2 having identical values for each input combination therefore they are identical functions.

Note : By an appropriate selection of complemented and uncomplemented input variables a function can be realized by a single T-gate having desired sign distribution. Therefore, it can be realized by only positive weights.

In this section, we have tried to find out the answer of questions: Is the threshold gate a universal gate? Can a single T-gate realize any switching function?

The answer of first question is *yes* while the answer to the second question is *no*. As we have seen in example A.3 that EXCLUSIVE OR function cannot be realized by a single T-gate.

In the subsequent section the condition that a given function must satisfy so that it can be realized by a single T-gate is found.

A.4 PROPERTIES OF THRESHOLD FUNCTIONS

(1) Isobaric Functions

Two threshold functions are said to be isobaric if they have different threshold values but the same set of weights.

(2) Unate Functions

A function is called a unate function, if the minimal sum of product (SOP) form contains each variables in complemented or uncomplemented form only. If all the variables appear in complemented form then the function is called a *negative unate* or *negative function* and if all the variables appear in uncomplemented form then the function is called a *positive function*. If a variable x_i appears only in complemented (uncomplemented) form, then the function is said to be *negative* (positive) *unate* in variable x_i.

An unate function can be represented by a cube. An n-cube contains 2^n vortices, each of which corresponds to a minterm of a n-variable function. Vertices corresponding to minterms for which function having value 1 are known as true vertices and for those function value 0 are called *false vertices*. To represent the function a line is drawn between every pair of vertices which differ in just one variable.

Example. *Check whether function*

$$F = A\overline{B} + B\overline{C}$$

is unate function.

Solution. From definition we know that function $F = A\overline{B} + B\overline{C}$ is positive in A and negative in C but not unate in B.

Example. *Find the three-cube representation of function* $y = \bar{x}_1\bar{x}_2 + \bar{x}_1x_3$ *implemented in example A.1.*

Solution. To represent the function, first draw a line between every pair of vertices which differs in just one variable as shown in Fig. A.6.

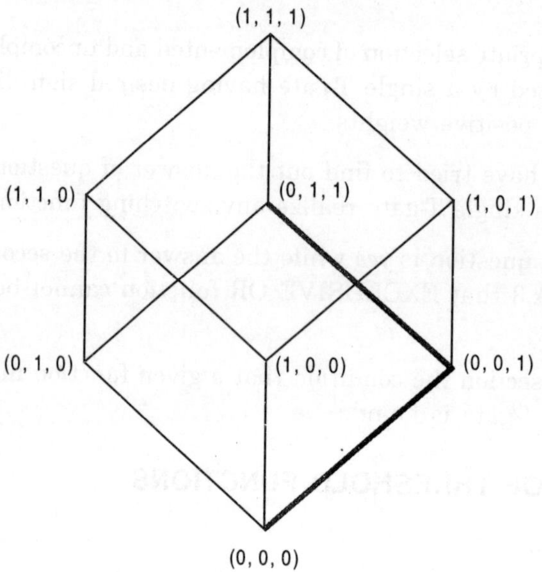

Fig. A.6 A three-cube representation of $y = \bar{x}_1\bar{x}_2 + \bar{x}_1x_3$

The heavier lines connecting the two pairs of the true vertices [(0, 0, 0) and (0, 0, 1) and (0, 0, 1) and (0, 1, 1)] represent the subcubes $\bar{x}_1\bar{x}_2$ and $\bar{x}_1x_3$.

(3) Linear Separability

Let $f(x_1, x_2, ..., x_n)$ be a threshold function represented in n-cube form, whose true vertices can be separated from the false ones by a hyperplane represented by linear equation

$$w_1x_1 + w_2x_2 + ... + w_nx_n = T$$

The output of a T-gate is dependent on whether the weighted sum is less than a certain constant (threshold value) or not.

Now, since $f = 0$ when

$$w_1x_1 + w_2x_2 + ... + w_nx_n < T$$

and $f = 1$ when

$$w_1x_1 + w_2x_2 + ... + w_nx_n \geq T$$

i.e., hyperplane separates the true vertices from the false ones.

Owing to this linear separability of true and false vertices, the threshold functions are called *linearly separable function*. Since all threshold functions are linearly separable, therefore can be realized by a single T-gate.

Note: Unateness is a necessary condition for linear separability.

A.5 SYNTHESIS OF THRESHOLD FUNCTIONS

The synthesis of threshold function utilizes the linear separability property, which determines whether or not their exists a hyperplane which separates the true vertices from the false ones. The synthesis of threshold function involved following steps:

(I) Test for unateners by minimal expression of the function.

(II) If unate, convert it into positive function in all its variables.

(III) Find the minimal true and maximal false vertices of positive function obtained in Step 2.

(IV) Determine linear separability of positive function and if it is, find an appropriate set of weights and threshold.

(V) Convert this weight-threshold vector to one which corresponds to the original function.

Example. *Determine whether the function*

$$F(A, B, C, D) = AB\overline{C}D + B\overline{C}\,\overline{D}$$

is threshold function, and if it is, find a weight-threshold vector.

Solution. Step 1: Test for unateness by minimal expression of the function

$$F(A, B, C, D) = AB\overline{C}D + B\overline{C}\,\overline{D}$$

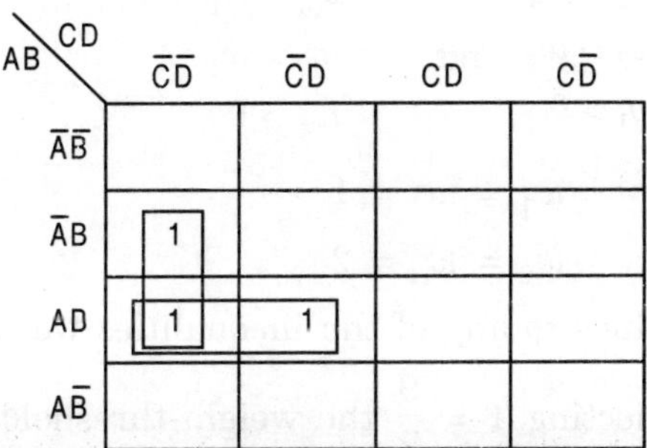

$$F = AB\overline{C} + B\overline{C}\,\overline{D}$$

This function is unate.

Step 2: Convert it into positive function F_2

$$F_2 = ABC + BCD$$

Step 3: Minimal true vertices (L) and maximal false vertices (U) of F_2. Minimal true vertices are (1, 1, 1, 0) and (0, 1, 1, 1). The maximal false vertices are found by determining all false vertices with two variables whose value is 0, and so on. Therefore, the maximal false vertices are (1, 1, 0, 1) and (1, 0, 1, 1) and (0, 1, 1, 0).

Step 4: To check linear separability, we have to solve LU inequalities, corresponding to the L minimal true and U maximal false vertices. For each pair of vertices A and B

$$\sum_{i=1}^{n} a_i w_i > \sum_{i=1}^{n} b_i w_i$$

where $A = \{a_1, a_2, ..., a_n\}$ = minimal true vertices

$B = \{b_1, b_2, ..., b_n\}$ = maximal false vertices

Therefore, for L = (1, 1, 1, 0) and (0, 1, 1, 1)

U = (1, 1, 0, 1), (1, 0, 1, 1) and (0, 1, 1, 0)

we get six inequalities as follows:

$$w_1 + w_2 + w_3 > w_1 + w_2 + w_4$$

$$w_1 + w_2 + w_3 > w_1 + w_3 + w_4$$

$$w_1 + w_2 + w_3 > w_2 + w_3$$

$$w_2 + w_3 + w_4 > w_1 + w_2 + w_4$$

$$w_2 + w_3 + w_4 > w_1 + w_3 + w_4$$

$$w_2 + w_3 + w_4 > w_2 + w_3$$

Since F_2 is a positive function, then if it is linearly separable the separating hyperplane will have positive co-efficients.

Solving the system of inequalities derive above, we get

$$w_3 > w_4 \qquad w_3 > w_1$$

$$w_2 > w_4 \text{ and } w_2 > w_1$$

$$w_1 > 0 \qquad w_4 > 0$$

Let $w_1 = w_4 = 1$

and $w_2 = w_3 = 2$

By substituting these values to any of the inequalities we find that T must be smaller than 5 but larger than 4. Selecting $T = \frac{9}{2}$ the weight-threshold vector for F_2

$$V_2 = \left\{\frac{9}{2}; 1, 2, 2, 1\right\}.$$

Step 5: To convert the weight-threshold vector corresponds to the original function F, result of example is used, where for every input x_j which is complemented in the original function, w_j must be changed to $-w_j$ and T to $T - w_j$. In this case, inputs x_3 = C and x_4 = D are complemented in F; thus, in the new weight-threshold vector V, the weights are 1, 2, –2 and –1, and the threshold is $\frac{9}{2} - 2 - 1 = \frac{3}{2}$, so

$$V = \left\{\frac{3}{2}; 1, 2, -2, -1\right\}$$

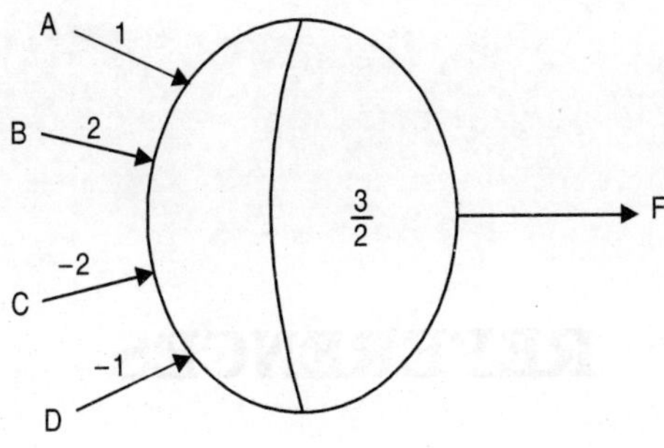

Fig. A.7

A.6 MULTI-GATE SYNTHESIS

So far we have been concerned mainly with the switching functions realizable with single T-gate. But how any arbitrary switching function can be realized using threshold gates? The general synthesis procedure for non-series parallel or multiple-output networks are not yet available. These problems are, however, solved for the particular case of networks specified by means of symmetric functions. One approach to synthesise any arbitrary switching function is to develop a procedure for the decomposition of non-threshold functions into two or more factors, each of which will be a threshold function.

A T-gate realization of any arbitrary switching function can be accomplished by selecting a minimal number of admissible patterns (A pattern of 1 cells which can be realized by a single T-gate). Such that each 1 cell of the map is covered by at least one admissible pattern.

A.7 LIMITATIONS OF T-GATE

The input-output relations of circuits constructed of conventional gates specified by switching algebra but till date no such algebraic means are developed for T-gate.

Other limitation of threshold logic is its sensitivity to variations in circuit parameters. Because of the variations in circuit parameters, the weighted sum may deviates from its designed value and cause a circuit malfunction with a large number of inputs. As we have seen in the physical realization of T-gate with resistor-transistor that threshold value depends on the resistance and supply voltage. Since resistance value may change up to 15-20 percent of their nominal value and supply as well input voltage may also vary therefore there is restriction on the threshold value T and on the maximum allowable number of inputs.

Another limitation is the lack of effective synthesis procedures.

A.8 EXERCISE

1. Show that the symmetric function F(A, B, C) is not linearly separable.
2. Show a threshold-logic realization of a Full adder requiring only two T-gates.
3. Realize $F(x_1, x_2, x_3, x_4) = \Sigma(3, 5, 7, 10, 12, 14, 15)$.
   Using (*a*) AND-OR realization.
   (*b*) T-gate realization.
4. A switching function $F(x_1, x_2, ..., x_n)$ is unate if and only if it is not a tautology and the above partial ordering exists, so that for every **pair of vertices** $(a_1, a_2, ..., a_n)$ and $(b_1, b_2, ..., b_n)$, if $(a_1, a_2, ..., a_n)$ is a true vertex and $(b_1, b_2, ..., b_n) \geq (a_1, a_2, ..., a_n)$ then $(b_1, b_2, ... b_n)$ is also a true vertex of F.
5. Realize $y = \Sigma(1, 2, 3, 6, 7)$ using T-gate.

REFERENCES

1. A.K. Singh, *Digital Logic Circuits,* New Age International Publishers, Delhi, 2007.
2. A.K. Singh, Manish Tiwari, *Digital Principles, Function of Circuit Design and Application*, New Age International Publishers, Delhi, 2006.
3. H. Taub, D. Schilling, *Digital Integrated Electronics*, McGraw-Hill, Koga Kusha, 1997.
4. A.S. Sedra, K.C. Smith, *Microelectronics Circuits*, 4th ed, Oxford University Press, New York, 1998.
5. J. Millman, H. Taub, *Pulse Digital and Switching Waveforms*, McGraw-Hill, Singapore.
6. M.M. Mano, *Digital Design*, 2nd ed, Prentice-Hall of India, 1996.
7. R.L. Tokheim, *Digital Electronics: Principles and Applications*, 6th ed, Tata McGraw-Hill, New Delhi 2004.
8. J. Millman, C.C Halkias, *Integrated Electronics: Analog and Digital Circuits and Systems*, Tata McGraw-Hill, New Delhi, 1994.
9. A.P. Malvino, D.P. Leach, *Digital Principles and Applications*, 4th ed, Tata McGraw-Hill, New Delhi, 1991.
10. R.P. Jain, *Modern Digital Electronics*, Tata McGraw-Hill, New Delhi, 1992.
11. Virendra Kumar, *Digital Technology; Principles and Practice*, New Age International Publishers, Delhi.
12. J.P. Hayes, *Computer Architecture and Organization*, 2nd ed, McGraw-Hill, Singapore, 1988.
13. V.C. Hamacher, Z.C. Vranesic, S.G. Zaky, *Computer Organization*, 4th ed, McGraw-Hill, 1996.
14. Gopalan, *Introduction to Digital Microelectronics* Circuits, Tata McGraw-Hill, 1998.
15. P.K. Lala, *Digital System: Design using Programmable Logic Devices*, BS Publication, Hyderabad, 2003.
16. J.M. Rabey, *Digital Intergrated Circuits: A Design Perspective*.
17. Charles H. Roth, Jr., *Fundamentals of Logic Design*, 4th ed, Jaico Publishing House, 2003.
18. ZVI Kohavi, *Switching and Finite Automata Theroy*, 12th ed, Tata McGraw-Hill, 1978.

INDEX

N

O

P

Q

R

S

T

U

W

X